Contributions from Biology Education Research

Contributions from Biology Education Research (CBER) is the international book series of the European Researchers in Didactics of Biology (ERIDOB). The series includes edited collections of state-of-the-art papers presented at the ERIDOB international conferences, and monographs or edited collections of chapters by leading international scholars of the domain. The aim of the series is to shed light on global issues and trends in the teaching and learning of biology by gathering cutting edge research findings, theoretical views, and implications or concrete suggestions for everyday school practice regarding biology. The books may serve as resources for (a) getting informed about the most recent findings of biology education research to possibly integrate them in new personal research, and (b) studying about the teaching and learning of biology as a pre-service or in-service biology teacher. So, the main audiences for the series range from senior to early career biology education researchers and pre- or in-service biology teachers working at all educational levels. Book proposals for this series may be submitted to the Publishing Editor: Claudia Acuna E-mail: Claudia.Acuna@springer.com

Nancy J. Pelaez • Stephanie M. Gardner
Trevor R. Anderson
Editors

Trends in Teaching Experimentation in the Life Sciences

Putting Research into Practice to Drive Institutional Change

Springer

eridob
european researchers
in didactics of biology

Editors
Nancy J. Pelaez
Biology Education Area, Department of Biological Sciences
Purdue University West Lafayette
West Lafayette, IN, USA

Stephanie M. Gardner
Biology Education Area, Department of Biological Sciences
Purdue University West Lafayette
West Lafayette, IN, USA

Trevor R. Anderson
Divisions of Chemical Education and Biochemistry, Department of Chemistry
Purdue University West Lafayette
West Lafayette, IN, USA

ISSN 2662-2319 ISSN 2662-2327 (electronic)
Contributions from Biology Education Research
ISBN 978-3-030-98591-2 ISBN 978-3-030-98592-9 (eBook)
https://doi.org/10.1007/978-3-030-98592-9

This Springer imprint is published by the registered company Springer Nature Switzerland AG
The registered company address is: Gewerbestrasse 11, 6330 Cham, Switzerland

Preface

The chapters in this volume are organized into six parts to show how faculty and instructors in post-secondary biology departments are using contributions from biology education research to help biology students learn about science research practices, including biological experimentation. All chapters are aimed at putting knowledge about biological research into pedagogical and classroom practice. The first chapter introduces and defines competencies for biological experimentation in a way that is intended to help advance the teaching, learning, and assessment of biological experimentation as a foundation for educating students at the tertiary level. This and the first four of the six parts build on work by members of a Research Coordination Network funded by the National Science Foundation in the United States to develop Assessments of Competence in Experimental Design in Biology (ACED-Bio). From May 2014 until September 2020, this project periodically brought together biology research scientists as well as biology faculty who were education specialists and university biology educators to work toward helping all post-secondary biology students to learn that experimentation in biology produces data that are useful for evidentiary reasoning, leading to findings that play a central role in our world to the benefit of society.

A goal of our project as well as this book has been to address this aim with new guidelines, instructional strategies, and norms for collaborative action involving both scientists and education specialists to inform educational innovations using contributions from biology education research and cutting-edge science. The scientists who were founding network members did not feel enthusiastic about belonging to a network on "Assessment." Furthermore, we agreed that it is not possible to separate experimental design from experimentation in general nor could assessment be considered separately from learning, so we decided to refer to our project as the Advancing Competencies in Experimentation – Biology (hereafter, ACE-Bio) Network.

Network members shared a common interest in teaching students about biology experimentation. Students struggle when it comes to developing hypotheses, thinking about the difference between dependent and independent variables, considering what control is appropriate for experiments, measuring variability and thinking about whether a claim is appropriate when they have some data to analyze as evidence, and so our network aimed to address these difficulties. An increasing number of studies in biology have been influencing the number of students who are engaged with biological research as a process, yet very little was known about what undergraduate students actually learn from these educational innovations.

To address these issues and after reflecting on our collective contribution, the first four chapters of this book present examples and a critical review of biology experimentation education informed by a framework from the ACE-Bio Network to address the various challenges that biology instructors face, whereas the last two parts extend this work. We grouped the chapters of this book into a logical sequence of parts to illustrate how four types of challenges faced by departments where science is taught as research are being addressed. These chapter contributions are useful, according to Akuma and Callaghan (2019), for establishing a shared vision (Part I), operationalizing a plan for instruction (Part II), engaging students with experimentation during instruction (Part III), and assessing what students learn about by doing biological research (Part IV). Although an ACE-Bio Competencies framework introduced in Chap. 1 was applied to these four types of challenges in the first four parts, several strengths and weaknesses are identified by chapter authors, so a fifth part addresses the limitations by introducing complementary frameworks for guiding students' biological research, including experimentation (Part V). The final part in this book (Part VI) moves beyond the ACE-Bio Competencies framework to consider factors of importance to biological experimentation instruction and its implementation that are also of major relevance to post-secondary biology education programs in general.

Most chapters could fit into a structure of this book in multiple sets. However, we opted to position chapters to illustrate how instructors address the above-mentioned challenges with teaching experimentation according to two key references that guided our decisions about the book structure (Anderson & Rogan, 2011; Akuma & Callaghan, 2019). Some other powerful frameworks which are drivers of reform are represented by individual chapters (which sometimes, but do not always, align themselves with the ACE-Bio Competencies framework). Next, we detail connections between the contributions from various chapter authors within each part to explain what valuable insight each chapter provides in terms of existing theoretical frameworks or what is currently known about the processes of learning with and about biological experimentation.

Part I: Vision and Initiation Phase: Envisioning What, When, and How Students Learn About Biological Experimentation

The chapters in Part I discuss what university biology students should learn as they progress through their undergraduate careers, and factors that interact with their learning of the desired concepts and skills. Pelaez et al. (Chap. 1) motivate the need for improving student understanding of the key role played by experimentation in our understanding of the natural world, as well as their competency with various aspects of experimental research. The process and product of the work done by a diverse group of instructors of the ACE-Bio Network are described, and the ACE-Bio Competencies framework is presented. Concept-skill statements describe what a competent biologist does when engaged in experimentation and can be used to guide instruction and assessment in undergraduate biology education. The ACE-Bio Competencies have served as an inspiration, guide, and point of comparison for the work described in the chapters of the first four book parts.

Bowe and Irby (Chap. 2) provide a process for, and examples of identifying anticipated learning outcomes (ALOs) for teaching and assessing student competence with science practices, including experimentation. They highlight the need for identifying a degree of specificity that will allow for the articulation of clear and measurable learning targets for assessments. The authors note that while the ACE-Bio Competencies are descriptive of what is done during experimentation, each statement is too broad to clearly guide instruction and assessment to generate evidence of student learning. Two frameworks that can assist instructors in the articulation of ALOs relevant for their courses are presented.

Cole and Beck (Chap. 3) used the ACE-Bio Competencies as a tool to assist in curriculum reflection and reform. The ACE-Bio Competencies served as the basis for surveying faculty in a biology department about what and when students should learn aspects of experimentation. This approach allowed the authors to identify areas of consensus and divergence within the faculty regarding student learning and target competence with aspects of experimentation. While there was agreement that all competencies were appropriate for teaching undergraduate biology students, some ACE-Bio Competencies were viewed as more easily developed than others (e.g., *Conduct* and *Conclude*). The findings described in Chap. 3 enable deeper discussions about content, sequencing, and expectations for student learning across the biology curriculum to lead to mastery by the time students graduate.

Finally, Chatzikyriakidou and McCartney (Chap. 4) use the ACE-Bio Competencies framework to explore its relationship with other frameworks as drivers of reform. They find that these multiple frameworks are both useful in ways that are independent but complement each other for envisioning what and how students learn about biological experimentation. Important philosophical and psychological precursors of contemporary learning theories are applied to education, and the links between learning theories and instruction are explained in this chapter. This also includes the epistemologies students bring with them and develop within biology classrooms.

Part II: Operationalizing and Planning: Designing Instruction to Promote Learning of Biological Experimentation

The chapters in Part II move from the "what" of teaching biological experimentation, which was a strong theme in the first part chapters, into outlining considerations for instructional designs to support students' learning of aspects of biological experimentation. These chapters discuss both large-scale and theoretical instructional considerations and speak to specific approaches that could be used. Large-scale and theoretical considerations include the selection of resources and contexts in which to engage students, and instructional frameworks to model experimentation practices to support students in their learning. Specific approaches include an extension of the traditional backward design approach according to Wiggins and McTighe (2011) and using the ACE-Bio Competencies as a starting point from which to organize instructional topics and assignments in a course.

Cheng et al. (Chap. 5) present an innovative biochemistry lab course with authentic research experiences aiming to promote students' gains in research abilities. The authors describe a reform approach focused on identifying anticipated learning outcomes (ALOs) and using backward design (Wiggins & McTighe, 2011) for reforming teaching and the curriculum. The role of the ACE-Bio Competencies for defining ALOs is described as the first of two stages of the lab course design: (1) using learning objectives (ALOs) to guide and navigate the design of assessments and teaching activities, and (2) using research questions to guide and navigate the administration of instruction where students are immersed in literature research, gap analysis, question identification, study design and planning, protocol development and troubleshooting, data acquisition, analysis, and reporting presentations.

Thomas (Chap. 6) used the ACE-Bio Competencies framework as a course planning tool to guide students' independent research. The competencies informed all aspects of course design from the topics, their sequencing within the courses, and assignments. While it emerged during the articulation of the ACE-Bio competencies that there was no pre-defined order of their enactment during biological experimentation (see Chap. 1), the author found it useful for course planning and student learning to use them in the order in which they are presented: *Identify*, *Question*, *Plan*, *Conduct*, *Analyze*, *Conclude*, *Communicate*. Having a set order allowed for synchronizing students in the course and provided structure for their learning. The author provides for instructors' valuable insights and reflections on the course, its design, and student learning including the duration and timing of student engagement of activities related to each competency and details of ACE-Bio competency-aligned assignments.

Linton et al. (Chap. 7) focus on experiments in data mining using digitized natural history collections to introduce students to data science. This work was initially informed by UC Berkeley's Museum of Paleontology multifaceted model of science (Thanukos et al., 2010) along with their suite of resources that were developed independently of the ACE-Bio Network project. The authors demonstrate how the Biodiversity Literacy in Undergraduate Education – Data Initiative (BLUE Data) project, which brought together communities of biodiversity, data, and education specialists, design lessons for promoting student understanding of what science is about and how it is done. This chapter expands the term "experiment" to show that the ACE-Bio Competencies can be applied to this type of inquiry that involves getting evidence for new discoveries from existing data by observation or comparison. Both the ACE-Bio Competencies framework and the UC Berkeley model of science focus on the iterative nature of science and the need for science processes to be flexible, with evidence-based reasoning, and linkages to the communities in which the science aims to address research questions.

Gardner et al. (Chap. 8) developed and implemented a framework for teaching and learning graphing in undergraduate biology, that is consistent with the ACE-Bio Competency areas of *Plan* (the ability to plan feasible and ethical experiments to answer research questions or test hypotheses), *Analyze* (the ability to analyze and process data), and *Conclude* (the ability to draw conclusions about data with inference that are limited to the scope inherent in the experimental design). In addition to diving deeper into each of the competency areas, the authors offer guidance for engaging students in an inclusive and authentic way. This includes considering tenets of universal design for learning to ensure that all students can engage with learning materials and activities, using data that are of interest to the students and that represent true, messy biological data, while encouraging students to be reflective and critical. The authors provide several recommendations for instructional designs including within the context of two case studies.

Part III: Implementation and Student Engagement: Guiding Learners to Do Experiments and Use Representations in Biological Research

One of the major challenges when implementing plans such as those outlined in the previous part is to engage students with experimentation in biology in a way that helps students integrate disciplinary knowledge with other scientific practices that give experiments purpose and meaning. Yet students tend to experience experiments as isolated and stand-alone activities. Implementation-phase challenges include not just engaging the learners but also persuading them to reflect on their experiences and findings. The chapters in Part III illustrate a range of approaches to engage and guide students. The examples show how scientific visualizations are used to help students while the chapters also demonstrate several approaches to peer review for feedback during instruction.

A productive way of using the ACE-Bio Competencies framework for engaging students is to look at one competence in detail in order to provide a fine-grained analysis that makes an under-recognized competence more visible. Kruchten and O'Brien (Chap. 9) focus on *Identify* as a key competence. The authors give a detailed account of how this competence can be fostered through visual signaling pathways that students use to identify a gap in the research literature. When students found no published information about a particular process in a cell signaling map, they added a question mark to denote a gap that could be a target for future research. They then wrote a research proposal to address the gap. Practitioners can refer to their step-by-step implementation process, examples of student work, and the discussion of how to motivate students and incorporate peer feedback on the research proposals by implementing a mock study section to award funding to selected proposals.

Casali et al. (Chap. 10 https://doi.org/10.1007/978-3-030-98592-9_10) found the ACE-Bio Competencies framework to be a suitable tool in the design of exercises to engage students in simulated research experiences with their online Virtual Microscope at Universidad de Buenos Aires in Argentina. First, they describe the simulation features of the virtual microscope that realistically support the use of this tool for helping students gain experience with experimental techniques and processes. Then they present examples to illustrate how they engaged students in designing a research investigation and to reflect on their learning. They discuss how the ACE-Bio Competencies encouraged them to improve their exercises and to create new ones to elicit rational experimentation research behaviors. Thus, the ACE-Bio Competencies framework helped them improve their virtual microscope platform for use as a simulation tool for biological experimentation and will guide future implementation of the tool.

A focus on applying the Wiggins & McTighe (2005, 2011) Understanding by Design framework (UbD) is presented by Spence (Chap. 11). Spence integrates the two independent UbD and ACE-Bio Competencies frameworks by first thinking through the desired outcomes before outlining step-by-step the guidance given to students through performance tasks and assessments, where students demonstrate scientific practices that are informed by the ACE-Bio Competencies framework. Examples in this chapter demonstrate that students from diverse backgrounds, including those who are not science majors, can actively engage in the scientific process through collaborative investigation activities including peer review. Scaffolding is provided to help students with experimentation practices such as formulating research questions, conducting hypothesis-driven experiments, and using and analyzing data as evidence for drawing conclusions, with exercises to guide students to reflect on these aspects of their research experiences.

The ACE-Bio Competencies were an important affirmation of an instructional sequence of lab modules in a series of courses where students performed and communicated authentic research in a program described by Batzli et al. (Chap. 12). This chapter presents an integrative lab sequence in which students repeatedly learned to use feedback as discourse as they engaged in a sequence of independent research projects where students were guided to solicit and use feedback as an important feature of the scientific enterprise. They present how feedback as a skill is particularly important for students as they plan their experiments and analyze their findings. The authors illustrate how to support students as they learn to give and receive feedback. Readers will see how feedback is strategically scaffolded within a multi-week laboratory curriculum to help students understand that scientific discourse and feedback are key components that serve as "glue" between the various ACE-Bio experimentation competency areas, creating "a culture of curiosity and humility that are at the heart of science."

Dasgupta et al. (Chap. 13) focus on experimental design as a key competence for undergraduate students in a large enrollment introductory biology laboratory course. This example is focused on a single course in which the changes are guided with "backward design" (Wiggins & McTighe, 2005, 2011). The authors first focus on experimental design learning outcomes that can be measured with a deliberate selection of assessment instruments that are aligned to the ACE-Bio Competency areas: identify a problem, ask a question and formulate a hypothesis, conduct an experiment to get data to analyze, and communicate the experimental design and expected findings. After the students engaged with a literature review, they were given a 24-well plate research tool as a visual aid to guide them toward detailing experimental design plans that included replicating experimental and control treatments of relevance to understanding the role of environmental chemicals in embryonic development, using Fetal Alcohol Spectrum Disorder (FASD) as an initial example. Throughout the course, information from the assessment of ACE-Bio Competencies informed modifications to the teaching including moving some activities to an online platform during the pandemic to engage students with planning experiments in their introductory biology lab course. This chapter bridges from student engagement to the assessment of student learning about research in Part IV.

Part IV: Assessment, Evaluation, and Grading What Students Learn About Biological Experimentation

The chapters in Part IV examine the evaluation phase of instruction to show how the ACE-Bio Competencies framework has been used by instructors to address concerns and difficulties linked to the grading of learning from biological experimentation as well as to evaluate the strength of a particular educational program or approach. These chapters illustrate how decisions are made based on assessment of student learning of relevance to foundational experimentation research abilities. They also illustrate how to identify gaps for future work. The process of reasoning with evidence from assessment of student learning can be informed by the assessment triangle (National Research Council, 2001; Pellegrino, 2012). Three key elements underly any assessment of learning about biological experimentation: ideas about the cognitive engagement involved in the research process, use of a measurement tool that elicits a performance to permit observation of cognition related to the expected competence, and an interpretation process for making sense of the expertise or difficulties the student has with that particular aspect of the research process. The ACE-Bio Competencies framework has been used by authors of chapters in Part IV to guide such inferences using evidence from assessments of relevance to the ACE-Bio Competencies areas.

Zelaya et al. (Chap. 14) used the ACE-Bio Competencies framework to categorize individual items that address aspects of experimentation in assessments currently used in undergraduate biology courses. Mapping of assessments on this framework helped instructors to better understand what cognition can actually be observed and assessed. The authors also identified gaps in the ACE-Bio Competencies (collaborative skills, statistical literacy). This critical reflection contributes to the readers' understanding of the ACE-Bio Competencies, its strengths and weaknesses, and the chapter helps education researchers to identify areas of need for future developing our arsenal of assessments related to experimentation.

Kinkade and Wilson (Chap. 15) adopted the claim from the AAAS (2010) report, *Vision and Change in Undergraduate Biology Education*, that effective communication is an essential skill of scientists, and therefore, formal methods of written communication should be a standard part of undergraduate biology education. The authors used the ACE-Bio Competencies framework to draft a Research Across Curriculum Rubric (RAC-R), which they modified in collaboration with instructors in their department to come up with an adaptable rubric for use at multiple levels to evaluate journal article style lab reports. RAC-R potentially benefits students by articulating the skills required to become an "accomplished" scientific writer. The RAC-R benefits the biology department by providing evidence from consistent assessment of student achievement and performance in biological research communication. Biology instructors could use RAC-R "as is," or they could independently or in collaboration with other members of a department modify criteria or levels of accomplishment as needed to meet instructional and assessment needs.

Rulfs and Caron (Chap. 16) found the level of detail and the elaboration of skills and concepts for each of the ACE-Bio Competencies to be useful for assessing student progress in the research process and as an important affirmation of their institutional expectations and department's instructional sequence where students do and communicate authentic research. They also evaluated their expected department outcomes in terms of the strength of their existing assessment plans that they aligned relative to foundational skills used in biological experimentation. As a result of their alignment study, their critical analysis of the ACE-Bio Competencies revealed that "the ability to function on multidisciplinary teams" (an ABET 2017 Criteria for Accrediting Engineering Programs), and to "function effectively in a collaborative scientific environment" (one of their department outcomes) are two critical components not included in the ACE-Bio Competencies. This chapter highlights for readers a way to use the ACE-Bio Competencies for planning a comprehensive program of student assessment to inform stakeholders at the department and institution level. By pointing out a limitation of the ACE-Bio Competencies, this chapter also introduces the importance of topics addressed in chapters of this book that are less closely related to the ACE-Bio framework.

Shiyao Liu and coauthors (Chap. 17) examined published assessments using the Conceptual Analysis of Disciplinary Evidence (CADE) framework as a lens. CADE is another framework of value to the teaching of the research process that is independent of the ACE-Bio Competencies framework. CADE is a more holistic framework than evidence-based reasoning because it explicitly examines both the disciplinary knowledge and epistemological considerations of relevance to students' use of evidence at all stages of the research process, including the planning and conducting of experiments. The authors conducted comprehensive literature searches in six databases to identify publications with assessments to measure what undergraduate biology students learn about science research practices. The searches were not just limited to the assessment of student learning about experimentation, because the chapter is broadly focused on the assessment of student learning about inquiry including research practices such as evolutionary biology tree-thinking, which thus expands the focus of our book to consider the use of evidence for biological research in general. A few existing assessments measure epistemic considerations of relevance to students' use of research evidence, and those that do are generally used for grading students' scientific writing. No assessments were found in the literature that reveal how students link epistemic and disciplinary knowledge as a key component of their understanding and use of scientific evidence in ways that could be incorporated into a written pre-test or final exam. The CADE framework was then used to illustrate the design of assessments to measure the use of evidence as a competence for evolutionary tree-thinking as a disciplinary research context. These examples demonstrate how CADE as a tool could be used to evaluate or guide the development of assessments of research practices to address other problems or in other disciplines. This chapter bridges from the assessment of learning about experimentation informed by the ACE-Bio Competencies framework to explore multiple alternative frameworks of relevance to teaching and learning about research in Part V.

Part V: Complementary Frameworks for Guiding Students' Experimentation Practice

The chapters in Part V introduce a range of other frameworks and scientific practices that both complement the ACE-Bio Competencies framework and provide important and useful guidelines for developing students' experimentation practice. These particularly focus on the use of so-called "soft" skills, such as motivation, meaning-making, innovation, creativity, scientific and evidentiary reasoning, and critical thinking, that complement and enhance the successful use of the so-called "hard" experimental competences like hypothesizing, analyzing, etc., advocated by the ACE-Bio Competencies framework. Clearly, all the above skills, and many others, constitute the repertoire of skills necessary for becoming a complete and competent scientist, and therefore need to be developed to an optimal extent in our biology students.

Experimentation in science is integrated with other scientific practices that give experiments purpose and meaning. Yet most students experience experiments as stand-alone activities. In Chap. 18, Gouvea et al. argue that this makes it challenging for students to motivate and make meaning from experimentation. One way to ameliorate these challenges is to put experiments in conversation with other scientific practices. The authors describe the design of an introductory biology laboratory curriculum in which students conduct investigations using both experiments and computational models. They explore the potential for interactions between computational models and experiments to expand how students experience experimentation. Specifically, they demonstrated how this interaction can support students in identifying and articulating questions that motivate experimentation and the interpretation of data to generate meaning.

Buckholt and Rulfs (Chap. 19) focus on the use of electronic laboratory notebooks and how they can facilitate good experimental practice. This skill is transferable from the formal teaching laboratory to less structured environments such as research labs in universities and industry. In comparing the pros and cons of electronic versus paper lab notebooks, they particularly highlight the importance of instructors developing their students' abilities to archive, share, and analyze data. This includes helping students develop data documentation and data curation skills by continually communicating expectations and course objectives through various feedback mechanisms.

The work by Beno and Tucker (Chap. 20) focuses on the development of students' "soft" skills such as innovation, creativity, collaboration, communication, and critical thinking. To achieve this, the authors propose a toolkit designed to assess these skills and thereby stimulate the focus of instruction and learning on such skills. The toolkit includes surveys, semi-structured interview questions, and feedback reflection questions. They also advocate the regular use of behavioral assessment surveys to track student self- and peer-perceived growth in the innovation skill areas, as well as the use of guided discussion and feedback by the instructor or an evaluator during the learning process to identify student strengths and weaknesses and to correct any deficiencies. All such soft skills are considered key to successful experimentation and without their utilization could render the implementation of the "hard" science research competencies less effective. Thus, the authors argue that these skills need to be developed in concert with ACE-Bio Competencies as a future perspective.

Chaonan Liu and others (Chap. 21) focus on scientific reasoning according to members of a project funded by the National Science Foundation in the United States, namely the Faculty Developer Network for Undergraduate Biology Education (FDN-UBE), and they provide insights from the Conceptual Analysis of Disciplinary Evidence (CADE) framework. CADE, which was developed by the authors independently from the ACE-Bio Competencies, explicitly points out epistemological considerations in current life science research, which is absent from the ACE-Bio Competencies. They take a comprehensive look at the use of evidence throughout the research process, which they reveal is intuitively valued by biology faculty who were trained as scientists. Quotes, from the FDN-UBE members, who were interviewed about their interests and expertise, also emphasize the role of disciplinary knowledge in the practice of formulating testable hypotheses, explanations, or rationale for an investigation. The coding of practices that are valued according to the

CADE framework reveals important practices for helping students understand and use scientific evidence.

Part VI: Approaches to Biological Experimentation Instruction of Relevance to Biology Education Programs in General

The chapters in Part VI highlight several different approaches to teaching about biological experimentation which could be usefully incorporated by readers in biology education programs in general. These include approaches that foster a strong focus on collaboration among researchers, on their use of multiple representations, and on the deployment of scaffolding, feedback, and constraint in the design of educational activities. This Part ends with a chapter aimed at reminding readers that all innovations in this book, no matter how potentially useful, might require special strategies by instructors to overcome departmental and institutional barriers before they can be successfully implemented.

In Chap. 22, Johnson and coworkers focus on the importance of collaboration in the performance of modern biological research. To model collaborative research in an undergraduate course, they expose students to a wide range of activities aimed at developing their appreciation of the importance of collaboration in research, and the use of inclusive collaboration practices that value both their own and other collaborator contributions, regardless of racial, ethnic, socioeconomic, and religious backgrounds. The chapter ends with a useful list of recommendations that educators could use to design collaboration components in a course, including student activities aimed at developing their collaborative skills in the context of biological experimentation.

In Chap. 23, Jefferies and Jefferies review and evaluate the usefulness of representations, including drawings, paintings, music, and media, for the teaching of biochemical concepts at primary, secondary, and post-secondary educational levels. They analyze the usefulness and limitations of each type of representation and propose future areas of research for each type in accordance with basic competencies of biological experimentation. They also offer guidelines for instructors for teaching with representations, including how to incorporate them into creative course curricula that take cognizance of student learning styles and educational backgrounds.

Meir (Chap. 24) focuses on the use of scaffolding, feedback, and constraint in the design of educational activities. The central argument in this chapter is based on extensive evidence that, of these three, constraint has been underappreciated. Constraint is defined by the author as the amount of freedom to make choices that is provided to the student by the environment in which the educational activity takes place. Thus, a high constraint activity or question would be close ended, such as multiple choice, and low constraint would correspond to more open-ended activities and therefore is better suited to students of higher educational levels engaging in experimentation where there is not necessarily one answer or way of performing an experiment. They use examples to propose an Intermediate Constraint Hypothesis that states that focusing on level of constraint in student exercises and assessments is a practical and powerful way to maximize student learning and instructors' ability to assess student understanding of complex skills such as those required in biological experimentation. The chapter ends with useful tips on how to implement the hypothesis, including ensuring that the degree of constraint in an activity or assessment is appropriate to the level of the target student population, and in some cases considering whether to replace feedback and/or scaffolding with constraint.

In Chap. 25, Anderson and Pelaez provide a range of practical ideas on how readers might approach the implementation of the various innovations presented in this book and other literature. They highlight the importance of identifying the different stakeholders affected by the implementation, what they would tolerate and find feasible, as well as any potential barriers that may impact the effective and successful implementation of an innovation. Toward this end, potential contextual forces that may support and oppose implementation are listed for readers' convenience, together with several strategies that could be deployed to overcome the negative forces and promote the positive forces to ensure that colleagues and students accept and value the curricular changes.

Cross-Cutting Trends

The structure of this book reflects the ways in which the different trends (first framed by ACE-Bio Competencies and then by other frameworks) drive reform. However, the chapters present a systematic comparison of additional trends that did not neatly fit into the six parts of the book. Here we point out four cross-cutting trends that are best addressed by comparing chapters from different parts of the book: (1) using backward design (Wiggins & McTighe, 2005, 2011) to align instruction and assessment of student learning to anticipated learning outcomes; (2) integrating promotion of soft skills with students' experimentation practices so that biology will be understood as a research science that informs collective decisions to meaningfully impact people's lives; (3) using technology and development of online approaches to experimentation education; and (4) engaging instructors with research scientists and education specialists to focus educational programs on helping biology students understand and use scientific evidence as a foundation for their knowledge of biology.

Multiple chapters address the alignment of the experimentation competencies that are taught with how they are taught and assessed as a driver of change toward developing students' competence for biological research in the educational process. Some authors provided exemplars of assignments of value for preparing students to do biological experimentation. An often-used strategy to align anticipated learning outcomes to course activities and assessments was backward design (Wiggins & McTighe, 2011), employed with a focus on providing students with opportunities for experimentation according to Bowe and Irby (Chap. 2), Cheng et al. (Chap. 5), Thomas (Chap. 6), and Gardner et al. (Chap. 8). Other chapters provide examples of assessments of student learning about experimentation since direct measures are needed to confirm ALOs as verified learning outcomes (VLOs) according to Bowe and Irby (Chap. 2) and S. Liu et al. (Chap. 17) or else the instructional activities or the goals for the course may need to be changed. Therefore, the alignment of assessments with specific experimentation competencies in chapters by Dasgupta et al. (Chap. 13) Zelaya et al. (Chap. 14), Kinkade and Wilson (Chap. 15), Rulfs and Caron (Chap. 16), Liu et al. (Chap. 17), Beno and Tucker (Chap. 20), and Johnson et al. (Chap. 22) are most useful for exploring assessment options.

So-called "soft skills" connect biology education to students' lives as a whole and help show the relevance of biology education to society. A strand on collaboration is found throughout the book, stated in particular by Zelaya et al. in Chap. 14 and reinforced by Beno and Tucker in Chap. 20 where the argument emerged that collaboration is missing from ACE-Bio Competencies and yet this is important in the future. Johnson et al. review the literature and how to explicitly address this goal in Chap. 22 (https://doi.org/10.1007/978-3-030-98592-9_22). Furthermore, other chapters also target helping students function effectively in a collaborative environment, including those by Cole and Beck (Chap. 3), Chatzikyriakidou and McCartney (Chap. 4), Linton et al. (Chap. 7), Gardner et al. (Chap. 8), Casali et al. (Chap. 10), Kinkade and Wilson (Chap. 15), Rulfs and Caron (Chap. 16), and Jefferies and Jefferies (Chap. 23), and all of that work would especially benefit from the Johnson et al. (Chap. 22 https://doi.org/10.1007/978-3-030-98592-9_22) report on explicit teaching of collaboration. These chapters focus on involving students with investigating "real world" issues by supporting collaborative group work. As another major contribution, detailed examples show how to promote constructive feedback throughout the scientific process using peer-feedback approaches described by Thomas (Chap. 6), Gardner et al. (Chap. 8), Kruchten and O'Brien (Chap. 9), Spence (Chap. 11), Beno and Tucker (Chap. 20), C. Liu et al. (Chap. 21), and especially Batzli et al. (Chap. 12). Finally, consider how Thomas (Chap. 6), Spence (Chap. 11), Beno and Tucker (Chap. 20), and C. Liu et al. (Chap. 21) highlight the role of science communication (Table 1.9 in Chap. 1 in this volume) throughout the research process.

The use of online technology is a way to extend learning about experimentation beyond the constraints of a typical laboratory classroom. Several chapters help educators understand how to guide students in developing competence with experimentation by addressing authentic research questions with a range of different online resources. Some chapters present examples of instruction with online tools used to optimize opportunities for doing experimentation according to Casali et al. (Chap. 10), Gouvea et al. (Chap. 18), and Meir (Chap. 24). In contrast, other chapters use tools in ways that reflect common professional scientific practices for data archiving, sharing, and analyses such as those by Linton et al. (Chap. 7) and Buckholt and Rulfs (Chap. 19), while Meir (Chap. 24) goes on to discuss a theoretical focus on tailoring instruction to learner expertise. These chapters also provide general principles for scaffolding and feedback to help students develop competence with experimentation in ways that could be done with remote learning but that also apply to a laboratory classroom.

Several of the chapters illustrate how to involve biology instructors and research scientists, who have engaged in discussions and the work needed to unpack at various levels of their institutional context, what it means to work with students who are developing competence with research in the life sciences. Particular examples by Cole and Beck (Chap. 3), Kinkade and Wilson (Chap. 15), and Rulfs and Caron (Chap. 16) target reform focused on experimentation instruction at a larger level than one single course. They do this by demonstrating how to generate discussion among members of the faculty in their own departments. However, to harness the complementary expertise of the range of biology faculty who must work together on improving student competence with biological experimentation, they should commit to engaging guiding principles for collaboration, collegiality, communication, and continuity as they work together to develop and improve undergraduate biology programs (Pelaez et al., 2018). With more sharing of the frameworks in this book to clarify the educational mission beyond a single course, imagine how efficient experimentation education of biology students could become! However, educators would need to debate and then agree on, or modify for their own use, one or more of these models of practice. Possible strategies for promoting the success of this collaborative process are presented by Anderson and Pelaez in Chap. 25.

Contextual and Practical Implications for Instructors

Chapters in this book include explicit evidence-based guidelines as a bulleted list summary to help educators discuss options for supporting students' development of skills that are indispensable to different aspects of biological experimentation. It should be noted that many of the contributing authors are basic research scientists with no specialized pedagogical or education research training. Author teams were paired with discipline-based education research biologists for a peer review process that was set up to organize and improve chapters and as a mechanism for science faculty professional development. Since this entire book was written during the COVID-19 pandemic, we acknowledge the hard work done under difficult conditions by contributing authors who took time to support each other despite the pandemic. We are aware of other investigators who are doing relevant work and who intended to contribute a chapter to this book but were unable to at such a difficult time. We hope to see their work in journals soon.

The practical implications in this book aim to reduce specific challenges by providing support to post-secondary educators linked to the design and implementation of biology instruction of relevance to students' development of evidentiary reasoning about biological research. However, biology educators need to decide what disciplinary contexts and influences are relevant to their own situation, which challenges need targeting, and in what order of priority in line with the constraints and affordances of their local educational environment. To accomplish this aim, we offer the following recommendations as points to discuss for educators and future education researchers who use any chapters or sections of this book:

- Delve deeply with students into discovering what biological sciences research can and is doing for society
- Equip students with the disciplinary knowledge of relevance to their experiments
- Explicitly teach and assess science epistemology considerations for experiments, i.e., the scientific skepticism and development of logical reasoning about evidence that is essential along with their technical skill development
- Study performance from various diverse populations of students and, if necessary, adapt curricula to optimize their performance
- Investigate to what degree skills learned in a course transfer from one context to another context of value to society
- Design and test assessments that align with competencies to measure the success of educational innovations that involve students in experimentation
- Most importantly, employ strategies to identify and address barriers and incentives for incorporating authentic biology experimentation into biology

Summary

In summary, this book is intended for use by people who are delving deeply into what the biological sciences can and are doing for society, and how to teach post-secondary students to understand biology as a research science. Today, as we write this Preface the world is gripped by a raging global pandemic. Society faces important questions that require people to understand the difference between speculation and sound inferences based on evidence from data gathered using rigorous experimentation. The biological sciences are a set of agreed-upon practices from a discipline that has continually been striving to improve research procedures since before the time of William Harvey when he wrote that teaching and learning were "not from books…, not from positions of philosophers, but from the fabric of nature" (Harvey, 1628, p. 42). There may be no trust in biology without knowledge that has been called the epistemology of science, i.e., the skepticism and rigor of experimentation that must be taught along with disciplinary knowledge.

Acknowledgments Our thanks go to all members of the ACE-Bio Network and BASIL for their wide-ranging contributions to advance the work supported by NSF grants #1346567 for the Advancing Competencies in Experimentation–Biology (ACE-Bio) Network and #1710051 for the Biochemistry Authentic Scientific Inquiry Lab (BASIL) project. All chapter authors received a stipend as an ACE-Bio Network collaborator or they attended an ACE-Bio Network workshop. Any opinions, findings, conclusions, or recommendations expressed are those of the authors and do not necessarily reflect the views of the NSF. In particular, we thank the research scientists and education specialists who shared ideas collected under IRB protocol #1506016184 for RCN-UBE: ACE-Bio Network Phase 2. Additionally, we thank the ACE-Bio Network external advisory board members (James

B. Bassingthwaighte, Cynthia M. Bauerle, Erin Dolan, Maria Araceli Ruiz-Primo, and James Pellegrino) for their advice and support.

West Lafayette, IN, USA

Nancy J. Pelaez

Stephanie M. Gardner

Trevor R. Anderson

References

Akuma, F. V., & Callaghan, R. (2019). A systematic review characterizing and clarifying intrinsic teaching challenges linked to inquiry-based practical work. *Journal of Research in Science Teaching*, *56*, 619–648, especially Figure 2 on page 627 and Table 3 on page 624 at https://doi.org/10.1002/tea.21516

American Association for the Advancement of Science. (2010). Brewer C, Smith D. (Editors). *Vision and Change in undergraduate biology education: A call to action*. AAAS. Retrieved December 23, 2020, from https://live-visionandchange.pantheonsite.io/wp-content/uploads/2013/11/aaas-VISchange-web1113.pdf

Anderson, T. R., & Rogan, J. M. (2011). Bridging the educational research-teaching practice gap. Curriculum Development, Part 1: Components of the curriculum and influences on the process of curriculum design. *Biochemistry and Molecular Biology Education*, *39*(1), 68–76, especially the Figure 1 diagram on page 69 at https://doi.org/10.1002/bmb.20470

Harvey, W. (1628). *An anatomical disquisition on the motion of the heart and blood in animals*. Retrieved April 11, 2022 from the London: Sydenham Society publication online at https://www.fadedpage.com/link.php?file=20130911-a5.pdf

National Research Council. (2001). *Knowing what students know: The science and design of educational assessment*. The National Academies Press. https://doi.org/10.17226/10019

Pelaez, N., Anderson, T. R., Gardner, S. M., Yin, Y., Abraham, J. K., Bartlett, E. L., Gormally, C., Hurney, C. A., Long, T. M., Newman, D. L., Sirum, K., & Stevens, M. T. (2018). A community-building framework for collaborative research coordination across the education and biology research disciplines. *Life Sciences Education*, *17*(2). https://doi.org/10.1187/cbe.17-04-0060

Pellegrino, J. W. (2012). Assessment of science learning: Living in interesting times. *Journal of Research in Science Teaching*, *49*, 831–841. https://doi.org/10.1002/tea.21032

Thanukos, A., Scotchmoor, J. G., Caldwell, R., & Lindberg, D. R. (2010). The real process of science: Understanding science at http://www.understandingscience.org and Science 101: Building the foundations for real understanding. *Science*, *330*(6012), 1764–1765. https://doi.org/10.1126/science.1186994

Wiggins, G. P., & McTighe, J. (2005). *Understanding by design*. Association for Supervision and Curriculum Development (ASCD).

Wiggins, G. W., & McTighe, J. (2011). *The understanding by design guide to creating high-quality units*. Association for Supervision and Curriculum Development.

Contents

Editors and Contributors

About the Editors

Nancy J. Pelaez Pelaez, Professor Emerita of Biology, Purdue University, was Associate Editor of *Advances in Physiology Education* and Retiring Chair of Section Q (Education) for the American Association for the Advancement of Science (AAAS). With a PhD in Physiology and Biophysics, Pelaez is a biology educator who started as a science teacher with K-12 California teaching credentials in life sciences and physical science. Pelaez has been honored with a Fulbright Award to Vienna, Austria, the American Physiological Society Guyton Educator of the Year Award, and as AAAS Fellow. Biological Sciences, Purdue University, West Lafayette, IN, USA

Stephanie M. Gardner Gardner, Associate Professor of Biological Sciences, Purdue University and Monitoring Editor of *CBE-Life Science Education*, aims to reveal student knowledge while engaging them in the practices of science as a foundation to design evidence-based instruction. Taking a qualitative research approach and drawing on theory and methods from education and the learning sciences, she examines students' exploration of biological evidence, systems thinking, and understanding of data and graphing, and teaches biological experimentation. Biological Sciences, Purdue University, West Lafayette, IN, USA

Trevor R. Anderson Anderson, Professor Emeritus of Chemical Education and Biochemistry, Purdue University, served on the IUBMB Education Committee and the editorial board of Biochemistry and Molecular Biology Education (BAMBEd) for 22 years. He does biochemistry education research into visual literacy, science inquiry and reasoning about experiments, assessment of reasoning with concepts and representations, and faculty development, including the Bridging-the-Gap series in *BAMBEd* aimed at the application of educational research to teaching practice. Chemistry, Purdue University, West Lafayette, IN, USA

Contributors

Deborah Allen Department of Biological Sciences, University of Delaware, Newark, DE, USA

Trevor R. Anderson Department of Chemistry, Purdue University, West Lafayette, IN, USA

Aakanksha Angra Department of Biology, Georgia State University, Atlanta, GA, USA
Office of Medical Education, University of Minnesota, Minneapolis, MN, USA

Janet M. Batzli Biology Core Curriculum (Biocore), University of Wisconsin-Madison, Madison, WI, USA

Christopher W. Beck Department of Biology, Emory University, Atlanta, GA, USA

Sarah M. Beno Department of Biology, Birmingham-Southern College, Birmingham, AL, USA

Lawrence S. Blumer Department of Biology, Morehouse College, Atlanta, GA, USA

Laura Bonofiglio Departamento de Microbiología, Biotecnología, Inmunología y Genética. Cátedra de Microbiología, Universidad de Buenos Aires. Facultad de Farmacia y Bioquímica, Buenos Aires, Argentina

Kathleen A. Bowe Department of Chemistry, University of New Hampshire, Durham, NH, USA

Michael A. Buckholt Biology and Biotechnology Department, Worcester Polytechnic Institute (WPI), Worcester, MA, USA

Chao Cai Department of Biological Sciences, Purdue University, West Lafayette, IN, USA

Jessica Caron Northeastern University, Boston, MA, USA

Cecilia I. Casali Departamento de Ciencias Biológicas. Cátedra de Biología Celular y Molecular, Universidad de Buenos Aires. Facultad de Farmacia y Bioquímica, Buenos Aires, Argentina

Kyriaki Chatzikyriakidou STEM Transformation Institute, Florida International University, Miami, FL, USA

Zhiyong Cheng Food Science and Human Nutrition Department, Institute of Food and Agricultural Sciences, University of Florida, Gainesville, FL, USA

Kari L. Clase Department of Agricultural and Biological Engineering, Biotechnology Innovation and Regulatory Science (BIRS) Center, Purdue University, West Lafayette, IN, USA

Megan F. Cole Department of Biology, Emory University, Atlanta, GA, USA

Annwesa Dasgupta STEM Education Innovation and Research Institute (SEIRI), Office of the Vice Chancellor for Research, Indiana University Purdue University, Indianapolis (IUPUI), Indianapolis, IN, USA

Natalie F. Douglas Departments of Biology and Communication Sciences & Disorders, Central Michigan University, Mt. Pleasant, MI, USA

Elizabeth Ellwood iDigBio, Florida Museum of Natural History, University of Florida, Gainesville, FL, USA

Nicolás O. Favale Departamento de Ciencias Biológicas. Cátedra de Biología Celular y Molecular, Universidad de Buenos Aires. Facultad de Farmacia y Bioquímica, Buenos Aires, Argentina

Stephanie M. Gardner Department of Biological Sciences, Purdue University, West Lafayette, IN, USA

Julia Gouvea Department of Education, Tufts University, Medford, MA, USA

Michelle A. Harris Biology Core Curriculum (Biocore), University of Wisconsin-Madison, Madison, WI, USA

Joseph A. Harsh Department of Biology, James Madison University, Harrisonburg, VA, USA

Robert Hayes Department of Education, Tufts University, Medford, MA, USA

Heidi A. Horn Department of Biology, Edgewood College, Madison, WI, USA

Stefan M. Irby Department of Chemistry, University of Central Florida, Orlando, FL, USA

Latisha R. Jefferies Department of Chemistry, United States Air Force Academy, USAFA, Colorado Springs, CO, USA

Shanae S. Jefferies Department of Sociology, Collin College, McKinney, TX, USA

Kathryn M. S. Johnson Trail Build, LLC, East Troy, WI, USA

Karla B. Kinkade Biology Department, College of Arts and Sciences, Marian University, Indianapolis, IN, USA

Anne E. Kruchten Biology Department, The College of St. Scholastica, Duluth, MN, USA

Dennis Lee BSCS Science Learning, Colorado Springs, CO, USA

Debra L. Linton Departments of Biology and Communication Sciences & Disorders, Central Michigan University, Mt. Pleasant, MI, USA

Chaonan Liu Department of Biological Sciences, Purdue University, West Lafayette, IN, USA

Shiyao Liu Department of Biological Sciences, Purdue University, West Lafayette, IN, USA

Namoonga M. Mantina Mel and Enid Zuckerman College of Public Health, University of Arizona, Tucson, AZ, USA

James A. Marrs Department of Biology, Indiana University Purdue University, Indianapolis (IUPUI), Indianapolis, IN, USA

Kathleen A. Marrs Department of Biology, Indiana University Purdue University, Indianapolis (IUPUI), Indianapolis, IN, USA

Melissa McCartney STEM Transformation Institute, Florida International University, Miami, FL, USA
Department of Biological Sciences, Florida International University, Miami, FL, USA

Eli Meir SimBiotic Software, Inc, Missoula, MT, USA

Anna K. Monfils Departments of Biology and Communication Sciences & Disorders, Central Michigan University, Mt. Pleasant, MI, USA

Rocio A. Moreira Szokalo Departamento de Bioquímica Clínica. Cátedra de Bioquímica Clínica II. Citogenética Humana y Genética Toxicológica, Universidad de Buenos Aires. Facultad de Farmacia y Bioquímica, Buenos Aires, Argentina

Jenean H. O'Brien Biology Department, The College of St. Scholastica, Duluth, MN, USA

Nancy J. Pelaez Department of Biological Sciences, Purdue University, West Lafayette, IN, USA

Heather R. Pelzel Department of Biological Sciences, University of Wisconsin-Whitewater, Whitewater, WI, USA

Lucila G. Pescio Departamento de Ciencias Biológicas. Cátedra de Biología Celular y Molecular, Universidad de Buenos Aires. Facultad de Farmacia y Bioquímica, Buenos Aires, Argentina

Daniela J. Romero Departamento de Ciencias Biológicas. Cátedra de Biología Celular y Molecular, Universidad de Buenos Aires. Facultad de Farmacia y Bioquímica, Buenos Aires, Argentina

Jill Rulfs Biology and Biotechnology Department, Worcester Polytechnic Institute (WPI), Worcester, MA, USA

Ala Samarapungavan Department of Educational Studies, Purdue University, West Lafayette, IN, USA

Bruno J. Santacreu Departamento de Ciencias Biológicas. Cátedra de Biología Celular y Molecular, Universidad de Buenos Aires. Facultad de Farmacia y Bioquímica, Buenos Aires, Argentina

Swapnalee Sarmah Department of Biology, Indiana University Purdue University, Indianapolis (IUPUI), Indianapolis, IN, USA

Matt Simon Department of Education, Tufts University, Medford, MA, USA

Porché L. Spence Department of Biological and Biomedical Sciences, North Carolina Central University, Durham, NC, USA
Department of Natural Resources and Environmental Design, North Carolina Agricultural and Technical State University, Greensboro, NC, USA

Aeisha Thomas Crown College, St. Bonifacius, MN, USA

Diane C. Tucker UAB Honors College and Department of Psychology, University of Alabama at Birmingham, Birmingham, AL, USA

Aditi Wagh Department of Comparative Media Studies & Writing, Massachusetts Institute of Technology, Cambridge, MA, USA

Lisa D. White University of California Museum of Paleontology, University of California at Berkeley, Berkeley, CA, USA

Kristy J. Wilson Biology Department, College of Arts and Sciences, Marian University, Indianapolis, IN, USA

Anna J. Zelaya Department of Biology, Emory University, Atlanta, GA, USA

Abbreviations

5CCs	Five core concepts of biology
AAAS	American Association for the Advancement of Science
AAC&U	American Association of Colleges and Universities
ABET	Accrediting Board for Engineering and Technology
ACE-Bio	Advancing competencies in experimentation–biology
AIM-Bio	Authentic inquiry through modeling
ALO	Anticipated learning outcomes
ANOVA	Analysis of variance
ARE	Apprentice-based research experiences
ASCD	Association for supervision and curriculum development
BASIL	Biochemistry Authentic Scientific Inquiry Lab
BEDCI	Biological experimental design concept inventory
Bio-TAP	Biology thesis assessment protocol
BioSQuaRE	Biological science quantitative reasoning exam
BioVEDA	Biological variation in experimental design and analysis
BISON	Biodiversity information serving our nation
BLAST	Basic local alignment search tool
BLUE	Biodiversity literacy in undergraduate education
BMB	Biochemistry and molecular biology
C-R-M	Conceptual knowledge-reasoning-mode of representation model
CADE	Conceptual analysis of disciplinary evidence
CAM	Cognitive apprenticeship model
cAMP	Cyclic adenosine monophosphate
CEAT	Claim and evidence assessment tool
COVID-19	Coronavirus disease of 2019
CRBS	Competency Rubric Bank for the Sciences
CRE	Course-based research experience
CSE	Council of Science Editors
CURE	Course-based undergraduate research experience
CYP3A4	Cytochrome P450 3A4
DAG	Diacylglycerol

DNA	Deoxyribonucleic acid
E-EDAT	Expanded experimental design ability tool
EBE	Exploring biological evidence
EDAT	Experimental design ability test
EDR	Experimental design rubric
ELN	Electronic laboratory notebook
ENaC	Epithelial sodium channel
EV	Evolution
FDN-UBE	Faculty Development Network for Undergraduate Biology Education
FFyB	Facultad de Farmacia y Bioquímica
FINERMAPS	Feasible, interesting, novel, ethical, relevant, manageable, appropriate, potential value and publishability, and systematic
GAPDH	Glyceraldehyde 3-phosphate dehydrogenase
GBIF	Global Biodiversity Information Facility
GDP	Guanosine diphosphate
GEA	Group effort analysis
GTP	Guanosine triphosphate
HDX	Hydrogen-deuterium exchange
HDX-MS	Hydrogen-deuterium exchange mass spectrometry
IACUC	Institutional Animal Care and Use Committee
ICT	Information and Communication Technology
iDigBio	Integrated digitized biocollections
IFES	Information flow, exchange, and storage
IMCA	Introductory molecular and cell biology assessment
IP3	Inositol triphosphate
IPSA	Individual problem-solving assessment
IRB	Institutional Review Board
IWCR	Inquiry and written communication rubric
jmuBERG	James Madison University Biology Education Research Group
K	Kindergarten
LB	Luria-Bertani
LCAS	Laboratory Course Assessment Survey
LMS	Learning Management System
MACH	Methods-analogy-context-how
MAtCH	Methods-analogy-theory-context-how
MBDAT	Molecular biology data analysis test
MRC	Meta-representational competence
NAS	National Academies of Science
NASEM	National Academies of Sciences, Engineering and Medicine
NEON	National Ecological Observatory Network
NGSS	Next generation science standards
NIH	National Institutes of Health
NOS	Nature of Science
NRC	National Research Council

NSF DUE	National Science Foundation Division of Undergraduate Education
NSF	National Science Foundation
OER	Open educational resources
PDB	Protein Data Bank
PEPCK	Phosphoenolpyruvate carboxykinase
PIBERG	Purdue International Biology Education Research Group
PICURA	Process to identify course-based undergraduate research abilities
PIP2	Phosphatidylinositol 4,5-bisphosphate
POGIL	Process oriented guided inquiry learning
POS	Process of science
PTEM	Pathways and transformations of energy and matter
QLR	Quantitative literacy rubric
QUBES	Quantitative undergraduate biology education and synthesis
RAC-R	Research across the curriculum rubric
RED	Rubric for experimental design
SDS-PAGE	Sodium dodecyl sulphate–polyacrylamide gel electrophoresis
SEA-PHAGES	Science Education Alliance-Phage Hunters Advancing Genomics and Evolutionary Science
SEGU	Stream ecology graphing unit
SERNEC	Southeast Regional Network of Expertise and Collections
SF	Structure and function
SPFA	Scientific process flowchart assessment
SRBCI	Statistical reasoning in biology concept inventory
Src	Src kinase
SRQ	Scientific reasoning questions
STEM	Science, technology, engineering, and mathematics
SURE	Summer undergraduate research experience
SYS	Systems
TA	Teaching assistant
TIED	Tool to assess interrelated experimental design
TIPS	Test of integrated process skills
TOSLS	Test of scientific literacy skills
UbD	Understanding by design
UC	University of California
URSSA	Undergraduate research student self-assessment
uTA	Undergraduate teaching assistant
VLO	Verified learning outcomes
VM	Virtual microscope
WPI	Worcester Polytechnic Institute
WR	Weighted relevance
WSI	Whole slide imaging
ZFI	Zone of feasible innovation
ZoT	Zone of tolerance

Part I
Vision and Initiation Phase: Envisioning What, When, and How Students Learn About Biological Experimentation

Chapter 1
The Problem with Teaching Experimentation: Development and Use of a Framework to Define Fundamental Competencies for Biological Experimentation

Nancy J. Pelaez, Stephanie M. Gardner, and Trevor R. Anderson

1.1 Scientific Rigor in Experimentation Is Integral to Trust in Science

Undergraduate students are increasingly learning about responsible conduct of research, often by doing biology investigations to meet more rigorous academic criteria, to gain a competitive employment edge upon graduation, or for various other reasons (National Academy of Sciences (NAS), 2009; American Association for the Advancement of Science (AAAS), 2011; Corwin et al., 2015). While observational studies provide valuable insights into what is happening, without experiments there would be no way of investigating the nature of mechanisms in living systems; for example, experiments revealed how a firefly glows and how cells "know" when to divide. Seymour and Hunter (2019) report that undergraduate students who felt well-prepared in terms of background knowledge or academic skills had experienced curriculum that focused on inquiry to make the material more interesting and open-ended. They also reported valuing creative exploration, exposure to scientific research or publications, and good teacher characteristics that included use of interactive inquiry approaches. However, many students struggle with experimentation (Ruiz-Primo et al., 2010; Shi et al., 2011; Dasgupta et al., 2014; Irby et al., 2018), despite numerous calls to involve them in authentic research experiences (AAAS, 2011; Corwin et al., 2015; Mulnix & Vandergrift, 2014;

N. J. Pelaez (✉) · S. M. Gardner
Department of Biological Sciences, Purdue University, West Lafayette, IN, USA
e-mail: npelaez@purdue.edu; sgardne@purdue.edu

T. R. Anderson
Department of Chemistry, Purdue University, West Lafayette, IN, USA
e-mail: ander333@purdue.edu

N. J. Pelaez et al. (eds.), *Trends in Teaching Experimentation in the Life Sciences*, Contributions from Biology Education Research,
https://doi.org/10.1007/978-3-030-98592-9_1

O'Mahony et al., 2019). Designing experiments in biology involves students framing research questions to investigate some component or the entirety of a living subject, being able to define and understand measurable variables, and processing, visualizing and interpreting results. All of these processes require students to engage with biology in a way that can help them understand how experimental research is done to yield all the knowledge we have about complex mechanisms in the natural world. But despite the obvious importance of such knowledge and skills building in the education of biology students, surprisingly little is known about what students actually learn from conducting biology investigations, compared to what they ought to learn to become knowledgeable and productive citizens.

The COVID-19 pandemic has illustrated the need for all students, not just those who will become scientists, to understand the value of rigor in scientific experimentation. Many people do not understand how rigorous experimentation in the life sciences relies on disciplinary norms that make it possible, for example, to discover safe and reliable vaccines that the public can trust, i.e. through application of scientific research methods to an unbiased sample; a well-controlled experimental design; transparency in communicating the full experimental details in terms of methodology, analysis, and interpretations; and, with reproducibility to document the variation so that others may trust the results of the research and understand their limitations. Unfortunately, in the United States part of this problem stems from a failure to provide equitable access for pre-college students to experience interactive inquiry-based science teaching methods known to make the material more interesting, drawing students' attention toward science as a creative, open-ended research endeavor. Data from a recent study done by the National Survey of Science and Mathematics Education were analyzed according to "equity factors" including community type (urban, suburban, rural). A study by Trygstad et al. (2020) at the pre-college level reported that classes in urban schools were more likely than classes in rural schools to have children learn science by doing science, yet few science classes in either category had students use data and reasoning to defend a claim or refute alternative claims (26% vs. 17%), and even fewer classes had students determine what details about an investigation might persuade a targeted audience about a scientific claim (17% vs. 10%). Therefore, it is important for all students to be given opportunities at the tertiary level to understand and experience how experiments are done. Unfortunately, most university students do not take a course that explicitly teaches rigorous experimentation. Therefore, those who receive no formal instruction in the principles of experimentation must infer what they know about it from reading the literature, which means that questions remain about how to build public confidence and reliability in science research even though this issue was already raised 35 years ago (Koshland, 1985).

News in the past decade has highlighted problems with the reproducibility of science research and the quality of experimentation. Yong (2017), in writing about reproducibility problems in cancer research, clarified why different labs attempting the same experiment may have found different results: the methods section of publications rarely reports step by step exactly what was actually done. Furthermore, investigators often fail to recognize hidden sources of variation, the need for additional quality control steps, and the importance of validating the findings with data that confirms the reliability of their experiments. Even today during the race to resolve the COVID-19 pandemic, according to another recent report by Yong (2020), although "many researchers had spent the past few decades transforming science from a plodding, cloistered endeavor into something nimbler and more transparent," still "flawed research made the pandemic more confusing, influencing misguided policies. Clinicians wasted millions of dollars on trials that were so sloppy as to be pointless." The *Economist* news magazine also reported on "How Science Goes Wrong," raising concerns based on reports that many published science research findings may not actually be true (Fig. 1.1). An October 2013 issue highlighted several causes for unreliable research under the headline "Trouble at the lab," concluding that young scientists must be taught technical skills and "imbued

Fig. 1.1 An October 2013 issue of *The Economist* news magazine focused on "How Science Goes Wrong." Articles in this issue exposed reasons why some scientists have been experiencing problems with reproducing research done by others. The suggestion that companies cannot rely on academic research was raised after attempts by Bayer HealthCare scientists to replicate 67 studies found that in only a quarter of the studies were they able to reproduce the original results. This problem was attributed not just to statistical mistakes, but rather to experimental designs that were poorly thought through, such as an inadequate control group, or describing the methods in a way that failed to expose "tacit knowledge," the detailed skills that were actually used. (*The Economist* cover image used with permission)

with skepticism towards their own results and those of others" (*The Economist*, October 19–25, 2013, p. 30).

To address the abovementioned problems, in 2014 we established a network of two types of biology faculty best qualified to collaboratively develop educational solutions. One group is faculty members who give instruction in biology or have an active research portfolio (hereafter called "scientists"); the other group is faculty who specialize in biology education including both tertiary instructors as well as those who conduct research into how students learn (hereafter called "educational specialists") (Pelaez et al., 2018). The scientists are essential because they engage undergraduate students in research and can convey to them their own research experiences. However, scientists typically have not received training in biology education and so may be challenged to develop effective pedagogical methods on their own. The contributions of the educational specialists, who are experts in science education, therefore, could complement their scientific skills and knowledge. Unfortunately, diffusion of knowledge about undergraduate science education has been slow because the faculty who are education specialists preferentially talk to each other and not to scientists who are at the cutting edge of changing research paradigms in biology (Lane et al., 2020). Paired teaching assignments was suggested as one way to encourage interaction between these two types of science faculty, but since most college and university biology departments do not house both scientists and educational specialists, there is a need to reach science faculty more broadly. Thus, our collaborative network was funded by the National Science Foundation in the US to synergistically unify expertise required in order to improve the teaching of experimentation in the life sciences. This project focused on Advancing Competencies in Experimentation–Biology (hereafter, "ACE-Bio") to enhance tertiary students' understanding of and proficiency in biological experimentation and to work toward widespread acceptance and pedagogical knowledge including assessments that directly measure what students learn about such experiments. Project members agreed to adopt guidelines (Pelaez et al., 2018) to facilitate ethical (Hanson, 2014) and productive collaboration among scientists and education specialists, drawing on overlapping areas of interest and complementary professional expertise across various sub-disciplines of biology, and serving diverse students, including those in small private colleges, minority-serving institutions, and major research universities. The sub-disciplines included ecology, plant biology, physiology, biochemistry, microbiology, and evolutionary biology. Since the chapter authors for this volume have been informed by our project in documenting their recent *Trends in Teaching Experimentation in the Life Sciences*, we aim for this volume to stimulate life science instructors at the tertiary level to reconsider and to assess key anticipated learning outcomes that must be identified and measured in order to know what students are learning when they actively engage with biological experimentation in their courses. Reports in the chapters of this volume are aimed at helping individual biology instructors evaluate and improve their own instructional innovations and diagnose and remediate students' difficulties with biology experiments, as well as help biology program directors characterize the quality of biology research opportunities provided to students by understanding how well the fundamentals of experimentation are being taught. Since these are the aims of this book, each chapter ends with a bulleted list of practical advice for life science programs and instructors.

1.2 How Do Competent Life Scientists Do Experimentation?

To investigate how a competent life scientist does experimentation, we established the ACE-Bio network project as a collaboration that included a diversity of life scientists who were recruited from different types of tertiary-level institutions, representing different biology research and teaching focus areas, who were experts based their published work or teaching different types of students biological experimentation. The collaborators shared a concern about the importance of strengthening biological experimentation education and assessment at the tertiary level (Table 1.1). With the aim to bridge the gap between education research and its application to teaching practice, the participants agreed to come up with a framework for evidence-based strategies for teaching biological experimentation.

1.2.1 Articulation of Competency Statements

Guided by an empirical theoretical framework, the Conceptual knowledge-Reasoning- Mode of representation (C-R-M) model of Anderson and colleagues (Schönborn & Anderson, 2009; Anderson et al., 2013; Dasgupta et al., 2016; Irby et al., 2018), the collaborators were tasked with a process to develop and structure a set of competencies, i.e. a description of actions that a competent life scientist performs while engaged in experimentation. The empirical C-R-M model (Fig. 1.2) was used to focus attention on prior knowledge of experimental, mathematical and biological concepts (C), the scientific reasoning skills/abilities (R) applied to the concepts (R-C) and representations (R-M) and the nature and quality of the relevant modes of representation (M) such as diagrams, graphs, and mathematical equations that are indispensable to their complete understanding, designing, and implementation of a biological experiment. In other words, these interdependent factors of the model were viewed as key components for framing experimentation competencies such that difficulties with all or part of one of the factors could lead to difficulties in the area of biological experimentation.

The C-R-M model emphasizes the crucial role of visualizations (M), whether they are of hypotheses, methodology, data or research outcomes, as a component for understanding the research process for any experiment (Schönborn & Anderson, 2009; Dasgupta et al., 2016). Therefore, any framework for understanding ways of reasoning about experimentation involves both representations (R-M) and experimentation concepts (R-C). Since concepts and representations are cast as nouns while reasoning abilities or competencies are indicated as verbs, this noun- verb

Table 1.1 Importance of strengthening biological experimentation education and assessment as determined by poll of initial collaborating ACE-Bio Network members[a]

Statement	Strongly agree 5	Agree 4	Neutral 3	Disagree 2	Strongly disagree 1	Mean (N = 13)	Std. Dev.
Undergraduate students' knowledge about experimental research in biology is NOT sufficiently assessed.	61.5	30.8	7.7			4.54	.66
Undergraduate students need to know experimental research better.	38.5	61.5				4.38	.51
Overall, students have insufficient training of experimental research in biology.	38.5	61.5				4.38	.51
It is very critical for biology students to know experimental research.	38.5	53.8	7.7			4.31	.63
It is unclear what the expected objectives are for students in learning about experimental design.		61.5	15.4	23.1		3.38	.87
It is unclear what are the critical skills students need to learn to conduct experimental research in biology.		38.5	53.8	7.7		3.31	.63
It is unclear how we should assess students' knowledge about experimental research in biology.	7.7	46.2	15.4	23.1	7.7	3.23	1.17
It is unclear what are the critical concepts about experimental research in biology.	7.7	38.5	23.1	23.1	7.7	3.15	1.14

[a]Percentage of respondents are shown for each Likert Scale category. Data are from Yue Yin, ACE-Bio project evaluator

structure, framed by the C-R-M model, was used to identify concepts or representations that are paired with actions (verbs) as competencies when a competent scientist engages in experimentation (Fig. 1.2).

The C-R-M framework then guided a multi-stage process for developing competency statements. In brief, personal expertise was shared among research scientists

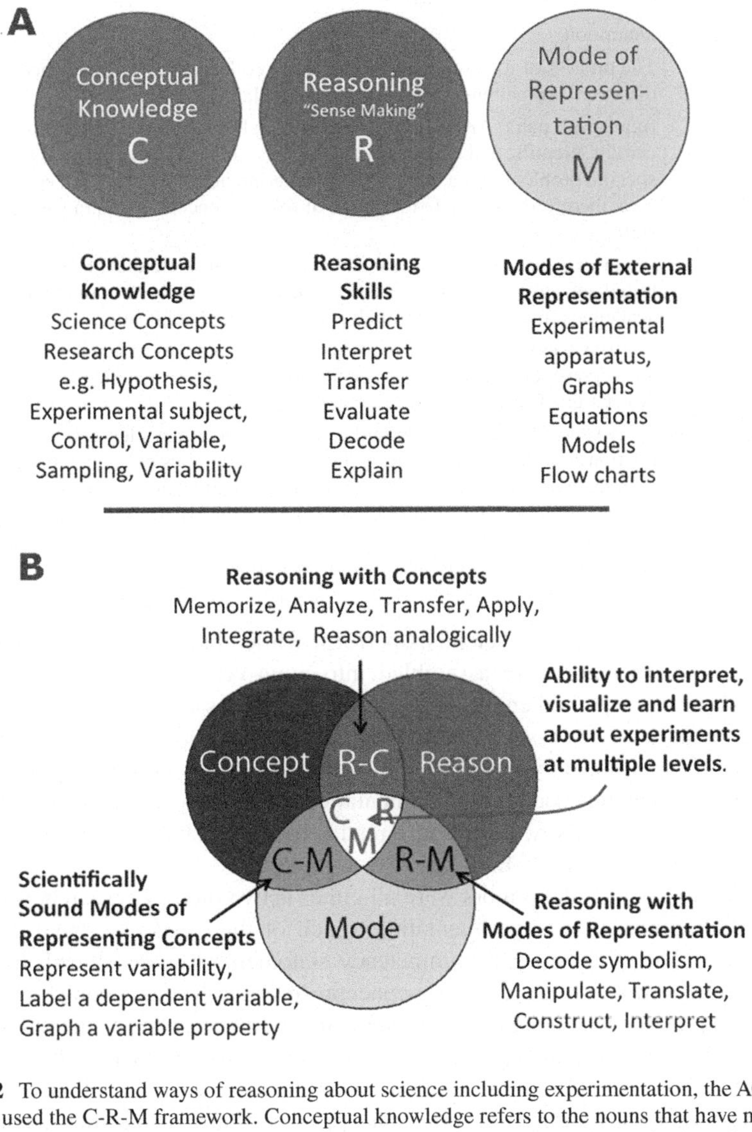

Fig. 1.2 To understand ways of reasoning about science including experimentation, the ACE-Bio project used the C-R-M framework. Conceptual knowledge refers to the nouns that have meaning agreed upon by experts in the discipline. Reasoning skills are the verbs or actions applied to the concepts. Furthermore, deep understanding involves reasoning with modes of external representation such as equipment that is calibrated, flow charts for research plans, data tables, graphs, and models for data analyses (modified from Schönborn & Anderson, 2009). With this framework, ACE-Bio project members examined the abilities to interpret and learn about experiments in terms of how these components were integrated in practice

and education specialists who were brought together to frame competency statements as noun-verb statements. This was initially approached in two parts: identification of the concepts essential to biological experimentation and development of

Table 1.2 Operational definitions of key terms

Term	Definition
Experiment	The process of research using methods for collecting data as evidence to discover if something works or to refute a claim.
Competency	(noun-verb pairs) An umbrella or overriding human ability to apply a skill to certain scientific and experimental concepts to target a specific goal or solve a specific problem in the target area of experimental research. Each competency area, therefore, may include a range of related concepts and a range of specific skills.
Concept	(noun) An idea, abstract or real, about the natural world that has been well-tested and substantiated by research and agreed upon by the community of academics in a discipline through peer review and publication.
Skill	(verb) A cognitive or reasoning ability used by humans to manipulate concepts and representations to target a specific goal or solve a specific problem in the target area of research.
Representation	(noun) Any abstraction or simplification that shows key elements and their relationships. The word "representation" is sometimes used interchangeably with the word "model" in scientific experimentation and, as shown in this book, means different things to different people. In our experience, the definitions of representations and models has raised many questions and divergent views among life scientists and warrants further attention.

verb actions that competent scientists perform while engaged in experimentation. These noun-verb pairs were assembled into noun-verb competency statements, organized into sets and taken through a validation processes that included reaching outside the network for essential feedback and input for validation. The product of these efforts was not only externally validated by experts but also by consulting existing resources for teaching experimentation. During the process of organizing this work, a glossary of key terms was agreed to by the collaborators and these terms are defined operationally in Table 1.2.

While the initial collaborators were all either active biology researchers or educators with expertise in experimentation, much of their expertise would be tacit knowledge. Therefore, a set of competency statements was developed through a process designed to articulate the key concepts and actions that a competent scientist would practice when engaged in biological experimentation. At the outset, time was dedicated for individual participants to crystallize for themselves the important elements of biological experimentation based on their own disciplinary publications, experiences, training, teaching, and practice. Participants then engaged in small group activities followed by a large group discussion for the sharing and synthesis of ideas to identify common elements, and areas of distinction, among individuals.

First, each ACE-Bio network member was asked to compile and discuss a set of primary literature articles from their field of study within biology. The objectives of this activity were to prompt ACE-Bio members to reflect on the features of experimentation within their field of research and to gather a diverse set of examples of biological experimentation practices from a variety of sub-disciplines within biology. The second way in which individuals were guided to contribute their knowledge and experience was to reflect on the concepts and skills that they themselves use in their research and teaching. This was accomplished by having ACE-Bio members complete an anonymous survey. In the survey, ACE-Bio members were asked to respond to the following: (1) What do scientists visualize when they design, conduct, or report experimental findings and how does their thinking about biology experiments compare with what we are asking of students? (2) What do students need to know, and be able to do, to design or conduct a biology experiment? (3) Where do students go wrong when they fail to design or conduct a biology experiment appropriately? and (4) What would the faculty like to measure? How do they currently assess student learning about biological experiments? Are they satisfied with the assessment? What are their concerns? How would they improve their assessments? The objective was to help individuals begin to articulate what concepts and skills they regard as important for biological experimentation according to their challenges and needs for teaching students.

The articulation of key concepts and skills was then documented in concept maps developed to represent the building blocks of essential competencies for biological experimentation. Four activities supported this conceptualization work. The first activity by each participant was to construct a concept map (Novak, 1990; Novak & Cañas, 2007) as a way to organize and visualize their thinking about biological experimentation. Each individual made a list of key concepts and skills in biological experimentation from their disciplinary perspective. The concepts were used as nodes in the creation of a visual representation with each concept linked to others in the map with uni- or bidirectional arrows over which a statement describing the relationship between the two concepts was written. Participants were not required to incorporate the hierarchical structure of a concept map, but were allowed to organize the concepts and their relationship in a manner that was representative of their thinking. As a second activity, individuals from different biology subdisciplines worked together in small groups. Each member of the small group identified and discussed a primary literature article that they had identified as an exemplar of experimentation in their field. The objective was to identify elements of experimentation, based on the methods described in papers that were chosen from the subfields of biology and the group members' research experience, to clarify ideas they had identified in the individual concept mapping activity. This discussion raised awareness of individual variation of experiences, thus broadening the perspectives on the types of research that compose "biology." For an example, it became clear that not all biology research is hypothesis-driven in the literal sense, and that it would be necessary to consider experimentation in varied contexts. A third small group activity involved using the individual concept maps to communicate ideas between individuals in a group and to create a single, consensus concept map based on the discussions of the primary literature articles and the individual concept maps. New groups of four to five people were formed, mixing up the original groups to include as much biology sub-discipline diversity as possible. These new groups developed consensus concept maps for two purposes: (1) to facilitate the exchange of ideas between members of the small groups and (2) to communicate a group synthesis of key concepts and skills for experimentation to all of the participants. During this work, group members compared their individual ideas and started to identify ideas that transcended disciplinary boundaries as well as those that were discipline-specific. As a final activity, a presentation by each group describing their process of consensus concept map construction and the resulting concept map was followed by discussion of competency statements.

The initial drafting of competency sets for biological experimentation was complex and involved an iterative process of individual and group reflection and work. Each group used its own concept maps to write statements of competencies. Using the C-R-M model as a guiding framework for the structure of these statements, concepts from the concept maps (nouns) were fleshed out with related skills (driven by verbs or action statements). Groups progressed through this difficult task and sometimes focused on different concepts. Each group then sent one or more team members to a mixed working group to synthesize the results from four small groups. Concept-skill (noun-verb) statements were further developed and then grouped into a smaller number of overarching competencies guided by the relationships between concepts within the concept maps. It became clear that some skills would overlap between competencies, but that was not considered to be problematic, because many concepts are used at multiple points in the process of experimentation but in slightly different ways. In addition, one important conversation involved how broadly the process of "experimental design" would be defined. The consensus was to not exclude certain biological disciplines (e.g. genomics) by assuming that particular skills were essential (e.g. hypothesis generation), and that certain skills, while related, were outside the scope of the work (e.g. literature searching). Additionally, in alignment with the C-R-M framework, the role of visual representations was considered and incorporated, where appropriate. The competency statements refer to what a competent scientist would do in their research. The statements are by no means an exhaustive list allowing, as illustrated in this book, for the addition of other competences of relevance to expert experimental practice. Tertiary students taking biology courses are the target population for eventual instruction and assessment design addressing the competencies.

1.2.2 Validity Evidence for Refining the Competency Statements

The articulated competency statement sets for biological experimentation were refined in a process that continued during a period of several years through internal and external validation. Initial sets of competency statements were revisited, reworked, and refined through asynchronous, individual consideration and synchronous work in both small and large group conference calls involving ACE-Bio network members. What emerged from this work are seven sets of statements describing what a competent scientist does when conducting experiments in biology (Tables

1.3, 1.4, 1.5, 1.6, 1.7, 1.8 and 1.9). Each set includes concepts and associated competency statements, which explain the ways in which each concept would be used. For example, under the *Question* Competency and the big idea of 'The ability to generate research questions and formulate hypotheses,' for the concept of 'Hypothesis' one competency statement is "Generate multiple explanations of the natural world that are testable and potentially falsifiable", while another is "Predict the associations between treatment conditions and outcome variables for the research target". All participants contributed to the competency statement editing so that the draft that finally emerged truly represented the thoughts, input, and consensus of all members. The inclusive process used in the generation of the competencies along with the diverse make-up of the participants allowed for an internal validation of an initial draft of the competency sets for biological experimentation.

External validation of the competence statements consisted of two tasks: (1) soliciting feedback from active biology research scientists as to the relevance of each competency to their own research and, (2) mapping the competency statements to published assessments and textbooks for teaching experimentation at the tertiary level. ACE-Bio network members recruited 23 research scientists to interview them about their use of each competence in their own use of experimentation in biology (IRB approval from Purdue University). Of the 23 life scientists, 14 identified as ecology or evolutionary biologists, seven as cell or molecular biologists, one as a

Table 1.3 Expert ratings (N = 23) of the importance of the *Identify* Competency to their own Biological Experimentation research: The ability to identify gaps or limitations in current research knowledge through the review, filtering and synthesis of relevant literature

Concepts	Skills: The ability to use the concept to…	Agree	Disagree	Unclear
A. Relevant background knowledge	1. Find appropriate sources of relevant scientific information (primary, secondary, etc.)	23	0	0
	2. Filter and evaluate the relevance of information from appropriate sources to the specific research focus	22	0	1
	3. Evaluate background information with critical scientific skepticism[a]	20	0	3
	4. Synthesize and apply current knowledge to generate a contextual foundation for the research problem.	21	0	2
	5. Reflect on the skills and knowledge needed in the relevant field before proceeding to do research.	21	1	1
B. A gap in current knowledge	1. Recognize a gap in current scientific knowledge that can be addressed with experimentation	22	0	0
	2. Reflect on limits of background knowledge related to the gap	21	0	1
	3. *Identify* a problem that is timely, relevant, and interesting, and, if addressed, could build on our foundational knowledge of science	22	0	0

[a]Statements in italics highlight those for which there was less expert agreement

Table 1.4 Expert ratings (N = 23) of the importance of the *Question* Competency to their own biological experimentation research: The ability to generate research questions and formulate hypotheses

Concepts	Skills: The ability to use the concept to…	Agree	Disagree	Unclear
A. Observations	*1. Apply systematic observations to discern variable properties of components of biological systems*[a]	19	2	2
	2. Compare observations to existing knowledge, models, or theories	21	1	0
B. Research questions	1. Develop novel, relevant, and testable research questions based on patterns or properties of components observed in biological systems or described in primary literature	22	0	1
	2. Evaluate ethical, theoretical, practical and cost constraints associated with a research question[a]	18	0	5
C. Models[b]	*1. Develop a model*[a] *to approximate or represent the behavior of a natural phenomenon*[a]	15	7	1
	2. Articulate the assumptions and limitations of a model[a]	15	7	0
	3. Evaluate a model to identify ways to improve it[a]	16	5	0
D. Hypotheses	*1. Use a model*[a] *to generate new hypotheses*[a]	13	8	1
	2. Generate multiple explanations of the natural world that are testable and potentially falsifiable[a]	18	3	1
	3. Predict associations between treatment conditions and outcome variables for the research target	19	2	0
	4. Determine whether multiple hypotheses are mutually exclusive and based on predictions of a model[a]	15	4	1

[a]Statements in italics highlight those for which there was less expert agreement
[b]Models in this competency refer to any abstraction or simplification that shows key elements and their relationships, including computational models to simulate natural phenomena

neuroscientist, and one as a biology education researcher. For each competency area, each life scientist was asked to indicate whether they used the skill statement in their own research (agree), did not use it in their research (disagree), or were unclear about the meaning of the statement. The results corresponding to each competency statement are presented in Tables 1.3, 1.4, 1.5, 1.6, 1.7, 1.8 and 1.9 and summarized in Fig. 1.3. In support of the importance of the competences to all the participant sub-disciplines, the results in Fig. 1.3 show high levels of agreement among the scientists in general, although slightly less consensus about the *Question* and *Plan* competency statement sets. Regarding the data in Tables 1.3, 1.4, 1.5, 1.6, 1.7, 1.8 and 1.9, most of the skills received greater than 90% agreement by research scientists. Although 23 life scientists participated, not every life scientist responded to all of the statements, therefore some items do not have 23 responses. The detailed responses reveal areas of diversity in the *Question* (Table 1.4) and *Plan* (Table 1.5)

Table 1.5 Expert ratings (N = 23) of the importance of the *Plan* Competency to their own biological experimentation research: The ability to plan feasible and ethical experiments to answer research questions or test hypotheses

Concepts	Skills: The ability to use the concept to...	Agree	Disagree	Unclear
A. Representations	1. Diagram a flow chart with steps of an experimental method	20	2	0
	2. Construct a visual representation (e.g. a graph or diagram) of predicted results[a]	17	5	1
	3. Diagram, label and title components for a proposal to conduct an experiment[a]	17	5	1
B. Experimental design	1. Identify assumptions and pros and cons of the different types of experiments (manipulative, observational/discovery, natural)	21	1	1
	2. Choose the most appropriate design approach to answer the research question(s) raised	23	0	0
	3. Propose measurable outcomes that would support or refute hypotheses	23	0	0
	4. Optimize treatments for efficiency[a]	19	2	2
	5. Identify potential sources of systematic and random error	22	1	0
	6. Draw a timeline of experimental procedures[a]	19	0	3
C. Variables	1. Identify relevant, measurable variables for testing the hypothesis	23	0	0
	2. Identify dependent and independent variables	23	0	0
	3. Identify confounding, and/or covariate variables aligned with experiment[a]	20	1	2
D. Controls (if relevant)	1. Design controls to anticipate likely sources of error to allow for comparison with experimental treatment groups in the context of the experiment.	20	1	1
	2. Select appropriate positive and negative controls to define an expected range of outcomes and to allow for comparison with outcomes from experimental treatments.[a]	17	2	3
	3. Consider what conditions are necessary to perform the experiments.	20	1	1
	4. Randomize the order in which experimental subjects or units experience treatment or control conditions as a way to reduce the chance of bias in the experiment.[a]	16	4	1
	5. Explain the implications of a control that did not show the expected result	18	3	0

(continued)

Table 1.5 (continued)

Concepts	Skills: The ability to use the concept to…	Agree	Disagree	Unclear
E. Measurement	1. Choose appropriate measurements based on available equipment, population/species, natural variation, and research question(s)	21	0	0
	2. Align variables appropriately with measurement tools/scale/instruments	19	1	1
	3. Recognize the limitations of measurement tools/equipment	21	0	0
F. Sample	1. Identify a target population(s) (might be molecules, cells, organisms, or populations) for the planned experiment	22	1	0
	2. Design the sampling strategy to expose and account for natural variation and measurement error	22	1	0
	3. Align sampling protocol with the research question or hypothesis	21	1	0
	4. Randomly sample subjects for control and treatment groups to be matched as closely as possible to equally reduce the effect of lurking variables on both groups.[a]	16	4	2
G. Variation	1. Differentiate between measurement variability and system variability (natural variation or heterogeneous populations)	19	1	1
	2. Determine replication or repeatability needed to quantify variation	20	1	0
H. Ethics	*1. Integrate professional and community ethics into research design*[a]	16	3	1
	2. Submit planned research to the institutional review board or animal care and use committee for evaluation, as appropriate[a]	16	4	1
I. Limitations	1. Evaluate assumptions in the experimental design	21	1	1
	2. Evaluate bias in the experimental design[a]	19	3	0
	3. Evaluate uncertainty in protocols (e.g. how we measure variables) analytical methods (e.g., assumptions of statistical tests), and interpretations of results[a]	19	2	0
	4. Evaluate limitations of methods	22	1	0
J. Iteration	*1. Model the research process to feedback loops (repeated experiments)*[a]	8	8	5
	2. Use feedback from preliminary results to improve protocols in new experiments			

[a]Statements in italics highlight those for which there was less expert agreement

sets since these had the largest number of skills that were not fully agreed on by experts, as indicated by the statements in italics. In particular, any statement that used the word "model" received a larger share of "disagree" or "unclear" ratings,

Table 1.6 Expert ratings (N = 23) of the importance of the *Conduct* Competency to their own biological experimentation research: The ability to conduct an investigation to achieve research goals

Concepts	Skills: The ability to use the concept to…	Agree	Disagree	Unclear
A. Measurement	1. Record observational data carefully and appropriately.	23	0	0
	2. Measure the response of the subjects to the treatment conditions carefully and appropriately.[a]	20	3	0
B. Variable outcomes	1. Monitor study for unexpected outcomes due to technical errors, equipment failure, subject characteristics, and unplanned factors.	22	0	0
	2. Evaluate potential for non-treatment causes for differences or similarities in research outcomes.[a]	19	3	0
	3. Troubleshoot technical errors.	23	0	0
C. Data documentation	1. Maintain a written or digital laboratory notebook or field journal that provides a record describing how, when, where, and why data were collected.	22	1	1
	2. Archive important and sensitive data in an accessible format that is intelligible, secure, and ethical.	20	2	0
	3. Record data in an organized and systematic way using appropriate tables, forms, etc.			
	4. Enter data with appropriate labels, units of measure, and levels of precision.			

[a]Statements in italics highlight those for which there was less expert agreement

including the outliers for Plan and Communicate. The external reviewers of the competency statements were able to expand on any ideas they found confusing or irrelevant, and to identify additional areas of difficulty – those that have no responses in Tables 1.5 and 1.6 – that they have observed or experienced in addition to those in the original ACE-Bio Competencies.

One-on-one interviews with the 23 research scientists made it possible to elicit additional feedback and suggestions regarding how to articulate skills considered most important to different scientists. Findings revealed that some scientists see themselves as focusing more attention on certain parts of the competency sets. For example, a cell biologist who was working with human blood samples to identify allergens, reported spending much more time planning and then concluding and communicating about their findings from a few careful experiments. In contrast, a molecular biologist runs lots of inexpensive assays on samples that were easy to get so that people in their lab first conduct many experiments and analyze the findings before deciding which experiments are most important to report. This approach of the molecular biologist expert is visually represented in Fig. 1.4 with the different sized shapes in the second panel. Another outcome of both the initial collaborative work as well as the interviews with experts was recognition of several "threads" across the entire process of experimentation. These threads included, for example, professional skepticism that involved evaluating limitations, finding corroborative

Table 1.7 Expert ratings (N = 23) of the importance of the *Analyze* Competency to their own biological experimentation research: The ability to apply analytical reasoning to data processing

Concepts	Skills: The ability to use the concept to…	Agree	Disagree	Unclear
A. Data curation	1. Construct appropriate ways to organize data (e.g. tables, figures)	23	0	0
	2. Explore and reduce raw data to discern trend and summarize relationships among variables	23	0	0
	3. Identify outliers and/or errant data by generating criteria for inclusion or rejection of data[a]	20	2	1
	4. Display appropriate comparisons (i.e. detect natural groupings)[a]	19	2	1
	5. Conduct transformations that facilitate statistical or other analytic tests	20	1	1
	6. Conduct computations for summarizing/interpreting findings	21	0	0
B. Data analysis	1. *Analyze* clean data using discipline-appropriate methods based on the measurements collected and the experimental questions.	22	1	0
C. Statistics	1. Choose and conduct statistical tests that are appropriate for the type/nature of data	22	0	1
	2. Choose and conduct statistical tests that are aligned with hypotheses and experimental research methods[a]	19	2	2
	3. Generate statistics for a sample to summarize and/or describe parameters for a whole population (e.g., mean, median, measures of variance).	21	1	1
D. Data summary	1. Appropriately identify a legend, label axes, and select appropriate scale to graph findings	22	0	0
	2. Considering the variables intended for comparisons, select an appropriate graphical type for the particular data type (e.g. contingency tables, bar graphs, histograms, scatterplots, etc.)	22	0	0
	3. Display findings with a representation that is effective in summarizing trends or major findings, including illustrating contrasts among categorical groups where relevant	22	0	0

[a]Statements in italics highlight those for which there was less expert agreement

evidence and continual evaluation of alignment of methods, data collection and analysis with the research question and/or hypothesis for a given research study. Discussions and interviews of experts also highlighted the iterative nature of research, which is often poorly represented in the classroom and teaching lab or in the classical model of the scientific method. Therefore, although the competency statement sets are listed in sequence for convenience, a figure was agreed upon to iconically show the seven areas of competence without any particular sequence.

A second approach to external validation of the competency statement sets involved mapping them to various textbooks about teaching experimentation in the life sciences. For example, the competencies were mapped to the contents of the *Experimental Design for the Life Sciences* book by Ruxton and Colegrave (2010) at

Table 1.8 Expert ratings (N = 23) of the importance of the Conclude Competency to their own biological experimentation research: The ability to draw conclusions about data that are limited to the scope inherent in the experimental design

Concepts	Skills: The ability to use the concept to...	Agree	Disagree	Unclear
A. Patterns and relationships	1. Describe trends in numeric and visual representations of data	23	0	0
	2. Interpret whether the results suggest a causal mechanism beyond simple correlation	22	0	1
	3. Distinguish biologically-meaningful trends from expected natural biological variability	22	1	0
B. Inferences and conclusions	1. Generalize results to an appropriate level (more than single experiment, less than universal)	23	0	0
	2. Connect analysis of results with valid claims or conclusion in a logical way	23	0	0
	3. Evaluate limitations of the findings and limitations that determine scope of inference (experimental and practical limitations)	22	0	1
	4. Compare results to other previously reported results and reconcile differences	23	0	0
	5. Align conclusion with analyses, hypotheses, research question(s), and existing knowledge	23	0	0
	6. Determine and articulate whether data support or refute hypotheses and predictions	23	0	0
	7. Express uncertainty by discussing limitations of data analysis (sources of error, inaccurate measurement, and sample bias, statistical significance vs. biological relevance)	23	0	0
	8. Identify future directions that will make conclusions more certain	23	0	0
	9. Understand that scientific knowledge is tentative[a]	20	1	2

[a]Statements in italics highlight those for which there was less expert agreement

two levels: first, the headings and subheadings from each content chapter (Chaps. 2, 3, 4, 5 and 6) were reviewed and mapped to a competency set and specific competency statement(s), if appropriate; second, the entire contents of a chapter were reviewed to determine the specific context of concepts covered to refine the initial mapping to the competencies. Concepts covered in this book mostly matched the competency sets for Plan (The ability to plan feasible and ethical experiments to answer research questions or test hypotheses) and Conduct (The ability to conduct an investigation to achieve research goals), as would be expected. In general, this textbook provided very specific examples and concepts that mapped onto our more general ideas. For example, this textbook elaborated on the selection and implications for choosing specific study designs and systems and it detailed random sample and random assignment to treatment groups, two areas with less consensus among our experts (Table 1.3). The book did not explicitly mention any visual representations or models, though these were included to illustrate points and concepts.

Table 1.9 Expert ratings (N = 23) of the importance of the <u>*Communicate*</u> Competency to their own biological experimentation research: The ability to communicate research work in professionally appropriate modes, including visual, written, and oral formats

Concepts	Skills: The ability to use the concept to…	Agree	Disagree	Unclear
A. Representations	1. Distill results into clear numeric and/or graphical forms that are aligned with the experimental objective/question/hypothesis	22	0	0
	2. Develop a predictive or explanatory model to summarize research findings.[a]	18	3	1
B. Scientific communication	1. Construct scientific communications using standard conventions.	22	0	0
	2. Distinguish typical structure and detail of an oral versus a written presentation	20	0	2
	3. Tailor structure and content of a presentation to the probable audience (e.g., scientific vs. public)	22	0	0
	4. Construct a wide range of representations such as tables, graphs, slides, diagrams, animations and simulations to present main points clearly in written and oral presentations	21	0	1
	5. Select the representation that best depicts the data to allow for appropriate inferences	22	0	0
C. Limitations	1. Articulate limitations, unanswered questions, and the tentative nature of results (both positive and negative)	21	0	1
	2. Contrast results and findings with previously published scientific work	22	0	0
	3. Offer alternative hypotheses	22	0	0
	4. Construct a justification and counter-justification argument for each alternative, if possible	21	0	0
D. Synthesis and reflection	1. Evaluate, analyze and explain the significance and implications of the research	22	0	0
	2. Revise an existing model based on observations or data[a]	17	4	0
	3. Articulate how findings contribute to new knowledge that can drive further inquiry	21	0	0
	4. Propose follow up experiments based on inferences from predicted or actual results of experiments.	19	0	1

[a]Statements in italics highlight those for which there was less expert agreement

Chapter 2 covered Hypotheses in detail but ignored important skills related to observations, research questions and models. Examples of experimentation processes that were missing from the book are the need to <u>*Identify*</u> relevant background knowledge in order to target a gap in current knowledge, and the focus on communication and conclusions. In addition to the missing coverage of <u>*Identify*</u> as background theory and observations to inform the experiments and <u>*Conclude*</u> and

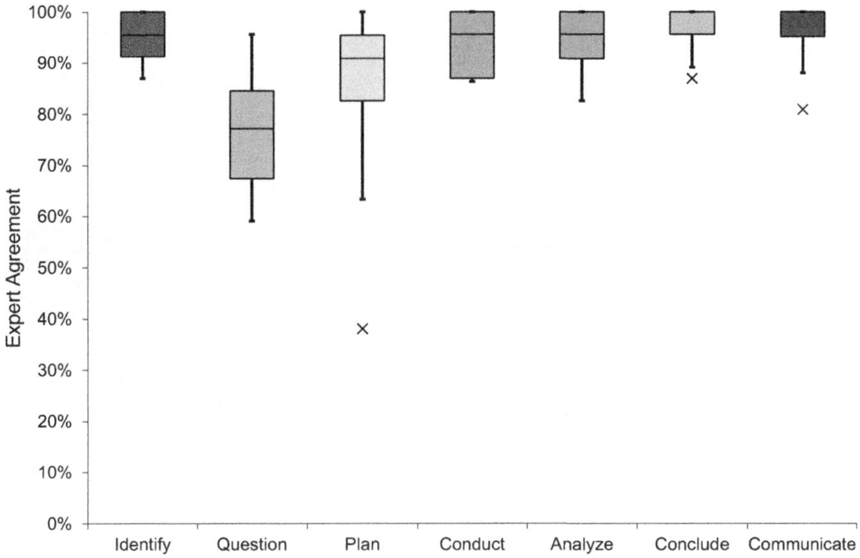

Fig. 1.3 Expert agreement with skill statements for each competency area. Life scientists (N = 23) were asked to rate their agreement with the importance of each skill to their own research. Median percent agreement for all skills within each competency is shown (box = middle two quartiles, whisker = 1.5 IQR, X = outlier). Each competency encompassed many associated skills, listed in Tables 1.3, 1.4, 1.5, 1.6, 1.7, 1.8 and 1.9

Communication as integral to building new knowledge from experimentation, data curation and data summary were ACE-Bio competency concepts found to be missing or underexplored in books. In fact, none of the textbook chapters or experts in our validation study articulated the purpose of visual representations in the process of study design or experimentation, which was a critical aspect of our work aligned with the C-R-M framework.

1.3 The ACE-Bio Competencies for Biological Experimentation

The biological experimentation competencies were published as an ePub by members of the ACE-Bio Network who identified the seven areas a competent biologist calls in when doing experimentation in biology (Pelaez et al., 2017). The researcher sits at the center of the process diagram model (Fig. 1.4), electing to start experimentation at any point given their understanding of the current state of knowledge, availability of preliminary data, and access to resources needed to fully execute experimental design and analysis of results. Researchers may choose to start with

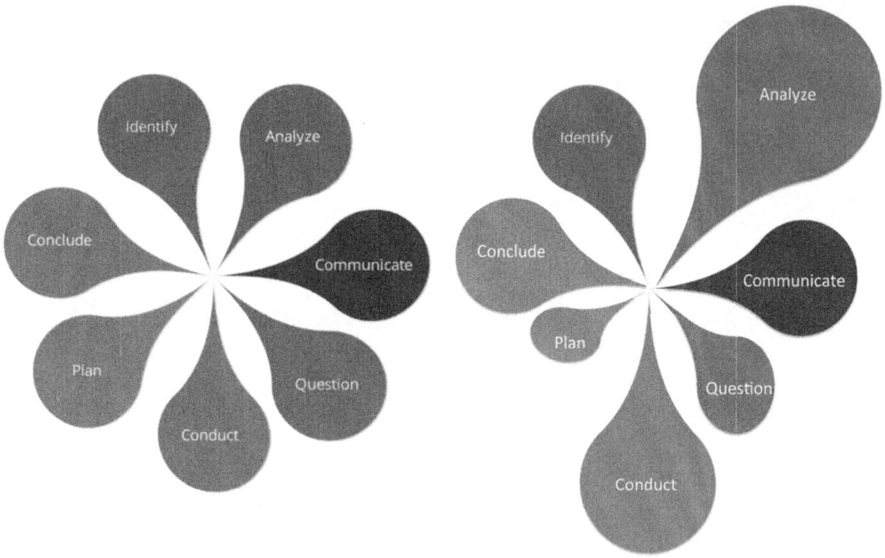

Fig. 1.4 This model shows seven areas a competent biologist calls in when doing experimentation in biology. Each competency is represented by a summary word on a uniquely colored segment of the model. Relevant skills are called on when needed and not necessarily in a linear sequence. Regardless of the sequence, an experimenter must *Identify* gaps or limitations in current research knowledge through the review, filtering and synthesis of relevant literature to understand how to *Conduct* experiments. They apply reasoning to data processing to *Analyze* their own findings as well as data reported by others. Even at the research proposal stage, scientists *Communicate* research work in professionally appropriate modes, including visual, written, and oral formats. They *Question* to generate research aims and formulate hypotheses based on their own work and evidence from others. They *Plan* feasible and ethical experiments to answer research questions or test hypotheses. A competent scientist is careful to *Conclude* with a narrow enough claim to align with the data and limitations to the scope inherent in the experimental design. The weight given to each area can vary among labs, as illustrated by the second panel to represent the research process of a molecular cell biologist who reported spending the most time *Conducting* many quick experiments and *Analyzing* results

Identifying the gaps in current research knowledge or they may begin planning (*Plan*) experiments that will inform the development of a testable hypothesis. Alternatively, they may generate research questions (*Question*) or models that they will explore through subsequent experimentation. Scientists indicate that their approach to experimentation is non-linear, which is consistent with more modern visual representations of the experimental process (Thanukos et al., 2010; Wilson & Rigakos, 2016). Unexpected results from their experiments, publication of new findings by other researchers, or other unanticipated circumstances contribute to the overall non-linear nature of their experimental schema. Presenting the Experimental Competencies as a process flow model to students might broaden their awareness of the elements of experimentation, debunk misconceptions that experimentation is linear, and offer opportunities for more robust skill development that maps back to all elements of the experimentation process.

For presentation convenience, the seven major areas within experimentation in biology are organized in tables in a linear manner. However, this is not meant to convey a particular order for approaches to experimentation. In fact, our studies revealed the flexible order and variable focus in terms of their use throughout the process of experimentation and that different investigators do not emphasize the seven areas equally. Because this work provides a framework for instructors or academic leaders in the biological sciences to study implementation of experimentation activities and assessments across diverse institutional and curricular contexts, apart from the document in pdf format, a link from the ePub also provides the file in MSWord format so that users can easily modify it (Pelaez et al., 2017). The aim is to support academics in making a tailored guide for assessment of student learning about experimentation, undergraduate biology instruction, curriculum development, professional faculty development, program evaluation, or review of research literature in a way that is appropriate to their own context.

While the average overall agreement, across the various sub-disciplines of biology, with each of the seven Experimentation Competencies was very high, there was less agreement on particular skills that included the terms model, representations, controls, and randomization. These disagreements could stem from differing language interpretations, sub-disciplinary contexts, or the nature of research (e.g., field studies vs. molecular studies). For example, while we carefully defined what we meant by model to be broad, the term "model" may trigger a narrower set of ideas such as a computational model or perhaps a computer-simulation. According to one biologist, "hypothesis and models are more or less the same thing," but then later the same person stated that "a model explains how the hypothesis works", while another participant reported that "the word 'model' is unclear." In all disciplinary fields, vocabulary is essential to communicating clearly, thus an area of future research could target the language differences used by disciplinary experts to describe their experimental practice. In the meantime, in this book, we have endeavored to define all the terms where relevant in the particular chapters.

In summary, in this chapter we have detailed how a competent life scientist does experimentation including the formal practices of observation, experimentation, hypothesis testing, and using modeling and simulation to focus on the study of biological phenomena, which are Core Competencies for Disciplinary Practice according to the *Vision and Change in Undergraduate Biology Education* report (AAAS, 2011; Clemmons et al., 2020). Although there had previously been only rare attempts to establish consensus about the details of experimentation practices for scientists and educators (Thanukos et al., 2010), here we have established a practical framework meant to guide decisions about teaching biological experimentation at the tertiary level to address the agreement in the literature that students have difficulty asking fruitful questions, planning and carrying out investigations, making predictions and observations, understanding experimental uncertainty, and constructing explanations from evidence (Ruiz-Primo et al., 2010; Shi et al., 2011; Dasgupta et al., 2014; Irby et al., 2018). Even for biology programs that rely on a class in statistics to assess what students know about experimentation, it is still important to understand how well students transfer what they learn in statistics to a biological research context. Experts reason about such basic research concepts by visualizing the experimental subject, the variable properties, and the amount of variation to predict a sample size needed for the overall research process. New assessments are still needed that will yield greater insight into the nature of these and other difficulties so that creative teaching interventions can be devised to address them (Auchincloss, et al., 2014). These efforts can be evaluated in terms of coverage of these ACE-Bio Experimentation Competencies. In the process of detailing the experimentation concepts, it was noted that many of the concepts of experimentation have an active verb form of the term. A scientist who comes up with a hypothesis has to hypothesize. To write a conclusion, an investigator must conclude from their findings. Experimentation is an active process. Our goal is to engage more biology students at the tertiary level in that type of research activity. It should be noted, though, that the concept of *measurement* under the *Plan* competency set is about planning and reflecting on taking and monitoring measurements, which may be mentally but not physically active. However, under *Conduct* where the focus is on the ability to conduct an investigation to achieve research goals, there is clearly an activity involved (the doing). Moreover, the Experimentation Competencies integrate skills that span multiple areas (e.g., construct visual representations, evaluate limitations, and explain implications), define skills associated with foundational concepts (e.g., observations, hypotheses, and measurement). In using this framework to guide decisions about teaching biological experimentation, it should be noted that the ACE-Bio Competencies reflect a non-linear approach to experimentation reflected in the analysis of expert concept maps and interview data (Figs. 1.3 and 1.4 and Tables 1.3, 1.4, 1.5, 1.6, 1.7, 1.8 and 1.9).

1.4 Practical Use of the ACE-Bio Competencies as a Framework

Many of the tertiary biology educators who wrote chapters for this volume used the ACE-Bio Competencies as drivers of educational reform to address challenges inherent to the teaching of experimentation in the life sciences. These challenges correspond to four categories of intrinsic teaching challenges found in published science education studies that reveal gaps in the knowledge, skills, and values of secondary school science teachers for teaching inquiry according to a systematic literature search by Akuma and Callaghan (2019): at an initiation phase the ACE-Bio Competencies influenced knowledge and beliefs about the biology curriculum to include explicit teaching of experimentation (Part I); at the planning phase they provided a fine-grained analysis of competencies pertinent to experimentation to inform the design of instruction (Part II); at the implementation phase the ACE-Bio Competencies were useful to persuade learners to engage in experimentation and to reflect on their experiences (Part III); and at the evaluation phase they were used by instructors to address concerns and difficulty linked to the grading of learning from biological experimentation as well as for evaluating the strength of a particular assessment relative to foundational experimental research skills (Part IV).

In addition to addressing these challenges, some authors explored relationships between the ACE-Bio Competencies framework and other frameworks or broad over-arching skills of relevance that still need to be addressed to contribute to the practical knowledge and values of post-secondary biology educators in order to advance their students' experimentation practices (Part V). Furthermore, authors of the final chapters wrote about topics that complement biological experimentation but that are also of relevance to biology education programs in general (Part VI).

In summary, by illustrating practical application of the ACE-Bio Competencies and other approaches, the book chapters that follow show practical ways to reduce specific challenges linked to the design and implementation of instruction as well as providing support to post-secondary educators who need to decide what influences are relevant to their own situation, which challenges need targeting, and in what order of priority in line with the constraints and affordances of their local educational environment. There is a need for reflection on the problem of how to focus post-secondary biology instruction on the understanding and use of disciplinary evidence and to guide students' development of reasoning with and about the process of experimentation, which is central to understanding biology for all students, not just for science students. Examples of approaches that highlight the practical strengths and limitations of the ACE-Bio Competencies should contribute to the readers' understanding of how to help students learn how anyone can possibly know something in biology, which is important because there is no trust in science without an understanding of the ways of knowing through experimentation.

Acknowledgments We gratefully acknowledge all members of the ACE-Bio Network for their wide-ranging contributions to our project. In particular, we thank the research scientists and education specialists who shared ideas collected under IRB protocol #1506016184 for RCN-UBE: ACE-Bio Network Phase 2: Joel Abraham, Edward Bartlett, Sue Ellen DeChenne-Peters, Cara Gormally, Mark A. Gallo, Jeff Hill, Mildred Hoover, Carol A. Hurney, Tammy Long, Dina Newman, Karen Sirum, Michael T. Stevens, and L. Kate Wright, who were willing to share their professional experiences through presentations and discussions with us, which catalyzed ideas for this book. In particular, thanks go to Joel Abraham, Cara Gormally, Mildred Hoover, Dina Newman, and Kate Wright for contributing to gathering of data from scientists for Fig. 1.3 and Tables 1.3, 1.4, 1.5, 1.6, 1.7, 1.8 and 1.9 and to Sue Ellen DeChenne-Peters for working on quantitative analyses of our data from scientists. Additionally, we thank our project evaluator, Yue Yin, and external advisory board members (Cynthia M. Bauerle, Erin Dolan, Maria Araceli Ruiz-Primo, and James Pellegrino) for their advice and support. This work was funded in part by NSF grant #1346567 for the Advancing Competencies in Experimentation–Biology (ACE-Bio) Network. Any opinions, findings, and conclusions or recommendations expressed are those of the authors and do not necessarily reflect the views of the NSF.

References

Akuma, F. V., & Callaghan, R. (2019). A systematic review characterizing and clarifying intrinsic teaching challenges linked to inquiry-based practical work. *Journal of Research in Science Teaching, 56*, 619–648. https://doi.org/10.1002/tea.21516

American Association for the Advancement of Science (AAAS). (2011). *Vision and change in undergraduate biology education: A call to action* (C. Brewer, & D. Smith, Eds.). Washington, DC. https://visionandchange.org/finalreport/. Accessed 18 Oct 2021.

Anderson, T. R., Schönborn, K. J., Plessis, L., Gupthar, A. S., & Hull, T. L. (2013). Multiple representations in biological education. In D. F. Treagust & C.-Y. Tsui (Eds.), *Multiple representations in biological education, series: Models and modeling in science education* (Vol. 7, pp. 19–38). Springer. https://doi.org/10.1007/978-94-007-4192-8

Auchincloss, L. C., Laursen, S. L., Branchaw, J. L., Eagan, K., Graham, M., Hanauer, D. I., Lawrie, G., McLinn, C. M., Pelaez, N., Rowland, S., Towns, M., Trautmann, N. M., Varma-Nelson, P., Weston, T. J., & Dolan, E. L. (2014). Assessment of course-based undergraduate research experiences: A meeting report. *CBE Life Sciences Education, 13*(1), 29–40. https://doi.org/10.1187/cbe.14-01-0004

Clemmons, A., Timbrook, J., Herron, J., & Crowe, A. (2020). BioSkills guide: Development and national validation of a tool for interpreting the vision and change core competencies. *CBE—Life Sciences Education, 19*, 4. https://doi.org/10.1187/cbe.19-11-0259

Corwin, L. A., Graham, M. J., & Dolan, E. L. (2015). Modeling course-based undergraduate research experiences: An agenda for future research and evaluation. *CBE Life Sciences Education, 14*(1), pii: es1.

Dasgupta, A. P., Anderson, T. R., & Pelaez, N. J. (2014). Development and validation of a rubric for diagnosing students' experimental design knowledge and difficulties. *CBE-Life Sciences Education, 13*, 265–284.

Dasgupta, A. P., Anderson, T. R., & Pelaez, N. J. (2016). Development of the neuron assessment for measuring biology students' use of experimental design concepts and representations. *CBE—Life Sciences Education, 15*(2). https://doi.org/10.1187/cbe.15-03-0077

Hanson, K. (2014). *Ethical decision making [mobile phone app]*. Markkula Center For Applied Ethics, Santa Clara University.

Irby, S. M., Pelaez, N. J., & Anderson, T. R. (2018). Anticipated learning outcomes for a biochemistry course-based undergraduate research experience aimed at predicting protein function from structure: Implications for assessment design. *Biochemistry and Molecular Biology Education, 46*(5), 478 492.

Koshland, D. E. (1985). Scientific literacy. *Science, 230*(4724), 391. https://doi.org/10.1126/science.230.4724.391

Lane, A. K., McAlpin, J. D., Earl, B., Feola, S., Lewis, J. E., Mertens, K., Shadle, S. E., Skvoretz, J., Ziker, J. P., Couch, B. A., Prevost, L. B., & Stains, M. (2020). Innovative teaching knowledge stays with users. *Proceedings of the National Academy of Sciences, 117*(37), 22665–22667.

Mulnix, A. B., & Vandergrift, E. V. H. (2014). A tipping point in STEM education reform. *Journal of College Science Teaching, 43*(3), 14–16.

National Academy of Sciences (NAS), National Academy of Engineering, and Institute of Medicine. (2009). *On being a scientist: A guide to responsible conduct in research* (3rd ed.). The National Academies Press. https://doi.org/10.17226/12192

Novak, J. D. (1990). Concept mapping: A useful tool for science education. *Journal of Research in Science Teaching, 27*, 937–950.

Novak, J. D., & Cañas, A. J. (2007). Theoretical origins of concept maps, how to construct them and uses in education. *Reflecting Education, 3*(1), 29–42.

O'Mahony, C., Deicke, W., Hasenknopf, B., Morgan, W., & Woodcock, L. (2019). *Towards successful learning: Controversies and common ground.* European Learning & Teaching Forum Organised by European University Association, University of Warsaw 14–15 Feb. 2019 https://eua.eu/downloads/content/2019.43_omahony_deicke_hasenknopf_morgan_woodcock.pdf

Pelaez, N., Anderson, T. R., Gardner, S. M., Yin, Y., Abraham, J. K. Bartlett, E. L., Gormally, C., Hill, J. P., Hoover, M., Hurney, C. A., Long, T. M., Newman, D. L., Sirum, K., & Stevens, M. T. (2017). The basic competencies of biological experimentation: Concept-skill statements. *PIBERG Instructional Innovation Material Paper 4*. https://docs.lib.purdue.edu/pibergiim/4/. Accessed 8 Nov 2021.

Pelaez, N., Anderson, T. R., Gardner, S. M., Yin, Y., Abraham, J. K., Bartlett, E. L., Gormally, C., Hurney, C. A., Long, T. M., Newman, D. L., Sirum, K., & Stevens, M. T. (2018). A community-building framework for collaborative research coordination across the education and biology research disciplines. *CBE—Life Sciences Education, 17*, 2.

Ruiz-Primo, M. A., Li, M., Tsai, S.-P., & Schneider, J. (2010). Testing one premise of scientific inquiry in science classrooms: Examining students' scientific explanations and student learning. *Journal of Research in Science Teaching, 47*(5), 583–608.

Ruxton, G., & Colegrave, N. (2010). *Experimental design for the life sciences* (3rd ed.). Oxford University Press.

Schönborn, K. J., & Anderson, T. R. (2009). A model of factors determining students' ability to interpret external representations in biochemistry. *International Journal of Science Education, 31*, 193–232.

Seymour, E., & Hunter, A.-B. (2019). *Talking about leaving revisited: Persistence, relocation, and loss in undergraduate STEM education.* Cham, Switzerland, Springer. https://doi.org/10.1007/978-3-030-25304-2

Shi, J., Power, J. M., & Klymkowsky, M. W. (2011). Revealing student thinking about experimental design and the roles of control experiments. *International Journal for the Scholarship of Teaching and Learning, 5*(2), 8. https://digitalcommons.georgiasouthern.edu/ij-sotl/vol5/iss2/8/

Thanukos, A., Scotchmoor, J.G., Caldwell, R., & Lindberg, D.R. (2010). The real process of science-understanding science at http://www.understandingscience.org and Science 101: Building the foundations for real understanding. *Science, 330* (6012), 1764–1765. https://doi.org/10.1126/science.1186994.

Trygstad, P. J., Malzahn, K. A., Banilower, E. R., Plumley, C. L., & Bruce, A. D. (2020). Are all students getting equal access to high-quality science education? *Data from the 2108 NSSME+*. Horizon Research, Inc.

Wilson, K. J., & Rigakos, B. (2016). Scientific process flowchart assessment (SPFA): A method for evaluating changes in understanding and visualization of the scientific process in a multidisciplinary student population. *CBE—Life Sciences Education, 15*(4), ar63. https://doi.org/10.1187/cbe.15-10-0212

Yong, E. (2017). How reliable are cancer studies? *Atlantic*, 319 (1), https://www.theatlantic.com/science/archive/2017/01/what-proportion-of-cancer-studies-are-reliable/513485/. January 18, 2017. Accessed 15 Dec 2020.

Yong, Ed. (2021). How science beat the virus. *Atlantic, 327* (1), 48–58, https://www.theatlantic.com/magazine/archive/2021/01/science-covid-19-manhattan-project/617262/. December 14, 2020. Accessed 8 Mar 2022.

Chapter 2
Using Data to Identify Anticipated Learning Outcomes for New and Existing Curricula

Kathleen A. Bowe and Stefan M. Irby

2.1 A Case for Data-Driven Curriculum Development

Within a science curriculum, students are expected to acquire a range of content knowledge and abilities. Ideally, a course should prompt students to engage in activities that reflect/model the application of knowledge and abilities characteristic of scientists. Past research has identified gaps in students' abilities to interpret data in light of scientific theoretical models (Ryder & Leach, 2000), and that inquiry-based laboratory experiences improve students' abilities to engage in activities important to experimentation like evaluation of data and recommending improvements to study design (Berg et al., 2003). Student mastery of disciplinary knowledge and know-how is guided by and measured against competencies such as these (e.g., data evaluation and model development). However, even with the best intentions there is a tendency in science curricula to focus on technical/procedural knowledge or on information-processing skills that are too general to be thoroughly assessed (as discussed in Irby et al., 2018a, b). While these broader long-term goals for disciplinary programs can help instructors and students see how a particular course outcome relates to their career aspirations (Dries et al., 2016; Elmgren et al., 2015), instructors' desire for students to advance their knowledge and abilities without anticipating competencies that are both specific and assessable is a problematic disconnect in curriculum design. Without specific anticipated learning outcomes (ALOs), a curriculum – which includes assessments – cannot properly target knowledge and

K. A. Bowe (✉)
Department of Chemistry, University of New Hampshire, Durham, NH, USA
e-mail: kathleen.jeffery@unh.edu

S. M. Irby
Department of Chemistry, University of Central Florida, Orlando, FL, USA

N. J. Pelaez et al. (eds.), *Trends in Teaching Experimentation in the Life Sciences*, Contributions from Biology Education Research,
https://doi.org/10.1007/978-3-030-98592-9_2

abilities and reveal if students have in fact developed competence (Anderson, 2007; NRC, 2001).

Identifying ALOs early when planning curriculum (Case 1) or as part of an evaluation strategy (Case 2) helps ensure that vision for instruction is aligned with intended pedagogical goals. This is especially important when an instructor desires to build a curriculum that goes beyond simple, routine laboratory experiments and instead develops research abilities and necessitates thinking like a scientist. The ACE-Bio competencies can be used for establishing a vision for curriculum design or evaluation in order to structure curriculum to promote the learning of biological experimentation. However, design challenges may persist in defining a content focus and in the unpacking and selecting of specific competencies to develop learning outcomes for students that are applicable, teachable, and assessable in the context(s) of interest.

Take, for example, "the ability to *Analyze* and process data." This is one of the most basic and indispensable competencies of biological experimentation (e.g., ACE-Bio Competencies in Pelaez et al., 2017; Chap. 1 in this volume and particularly Fig. 1.4), and of scientific research more broadly. However, while it helps focus curriculum design, an ALO this broad is difficult to assess and means little to students new to a discipline. A more specific sub-competency is to "*Analyze* clean data using discipline-appropriate methods based on the measurements collected and the experimental questions" (Pelaez et al., 2017). This ALO is more assessable, but it is not associated with any particular activity within a course, which makes it more difficult for instructors to create targeted assessments and much harder for students to self-assess.

We argue that ALOs should, at a minimum, adhere to the following three criteria: (1) active and contextualized; (2) supported by evidence; and (3) assessable. One useful guide for identifying ALOs that meet these criteria is the assessment triangle (NRC, 2001). At a fundamental level, the research examples presented in this chapter consider the components of the assessment triangle – cognition, observation, and interpretation – which together act as a scaffold for reasoning from evidence. The cognition component requires that an instructor has an explicit and clearly conceptualized model of how learners represent knowledge and develop competence in a discipline. This is similar to Anderson's (2007) argument that the first step in assessment is identifying the specific cognitive abilities students should learn. This relates to criteria one and two: to identify active and contextualized ALOs, instructors and designers should seek out evidence of how (active verb) and when (specific context) experts apply relevant knowledge and abilities first-hand. Figure 2.1 shows ALOs written with increasing specificity. As illustrated in this chapter, analyzing and reflecting on expert ways of knowing within a discipline enables instructors to identify ALOs and develop curricula that deliberately focus assessment – and therefore students – on the representations, tools, and activities necessary when engaging in experimentation like (future) scientists (Airey & Linder, 2009). These representations, tools, and activities include resources such as diagrams, ribbon diagrams of proteins, computer databases like BLAST, analytical routines, and modeling.

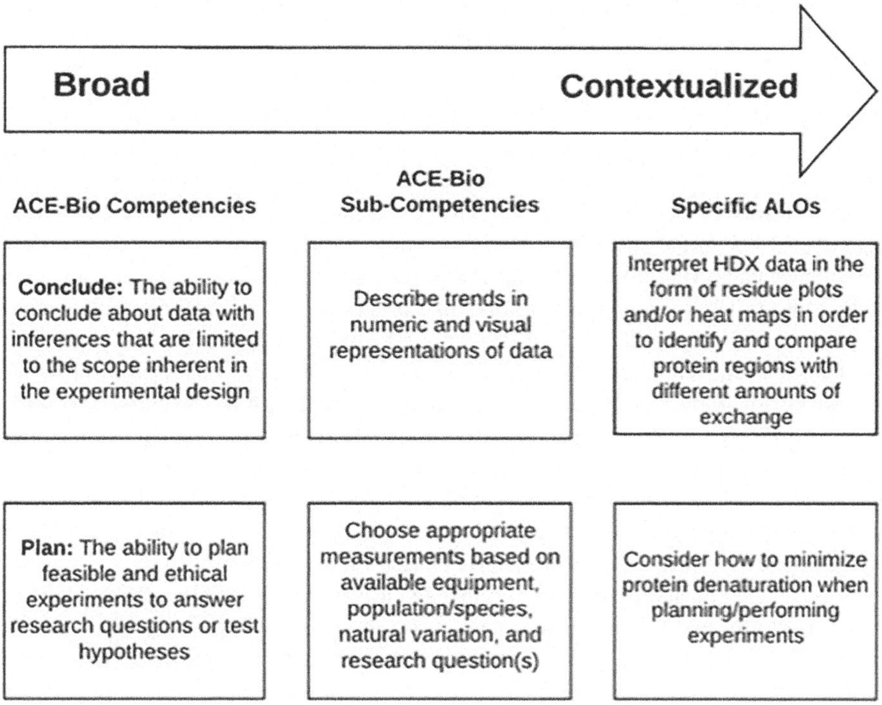

Fig. 2.1 Examples from both cases (Case 1, top row; Case 2, bottom row) of increasing specificity going from the ACE-Bio Competencies to specific course-level ALOs

The observation component requires that assessment tasks are carefully designed to elicit useful responses (i.e., demonstrative of competency). Given their intimate connection, so too should ALOs, rather than being arbitrary objects contrived by instructors. Relatedly, the interpretation component requires consideration of how observations of student competency derived from assessment tasks will be analyzed and what steps will be taken to ensure the interpretation is valid. When we say that an ALO should be assessable we mean that its structure should frame the design of assessment tasks that produce responses that can be interpreted reliably through cognitive frameworks or analytical lenses (also see Laverty et al., 2016). Sound curriculum design should explicitly align teaching, learning, and assessment (Anderson, 2007; NRC, 2001).

To support practitioners, researchers, and curriculum designers in rigorously identifying ALOs, we highlight here how different data sources can provide evidence to support ALO identification for the purposes of developing new curricula through backwards design (Case 1) and informing and refining existing curricula (Case 2). To provide practical illustrations, we report on aspects of these two cases, which are described briefly below.

2.2 Expert Sources as Contexts for Identifying ALOs for New and Existing Curricula

The most obvious instance in which to identify ALOs is in the development of new curricula. Curriculum development that starts with ALOs is said to follow a backward design approach (Cooper et al., 2017; Wiggins & McTighe, 2005). In this approach, instructors first identify what they want students to be able to do (i.e., identify ALOs), what student competency would look like, and then design curriculum materials and activities based on that information. For instructors and designers that want students to engage in expert-like practices and acquire competencies in experimentation, expert sources are an under-utilized resource for identifying evidence-based ALOs (Trujillo et al., 2016b). Previous work has used information gleaned from studying expert knowledge and reasoning to develop classroom activities, resources, and/or guidelines for connecting levels of biological organization (Van Mil et al., 2013, 2016), developing representational competence in chemistry (Kozma & Russell, 1997), and supporting students in monitoring their explanations of biological mechanisms (Trujillo et al., 2015, 2016a). We employ two cases to show how expert sources can inform ALO identification.

Case 1 used backward design to translate cutting-edge biochemistry research on protein folding and dynamics into new curricula for an introductory biochemistry course for allied health science majors. Experts' explanations of their research (Jeffery et al., 2018, 2019) served as the primary source of data to inform the development of curriculum materials about the use of hydrogen-deuterium exchange mass spectrometry (HDX-MS) to study protein structure and dynamics in the context of biopharmaceutical and small-molecule drug development (Bowe et al., in preparation; Jeffery, 2019; Jeffery et al., 2019). Primary literature, textbooks, and recent disciplinary resources from educators in the field served as secondary sources (Fig. 2.2). In the developed materials, students are introduced to the basics of HDX-MS and walked through interpretation of several relevant data representations in a short pre-activity reading and lecture. This is followed by the completion of two short activities, each based on a different primary research study. In the activities, students are guided through analysis of several different figures and tables containing HDX-MS data, and ultimately asked to draw conclusions about protein structure and dynamics within the context of each study's research goal.

Case 2 developed and used a process to identify evidence-based ALOs for an existing curriculum in order to contribute to its refinement and assessment strategy. The curriculum of interest was the Biochemistry Authentic Scientific Inquiry Lab (BASIL) curriculum (Craig et al., 2018; Irby et al., 2018a, b, 2020; McDonald et al., 2019; Roberts et al., 2019; Sikora et al., 2020) and is freely accessible online (https://doi.org/10.35071/m89q-fa20). The BASIL curriculum was designed as a Course-based Undergraduate Research Experience (CURE) and was modeled off the experimentation and activities conducted in biochemistry, structural biology, bioinformatics, and computational biology and chemistry research (e.g., McKay et al., 2015). The students who partook in the BASIL curriculum used a

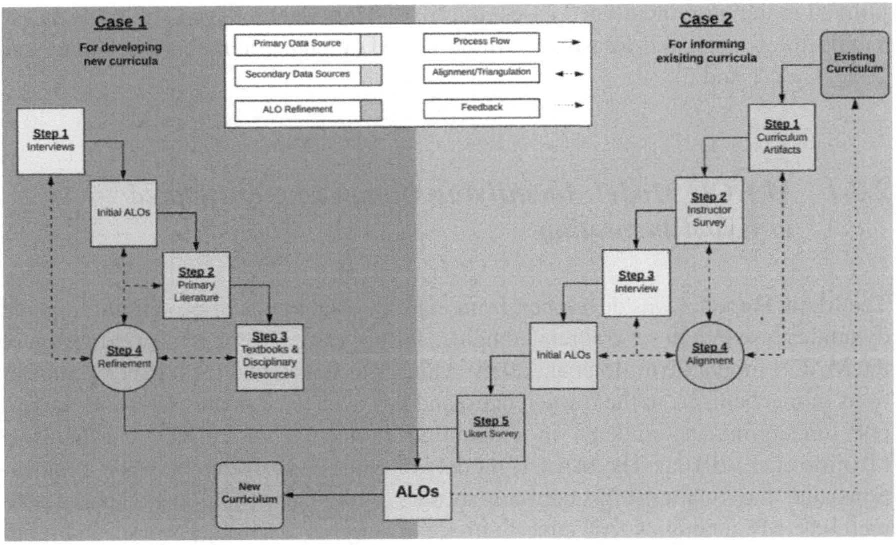

Fig. 2.2 Outline of the data sources used and the order in which they were used by the two cases to identify evidence-based ALOs to be used to develop a new curriculum (Case 1) and to inform an existing curriculum (Case 2)

combination of computational and biochemical wet-lab techniques to investigate a protein of interest that had been structurally characterized and uploaded to the Protein Data Bank (PDB), but its function remained unknown. Students used computational programs to predict the function of their protein of interest, and then conduct wet-lab experiments to begin to test the function of their protein and to evaluate their prediction. As a CURE, BASIL is therefore particularly concerned with student development of research competencies. Case 2 developed the Process to Identify Course-based Undergraduate Research Abilities (PICURA; Irby et al., 2018b), which triangulated the results of a content analysis of curriculum artifacts, an open-ended survey, and an interview to identify evidence-based ALOs. A final Likert survey was also used to refine and rank the identified ALOs so that initial assessment and course evaluation efforts could be focused. These steps are outlined in Fig. 2.2.

2.3 Useful Frameworks for Identifying Evidence-Based ALOs

While principles of sound curriculum and assessment design informed the overall approach of both cases, additional frameworks were used to identify ALOs. Frameworks are necessary because they focus attention on seeking out data that reveals specific knowledge or skills. In this section we highlight two frameworks utilized in the aforementioned cases: the MAtCH model (Jeffery et al., 2018; Case 1) and the Conceptual-Reasoning-Mode (CRM) model (Schönborn & Anderson, 2009; Case 1 and 2).

2.3.1 MAtCH Model: Identifying Competencies Related to Experimentation

The MAtCH model was developed from expert explanations of protein folding and dynamics research in several related fields (Jeffery et al., 2018). It is an extension of the MACH model (Trujillo et al., 2015) which was developed from expert explanations of mechanisms in the life sciences, and has been used in the classroom to support undergraduate students in explaining molecular or cellular mechanisms (Trujillo et al., 2016a). The MAtCH model identifies components of expert explanations and the relationships between those components, including experimental methods (M), analogies (A), underlying knowledge of scientific theories and models (t), social or biological research context (C), and descriptions of how a phenomenon operates (H). The MAtCH model focuses attention on the use of experimental methods and representations to understand phenomena in various research contexts. This is useful for both identifying ALOs targeted at experimentation as well as defining potential contexts for curricular activities, modules, or laboratories.

MAtCH can be used in two different ways to explore data and find evidence of expert knowledge and skills. First, it can be used to identify components of a research explanation (M, A, t, C, and H). For example, analysis of primary literature in Case 1 identified deuterium uptake plots and heat maps as representations (A) with which an individual studying protein folding or using HDX should be competent. Second, the MAtCH model can be used to attend to relationships between components in order to structure ALOs and specific curricular activities, to focus students on applying theory to navigate between the experimental methods' means of exploring physical systems and representing data outputs to describe systems. Focusing on these connections is important as Trujillo et al. (2016a, b) found that undergraduate students using the MACH model to think about components of mechanistic explanations still struggled to make connections – especially between phenomena (H) and how they are measured (M) – and thus produced disjointed explanations. In order to do this, an instructor or designer can, for example, analyze expert data sources with a focus on A-H or A-C connections in order to identify how experts analyze data representations to draw conclusions about physical systems and/or to refute/support initial hypotheses relevant to the research context. An A-H/C focused ALO from Case 1 would therefore be "use HDX data in order to draw conclusions about the spatial organization, stability, or function of different regions of a protein structure" (Bowe et al., in preparation; Jeffery et al., 2019). A specific curricular question focused on the A-H/C may ask students to "predict which regions are part of the binding site, given the change in deuteration level for the different regions and your general knowledge of enzyme structure." Note that both the ALO and questions here are active (lead with a verb) and contextualized. By using the MAtCH model to systematically check for the presence of components and connections, instructors can critique curricula based on expert practice in order to ensure that it exposes students to experimentation in a more holistic manner.

2.3.2 *Conceptual-Reasoning-Mode (CRM) Model: Identifying Competencies Related to Concepts and Representations*

The Conceptual-Reasoning-Mode (CRM) model has been used to frame how people interpret external representations (Schönborn & Anderson, 2009). CRM serves as a scaffold for understanding how scientists and students construct and use representations to rationalize experimental conclusions by incorporating conceptual knowledge (e.g., biochemical or experimental concepts and theories) and reasoning skills (e.g., problem solving or data analysis) to generate their understanding of a phenomenon (Schönborn & Anderson, 2010).

CRM is a useful tool in ALO identification for two reasons. First, the model provides a convenient way to look for evidence of reasoning with concepts or representations by looking for verb (reasoning) and noun (concept or mode of representation) pairs within data sources (Anderson et al., 2013; Irby et al., 2018b). The resulting verb-noun pairs capture reasoning with concepts (RC) and reasoning with representations (RM). Case 1 employed this analytical framework to review expert interviews and primary research literature, with a specific focus on how experts used representations of experimental methods to discuss physical systems, and how representations of research data were used to *Communicate* evidence of a model of a physical system (Jeffery et al., 2019). For Case 2, CRM was used at all steps of data analysis (Irby et al., 2018a, b) to guide the identification of concepts and representations in curriculum materials; of representations and instructor reasoning about the research students would be conducting (RC and RM); and of detailed examples of expert reasoning (RC and RM) during an interview with the lead curriculum designer. Second, CRM verb-noun pairs can be used to structure ALOs, a format which makes them necessarily both active and contextualized. For example, an ALO from Case 2 is "*translate* or *map* features *between* **2D and 3D representations of proteins**" (more examples in Irby et al., 2018a). In this case, the verbs 'translate' or 'map' (hereafter *italicized*) describe an expected action and the noun phrase '2D and 3D representations of proteins' (hereafter **bolded**) provides a specific context in which an action is expected to take place. Instructors and designers should note that verb-noun pairs do not necessarily have to be a singular verb and noun. Both Case 1 and 2 used CRM verb-noun pairs to inform the structure of their ALOs.

2.4 Reasons for and Examples of How to Use Data Sources in Curriculum Development

This section provides a practical look into how an instructor or designer might use data sources to identify ALOs for courses involving research experiences (e.g., CUREs) or exploring research or experimental methods outside of a laboratory setting. The data sources and the order in which they were used by each case is shown in Fig. 2.2. Both cases used interviews with experts as their primary data source, albeit at different steps of the ALO identification process. Case 1 employed secondary sources (primary literature, textbooks, and disciplinary resources) to refine initial ALOs and situate them within content and skill expectations for biochemistry undergraduates as defined by traditional curriculum and by more recent efforts of the broader biochemistry and molecular biology (BMB) educator community. Case 2 employed curriculum artifacts and surveys as secondary data sources to inform the scope of the interviews and to check alignment between initial ALOs and evidence of their relevance to the curriculum of interest (Irby et al., 2018a, b). We discuss the purpose and use of these sources in developing specific competencies and curriculum materials, and provide practical examples of what can be gained from each data source.

2.4.1 Interviews

Interviews are a valuable source for ALO identification and curriculum development. In general, they can provide examples of relevant social and experimental contexts for course content (Jeffery et al., 2018); simple language and representations to explain complex experimental methods; and examples of reasoning abilities used by scientists to conduct research and communicate with representations of experimental data (Irby et al., 2018b; Jeffery et al., 2019). Interviews can be conducted with different people and used for different purposes and at different time points (Fig. 2.2). In creating a new curriculum in Case 1, interviews with practicing research scientists were the first step and they provided the seed and expectations (i.e., initial ALOs) for designing dedicated materials and activities related to course content (e.g. biopharmaceuticals to study proteins). In Case 2, an interview with the lead curriculum designer of the BASIL curriculum occurred toward the end of the process to follow up on the lead designer's responses to the open-ended survey and to discuss additional representations that the lead designer mentioned during the open-ended survey and provided at the interview. To illustrate how interviews support ALO identification, we provide an example from each case.

Viewed through the lens of the MAtCH and CRM models, the interviewees in Case 1 provided authentic research contexts and experimental methods that could serve as the foundation for curriculum materials about protein folding and dynamics, as well as examples of how scientists reasoned and communicated with representations (Jeffery et al., 2018, 2019). For example, one Case 1 interviewee explained the following about their research:

> [...we're interested] in what happens to proteins when you [freeze dry] them. [...] to help preserve them. To help make them more stable than they would be in solution [...] degradation reactions and bad things happen at a much slower rate in solid forms [...] but unfortunately, one of the things that can happen when you're getting proteins into the solid state is that the process [...] can damage them. [...] so we wanted to use this idea to try and understand what happens to proteins in the solid state. And so in order to do that it's a pretty simple experiment actually: you just take the protein and the things you want to mix with the protein, you [freeze dry] the protein, and then you take the powder and expose it to D_2O in the vapor phase at controlled temperature, relative humidity [...] we can get [...] a heat map that shows extent of deuterium incorporation as a function of the, in this case, the excipient that we chose. So cool colors are less deuterium incorporated; hot colors are more. And so now with this HDX method, we can *map* **what parts of the molecule incorporated a lot of deuterium** [...] you can *see* right here **these two alpha helices** are really closely opposed [...] so they're protected from exchange. Whereas this loop over here is brighter colored, hotter colored so it incorporates a lot more deuterium. And [...] you *can see differences* in formulation [...] And it turns out all of that is pretty highly correlated with how stable the molecule is when you store it [...] So now this can be a tool that people developing formulations for proteins and other biologics can use.

Viewed through a MAtCH lens, the above excerpt provides evidence of a compelling research context (C; development and storage of protein drugs); of experimental methods (M; solid state HDX-MS); of representations used in this kind of research (A; heat maps); and of a particular phenomenon (H; protein drug dynamics and degradation). All of this is underpinned with implicit theoretical knowledge (t; higher-order protein structure and chemical equilibrium). Furthermore, this excerpt provides evidence of how this scientist would interpret experimental data to draw conclusions. In the excerpt, RM abilities are engaged when they talk about the colors on the 3D representation of a protein and combined with RC abilities in how they relate the amount of deuterium incorporation to protein structure. From this, an ALO statement such as the following could be produced: "interpret HDX data in the form of a heat map in order to *Identify* and compare protein regions with different amounts of exchange."

For Case 2, interviews also provided detailed examples of RC and RM abilities a scientist would need to have as part of experimental competency in a field of research. For example, when discussing protein function and structural characteristics the interviewee stated:

> [...] just because you have a **catalytic triad** that doesn't mean that an enzyme will cut proteins, maybe it will cut lipids, maybe it will cut something else. I'm hoping that they'll have *some sort of grasp* on the **physical nature of proteins** you know like the molecular weight of proteins, how they behave. Also, some idea of how fragile life is. They work to purify this protein in-lab and they come back the next week and it's dead. What happened? You know, so them to get a better understanding of that is important to me.

Viewed through a CRM lens, the above excerpt provides evidence of RC abilities pertaining to the consideration of how protein homology and the structure of a protein may impact protein function and how to handle a protein *in vitro*. From this, an

ALO statement using a verb-noun format would be "grasp the limitations of research methods based on homology."

Both MAtCH and CRM revealed how experts engage in various aspects of experimentation. For example, the ALO above from Case 2 analyzed using the CRM framework would fall under the ACE-Bio competency of plan and the sub-competency of "choose appropriate measurements based on available equipment, population/species, natural variation, and research question(s)" (Fig. 2.1). Interviews are capable of capturing many aspects of experimentation, including how scientists plan investigations, propose a testable research *Question*, and *Communicate* limitations. In both Case 1 and Case 2, the interviews shed light on aspects of experimentation that are not typically emphasized in curricula, such as *Plan*, *Question*, and *Communicate*. The cases described here employed semi-structured interviews as their primary data source. We believe a semi-structured interview is the most useful interview format as it offers guidance concerning what the interviewer wants to learn from the interview, but allows for flexibility to discover new, relevant, ideas being shared by the interviewee (Cohen et al., 2002). Thus, if an instructor is interested in highlighting a particular aspect of experimentation in their course, they can design the interview to focus on that aspect (e.g., ACE-Bio competency set, *Question*).

2.4.2 Case 1 Supporting Data: Primary Literature & Other Disciplinary Resources

For Case 1, supporting sources were used after conducting and analyzing interviews in order to refine initial ALOs and make them both realistic for the target course and relevant to the student audience.

2.4.2.1 Primary Literature

Primary literature can support the development of curriculum in several ways. First, it provides an additional way to identify and understand phenomena, experimental methods, and contexts related to the content area of interest that may have been minimally discussed or absent in interviews. For example, Case 1 reviewed research articles employing HDX-MS to study proteins (some of which had been recommended by interviewed experts) and applied the MAtCH model as a lens to understand in broad strokes where and how the articles talked about research context (C) and described how experimental methods and data revealed information about protein folding or structure (M-t-H or A-t-H connections). Similarly, Case 1 curriculum activities were based on HDX in developing small molecule drugs and biopharmaceuticals (Hsu et al., 2013; Moorthy et al., 2014), but could have been based on computational modeling and docking of potential coronavirus inhibitors (e.g., Ton et al., 2020). This process can simultaneously identify relevant literature to bring into a curriculum and deepen the knowledge of the instructor or designer. For example, numerous articles summarizing the theory behind and application of HDX were found during Case 1 curriculum development and informed curriculum development (Jeffery, 2019; Jeffery et al., 2019: in prep). Primary literature can also be used to identify common types of representations, which students need to be competent in interpreting when conducting experiments in a particular research context or using a particular method. Application of the CRM model to article narratives and captions can support the identification of how authors reason with representations and use them to draw conclusions from data or communicate research findings. This information can be used to identify ALOs or write activity questions. For example, an article may direct readers to attend to HDX-MS data for a protein region of interest and make a statement about the stability of that region. An ALO based on this data might be "use HDX data to draw conclusions about the spatial organization, stability or function of different regions of a protein structure."

2.4.2.2 Learning Outcomes Defined by Other Disciplinary Resources

Case 1 also utilized the learning outcomes defined in relevant chapters of biochemistry textbooks and expectations developed by current educators in the BMB community in order to identify realistic and relevant ALOs for the particular student audience. This is an important step when using expert data sources in ALO identification for two reasons. First, it enables the instructor or designer to confirm what knowledge is prerequisite in order to engage with the material, and thus what can realistically be addressed within a course. For example, students must have a basic understanding of higher-order protein structure, particularly the role of hydrogen-bonding, to explore how HDX-MS investigates protein structure (Jeffery, 2019). This knowledge is likely implicit in explanations found in expert interviews and primary literature, which is why it is important to look at learning outcomes from external educational sources if one starts ALO identification with expert sources. Students will need to be taught this information explicitly. Relatedly, this analysis supports the positioning of particular activities (e.g., using HDX-MS to explore protein dynamics) within a broader course context (after learning about protein structure) and in relation to other activities (e.g., before exploring how environmental conditions and site-directed mutagenesis can affect interactions in proteins and their ligands). In this way, considering the learning outcomes of external educational sources helps ensure that materials reinforce course competency goals through rich research and social contexts, rather than merely generate additional information that students are expected to digest and regurgitate.

Second, this process enables an instructor/designer to identify ALOs and design curricula that are relevant and cognizant of current disciplinary expectations for graduates. For example, one consensus expectation from the American Society for Biochemistry and Molecular Biology (ASBMB, 2020; Tansey et al., 2013; White et al., 2013) is that graduates should be able to "discuss the time scales of various conformational effects in biological macromolecules and design appropriate experiments to investigate ligand induced changes in conformation and dynamics." When faced with trying to design curricula that engage students in research competencies (e.g., 'plan feasible experiments' from ACE-Bio), expectations laid out in discipline-specific consensus documents can support the selection of research contexts already aligned with known content goals. This can make it easier to integrate cutting-edge research into a traditional curriculum or update a curriculum. For example, the Case 1 curricular materials focus on an experimental method traditionally beyond introductory biochemistry and are designed for a non-laboratory setting, but the activities extend course outcomes beyond an overview of protein structure and folding (i.e., traditional course expectations) towards developing competency in interpreting data pertaining to protein structure, as well as building knowledge of how experimental methods can be used to investigate the physical word (i.e., experimental expectations relevant to graduates).

Comparisons to multiple external educational sources may not be necessary, but each source provides different insight. The authors recommend comparisons to expectations outlined in disciplinary resources from a particular educator community (if they exist) over the use of a textbook as textbooks may be out-dated, include de-contextualized representations and references to studies, and contain third-hand information. Those expectations put out by educator communities are more responsive to current disciplinary trajectories and tend to be more competency-focused than textbook objectives.

2.4.3 Case 2 Supporting Data: Curriculum Artifacts & Surveys

For Case 2, supporting sources were analyzed prior to and after conducting interviews. This was because the curriculum already existed and there were several instructors and curriculum designers working on the BASIL curriculum. By first conducting a content analysis of the curriculum artifacts, it allowed for an understanding of the information contained within the documents and how it was presented to students. The content analysis also informed the design of an open-ended survey to elicit responses from instructor and curriculum design stakeholders about how scientists ***Conduct*** research similar to what students will be conducting. The supporting data sources (Step 1 and 2) informed the interview (Step 3) and were used to triangulate the initial ALOs identified (Step 4; Fig. 2.2). A Likert survey (Step 5) was conducted at the end of the process in order to refine and achieve consensus on the most relevant ALOs (Fig. 2.2).

2.4.3.1 Curriculum Artifacts

In education research, it is common to perform a content analysis on a variety of course documents (e.g., lab protocols and student lab reports) to identify what students should be learning and to evaluate the content covered in a course. A content analysis helps interpret and better understand what is being conveyed in text or other forms of communications, resulting in exact data about the information contained within a document (Cohen et al., 2002; Hsieh & Shannon, 2005).

For Case 2, the existing BASIL curriculum materials were analyzed. These materials largely consisted of laboratory protocols, and were the unifying documents that were created by the BASIL development team to be used for all implementations of BASIL (McDonald et al., 2019). The content analysis focused on identifying the concepts and representations (Schönborn & Anderson, 2009, 2010) presented in the curriculum materials. For example, the introductions of each module presented background information on biochemical and experimental concepts and theories, as well as some example representations of data outputs and of protein simulations. In addition, the content analysis revealed the activities students would be exposed to and thus the abilities they would be expected to develop or demonstrate. The identified concepts and representations were used to inform subsequent data sources (Fig. 2.2) and to ensure that the concepts and representations that were contained within the initial ALOs were present in the curriculum materials (Irby et al., 2018a, b). Ensuring alignment between goals and content is an important step when evaluating a curriculum (Anderson, 2007; NRC, 2001), particularly when it comes to complex competencies (e.g., the ACE-Bio Competencies).

2.4.3.2 Surveys

Surveys are a versatile tool for ALO identification. They can gather a range of input efficiently, assess consensus (e.g., Abualrob & Daniel, 2013), and are particularly useful if an instructor or designer is limited on time, funds, location, etc. Case 2 utilized both an open-ended survey and Likert-scale survey (Fig. 2.2) in order to, respectively, gather initial impressions of expected experimental competencies for interview development and to assess consensus and relevance of initial ALOs for further refinement.

The purpose of the open-ended survey in Case 2 was to begin to develop an understanding of how instructors and designers of the BASIL curriculum, as researchers, would perform the type of activities that students would have to conduct, and how they would reason with the concepts and representations that the students be provided with or would have to produce. The open-ended survey asked a range of questions about how the participants, as scientists, would approach and explain the research that is conducted as a part of the BASIL curriculum (see Irby et al., 2018b for full details). Many of the participants discussed their use of different computational and biochemical techniques, skills needed, and sequences of steps. For example, according to one instructor:

> We use **protein sequence alignment** *to find* similar proteins with known function, we use **domain analysis** – *to find* proteins with *similar* **domain composition**, we use **structure alignment** *to find similar* structures with known functions, we use **docking** *to simulate interactions* between **enzyme** and possible **substrate** to try *to choose* more likely **substrate.** Each type of computational evidence does not generate one answer, but rather a list that can be ordered. We then *look for patterns* and **common functions** across different types of data – *to try to come up with* a short list of **functions/substrates** to try in [the] lab.

In this excerpt, the participant discussed several computational methods, and how they are used and the order they may be employed, but does not discuss specific steps or provide detailed reasoning for each method. At this stage there is only a basal level of evidence for RC and RM abilities so it did not provide sufficient detail to write ALOs. Thus, the results of this survey informed the interview protocol by showing what areas to probe further to reveal more detailed reasoning behavior (see interview excerpt above).

Case 2 also employed a Likert survey as a final step to determine the degree of relevance of an ALO to the BASIL CURE. This was based on the level of agreement among BASIL curriculum instructors and designers with regard to how unique each ALO was to the BASIL curriculum and how important gaining competence with that ALO was to being a scientist in that field of research. This Likert survey was formatted so that each initial ALO was rated on two tiers: where an ability should be acquired (not acquired in this course, both this course and some other course, or only in this course) and how important the ability is to students functioning as a scientist (unimportant, undecided, or important). There was also space for the participants to leave comments on the initial ALOs, which were reviewed when making refinements for the final set of ALOs.

To analyze the resultant survey data, a metric called weighted-relevance (WR) was devised (Irby et al., 2018a, b). A WR score is calculated by summing up the responses on both tiers of the survey for each ALO (such that 'important' and 'only in this course' had a weight of +1; 'undecided' and 'both' had a weight of +0.5; and 'unimportant' and 'not acquired' had a weight of −1). The use of WR helped determine which of the identified ALOs had the most agreement of importance to the BASIL curriculum, as well as whether or not any ALOs should be revised, removed, or added. For example, one initial ALO was worded as "Use SDS-PAGE gels for interpreting information about protein and plasmid integrity." However, despite a fairly high WR (Irby et al., 2018a), instructor comments in the comment prompts revealed confusion about the use of the word "integrity" and what information can actually be gained from an SDS PAGE gel. The phrasing of this ALO was therefore altered to "Use SDS-PAGE gels for interpreting information about a protein and its expression from a plasmid." Taking into account this additional input, each of the initial ALOs could be refined and those considered most relevant to the curriculum could be gathered to generate a final set of ALOs (Fig. 2.2; Irby et al., 2018a).

2.5 Conclusion

Rigorous identification of ALOs, which are (1) active and contextualized; (2) supported by evidence; and (3) assessable, is an essential component of creating curricula that support students in the development of expert-like competency in experimentation. We have demonstrated through two cases how the MAtCH (Jeffery et al., 2018) and CRM (Anderson et al., 2013; Irby et al., 2018a, b; Jeffery et al., 2019; Schönborn & Anderson, 2009, 2010) frameworks can be applied to expert data and other sources to productively frame experimentation, as well as identify specific and relevant applications of biochemical knowledge and skills to inform both new and existing curriculum development.

Each data source described here was found to have certain advantages. Surveys captured a variety of perspectives and consensus, but were limited in terms of response detail. Curriculum artifacts and other disciplinary resources provided a quick sense of content, but they did not provide much direction. Primary literature was used as a course resource and provided insight into current research trends and experimental approaches, but as a report it provided only an overview of the experimental process, which obscured the recursive nature of problem-solving in experimentation. Interviews provided rich detail with the capability of real-time follow-up, but they took more active engagement to prepare, conduct, and analyze. It may not be feasible for all instructors or designers to use multiple data sources as in the two cases described here. While this is understandable, the authors advance that evidence is vital in constructing active, contextualized, and assessable ALOs. The processes and frameworks outlined in this chapter are intended to make using multiple data sources to identify ALOs more feasible.

Curriculum development also does not end with creating ALOs. There is an intimate link between learning outcomes, teaching, learning, and assessment (Anderson, 2007). Although the ALOs presented in the two cases here emerged from data and are supported by evidence, ALOs must also be evaluated as to whether or not students partaking in the curriculum achieve them. This can be done through targeted research studies or well-designed classroom assessments. If there is evidence of student competency gains, then they can be deemed verified learning outcomes (VLOs; Irby et al., 2018a, b). For those creating a new curriculum, as in Case 1, activities and assessments must be developed, piloted, and student work evaluated to identify specific evidence of student competence (Bowe et al., in preparation; Jeffery et al., 2019; Trujillo et al., 2016b). The same is true for those revising or refocusing an existing curriculum (Case 2). Studies related to the Case 2 BASIL curriculum have investigated how students perform on open-ended assessments for specific ALOs (Irby et al., 2018a; Irby, 2019) and student perceptions of their knowledge, experience, and confidence for a subset of ALOs (Irby et al. 2020; Sikora et al., 2020) as part of the verification process.

Through this process, ALOs and VLOs can impact curricula. Thus we offer the following advice for instructors:

- Write specific course-level ALOs to bolster macro-level goals including institutional goals for graduates and disciplinary community goals for future scientists, by moving beyond generic content and skill goals to focus on explicit development of student competency in experimentation.
- Use ALOs to guide implementation of daily course activities, guide course assessment choices, and/or structure adaptive versions of the materials to new contexts.

We believe these steps are critical not only to enhance the impact of CUREs, but also to guide creation of new types of educational experiences. For example, during the pandemic there was a need to deliver online versions of wet-lab activities (Irby et al., 2018c; Sikora et al., 2020). ALOs will assist faculty who are working to fill the gap for students who cannot have CURE-like experiences due to lack of capital or institutional resources.

Acknowledgments This work was funded by the National Science Foundation (NSF) Division of Undergraduate Education (DUE) grants #1503798 and #1710051. Additionally, this paper was informed by discussions at meetings of the ACE-Bio Network (NSF grant #1346567). Any opinions, findings, and conclusions or recommendations expressed in this material are those of the authors and do not necessarily reflect the views of the National Science Foundation.

References

Abualrob, M. M. A., & Daniel, E. G. S. (2013). The Delphi technique in identifying learning objectives for the development of science, technology and society modules for palestinian ninth grade science curriculum. *International Journal of Science Education, 35*(15), 2538–2558. https://doi.org/10.1080/09500693.2011.610381

Airey, J., & Linder, C. (2009). A disciplinary discourse perspective on university science learning: Achieving fluency in a critical constellation of modes. *Journal of Research in Science Teaching, 46*(1), 27–49.

American Society for Biochemistry and Molecular Biology [ASBMB]. (2020). *Structure & function: Macromolecular structure determines function and regulation.* Retrieved July 24, 2020 from https://www.asbmb.org/education/teachingstrategies/foundationalconcepts/MacromolecularStructureFunction/

Anderson, T. R. (2007). Bridging the educational research-teaching practice gap: The power of assessment. *Biochemistry and Molecular Biology Education, 35*(6), 471–477. https://doi.org/10.1002/bambed.20136

Anderson, T. R., Schönborn, K. J., du Plessis, L., Gupthar, A. S., & Hull, T. L. (2013). Identifying and developing students' ability to reason with concepts and representations in biology. In D. F. Treagust & C.-Y. Tsui (Eds.), *Multiple representations in biological education, series: Models and modeling in science education* (Vol. 7, pp. 19–38). Springer. https://doi.org/10.1007/978-94-007-4192-8

Berg, C. A. R., Bergendahl, V. C. B., Lundberg, B., & Tibell, L. (2003). Benefiting from an open-ended experiment? A comparison of attitudes to, and outcomes of, an expository versus an open-inquiry version of the same experiment. *International Journal of Science Education, 25*(3), 351–372. Retrieved from https://doi.org/10.1080/09500690210145738.

Bowe, K. A., Pelaez, N., & Anderson, T. R. (in preparation). Using expert research knowledge and representations to inform design of a module on the use of hydrogen-deuterium exchange to study protein structure and dynamics. In progress.

Cohen, L., Manion, L., & Morrison, K. (2002). *Research methods in education*. Routledge. https://doi.org/10.4324/9780203224342

Cooper, K. M., Soneral, P. A. G., & Brownell, S. E. (2017). Define your goals before you design a CURE: A call to use backward design in planning course-based undergraduate research experiences. *Journal of Microbiology & Biology Education, 18*(2), 1–7. https://doi.org/10.1128/jmbe.v18i2.1287

Craig, P. A., Anderson, T., Bernstein, H. J., Daubner, C., Goodman, A., Irby, S. M., Koeppe, J., Mills, J. L., Pikaart, M., Ringer McDonald, A., O'Handley, S. F., Roberts, R., & Stewart, R. (2018). Using protein function prediction to promote hypothesis-driven thinking in undergraduate biochemistry education. *The Chemist, 91*(1), 1–8. Retrieved from http://theaic.org/pub_thechemist_journals/index.html

Dries, D. R., Dean, D. M., Listenberger, L. L., Novak, W. R. P., Franzen, M. A., & Craig, P. A. (2016). An expanded framework for biomolecular visualization in the classroom: Learning goals and competencies. *Biochemistry and Molecular Biology Education, 69–75*. https://doi.org/10.1002/bmb.20991

Elmgren, M., Ho, F., Åkesson, E., Schmid, S., & Towns, M. (2015). Comparison and evaluation of learning outcomes from an international perspective: Development of a best-practice process. *Journal of Chemical Education, 92*(3), 427–432. https://doi.org/10.1021/ed500542b

Hsieh, H. F., & Shannon, S. E. (2005). Three approaches to qualitative content analysis. *Qualitative health research, 15*(9), 1277–1288. https://doi.org/10.1177/1049732305276687

Hsu, Y. H., Bucher, D., Cao, J., Li, S., Yang, S. W., Kokotos, G., Woods, V. L., McCammon, J. A., & Dennis, E. A. (2013). Fluoroketone inhibition of Ca2+ – independent phospholipase A2 through binding pocket association defined by hydrogen/deuterium exchange and molecular dynamics. *Journal of the American Chemical Society, 135*(4), 1330–1337.

Irby, S. M. (2019). *Evaluation of a novel biochemistry course-based undergraduate research experience (CURE)* (Doctoral dissertation). Purdue University. https://doi.org/10.25394/PGS.7813796.v1

Irby, S. M., Pelaez, N. J., & Anderson, T. R. (2018a). Anticipated learning outcomes for a biochemistry course-based undergraduate research experience aimed at predicting protein function from structure: Implications for assessment design. *Biochemistry and Molecular Biology Education, 46*(5), 478–492. https://doi.org/10.1002/bmb.21173

Irby, S. M., Pelaez, N. J., & Anderson, T. R. (2018b). How to identify the research abilities instructors anticipate students will develop in a biochemistry course-based undergraduate research experience (CURE). *CBE—Life Sciences Education, 17*(2), es4,1–14. https://doi.org/10.1187/cbe.17-12-0250

Irby, S. M., Borda, E. J., & Haupt, J. (2018c). Effects of implementing a hybrid wet lab and online module lab curriculum into a general chemistry course: Impacts on student performance and engagement with the chemistry triplet. *Journal of Chemical Education, 95*(2), 224–232. https://doi.org/10.1021/acs.jchemed.7b00642

Irby, S. M., Pelaez, N. J., & Anderson, T. R. (2020). Student perceptions of their gains in course-based undergraduate research abilities identified as the anticipated learning outcomes for a biochemistry CURE. *Journal of Chemical Education, 97*(1), 55–65. https://doi.org/10.1021/acs.jchemed.9b00440

Jeffery, K. A. (2019). *Expert explanations of protein-folding and dynamics research: Implications for biochemistry instruction* (Doctoral dissertation). Purdue University. https://doi.org/10.25394/PGS.7805984.v1

Jeffery, K. A., Pelaez, N., & Anderson, T. R. (2018). How four scientists integrate thermodynamic and kinetic theory, context, analogies, and methods in protein-folding and dynamics research: Implications for biochemistry instruction. *CBE—Life Sciences Education, 17*(1), ar13.

Jeffery, K. A., Pelaez, N., & Anderson, T. R. (2019). Using expert data to inform the use of research methods and representations to enhance biochemistry instruction and textbook design. *Biochemistry and Molecular Biology Education, 47*(5), 513–531.

Kozma, R., & Russell, J. (1997). Multimedia and understanding: Expert and novice responses to different representations of chemical phenomena. *Journal of Research in Science Teaching, 34*(9), 949–968.

Laverty, J. T., Underwood, S. M., Matz, R. L., Posey, L. A., Carmel, J. H., Caballero, M. D., Fata-Hartley, C. L., Ebert-May, D., Jardeleza, S. E., Cooper, M. M., & Cooper, M. M. (2016). Characterizing college science assessments: The three-dimensional learning assessment protocol. *PLoS One, 11*(9), 1–21. https://doi.org/10.1371/journal.pone.0162333

McDonald, A. R., Bernstein, H. J., Daubner, S. C., Goodman, A. L., Irby, S. M., Koeppe, J. R., Mills, J. L., O'Handley, S. F., Pikaart, M. J., Roberts, R. A., Sikora, A., & Craig, P. A. (2019). *BASIL Biochemistry Curriculum*. Retrieved from https://doi.org/10.35071/m89q-fa20. Accessed 3 Jan 2021.

McKay, T., Hart, K., Horn, A., Kessler, H., Dodge, G., Bardhi, K., Bardhi, K., Mills, J. L., Bernstein, H., & Craig, P. A. (2015). Annotation of proteins of unknown function: Initial enzyme results. *Journal of Structural and Functional Genomics, 16*(43), 43–54. https://doi.org/10.1007/s10969-015-9194-5

Moorthy, B. S., Schultz, S. G., Kim, S. G., & Topp, E. M. (2014). Predicting protein aggregation during storage in lyophilized solids using solid state amide hydrogen/deuterium exchange with mass spectrometric analysis (ssHDX-MS). *Molecular Pharmaceutics, 11*, 1869–1879.

NRC. (2001). *Knowing what students know: The science and Design of Educational Assessment*. National Academies Press.

Pelaez, N., Anderson, T., Gardner, S. M., Yin, Y., Abraham, J. K., Bartlett, E., Gormally, C., Hill, J. P., Hoover, M., Hurney, C., Long, T., Newman, D. L.. Sirum, K., Stevens, M. (2017). The basic competencies of bio-logical experimentation: concept-skill statements. *PIBERG Instructional Innovation Materials*. Paper 4. http://docs.lib.purdue.edu/pibergiim/4

Roberts, R., Hall, B., Daubner, C., Goodman, A., Pikaart, M., Sikora, A., & Craig, P. (2019). Flexible implementation of the BASIL CURE. *Biochemistry and Molecular Biology Education, 47*(5), 498–505. https://doi.org/10.1002/bmb.21287

Ryder, J., & Leach, J. (2000). Interpreting experimental data: The views of upper secondary school and university science students. *International Journal of Science Education, 22*(10), 1069–1084. https://doi.org/10.1080/095006900429448

Schönborn, K. J., & Anderson, T. R. (2009). A model of factors determining students' ability to interpret external representations in biochemistry. *International Journal of Science Education, 31*(2), 193–232. https://doi.org/10.1080/09500690701670535

Schönborn, K. J., & Anderson, T. R. (2010). Bridging the educational research-teaching practice gap: Foundations or assessing and developing biochemistry students' visual literacy. *Biochemistry and Molecular Biology Education, 38*(5), 347–354. https://doi.org/10.1002/bmb.20436

Sikora, A., Irby, S. M., Hall, B. L., Mills, S. A., Koeppe, J. R., Pikaart, M. J., Wilner, S. E., Craig, P. A., & Roberts, R. (2020). Responses to the COVID-19 pandemic by the biochemistry authentic scientific inquiry lab (BASIL) CURE consortium: Reflections and a case study on the switch to remote learning. *Journal of Chemical Education, 97*(9), 3455–3462. https://doi.org/10.1021/acs.jchemed.0c00729

Tansey, J. T., Baird, T., Jr., Cox, M. M., Fox, K. M., Knight, J., Sears, D., & Bell, E. (2013). Foundational concepts and underlying theories for majors in biochemistry and molecular biology. *Biochemistry and Molecular Biology Education, 41*(5), 289–296.

Ton, A. T., Gentile, F., Hsing, M., Ban, F., & Cherkasov, A. (2020). Rapid identification of potential inhibitors of SARS-CoV-2 main protease by deep docking of 1.3 billion compounds. *Molecular Informatics, 2000028*, 1–8. https://doi.org/10.1002/minf.202000028

Trujillo, C. M., Anderson, T. R., & Pelaez, N. J. (2015). A model of how different biology experts explain molecular and cellular mechanisms. *CBE—Life Sciences Education, 14*(2), ar20.

Trujillo, C. M., Anderson, T. R., & Pelaez, N. J. (2016a). Exploring the MACH model's potential as a metacognitive tool to help undergraduate students monitor their explanations of biological mechanisms. *CBE—Life Sciences Education, 15*(2), ar12.

Trujillo, C. M., Anderson, T. R., & Pelaez, N. J. (2016b). An instructional design process based on expert knowledge for teaching students how mechanisms are explained. *Advances in Physiology Education, 40*(2), 265–273.

van Mil, M. H. W., Boerwinkel, D. J., & Waarlo, A. J. (2013). Modelling molecular mechanisms: A framework of scientific reasoning to construct molecular-level explanations for cellular behaviour. *Science and Education, 22*(1), 93–118.

van Mil, M. H. W., Postma, P. A., Boerwinkel, D. J., Klaassen, K., & Waarlo, A. J. (2016). Molecular mechanistic reasoning: Toward bridging the gap between the molecular and cellular levels in life science education. *Science Education, 100*(3), 517–585.

White, H. B., Benore, M. A., Sumter, T. F., Caldwell, B. D., & Bell, E. (2013). What skills should students of undergraduate biochemistry and molecular biology programs have upon graduation? *Biochemistry and Molecular Biology Education, 41*(5), 297–301.

Wiggins, G. P., & McTighe, J. (2005). *Understanding by design*. Association for Supervision and Curriculum Development (ASCD).

Chapter 3
ACE-Bio Experimentation Competencies Across the Biology Curriculum: When Should We Teach Different Competencies and Concepts?

Megan F. Cole and Christopher W. Beck

3.1 Introduction

Calls for exposure to research for undergraduates at all levels have been issued by the National Academies (National Research Council, 1999) and the AAAS (2011). More recently, "doing research" has been defined as an important program-level learning outcome for the "Process of Science" (Clemmons et al., 2020), one of the core competencies in the *Vision and Change* report (AAAS, 2011). As experimentation is fundamental to biological research, teaching experimentation is essential in the process of introducing undergraduate students to research.

To further the teaching and assessment of experimentation, a National Science Foundation-funded Research Coordination Network on Advancing Competencies in Experimentation (ACE-Bio) has developed a set of *Basic Competencies of Biological Experimentation* (Pelaez et al., 2017; Chap. 1 in this volume). The Competencies are organized into seven thematic areas: *Identify* (the ability to identify gaps or limitations in current research knowledge through the review, filtering, and synthesis of relevant literature); *Question* (the ability to generate a research question and formulate hypotheses); *Plan* (the ability to plan feasible and ethical experiments to answer research questions or test hypotheses); *Conduct* (the ability to conduct an investigation to achieve research goals); *Analyze* (the ability to analyze and process data); *Conclude* (the ability to draw conclusions about data with inferences that are limited to the scope inherent in the experimental design); and, *Communicate* (the ability to communicate research work in professionally-appropriate modes, including visuals, written, and oral formats). Each set of Competencies is includes a list of Concepts and then Skill statements, totaling 103

M. F. Cole (✉) · C. W. Beck
Department of Biology, Emory University, Atlanta, GA, USA
e-mail: megan.f.cole@emory.edu; cbeck@emory.edu

N. J. Pelaez et al. (eds.), *Trends in Teaching Experimentation in the Life Sciences*, Contributions from Biology Education Research,
https://doi.org/10.1007/978-3-030-98592-9_3

specific Skill statements. Other similar frameworks exist for understanding experimentation (Harwood, 2004; Thanukos et al., 2010; Clemmons et al., 2020). For example, Harwood (2004) proposed an "activity" model with 10 activities in which scientists participate, including "form the question", "carry out the study", "examine the results", and "communicate with others" that are similar to ACE-Bio Competencies (Harwood, 2004). More recently, the BioSkills guide was developed, and it includes a subset of the ACE-Bio Competencies, Concepts, and Skills (Table 3.1) in course-level learning outcomes across a range of *Vision and Change* competencies (Clemmons, et al., 2020).

Given the number and breadth of Competencies, Concepts, and Skills in the ACE-Bio framework, they cannot all be addressed in a single course or likely in an undergraduate STEM career. The order and depth in which different aspects of experimentation should be taught to students at various stages in their academic careers is not clear (Clemmons et al., 2020), hindering effective course and programmatic design as well as assessment to advance experimentation skills. Thus, research is needed into how student learning of experimentation should be scaffolded and progress through an undergraduate curriculum.

3.2 Methods

We sought to align the 103 experimentation Skill statements with an undergraduate biology curriculum by measuring experts' expectations of student proficiency in each competency at two points in an undergraduate degree – after one year of introductory level coursework and upon receiving a bachelor's degree. To this end, we developed a survey that divided the seven Competencies into separate survey blocks, each with text that identified and defined the Competency using the definition from Pelaez et al. (2017). Within each block, Concepts and Skill statements were presented as matrix questions with each skill within the Concept presented as a row and the columns comprised of a 4-point Likert scale for both "After completing Introductory Biology students should …" and "After completing an undergraduate degree students should …". The Likert scale items were: "Not important", "have knowledge of this", "have first-hand experience with this", and "perform with little or no guidance," in order to capture the range of experience with each skill that faculty members might expect students to have.

Because of the seven areas of Competencies with the large number of Concepts, and Skill statements, survey respondents were presented with only half of the Concepts within each Competency to reduce survey fatigue and encourage completion of the survey. The Concepts within each Competency that were presented to a respondent were determined at random by the survey software. The final survey block, which was presented to all respondents, included demographic questions on institution type; the percentage of time a respondent spends on disciplinary research, education research, education practice (e.g., curriculum development) or other activities; which sub-discipline in biology best describes their primary disciplinary

Table 3.1 Mapping of BioSkills Guide Program and Course-level Learning Outcomes on to the ACE-Bio Basic Competencies for Biological Experimentation

	BioSkills Guide		ACE-Bio Competencies	
Vision & Change Core Competency	Program-level learning outcome	Course-level learning outcome	Competency	Concept
Process of science	Information literacy	Find and evaluate the credibility of a variety of sources of scientific information, including popular science media and scientific journals.	Identify	Relevant background knowledge
		Interpret, summarize, and evaluate evidence in primary literature.		
	Question formulation	Recognize gaps in our current understanding of a biological system or process and identify what specific information is missing.		A gap in current knowledge
		Develop research questions based on your own or others' observations.		
		Formulate testable hypotheses and state their predictions.	Question	Hypotheses
	Study design	Compare the strengths and limitations of various study designs.	Plan	Experimental design
		Design controlled experiments, including plans for analyzing the data.		Variables, controls, variation
		Execute protocols and accurately record measurements and observations.	Conduct	Measurement
		Identify methodological problems and suggest how to troubleshoot them.		Variable outcomes
		Evaluate and suggest best practices for responsible research conduct (e.g., lab safety, record keeping, proper citation of sources).		Data documentation
	Data interpretation and evaluation	Analyze data, summarize resulting patterns, and draw appropriate conclusions.	Analyze & Conclude	Multiple
		Describe sources of error and uncertainty in data.	Conclude	Inferences & Conclusions
		Make evidence-based arguments using your own and others' findings.		
		Relate conclusions to original hypothesis, consider alternative hypotheses, and suggest future research directions based on findings.		
	Doing research		All	Most

(continued)

Table 3.1 (continued)

	BioSkills Guide		ACE-Bio Competencies	
Quantitative reasoning	Quantitative & Computational Data Analysis	Record, organize, and annotate simple data sets.	*Analyze*	Data curation
		Create and interpret informative graphs and other data visualizations.		Data summary
		Select, carry out, and interpret statistical analyses.		Statistics
		Interpret the biological meaning of quantitative results.	*Conclude*	Patterns & relationships
Modeling	Model application	Use models and simulations to make predictions and refine hypotheses.	*Question*	Hypotheses
	Modeling	Build and revise conceptual models to propose how a biological system or process works.	*Question*	Models
			Communicate	Synthesis & Reflection
Communication & Collaboration	Communication	Use appropriate language and style to communicate science effectively to targeted audiences (e.g., general public, biology experts, collaborators in other disciplines).	*Communicate*	Scientific communication
		Use a variety of modes to communicate science (e.g., oral, written, visual).		
Science & Society	Ethics	Identify and evaluate ethical considerations (e.g., use of animal or human subjects, conflicts of interest, confirmation bias) in a given research study.	*Plan*	Ethics

research area; and all sub-disciplines in biology that describe their disciplinary research area. Our initial sample included faculty experts drawn from principal investigators of National Science Foundation grants related to course-based research experiences (CREs) or laboratory curriculum development in biology over the past 5 years, and corresponding authors of articles on CREs or laboratory curriculum in major biology education journals. However, response rates from this sample were too low to draw meaningful conclusions. As a result, we widened our survey pool by sending the survey invitation and link to the membership of the Association for Biology Laboratory Education (ABLE) and the listserv for the Society for the Advancement of Biology Education Research.

We received 104 completed surveys and 40 incomplete surveys, in which respondents did not proceed through to the end of the survey. For incomplete surveys, we included responses in our analysis only for Competencies that were completed. Because of incomplete surveys and the random assignment of Concepts within each Competency, the final sample size varied among Competencies and Concepts. The number of responses for Concepts ranged from 48 to 71.

3.3 Results and Discussion

3.3.1 Sample

One-hundred and forty-four faculty completed at least a portion of the survey with 72% of respondents completing the entire survey. Of the faculty who completed the demographics portion of the survey, those from research universities were the most frequent respondents (48%), followed by faculty from liberal arts colleges (28%), comprehensive universities (19%), and 2-year colleges (5%). In terms of disciplinary research areas, respondents most commonly identified ecology (24%), microbiology (10%), cell biology (10%), plant biology (9%), and genetics (9%) as their primary research areas (Table 3.2).

3.3.2 Competency Expectations of Introductory Students

Overall, most faculty expected students after one year of coursework to have at least knowledge of all seven Competencies (Fig. 3.1) with at least 82% of Skill ratings within each Competency being rated at 'have knowledge of this' or higher. Faculty had the highest expectation levels for *Conclude*, with 93% of faculty ratings within this competency being at least 'have knowledge of this' skill. Results also indicated that faculty most frequently scored *Conduct* and *Conclude* skills at the level of 'have first-hand experience with this.' In contrast, for skills within the other Competencies, faculty most frequently scored these at the level of 'have knowledge of this' or lower. This suggests that faculty hold higher expectations for student skills in *Conduct* and *Conclude* after one year of biology training.

Table 3.2 Distribution of disciplinary research areas of survey respondents

Disciplinary Area	Number of respondents	Percentage of respondents
Animal behavior	7	7%
Animal physiology	5	5%
Biochemistry	7	7%
Cell biology	10	10%
Developmental biology	2	2%
Ecology	23	24%
Genetics	9	9%
Microbiology	10	10%
Molecular biology	8	8%
Neurobiology	2	2%
Plant biology	9	9%
Other	4	4%
No response	48	

Percentages are based on those who provided a response

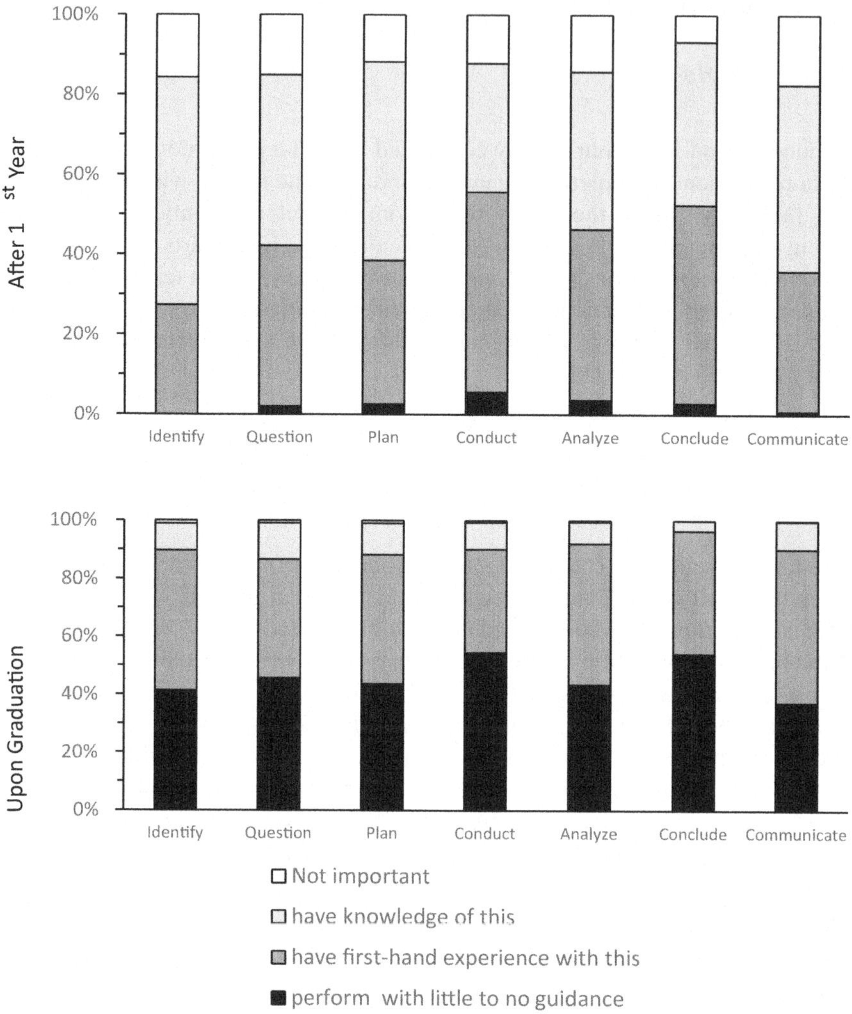

Fig. 3.1 Faculty competency expectations. Percentage of total skill ratings within each Competency shown. Number of ratings varied based on faculty participants and number of individual skills within each competency

At the level of having first-hand experience with the Competency skills, faculty rated *Conduct* skills highest (56% *Conduct* skill ratings were 'have first-hand experience of this' or higher) and rated *Identify* the lowest (27% of *Identify* skill ratings were 'have first-hand experience of this' or higher). This is in contrast to the fact that these skills were rated similarly in terms of expecting students to at least 'have knowledge of this' (88% and 84%, respectively). Together, these results suggest that faculty value exposing students to both of these Competencies. However, faculty might perceive that having first-hand experience with *Identify* skills may be less important for introductory students, more challenging for these students to perform, or more difficult for faculty to incorporate into first-year coursework.

Few faculty rated Competency skills at the level of 'perform with little to no guidance' for students after one year of coursework. At most, 6% of skills were rated at this level for any of the Competencies. This suggests that faculty do not expect high school students to have mastered any of these Competencies and that none should be expected to be fully learned within the first year of college-level coursework. This supports use of the Competencies to guide college-level appropriate course objectives (e.g. Irby et al., 2018a).

While most faculty expected students to have at least knowledge across all Competencies, a non-negligible number of skills were rated as 'not at all important' for introductory-level students. Six of the seven competencies had skill ratings within the range of 12–18% scored as 'not at all important' for introductory-level students (Fig. 3.1). However, *Conclude* only had 7% of ratings at this level, suggesting that faculty highly value student exposure to drawing conclusions from data in introductory coursework.

3.3.3 Competency Expectations of Degree Students

Faculty reported higher expectations for degree students compared to introductory students across all Competencies (Fig. 3.1). Competencies were almost universally rated at least at the level of 'have knowledge of this' (99–100% of skills) for graduating students. This suggests that faculty expect growth across all Competencies in post-introductory coursework. We did not specifically ask faculty how students should gain experience in experimentation. Thus, faculty may be including expectations of students' learning from upper-level lecture or laboratory courses, honors or other academic-year research, summer research experiences, or other experiences.

Substantial increase in expectations were also seen at the level of 'have first-hand experience with this' with 87–96% ratings at least at this level for degree students. The largest growth between introductory and degree students was seen in the increase from 27% to 89% for the *Identify* skills ratings of at least 'have first-hand experience with this.' This may speak to the fact that *Identify* skills require a background understanding of biology that faculty perceive that upper-level students have, but might be absent in introductory students.

Nearly half of all skills ratings for Competencies were at the level of 'perform with little to no guidance' (37–54%). This suggests that faculty have confidence in the ability of undergraduate students to master many of the competency skills. Given that ratings at this level were at or below 6% for students after the first-year coursework, it is clear that faculty feel undergraduate experiences provide students with adequate opportunity to practice and improve upon their experimentation skills.

While the lowest-rated Competency after the first-year coursework was *Identify*, faculty had the lowest expectations for *Communicate*, upon degree completion. This may align with faculty scoring *Communicate* skills more frequently than other

Competencies as 'not at all important' after the first-year coursework. Therefore, faculty may view skills within Communicate as less important throughout the undergraduate level than the other Competencies. However, this view conflicts with "Communication and Collaboration" as a core competency in the *Vision and Change* report (AAAS, 2011) and "Communication" as a program-level learning outcome in the BioSkills guide (Clemmons et al., 2020).

3.3.4 Variation Between Concepts and Skills Within Competencies

To better interpret faculty expectations of students' competencies in experimentation, it is important to examine Concepts and Skills within the Competencies, as faculty evaluated individual Skill statements in our survey and this allows us to identify more specific areas to focus on within the undergraduate curriculum. To examine variation between Concepts and Skills, we first converted each Concept or Skill to a single value by assigning numeric values to Likert ratings ('not important' = 0, 'have knowledge of this' = 1, 'have first-hand experience with this' = 2, 'perform with little to no guidance' = 3), tallying up total points, and converting to a percentage score based on the maximum score possible given the number of ratings. As can be seen in Fig. 3.2, Concepts and Skill ratings varied across a wide range of scores. This indicates that faculty can identify differences in their expectations between Concepts for a particular area of Competencies and Skills for applying an experimentation Concept.

3.3.4.1 Identify

The two Concepts within *Identify* were rated differently, with higher faculty expectations for identifying "Relevant Background Knowledge" compared to identifying "A Gap in Current Knowledge (Fig. 3.3). This difference helps explain the low scores for *Identify* at the level of 'have knowledge of this' or higher compared to the other Competencies. While, overall, *Identify* was rated low at the level of 'have first-hand experience with this' for introductory students (27%) compared to scores for the other Competencies (36–56%), identifying "Relevant Background Knowledge" ratings at this level were 35% while identifying "A Gap in Current Knowledge" was rated at only 14%. Thus, faculty may feel that introductory-level students can build the skills of finding and reading scientific literature, but do not yet have the breadth of knowledge within a field to *Identify* gaps in knowledge.

By the time students complete their undergraduate degree, faculty largely expect students to have experience with both *Identify* Concepts (83–93% ratings at least at the level of 'have first-hand experience with this'). This is consistent with particular course-level learning outcomes within the program-level learning outcomes of

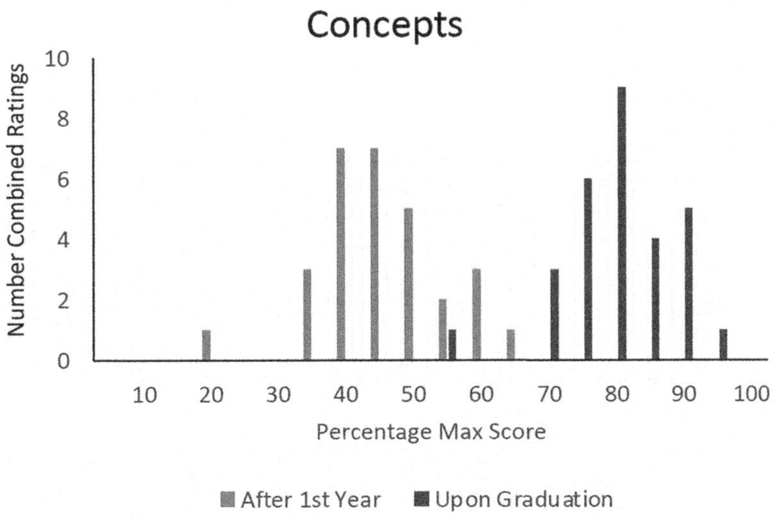

Fig. 3.2 Distribution of concept and skill ratings. Histogram of combined rating scores for each Concept (left) and Skill (right) shown. Combined rating scores were calculated by assigning 0 points for each 'not important' rating, 1 point for each 'have knowledge of this' rating. 2 points for each 'have hands-on experience with this' rating, and 3 points for each 'perform with little to no guidance' rating. Scores for each Concept or Skill were converted to percentages based on a maximum score of all 'perform with little to no guidance' ratings. Total Concepts = 29 for each of 'After 1st Year' and 'Upon Graduation' and total Skills = 103 for each of 'After 1st Year' and 'Upon Graduation'

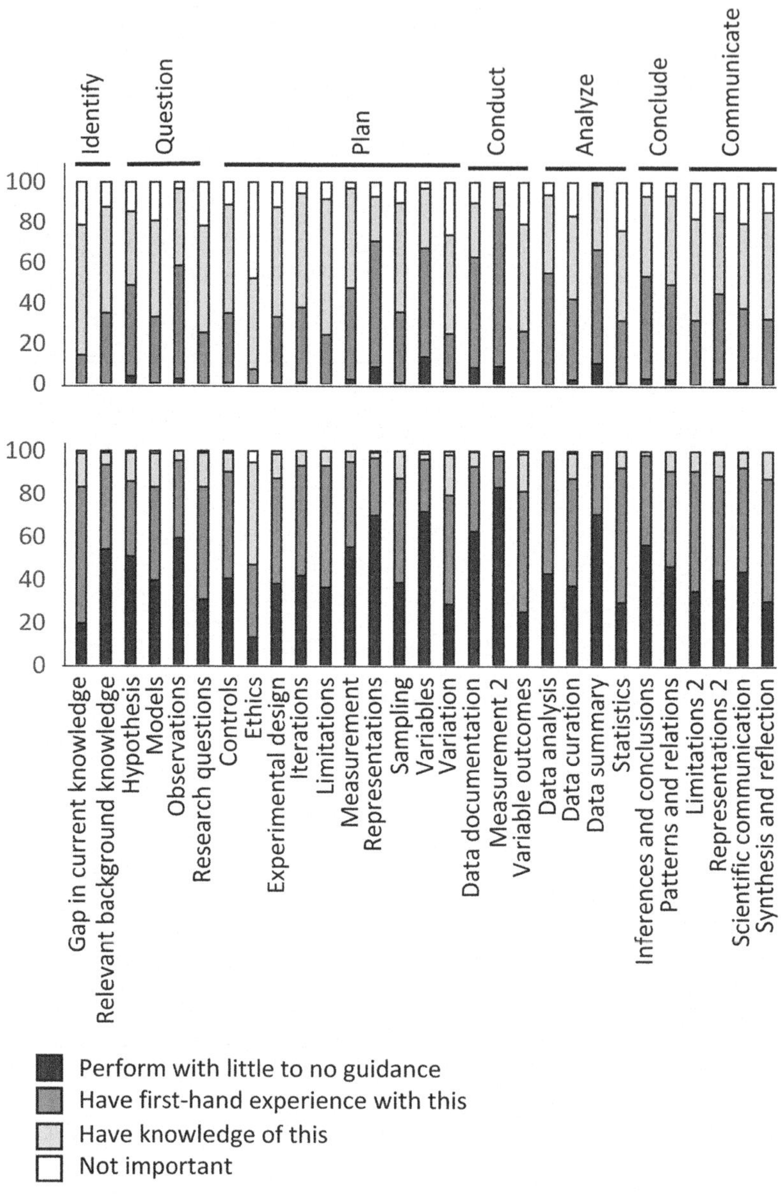

Fig. 3.3 Faculty concept expectations. Percentage of total Skill ratings within each Concept shown. Number of ratings varied based on faculty participants and number of individual skills within each concept

"Information Literacy" and "Question Formulation" in the BioSkills guide (Clemmons et al., 2020). However, expectations for the two *Identify* Concepts differ at the level of being able to 'perform with little to no guidance' (54% for "Relevant Background Knowledge" and 20% for "A Gap in Current Knowledge"). This further supports that faculty view skills needed to identify a gap in current knowledge as highly challenging for undergraduates and that these skills may not be obtained until graduate degree training.

Examining individual Skills within the Concept of "Relevant Background Knowledge" reveals differing expectations while skills within *Identifying* "A Gap in Current Knowledge" had relatively little variation. Ratings of degree students' ability to 'perform with little to no guidance' ranged from 38–80% across all Skills in both Concepts. Skills relating to finding and understanding literature (find appropriate sources of relevant scientific information, filter and evaluate the relevance of information of appropriate sources to the research focus, and evaluate background information with critical scientific skepticism) rated higher (51–80%) than skills focused on contextualizing background knowledge to guide research (synthesize and apply current knowledge to generate a contextual foundation for the research problem, and reflect on the skills and knowledge needed in the relevant field before proceeding to do research; 38–39%). This variation can be used by faculty to scaffold learning outcomes and focus on skills for finding and understanding literature for earlier coursework. After students have some mastery of finding and understanding literature in introductory courses, upper-level courses could then incorporate application of background knowledge to a specific research project. Approaches such as the C.R.E.A.T.E. method might be particularly helpful in this regard (Beck, 2019; Gottesman & Hoskins, 2013; Hoskins et al., 2011; Kenyon et al., 2016; Stevens & Hoskins, 2014).

3.3.4.2 Question

The *Question* Competency consists of four Concepts: Hypotheses, Models, Observations, and Research Questions. Amongst the Concepts, expectations are slightly higher for Hypotheses and Observations than for Models and Research Questions (Fig. 3.3). All Concepts showed increased expectations between completion of the first year and graduation. The higher-rated Concepts of Hypotheses and Observations showed the most increase at the level of 'perform with little to no guidance,' while the Concepts of Models and Research Questions had a larger increase at the 'have first-hand experience in this' level. These results suggest that faculty expect skill in Hypotheses and Observations to develop earlier or to a higher level than Models and Research Questions. Both hypotheses and research questions are course-level learning outcomes within the program-level learning outcome of "Question Formulation" in the BioSkills guide (Clemmons et al., 2020), confirming the importance of these experimentation concepts. Modeling (and hence models) is a core competency in the *Vision and Change* report (AAAS, 2011). However, within the ACE-Bio community, we struggled with differences in how faculty from different sub-disciplines understand the term “model.” Clemmons et al. (2020) found similar difficulties when working with faculty on the BioSkills guide. This disparate understanding and use of models might explain why models are considered less important.

3.3.4.3 Plan

Plan is the largest Competency and is composed of 10 unique Concepts and 70 individual Skills. Not surprisingly, there is variation between the 10 Concepts (Fig. 3.3).

Ratings for Ethics were surprisingly much lower than other Concepts with only 52% of ratings at the level of ‘have knowledge of this’ or higher after one year of coursework compared to the other Concepts which ranged from 74–97% (average 90%). Expectations remained lower for Ethics than for other Plan Concepts for graduating students with only 13% ratings at ‘perform with little to no guidance’ compared to other Concepts (39–72%, average 47%). However, the low ratings may be caused by the focus of the two Ethics skills on professional values and on regulatory committee submission. Other Concepts, which may relate to Ethics such as bias and efficiency in experimental design, appropriate data collection and accurate representation are incorporated into other Concepts or Competencies. Of the two Skills in Ethics, “Integrate Professional and Community Ethics into Research Design” is rated much higher than ‘Submit Planned Research to the Institutional Review Board or Animal Care and use Committee for Evaluation, as Appropriate’. Expectations for the first Skill were 81% at the level of ‘have knowledge of this’ after the first year and 71% ‘have first-hand experience with this’ for degree students, but only 24% and 23% at these levels regarding IRB and IACUC submission. Thus, it seems likely that faculty value Ethics Concepts for undergraduate students. Indeed, a research coordination network in undergraduate biology education (RCN-UBE) has been established to explore teaching ethics and the responsible conduct of research in the context of course-based undergraduate research experiences (Diaz-Martinez et al., 2019). However, specific knowledge or experience with research ethics regulatory committees may not be considered important for most undergraduate students (though in theory would be considered appropriate for students doing mentored research regulated by these committees).

Two Concepts within Plan stand out due to their higher expectations: Representations and Variables. These Concepts scored higher than others at the level of at least ‘have first-hand experience of this’ (71% and 67%, respectively, for introductory students compared to 31% average for other Concepts, and 97% and 96%, respectively, for graduating students compared to 84% average for other Concepts). They also scored higher at the level of ‘perform with little to no guidance’ (8% and 14%, respectively, for introductory students compared to 1% for other Concepts, and 70% and 72%, respectively, for graduating students compared to 37% for other Concepts). Although these data suggest that Representations is an important concept and that faculty expect student gains are made in it, few tools exist for assessing for Representations (Chap. 14 in this volume).

3.3.4.4 Conduct

Faculty had relatively high expectations for skills in *Conduct* for both early and graduating students (Fig. 3.3). One of the three Concepts within *Conduct* ("Variable Outcomes") was consistently rated with lower expectations than the other two Concepts ("Data Documentation" and "Measurement"). Within Data Documentation, however, the Skill 'archive important and sensitive data in an accessible format that is intelligible, secure and ethical' was rated with lower expectations. This skill may have been viewed similarly by faculty to Ethics in Plan, which also had low ratings.

By the time of graduation, 63% and 83% of skill ratings for "Data Documentation" and "Measurement" were at the level of 'perform with little to no guidance' while "Variable Outcomes" had only 25% of its skill ratings at this level. This suggests that faculty largely expect students to handle "Data Documentation" and "Measurement" appropriately without need for much assistance, but that responding to unforeseen outcomes might be considered less important or considered a more challenging concept. Although faculty largely expect students to have encountered this concept by the time they graduate, they also expect students to still need guidance when handling it. In courses where unexpected outcomes may occur, faculty should consider including scaffolding or mentoring to help students work through these challenges appropriately. Henry et al. (2019) provide a framework for understanding how students might deal with unexpected outcomes, which students might view as "failures."

3.3.4.5 Analyze

Analyze consists of four Concepts (Data Analysis, Data Curation, Data Summary, and Statistics) and 13 Skills. Overall, Analyze was rated similarly to the other Competencies (Fig. 3.1). Within the Analyze Concepts, faculty held the highest expectations for Data Summary and the lowest for Statistics (Fig. 3.3). In fact, of Concepts across all the Competencies for introductory-level students, Data Summary had the second highest rating (10%) for 'perform with little to no guidance' and Statistics had the third highest rating for 'not important' ratings (24%). It is possible faculty feel use of statistical tests is too challenging for introductory-level students or that this skill should be taught after students acquire more basic experimentation skills. However, while the importance of each of the Concepts increased between the end of introductory courses and the end of a degree program, Statistics was still considered less important than Data Summary and Data Analysis.

Within the three Statistics Skills ("Choose and conduct statistical tests that are appropriate for the type/nature of data", "Choose and conduct statistical tests that are aligned with hypotheses and experimental methods", and "Generate statistics for a sample to summarize and/or describe parameters for a whole population (e.g. mean, median, measures of variance)"), the two related to statistical tests were rated with lower expectations compared to the Concept on summarizing or describing the dataset or population. This was the case at both the introductory level and after degree completion. Whether this lack of emphasis on statistical analysis reflects differences among sub-disciplines in biology in the use of statistics is unclear. Unfortunately, our sample size in any sub-discipline in biology was not sufficient to explore this idea in further detail.

3.3.4.6 Conclude

Conclude, overall, had high expectations from faculty (Fig. 3.1) and consists of two Concepts ("Inferences and Conclusions", and "Patterns and Relationships") that are rated similarly at both the introductory and degree levels (Fig. 3.3). At the introductory level, Conclude was the highest rated in terms of students at least having knowledge of it, and about half the ratings were at the level of 'have first-hand experience with this.' Upon graduation, 96% of ratings for Conclude were at the level of at least 'have first-hand experience with this' with about half at the level of 'perform with little to no guidance' (Fig. 3.1). The two Concepts within Conclude were scored similarly by faculty (Fig. 3.3). Interestingly, the skill "Understand that Scientific Knowledge is Tentative" within "Inferences and Conclusions" was the third highest skill at the introductory level for 'perform with little to no guidance' (17%) out of all 103 Skills. This suggests the importance of reinforcing the tentative nature of science, especially in student presentations and writing.

3.3.4.7 Communicate

Communicate consists of four Concepts (Limitations, Synthesis and Reflection, Representations, and Scientific Communication) with 15 total Skills. Communicate had low faculty expectations compared to other Competencies at both the introductory and degree levels (Fig. 3.1). These lower ratings could not be entirely attributed to particular Concepts with much lower scores (Fig. 3.3). Although "Limitations" and "Synthesis and Reflection" had lower expectations than those for "Representations" and "Scientific Communication" the difference was not large. Low faculty expectations for Communicate were most prominent at the 'perform with little to no guidance' level for degree students. This may speak to the difficulty of the Skills or to the nature of scientific communication, which often involves feedback from collaborators, peers, or mentors.

Two Skills scored lower than others within Communicate: within Limitations, "Construct a Justification and Counter-justification Argument for each Alternative, if Possible" and within Synthesis and Reflection, "Revise an Existing Model Based on Observations or Data." These skills may be thought to be more challenging for students so may require more scaffolding when teaching these Concepts.

Two Skills also stood out due to relatively high scores: "Distill Results into Clear Numeric and/or Graphical Forms that are Aligned with the Experimental Objective/Question/Hypothesis" within Representations, and "Construct Scientific Communications Using Standard Conventions" within Scientific Communication.

These Skills had 68% and 61% ratings respectively of at least 'have first-hand experience with this' for introductory students (range of other Communication Skills 16–42%) and ratings of 55% and 57% ratings respectively of at least 'perform with little to no guidance' for graduating students (range of 15–49% for other Communication Skills). These ratings largely align with common assignments in biology laboratory courses. In addition, these Skills can be rigorously assessed with existing rubrics (Angra & Gardner, 2018; Kishbaugh et al., 2012; Timmerman et al., 2011).

Interestingly, faculty had higher expectations for Representations than Scientific Communication for introductory students, but this pattern was reversed for graduating students for all rating levels. Thus, Scientific Communication Skills had larger growth in expectations between first-year and graduating students. This seems to largely be caused by smaller growth in the Skill "Develop a Predictive or Explanatory Model to Summarize Research Findings in Representations." This suggests that this Skill, in particular, may be difficult to learn or that different faculty interpret this Skill differently.

3.4 Conclusions and Recommendations

Overall, faculty experts suggest that undergraduate students should develop to a certain level in all experimentation competencies during their undergraduate education rather than focusing on particular competencies in introductory courses and other competencies in upper-level courses. More specifically, faculty suggest greater experience with *Conduct* and *Conclude* as compared to the other Competencies after completing introductory courses. However, the expectations are more similar across Competencies at the completion of an undergraduate degree with students having at least "first-hand experience" with most of the experimentation Skills. Similarly, in their survey of science process skills, Coil et al. (2010) found that faculty considered most science process skills to be important or very important for graduating students (Coil et al. 2010)). Our results suggest that the ACE-Bio Basic Competencies for Biological Experimentation seem appropriate for teaching and a providing a framework for assessing experimentation at the undergraduate level.

Introducing all of the Competencies, although not all of the Skills, at the introductory level and providing students with more experience with the Competencies throughout their undergraduate careers is in line with the idea of scaffolding instruction in experimental design that is supported at the course level (e.g., D'Costa & Schlueter, 2013; Großmann & Wilde, 2019). As a result, faculty should consider how to appropriately scaffold student learning in experimentation throughout the curriculum. They should pose questions such as: "Where should students gain experience in experimentation between the end of introductory courses and graduation?" and "What is the relative importance of lecture and discussion courses, upper-level laboratory courses, course-based and mentored research, and summer research programs?"

In addition to considering where in the curriculum students should gain experience in experimentation, the results of our survey lead to several additional recommendations for both instructors and education researchers. These are summarized as bulleted lists and briefly discussed below.

3.4.1 Recommendations for Instructors

- Use the results to develop learning objectives appropriate for a particular course level;
- Based on each course-level learning objective, choose or design assessments for determining competency in experimentation at the appropriate level;
- Use the results to develop program-level learning objectives to determine how experimentation is taught across your curriculum; and,
- Engage your department in a similar exercise to determine whether your department's expectations follow those presented here.

3.4.2 Recommendations for Education Researchers

- Explore learning progressions in experimentation competencies;
- Perform follow-up surveys or focus groups to better understand why certain Skills are rated low;
- Perform follow-up surveys to examine whether what faculty teach and how they teach influence their perspectives on teaching experimentation; and,
- Develop assessments of experimentation that are appropriate for the course level and faculty expectations.

3.4.3 Discussion of Recommendations

For **instructors**, the proposed levels of experience for students in different Competencies and Concepts can provide a framework for developing course-level learning outcomes for courses at different levels (e.g. Irby et al., 2018b). Although learning outcomes related to experimentation can be determined after curriculum development (Irby et al., 2018a), defining learning objectives at the beginning in a backward design approach according to Wiggins and McTighe (2005) will allow instructors to better align assessments with learning outcomes in order to determine if student competencies in experimentation are indeed increasing (Irby et al., 2018a). Defining course-level learning objectives and articulating the level of expertise expected also allows instructors to better select appropriate assessments to examine gains in student competencies (see Chaps. 14 and 17 in this volume for a review of assessments in biological experimentation). It should be noted that although competencies in the *Identify* area were prioritized less by faculty in this study, chapters in this volume illustrate how those competencies are brought into focus for courses at other institutions (chapters by Cheng (5), Thomas (6), and Kruchten & O'Brien (9) in this volume). Similar to defining and aligning course-level learning objectives to instructional activities, the proposed levels of experience for students in different Competencies and Concepts can be used by departments to define program-level learning outcomes. In addition, departments can map existing courses on the Competencies and Concepts to see how they align with the proposed development of experimentation skills throughout a curriculum. In a similar vein, faculty could conduct a similar survey with their faculty to determine if their department views the development of experimentation skills in the same way. We did this with the faculty in the Department of Biology at Emory University and found patterns that largely correspond with those presented here.

Our results also suggest future directions for **education researchers** interested in experimentation competencies. First and foremost, our results represent the expectations of faculty experts. While our results align with ideas of scaffolded instruction, they do not necessarily reflect the way in which students best learn about experimentation. Learning progressions, which incorporate a fuller understanding of how students think about experimentation and reason while involved in different aspects of experimentation, will allow us to better understand how students learn experimentation and, therefore, how we should teach experimentation (Duncan & Hmelo-Silver, 2009; Scott et al., 2019). Second, for particular Skills that were rated lower, it is unclear whether faculty consider these Skills to be less important, more difficult to implement, or more challenging for students. In their survey, Coil et al. (2010) found that the most common reasons for not teaching particular science process skills were the amount of time needed to teach the skill and the need for students to have particular content knowledge before a skill is taught (Coil et al., 2010). Follow-up surveys or focus groups with faculty would allow us to better understand why certain Skills are rated low. Third, our survey and sample size of responses did not allow us to examine factors that might influence survey responses. For example, faculty who teach laboratory courses using CREs or whose department's incorporate CREs might have different perspectives than faculty who teach courses using other approaches. Future studies might consider in more detail whether what faculty teach (both area of biology and course level) and how they teach influence their perspectives on teaching experimentation. Finally, assessment is key to teaching, as it allows us to determine what students are learning. Our results suggest that assessments for competencies in experimentation should vary for students at different levels, as the expectations are different for students at different levels. However, assessments designed for students at a particular level may not emphasize the Competencies that are expected at that level. For example, many assessments of experimentation, including those designed for introductory students, focus on *Plan* (Chaps. 13 and 14 in this volume) whereas faculty do not have the expectation that students at the introductory level will have at least "first-hand experience" with many of the Skills within this competency (Fig. 3.1). In addition, assessments are lacking for certain Concepts that are deemed important. For instance, faculty expected students to have a high level of experience with Skills in the Concept Representations within the *Plan* Competency at both the introductory level and at graduation (Fig. 3.3). However, assessments of student competency with respect to Representations is wholly lacking (see Chap. 14 in this volume).

The ACE-Bio Basic Competencies for Biological Experimentation provide an important framework for curriculum development and assessment related to experimentation (Pelaez et al., 2017). Our results advance this work by clarifying the expectations of faculty experts on how these competencies should be developed over the course of an undergraduate career. We hope that instructors and education researchers leverage these results in their own curriculum development and research on student learning of experimentation.

Acknowledgments This work was partially funded by a stipend from the National Science Foundation (NSF) grant #1346567 for the Advancing Competencies in Experimentation–Biology (ACE-Bio) Network. Any opinions, findings, and conclusions or recommendations expressed are those of the author and do not necessarily reflect the views of the NSF. We thank Nancy Pelaez, Trevor Anderson, and Stephanie Gardner for facilitating the work of the ACE-Bio Network. Preliminary feedback from Trevor Anderson, Jeff Olimpo, and others in the ACE-Bio Network helped to improve this chapter. For this, we are grateful. This study was approved by the Emory University Institutional Review Board; Study 00000944.

References

American Association for the Advancement of Science (AAAS). (2011). *Vision and change in undergraduate biology education: A call to action* (C. Brewer & D. Smith Eds.). Washington, DC. https://visionandchange.org/finalreport/. Accessed 18 Oct 2021.

Angra, A., & Gardner, S. M. (2018). The graph rubric: Development of a teaching, learning, and research tool. *CBE—Life Sciences Education, 17*(4), ar65. https://doi.org/10.1187/cbe.18-01-0007

Beck, C. W. (2019). Integrating primary literature in a lecture course using a modified version of the C.R.E.A.T.E Approach. *CourseSource*, https://doi.org/10.24918/cs.22019.24925

Clemmons, A., Timbrook, J., Herron, J., & Crowe, A. (2020). BioSkills guide: Development and national validation of a tool for interpreting the vision and change core competencies. *bioRxiv*, 2020.2001.2011.902882. doi:https://doi.org/10.1101/2020.01.11.902882

Coil, D., Wenderoth, M. P., Cunningham, M., & Dirks, C. (2010). Teaching the process of science: Faculty perceptions and an effective methodology. *CBE—Life Sciences Education, 9*(4), 524–535. https://doi.org/10.1187/cbe.10-01-0005

D'Costa, A. R., & Schlueter, M. A. (2013). Scaffolded instruction improves student understanding of the scientific method & experimental design. *The American Biology Teacher, 75*(1), 18–28. https://doi.org/10.1525/abt.2013.75.1.6

Diaz-Martinez, L. A., Fisher, G. R., Esparza, D., Bhatt, J. M., D'Arcy, C. E., Apodaca, J., Brownell, S., Corwin, L., Davis, W. B., Floyd, K. W., Killion, P. J., Madden, J., Marsteller, P., Mayfield-Meyer, T., McDonald, K. K., Rosenberg, M., Yarborough, M. A., & Olimpo, J. T. (2019). Recommendations for effective integration of ethics and responsible conduct of research (E/RCR) education into course-based undergraduate research experiences: A meeting report. *CBE—Life Sciences Education, 18*(2), mr2. https://doi.org/10.1187/cbe.18-10-0203

Duncan, R. G., & Hmelo-Silver, C. E. (2009). Learning progressions: Aligning curriculum, instruction, and assessment. *Journal of Research in Science Teaching, 46*, 606–609. https://doi.org/10.1002/tea.20316

Gottesman, A. J., & Hoskins, S. G. (2013). CREATE cornerstone: Introduction to scientific thinking, a new course for STEM-interested freshmen, demystifies scientific thinking through analysis of scientific literature. *CBE—Life Sciences Education, 12*, 59–72.

Großmann, N., & Wilde, M. (2019). Experimentation in biology lessons: Guided discovery through incremental scaffolds. *International Journal of Science Education, 41*(6), 759–781. https://doi.org/10.1080/09500693.2019.1579392

Harwood, W. S. (2004). A new model for inquiry. *Journal of College Science Teaching, 33*(7), 29–33.

Henry, M. A., Shorter, S., Charkoudian, L., Heemstra, J. M., & Corwin, L. A. (2019). FAIL is not a four-letter word: A theoretical framework for exploring undergraduate students' approaches to academic challenge and responses to failure in STEM learning environments. *CBE—Life Sciences Education, 18*(1), ar11. https://doi.org/10.1187/cbe.18-06-0108

Hoskins, S. G., Lopatto, D., & Stevens, L. M. (2011). The C.R.E.A.T.E. approach to primary literature shifts undergraduates' self-assessed ability to read and analyze journal articles, attitudes about science, and epistemological beliefs. *CBE—Life Sciences Education, 10*, 368–378.

Irby, S. M., Pelaez, N. J., & Anderson, T. R. (2018a). Anticipated learning outcomes for a biochemistry course-based undergraduate research experience aimed at predicting protein function from structure: Implications for assessment design. *Biochemistry and Molecular Biology Education, 46*(5), 478–492. https://doi.org/10.1002/bmb.21173

Irby, S. M., Pelaez, N. J., & Anderson, T. R. (2018b). How to identify the research abilities that instructors anticipate students will develop in a biochemistry course-based undergraduate research experience (CURE). *CBE—Life Sciences Education, 17*(2), es4. https://doi.org/10.1187/cbe.17-12-0250

Kenyon, K. L., Onorato, M. E., Gottesman, A. J., Hoque, J., & Hoskins, S. G. (2016). Testing CREATE at community colleges: An examination of faculty perspectives and diverse student gains. *CBE—Life Sciences Education, 15*, ar8.

Kishbaugh, T. L. S., Cessna, S., Jeanne Horst, S., Leaman, L., Flanagan, T., Graber Neufeld, D., & Siderhurst, M. (2012). Measuring beyond content: A rubric bank for assessing skills in authentic research assignments in the sciences. *Chemistry Education Research and Practice, 13*(3), 268–276. https://doi.org/10.1039/C2RP00023G

National Research Council. (1999). *Transforming undergraduate education in science, mathematics, engineering, and technology*. The National Academies Press.

Pelaez, N.J., Anderson, T.R., Gardner, S. M., Yin, Y., Abraham, J. K., Bartlett, E., Gormally, C., Hill, J. P., Hoover, M., Hurney, C., Long, T., Newman, D.L., Sirum, K., & Stevens, M. (2017). The basic competencies of biological experimentation: Concept-skill statements. *PIBERG Instructional Innovation Material Paper 4.* Retrieved from http://docs.lib.purdue.edu/pibergiim/4/. Accessed 18 Oct 2021.

Scott, E., Wenderoth, M., & Doherty, J. (2019). Learning progressions: An empirically grounded, learner-centered framework to guide biology instruction. *CBE—Life Sciences Education, 18*, es5. https://doi.org/10.1187/cbe.19-03-0059

Stevens, L. M., & Hoskins, S. G. (2014). The CREATE strategy for intensive analysis of primary literature can be used effectively by newly trained faculty to produce multiple gains in diverse students. *CBE—Life Sciences Education, 13*, 224–242.

Thanukos, A., Scotchmoor, J. G., Caldwell, R., Lindberg, D. R. (2010). The real process of science-understanding science at http://www.understandingscience.org and science 101: Building the foundations for real understanding. *Science, 330* (6012), 1764–1765. https://doi.org/10.1126/science.1186994

Timmerman, B. E. C., Strickland, D. C., Johnson, R. L., & Payne, J. R. (2011). Development of a 'universal' rubric for assessing undergraduates' scientific reasoning skills using scientific writing. *Assessment & Evaluation in Higher Education, 36*(5), 509–547. https://doi.org/10.1080/02602930903540991

Wiggins, G. P., & McTighe, J. (2005). *Understanding by design*. Association for Supervision and Curriculum Development (ASCD).

Chapter 4
Integrating the Five Core Concepts of Biology into Course Syllabi to Advance Student Science Epistemology and Experimentation Skills

Kyriaki Chatzikyriakidou and Melissa McCartney

4.1 Introduction

A competency is defined as concept-based ability that informs professional practices during experimentation according to the American Association for the Advancement of Science (AAAS, 2011). It has been well-documented that college students are lacking in critical thinking, experimental hypothesis formation, and data interpretation (Barnett & Francis, 2012; Butler et al., 2012; Flores et al., 2012; Thompson & Blankinship, 2015), which are fundamental competencies used by professional biologists (AAAS, 2011; Pelaez et al., 2017). The reason for these competencies being challenging for students to acquire may have to do with current curricula offering information through separate course sections that students have to combine as they move through their college studies (Nehm, 2019). Recently it has been stated that science teaching practices are based on compartmentalized, not contextualized, learning experiences (Faria et al., 2014). Perhaps, such education practices are grounded on the assumption that students are able to combine the knowledge they acquire across courses, but such practices may actually impede learning (Southard et al., 2016).

In addition to above-mentioned lack of experimentation competencies, it has also been reported that students usually face difficulty in organizing their ideas and structuring a progressive sequence of events (Faria et al., 2014; Kang et al., 2014;

K. Chatzikyriakidou (✉)
STEM Transformation Institute, Florida International University, Miami, FL, USA
e-mail: chatzikyriak@wisc.edu

M. McCartney
STEM Transformation Institute, Florida International University, Miami, FL, USA

Department of Biological Sciences, Florida International University, Miami, FL, USA
e-mail: mmccartn@fiu.edu

N. J. Pelaez et al. (eds.), *Trends in Teaching Experimentation in the Life Sciences*, Contributions from Biology Education Research,
https://doi.org/10.1007/978-3-030-98592-9_4

Zangori et al., 2015). These findings point to the research need to invent in-class activities that could help students synthesize course content, and experimentation skills with disciplinary epistemic beliefs. Biology learning experiences would be considered authentic only when they would immerse students into the constructs and values intrinsic to science (epistemic beliefs), at the same time that they introduce students to content knowledge and experimental procedures (Gouvea et al., 2019).

Considering the time it takes a learner to develop good experimentation skills and also shift their initial epistemic beliefs to more sophisticated ones (Hall, 2013; Semsar et al., 2011), students should be immersed into authentic science learning experiences as soon as possible, starting at the freshman year of college education. Initial reform efforts in biology education have brought in consensus on the 5 Core Concepts in biology (5CCs; Evolution, Structure and Function, Information Flow, Exchange, and Storage, Pathways of Transformation of Energy and Matter, and Systems). The 5CCs provided a succinct conceptual framework, describing all potential biology knowledge summarized in five biological scales (molecular, cellular, organismal, population, and ecology) and five overarching concepts that dictate natural biological phenomena or processes (AAAS, 2011). Understanding that every biological process or phenomenon can be interpreted through five different perspectives, as suggested by the 5CCs, could help student improve their conceptual understanding and consequently several of the experimentation competencies, as described in the ACE-Bio Competences, that are currently challenging for them.

There has been little research on the effect of learning with the 5CCs on undergraduate biology student content, procedural and epistemic knowledge. In this manuscript, we aim to motivate the integration of the 5CCs into introductory biology syllabi and describe how teaching with the 5CCs could offer integrated learning experiences to undergraduate biology majors. By providing preliminary data along with current theoretical frameworks of how people form scientific explanations, we argue that students who learn with the 5CCs could have improved conceptual understanding and more advanced epistemological beliefs by the end of an academic semester. In addition, we present the argument that improvement of epistemic knowledge should be linked to improvement of content knowledge and procedural knowledge (experimentation competencies) as well.

4.2 Connections Between Concept-Based Knowledge and Experimentation Skills

At the undergraduate level, the core concepts of biology were first produced after collaboration efforts between the National Science Foundation (NSF) and the American Association for the Advancement in Science (AAAS), with the report *Vision and Change in Undergraduate Biology Education: A Call to Action* (AAAS, 2011). *Vision and Change* introduced five overarching Core Concepts in biology

Table 4.1 Core concepts outlined in *Vision and Change: A Call to Action* (AAAS, 2011)

Core Concept	Main description
1. Evolution (EV)	The diversity of life evolved over time by processes of mutation, selection, and genetic change.
2. Structure and function (SF)	Basic units of structure define the function of all living things.
3. Information flow, exchange, and storage (IFES)	The growth and behavior of organisms are activated through the expression of genetic information in context.
4. Pathways and transformations of energy and matter (PTEM)	Biological systems grow and change by processes based upon chemical transformation pathways and are governed by the laws of thermodynamics.
5. Systems (SYS)	Living systems are interconnected and interacting.

(5CCs; Table 4.1) as an effort to reform biology education and reach consensus on those concepts that are important for undergraduate biology majors to understand by the time they graduate.

The potential of providing these generic 5CCs in the first place has been adopted by biology education researchers who helped expand their use in general biology courses. The BioCore Guide (Brownell et al., 2014) is a nationally validated set of general principles and specific statements for each one of the 5CCs. On a similar note, the Conceptual Elements Framework offers a list of sub-concepts for each one of the 5CCs that can be used as a tool to make connections across sub-disciplines and scales (Cary & Branchaw, 2017). Both of these studies supported the initial efforts of reforming undergraduate biology education in accordance to the 5CCs. In addition, instructional approaches (Branchaw et al., 2020) and specific syllabi (Chatzikyriakidou et al., 2021a) have been developed to help educators integrate the 5CCs into their courses. Assessment of student understanding of the 5CCs can be measured with the BCCIs (Biology Core Concept Instrument) tool which was specifically designed to measure first–year students' ability to identify and describe concepts represented in biological phenomena, as well as to make connections between concepts (Cary et al., 2019). The BCCIs tool is composed of four narratives, each including a series of true-false/identify (TF/I) and open-ended questions, which have been aligned with the Conceptual Elements Framework.

According to recent literature in science education, the ability of a student to form scientific explanations is supported by two models: the unification model and the causal explanation model (de Andrade et al., 2019). The unification model assumes a direct relationship between explanations and promotion of understanding, and supports the idea that understanding increases as several distinct phenomena are linked by big ideas. In other words, scientific explanations are made out of comprehensive ideas that connect various aspects of the universe (de Andrade et al., 2019). The causal explanation model supports the idea that scientific explanations are based on the sequence of relevant causes that produce the phenomenon and the relationships between them (de Andrade et al., 2019).

According to the above-mentioned scientific explanation models, instruction with the 5CCs seems of great benefit to introductory biology students. The 5CCs compose a universal scaffold of thought process, regardless of someone's prior knowledge and personal means of understanding, and seem ideal for instruction at the undergraduate level (Chatzikyriakidou et al., 2021a). Having biology learners analyze a phenomenon into five different but interrelated perspectives (5CCs) is crucial for developing their understanding of the inherent relationships among biology concepts and realizing the coherent nature of scientific knowledge. Integrating the 5CCs in course syllabi, serves as a means to forming the epistemic belief that scientific knowledge is a coherent network of facts, and could never be composed of groups of unrelated facts. In addition, the 5CCs can be used for analysis of a biological phenomenon in multiple biological scales (molecular, cellular, organismal, population, or ecology) at the same time, which allows explanatory transitions to be made either between concepts or between biological scales. These explanatory transitions can be considered as elements of causal relationships.

Regarding procedural knowledge, we consider the set of experimentation competencies a student has acquired. Students have to experience concept-based abilities, known as competencies, in order to become biologically literate and practice science (AAAS, 2011). The BioSkills Guide provides both program-level and course-level learning outcomes for each of the 5CCs (Clemmons et al., 2020), however research findings on college biology students understanding of and assessment in these specific competencies have not been as expanded as in the 5CCs.

A more detailed and coherent guide on experimentation competencies is provided by the RCN-UBE Advancing Competencies in Experimentation–Biology (ACE-Bio) Network (Pelaez et al., 2017; Pelaez et al., 2018; Chap. 1 in this volume). According to the ACE-Bio Network, there are seven areas a competent biologist calls in when doing experimentation in biology: *Identify*, *Analyze*, *Communicate*, *Question*, *Conduct*, *Plan*, and *Conclude*. Each competency is accompanied by a list of specific learning outcomes to help biology educators with implementation of experimentation activities and assessments across diverse institutional and curricular contexts (Pelaez et al., 2017).

We provide some examples on how familiarity with the 5CCs of biology is directly linked to several of the ACE-Bio Competencies. The experimentation competency *Identify* is defined as the ability to identify the gaps or limitations in current research knowledge and synthesize information read in the literature. The experimentation competency *Question* is defined as the ability to formulate explanations or hypotheses that are compatible to current knowledge, testable and potentially falsifiable. In addition, the experimentation competency *Conclude* requires the understanding of patterns and relationships relevant to the information in hand (experimenting measures) in order to conclude or infer about research findings. Lastly the experimentation competency *Communicate* requires the ability to synthesize and reflect on the research findings, as well as to articulate the contribution of findings to what is already known. The learning outcomes of these experimentation competencies can be taught to all students of an introductory biology course when they learn biology using the 5CCs framework.

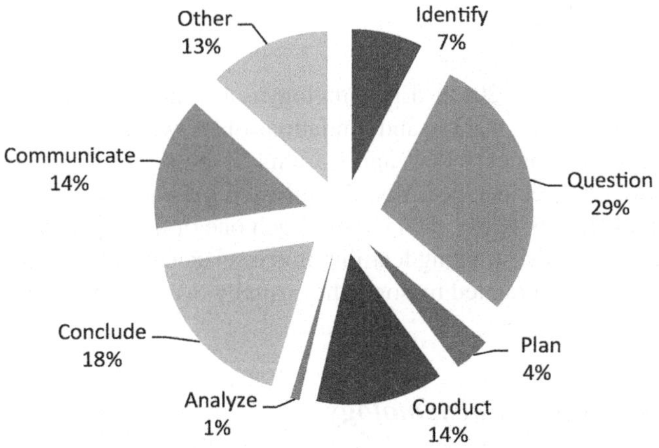

Fig. 4.1 Introductory biology student distribution (n = 137) of open-ended responses to the prompt "Please describe what doing science means to you." Student short answer responses were analyzed using thematic analysis (Braun & Clarke, 2006). The ACE-Bio Competencies were used as pre-defined themes, and each student response was coded with the theme(s) mostly referred to. Two researchers first independently coded student responses and then reviewed their coding together until a consensus was reached. Consensus is defined by achieving an inter-rater reliability (kappa value) of >0.9, which is higher than what is recommended in the literature (Syed & Nelson, 2015)

To support our opinion that 5CCs of biology should be incorporated into introductory biology syllabi to advance experimentation competencies, we present pilot data collected during Spring 2020. In a large enrollment introductory biology course, early in the semester, we asked students to briefly describe what doing science means to them (Fig. 4.1). We used the ACE-Bio Competencies (as pre-defined themes) to tag each student response and organize the dataset into the given seven categories. As seen in Fig. 4.1, the majority of responses were related to the ACE-Bio competency *Question* and most students referred to the concepts of Observations and Hypotheses (Pelaez et al., 2017; Table 1.4 in Chap. 1 of this volume). The second most chosen response was related to the ACE-Bio competency *Conclude* and most students referring to the concept of Inferences and Conclusions (Pelaez et al., 2017; Table 1.8 in Chap. 1 of this volume). These responses were collected from two different lecture sections of introductory biology with 75–87% freshmen and sophomores thus, they set a promising research direction regarding the effect of 5CCs instruction on advancing student experimentation skills in early college years.

Students must have real opportunities at being involved in the construction of their own scientific explanations in order to improve them (NRC, 2012; Tang, 2016; Zangori et al., 2015). Authentic science experiences can be offered to students when incorporating the 5CCs in course syllabi, even when the classroom is not a laboratory room. Student ability to form a causal and coherent story when explaining a phenomenon is a fundamental aspect of both conceptual understanding and science epistemology. The latter is further elaborated in the following section.

4.3 Epistemology

According to Kelly et al. (2012) "Epistemology is a branch of philosophy that investigates the origins, scope, nature, and limitations of knowledge" (pp. 281). Research concerning epistemological views and learning in the scientific disciplines can be described through three perspectives: disciplinary, personal ways of knowing, and social practices (Hayes-Klosteridis, 2019). Each one of these perspectives provides theoretical basis for investigating learning in the sciences, and even if they are used separately for research related reasons, they usually overlap and inform each other.

4.3.1 Disciplinary Epistemology

Although there is no specific set of beliefs about knowledge construction within the field of biology research, exploration of (general) science epistemology has been recorded in the literature for more than 60 years. Nature of Science (NOS) publications emerged from investigating K-12 students and teachers NOS conceptions, and has since extended to college level education.

Typically, NOS refers to the epistemology and sociology of science, science as a way of knowing, or the values and beliefs inherent to scientific knowledge and its development (Lederman, 1992). NOS views are considered a subset of epistemological beliefs about the nature of knowledge, and specifically about the development and justification of knowledge (Borgerding et al., 2017). The consensus view of NOS (Lederman, 1999; Schwartz & Lederman, 2008) has been an influential model describing the elements of the nature of scientific knowledge although the challenges of applying this model to various scientific disciplines have been discussed in the literature (Hodson & Wong, 2017).

Research framed by the disciplinary perspective of knowing, values the relationship between NOS and students' views of NOS. The core emphasis is given on students understanding of the elements of scientific knowledge and inquiry. An example of naïve to sophisticated disciplinary epistemology, regarding empirical NOS would be students believing that science is an entirely rational and orderly activity to students believing that scientific knowledge involves human imagination and creativity when observations are made. Another example of naive epistemology that has also been seen in biology majors is the challenge students have in distinguishing between theories and laws (Desaulniers Miller et al., 2010) usually believing that laws have a higher status than theories or theories could become laws with more evidence. However, professional scientists would keep theories separate from laws in the meaning that theories are inferred explanations for observable phenomena, while laws are descriptive statements of relationships among observable phenomena (Lederman et al., 2002).

4.3.2 Personal Epistemology

Personal epistemology is the study of people's thinking about knowledge and about how people know (Hofer & Pintrich, 1997; Hofer & Bendixen, 2012). Research in college students' personal epistemology has been actively debated since Perry's original work in Perry, 1970. Perry's model suggested that students' epistemological beliefs progress from naive to sophisticated ones throughout nine stages, which were later simplified to four stages by Moore (2002) as: dualism, multiplicity, relativism, and commitment to relativism (dialectical). Students were assumed to begin college with dualistic (right versus wrong) epistemological beliefs and shift to some of the other stages as they got exposed to conflicting views regarding the same issues.

Informed by Perry's model, research on personal epistemology has been branched out to three main approaches: the developmental, the beliefs, and the resources approaches. Although there are overlapping elements among the three approaches, the last two are the most relevant to this manuscript, and we focus on elaborating these further. Integrating the 5CCs in introductory biology syllabi would facilitate an authentic scientific context and help activate those epistemological resources students need to advance their intellectual experimentation skills.

4.3.2.1 Beliefs Approach

Schommer-Aikins (Schommer-Aikins, 2004; Schommer, 1990) proposed that there are multiple epistemological beliefs that develop independently in an individual's mind. In her model, epistemological beliefs included the "structure of knowledge (ranging from simple to complex), the stability of knowledge (certain to uncertain), the source of knowledge (omniscient authority to reason and evidence), the speed of learning (quick to gradual) and the ability to learn (fixed to improvable)" (Schommer-Aikins & Duell, 2013). These beliefs reflect assumptions, expectations, and attitudes that may affect reasoning processes (Hofer & Pintrich, 1997). Schommer-Aikins (2004) described beliefs as overlapping with knowledge, meaning that they are possessed by the individual, and they can be emotion-laden and context-independent.

College students' epistemological beliefs have been found to affect their learning with specific relationships recorded between personal epistemologies and academic performance (Schommer-Aikins & Easter, 2006), motivation (Muis, 2004), student learning (Hofer, 2000), and engagement (Kardash & Howell, 2000). Regarding the link between disciplinary and personal epistemology, Abd-El-Khalick and Ackerson (Abd-El-Khalick & Akerson, 2004) found that holding right-versus-wrong dualistic epistemological beliefs (low-level of Perry's model) interfered with acquiring informed NOS views, and Ackerson et al. (Akerson et al., 2006) found that epistemological beliefs played an important role in the retention of informed NOS views in preservice teachers.

4.3.2.2 Resources Approach

An epistemic resource is an individuals' perception of the source of their own knowledge, in other words the understanding of "how do I know what I know." An individual's understanding of "how they know what they know" helps them develop their personal epistemology (Hammer & Elby, 2002; Hofer, 2001, 2006), described above. Epistemological resources are context-specific, fine-grained cognitive resources that people use to reflect on their epistemic knowledge, activities, forms and stances (Barzilai & Zohar, 2014). Context-specific means that an individual will not always apply the same epistemological resources to every situation. The result is that a person can view a cognitive construct as "known" for different reasons at different times.

Designing instructional techniques, solely based on the beliefs approach, would aim to have an individual acquire the correct beliefs, concepts, ideas etc. in order to able to use it again, later in another context (Prince, 2004). From a resources perspective, however, learning is not conceptualized as the acquisition or formation of a particular cognitive object, but rather as a cognitive state the learner enters in the moment by activating multiple resources (Hammer & Elby, 2003). In these terms, successful learning would mean the learner entering a similar state later in a different context, likewise from the classroom to the real lab room.

Borgerding et al. (2017) suggested that the relationship between overall epistemic views (Perry's model) and context-dependent epistemological beliefs (Hofer's multidimensional model) should be considered as dynamic. In other words, the authors hypothesized that "overall epistemic beliefs can influence to what extent one can improve their context-dependent epistemological beliefs, and in turn, improved context-dependent epistemological beliefs can facilitate one's transition from a lower to a higher epistemological position" (pp. 495). It has also been stated that epistemological resources may gradually evolve into beliefs as they become articulated and more stable, although stability is usually a characteristic of expert-like epistemology (Louca et al., 2004). Epistemological resources are activated within epistemological frames defined as "locally coherent activation of a network of resources that may look like a stable belief or theory" (Elby & Hammer, 2010).

4.3.3 The Effect of Classroom's Epistemic Climate on Student Learning – Social Practices

Hall (2013) argued that students bear classroom expectations; a predictive set of ideas or assumptions students make regarding the nature of their classroom experience. In other words, these expectations are a student's answer to the question: "What do I think is the nature of the knowledge I am learning and what is it that I have to do in order to learn it?"

When a student's epistemic resources are activated in the right context, (i.e a college classroom) they can be productive reasoning tools that students use to understand disciplinary concepts (Hammer, 2000). In general, epistemological framing refers to the interpretations and adjustment that occurs as an individual makes sense of how to behave in particular situations or settings (Goffman, 1974; Tannen, 1993). Epistemological framing is a subcategory of framing that involves interpreting the knowledge forms and knowledge activities that are valued or appropriate (Hammer et al., 2005; Bing & Redish, 2009).

How a learner makes the decision about what (learning) activity is valuable or appropriate in a specific context is attributed to two factors: prior knowledge and physical and social cues (Gouvea et al., 2019). Preparing for a course exam composed of multiple-choice questions might trigger the epistemic resource that memorization would serve the learner best in this examination. Students may have prior experiences of success with multiple-choice exams and memorization of lecture notes. Students situated in a lecture hall with the instructor lecturing and students taking notes may reinforce a framing about knowledge transmission. However, in the scenario where the instructor facilitates an open-end discussion with the students, different resources will be activated, such as those students have about collaboration, argumentation and so on.

Previous work has shown that students may demonstrate multiple, contradictory, or context-dependent sets of expectations (Hall et al., 2011; Watkins et al., 2010), therefore, it is important to measure the nature of student expectations in a specific context (classroom) and study the alignment between students' and instructor's expectations, so syllabus adjustments can be made to help students improve from beginning to end of the semester, and consequently throughout their degree programs.

4.3.4 Assessment Tools for Biology Student Epistemology

Within the last decade, various efforts have been set to design measurement tools that focus on biology student epistemological beliefs at the college level, in addition to the reported efforts in NOS assessment. Two such tools have been developed for use in undergraduate biology populations, with either lower- or upper-level biology courses. However, due to the complex effect of a learning environment on student epistemology, validation of these tools would be required before drawing any conclusions. We briefly describe the structure and theoretical framework of each of these measurement tools.

4.3.4.1 The Colorado Learning Attitudes About Science Survey for Biology (CLASS-Bio)

CLASS-Bio was one of the first questionnaires developed to measure biology students' epistemological beliefs (Semsar et al., 2011). The survey consists of four categories of questions, based on four major perceptions about biology that are known to vary between experts and novices: 1. enjoyment of the discipline, 2. propensity to make connections to the real world, 3. recognition of conceptual connections underlying knowledge, and 4. problem-solving strategies (Semsar et al., 2011).

CLASS-Bio was designed based on Hammer (1994) who proposed that differences between how experts and novices view a discipline can be characterized into three main areas: (1) content and structure of knowledge, (2) source of knowledge, and (3) problem-solving approaches. The 31 Likert-scale statements of the survey aim to distinguish between the specific attitudes and beliefs of biologists from those of introductory students, offering educators the ability to track students' perceptions across a curriculum and across courses.

4.3.4.2 Maryland's Biology Expectations Survey (MBEX)

MBEX[1] was designed to measure students' epistemological beliefs about biology knowledge and biology learning in a reformed lower-level organismal biology course that integrated multidisciplinary concepts in its syllabus (Hall, 2013). This survey's theoretical framework is focused on epistemological framing, however it has only been qualitatively (student-expert interviews) validated, and there is currently no record of additional administration in other biology courses. It is composed of 32 Likert-scale statements, separated into four clusters: (I) Facts v. Principles, (II) Independence v. Authority, (III) Interdisciplinary Perspectives v. Silo Maintenance, and (IV) Connected v. Isolated.

4.4 Student Epistemological Beliefs and Learning Biology with the 5 CCs: A Case Study

Since the publication of the 5CCs, research has mostly focused on developing instructional techniques that incorporate the 5CCs (Branchaw et al., 2020; Chatzikyriakidou et al., 2021a), however there is no record yet about how and what students learn when they are taught with the 5CCs. In a first-year bio seminar class,

[1]The original MBEX survey was designed for an interdisciplinary biology class, thus we conducted exploratory (EFA, n = 318) and confirmatory (CFA, n = 211) factor analyses with lecture-based introductory biology student responses and resolved a different four-factor structure of the questionnaire: (a) Interdisciplinary knowledge, (b) Application of interdisciplinary knowledge, (c) Real-world connections, and (d) Learning facts v. principles in class. This new questionnaire structure was used for comparative analysis between the two groups of introductory biology students presented in this manuscript.

students used the 5CCs to analyze, compare and contrast various biological phenomena, before they took introductory biology and were found to perform better than students who had not taken the seminar. Although not directly measured, the authors hypothesized that the better performance was related to the 5CCs activities the students had completed in the first-year seminar (Wienhold & Branchaw, 2018).

Student ability to form scientific explanations seems to be interdependent on someone's ability in forming coherent and causal explanations of phenomena as well as their belief that coherence is part of the nature of scientific knowledge. Epistemological beliefs have been found significantly correlated to students' conceptual change when learning evolution theory (Borgerding et al., 2017) as well as physics students conceptual gain, both at the beginning and end of the semester (Perkins et al., 2005; Coletta & Philips, 2010). Similarly to these studies, we hypothesized that there would be a positive relationship between student 5CCs understanding and science epistemology, and we present preliminary data to explore the relationship between introductory biology students BCCI and MBEX scores.

Students of General Biology I consented to participate in an end-of-the-semester survey, which included the first narrative of the BCCIs tool (Cary et al., 2019) and the MBEX statements that have been shown to express student epistemological beliefs in a conventional introductory biology class. The first narrative of the BCCIs tool is about Recombinant Humulin production and asks students to identify the core concepts: IFES, SF, and PTEM (Table 4.1). Student BCCI scores were calculated as the percentage of correct answers out of the total 20 questions of the instrument and correlated to their average MBEX scores (Fig. 4.2). There were two Gen Bio I sections: an experimental section ($n = 37$) and a control section ($n = 31$). The experimental section had a semester-long experience of analyzing three course topics: Aquaporins, Aerobic respiration and DNA transcription with the use of the 5CCs (Chatzikyriakidou et al., 2021b), while the control section did not engage with the 5CCs.[2] In addition, at the beginning of the semester, and prior to any 5CCs activity, we asked students to rank their familiarity with each of the 5CCs on a percentage scale (0–100%). In order to gather accurate self-reported data, no description about the 5CCs was provided other than the full name of each core concept.

Comparing the two introductory biology sections, we found that students who engaged in the 5CCs activities scored significantly ($p < 0.05$) higher than the students of the control group and the number of students who answered at least half of the 20 questions correctly was twice higher than the number of students in the control group. The Recombinant Humulin production narrative targets students' conceptual understanding of the CCs: SF, IFES, and PTEM (Table 4.1). Based on the self-reported prior knowledge of these concepts, the control group reported higher familiarity with all three concepts as compared to the 5CCs group, with

[2] The 5CCs group participated in three in-class activities (one each month) that were completed during periodic exam review sessions. In each review session, students were provided with a single page 5CCs worksheet (Chatzikyriakidou et al., 2021a) modified to include auxiliary questions for each CC (Chatzikyriakidou et al., 2021b). Students were asked to fill in the worksheet with a short answer for each CC about the topic recently covered in lecture. The three topics selected for each of the three 5CCs activities were, in order: *Aquaporins, Aerobic respiration*, and *DNA transcription*. We chose these topics based on the course timeline and learning goals.

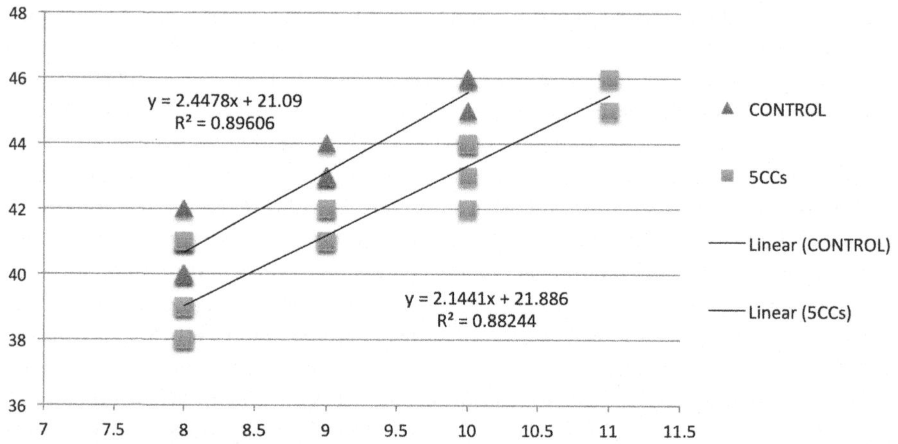

Fig. 4.2 Correlation values of introductory biology student BCCI and MBEX scores for the 5CCs and control groups. Pearson r correlations were significant ($p < 0.05$) for both groups. During an academic semester, the 5CCs group of students completed three analysis of biological phenomena with the use of the 5CCs, whereas the control group did not engage with the 5CCs

29–49% and 17–29% respectively. To the best of our knowledge there are no other records of BCCI scores in the literature, so these findings are promising for increasing student understanding of the 5CCs with regular in-class activities.

Regarding student epistemological beliefs, no significant differences were found between the two groups of students, with a range of scores between 23 and 58 points for the control group and between 24 and 56 points for the 5CCs group. An absolute expert-like (based on strongly agree/strongly disagree choices) score, based on the 11 MBEX items that were given to students, would total 41 points. However, due to the Likert-type 6-point scale an expert-like score could deviate between 35 and 46 points. These limits of the MBEX score deviation were used to select the subgroup of students expressing an overall sophisticated epistemological profile, while those with a lower or a higher MBEX score were excluded from analysis. Figure 4.2 shows those students who had both a sophisticated epistemological profile and a BCCI score of at least 8/20 correct answers, resulting in the 5CCs and control subgroups of 14 and 15 students respectively. Ideally, we would select students with a minimum of 10/20 BCCI score, however the small sample size did not allow for such selection.

In the subgroups of selected students (Fig. 4.2) there were significant ($p < 0.05$) and positive correlations between their MBEX and BCCI scores of 88.2% and 89.6% for the 5CCs and control groups respectively. Considering that students' scores were selected at the end of the semester, this finding may be attributed to a real relationship between conceptual understanding and science epistemological beliefs. In addition, no significant correlations were seen for either of the two groups of students, for those with lower or higher BCCI and MBEX scores than the selected ones. This finding implies a correlation between conceptual understanding and sophisticated epistemological beliefs. However, without a pre-MBEX score at the beginning of the semester, we cannot conclude on the effect of the specific 5CCs activity, on advancing student epistemological beliefs.

The results of this case study are preliminary and aim to increase motivation in gathering additional data on the relationships between student conceptual understanding of biology and science epistemology in introductory courses. Based on videotaped classroom scenarios, and interviews, Lising and Elby (2005) argued that a more expert-like epistemology indeed leads to better learning, and thus, curricular materials and teaching techniques should explicitly attend to students' epistemological beliefs. Because correlations between epistemologies and learning alone do not imply causal relationships, curricular material and teaching techniques have to be designed in ways that reflect the disciplinary epistemic knowledge that majors need to acquire upon completion of their degree programs.

4.5 Designing Epistemic Learning Environments in the Classroom

Each classroom has its own epistemic climate fostered by its main instructional activities and the type of notions these offer to students about construction of knowledge. In a recent study, hypothesizing that the classroom climate perceptions are indeed an important factor in student epistemology development, differences were found when the same instructor taught a traditional lecture-based section, and a constructivist-based section of organic chemistry (Barger et al., 2018). Two main findings were concluded from this study: 1. student perceptions of a complex learning environment predicted changes in their epistemology, and 2. student initial epistemological beliefs predicted how they perceived the classroom environment (Barger et al., 2018). These findings are not surprising if we recall the fundamentals of epistemological framing supporting that the physical cues of a class, along with student prior experiences and knowledge, inform those epistemological resources that the student finds appropriate to use within a specific class (context).

An epistemologically reformed course would look like a place where students are supported towards building their own network of coherent and sensible ideas about the world and about their learning process (Berland et al., 2015). Russ (2014) has previously argued that students need to negotiate scientific content and construct knowledge within a context informed by their past experiences and knowledge, in order to advance their epistemological beliefs. As a result of taking a biology class, we want students to recognize that they can use experimentation and reasoning to learn about the world, however this is challenging to implement in large enrollment courses.

Results from a meta-analysis of 24 studies in physics education which measured student epistemologies either with the CLASS (Colorado Learning Attitudes about Science Survey) or MPEX (Maryland Physics Expectations Survey) surveys, found that teaching method was a significant predictor in shifting student's beliefs (Madsen et al., 2015). Furthermore, the authors mentioned a pattern between teaching method, class size and student population, although non-significant at the moment (Madsen et al., 2015). Large positive shifts were seen in courses with small class size, explicit focus on model building and taught to non-science majors (Madsen et al., 2015). Relevant findings to these meta-analysis conclusions have been reported in biology education lab-based curricula. For example, AIM-Bio students were found to have gains in nature of science understanding (Hester et al., 2018) and C.R.E.A.T.E. students significantly shifted some of their epistemological beliefs (Gottesman & Hoskins, 2013). Both of these findings refer to freshmen in small size classes with explicit teaching of scientific method skills, both intellectual and procedural.

Although it may seem challenging to create a similar epistemic climate in lecture-based courses, the potential of implementing 5CCs activities could offer the ground for reforms in the current large enrollment biology courses. Multiple-choice testing (often used in large enrollment lecture courses) has been found to be unproductive for students' study habits and critical thinking (Stanger-Hall, 2012). Furthermore, it is paradoxical that the majority of introductory biology instructors want students to learn higher-order skills, but the course exams tend to focus at lower levels (Momsen et al., 2010). The already developed 5CCs activities (Branchaw et al., 2020; Chatzikyriakidou et al., 2021a) or newly designed material should be further explored along with measures of student conceptual understanding and epistemological beliefs.

4.6 Conclusion and Recommendations

Every learning environment comes with its own context, thus students will activate epistemic resources based on their prior knowledge on similar learning environments, as well as the current physical and social cues they interpret in the class which they are situated. Additional findings on student understanding of the 5CCs are of value for designing epistemologically informed curricula, teaching practices and assessment tools, and consequently help reform undergraduate biology education.

Researchers and educators need to understand how portrayals of knowledge in the classroom shape personal epistemology development. We support the argument that integrating the 5CCs in lecture-based curricula would be an important factor setting the appropriate epistemic climate of a classroom, even in large enrollment courses. Through implementation of 5CCs activities, biology instructors could adapt their course learning goals towards conceptual understanding and sense-making of biology knowledge, which would foster alignment of student epistemological beliefs with those of professional scientists. In addition, 5CCs activities can be an innovative teaching material to help educators gauge how well their students understand the biological concepts covered in the course material.

Three recommendations are provided for educators and education researchers:

- Teaching practices that apply the 5 Core Concepts in Biology (5CCs) could potentially foster integrated learning of content, procedural and epistemic knowledge. Such integrated learning practices are of primary importance in undergraduate biology education.
- Further research is needed to confirm this preliminary evidence about the effect of learning with the 5CCs on student conceptual understanding, experimentation skills, as well as science epistemology.
- The existence and type of casual relationships between conceptual, procedural and epistemic knowledge learning should be further investigated in undergraduate biology.

Acknowledgements This work was supported by the State of Florida through the 2018–2019 State University System of Florida Legislative Budget Request and a Florida International University start-up package.

References

Abd-El-Khalick, F., & Akerson, V. L. (2004). Learning as conceptual change: Factors that mediate the development of preservice elementary teachers' views of nature of science. *Science Education, 88*(5), 785–810. https://doi.org/10.1002/sce.10143

Akerson, V. L., Morrison, J. A., & Roth McDuffie, A. (2006). One course is not enough: Preservice elementary teachers' retention of improved views of nature of science. *Journal of Research in Science Teaching, 43*(2), 194–213. https://doi.org/10.1002/tea.20099

American Association for the Advancement of Science (AAAS). (2011). *Vision and change in undergraduate biology education: A call to action* (C. Brewer & D. Smith Eds.). Washington, DC. https://visionandchange.org/finalreport/. Accessed 18 Oct 2021.

Barger, M. M., Perez, A., Canelas, D. A., & Linnenbrink-Garcia, L. (2018). Constructivism and personal epistemology development in undergraduate chemistry students. *Learning and Individual Differences, 63*, 89–101. https://doi.org/10.1016/j.lindif.2018.03.006

Barnett, J. E., & Francis, A. L. (2012). Using higher order thinking questions to foster critical thinking: A classroom study. *Educational Psychology, 32*(2), 201–211. https://doi.org/10.1080/01443410.2011.638619

Barzilai, S., & Zohar, A. (2014). Reconsidering personal epistemology as metacognition: A multifaceted approach to the analysis of epistemic thinking. *Educational Psychologist, 49*(1), 13–35. https://doi.org/10.1080/00461520.2013.863265

Berland, L., Schwartz, V., Krist, C., Kenyon, L., Lo, S., & Reiser, J. (2015). Epistemologies in practice: Making scientific practices meaningful for students. *Journal of Research in Science Teaching, 53*(7), 1082–1112. https://doi.org/10.1002/tea.21257

Bing, T., & Redish, E. F. (2009). Analyzing problem solving using math in physics: Epistemological framing via warrants. *Physical Review Physics Education Research, 5*, 020108. https://doi.org/10.1103/PhysRevSTPER.5.020108

Borgerding, L. A., Deniz, H., & Shevock Anderson, E. (2017). Evolution acceptance and epistemological beliefs of college biology students. *Journal of Research in Science Teaching, 54*(4), 493–519. https://doi.org/10.1002/tea.21374

Branchaw, J. L., Pape-Lindstrom, P. A., Tanner, K. D., Bissonnette, S. A., Cary, T. L., Couch, B. A., Crowe, A. J., Knight, J. K., Semsar, K., Smith, J. I., Smith, M. K., Summers, M. M.,

Wienhold, C. J., Wright, C. D., & Brownell, S. E. (2020). Resources for teaching and assessing the vision and change biology core concepts. *CBE—Life Sciences Education, 19*(1), 1–9. https://doi.org/10.1187/cbe.19-11-0243

Braun, V., & Clarke, V. (2006). Using thematic analysis in psychology. *Qualitative Research in Psychology, 3*(2), 77–101. https://doi.org/10.1191/1478088706qp063oa

Brownell, S. E., Freeman, S., Wenderoth, M. P., & Crowe, A. J. (2014). BioCore guide: A tool for interpreting the core concepts of vision and change for biology majors. *CBE Life Sciences Education, 13*(2), 200–211. https://doi.org/10.1187/cbe.13-12-0233

Butler, H. A., Dwyer, C. P., Hogan, M. J., Franco, A., Rivas, S. F., Saiz, C., & Almeida, L. S. (2012). The Halpern critical thinking assessment and real-world outcomes: Cross-national applications. *Thinking Skills and Creativity, 7*(2), 112–121. https://doi.org/10.1016/j.tsc.2012.04.001

Cary, T., & Branchaw, J. (2017). Conceptual elements: A detailed framework to support and assess student learning of biology Core concepts. *CBE-Life Sciences Education, 16*(2), 1–10. https://doi.org/10.1187/cbe.16-10-0300

Cary, T. L., Wienhold, C. J., & Branchaw, J. (2019). A biology core concept instrument (BCCI) to teach and assess student conceptual understanding. *CBE Life Sciences Education, 18*(3), 1–17. https://doi.org/10.1187/cbe.18-09-0192

Chatzikyriakidou, K., Manrique, C., Tacloban, M. J., & McCartney, M. (2021a). Exploring primary scientific literature through the lens of the 5 core concepts of biology. *Course Source*. https://doi.org/10.24918/cs.2021.5

Chatzikyriakidou, K., Tacloban, M. J., Conception, K., Geiger, J., & McCartney, M. (2021b). Student Association of Lecture Content with the Five Core Concepts of Biology: Novel Results from an Introductory Biology Course. *Journal of Microbiology and Biology Education*, 22:e00105-21. https://doi.org/10.1128/jmbe.00105-21

Clemmons, A., Timbrook, J., Herron, J., & Crowe, A. (2020). BioSkills guide: Development and national validation of a tool for interpreting the *vision and change* Core competencies. *CBE—Life Sciences Education, 19*, 4. https://doi.org/10.1187/cbe.19-11-0259

Coletta, V. P., & Philips, J. A. (2010). Developing thinking & problem solving skills in introductory mechanics. *AIP Conference Proceedings, 1289*, 13. https://doi.org/10.1063/1.3515181

de Andrade, V., Freire, S., & Baptista, M. (2019). Constructing scientific explanations: A system of analysis for students' explanations. *Research in Science Education, 49*, 787–807. https://doi.org/10.1007/s11165-017-9648-9

Desaulniers Miller, M. C., Montplaisir, L. M., Offerdahl, E. G., Cheng, F., & Ketterling, G. L. (2010). Comparison of views of the nature of science between natural science and nonscience majors. *CBE—Life Sciences Education, 9*(1), 45–54. https://doi.org/10.1187/cbe.09-05-0029

Elby, A., & Hammer, D. (2010). Epistemological resources and framing: A cognitive framework for helping teachers interpret and respond to their students' epistemologies. In L. D. Bendixen & F. C. Feucht (Eds.), *Personal epistemology in the classroom: Theory, research, and implications for practice* (pp. 409–434). Cambridge University Press.

Faria, C., Freire, S., Baptista, M., & Galvão, C. (2014). The construction of a reasoned explanation of a health phenomenon: An analysis of competencies mobilized. *International Journal of Science Education, 36*(9), 1476–1490. https://doi.org/10.1080/09500693.2013.783723

Flores, K. L., Matkin, G. S., Burbach, M. E., Quinn, C. E., & Harding, H. (2012). Deficient critical thinking skills among college graduates: Implications for leadership. *Educational Philosophy and Theory, 44*(2), 212–230. https://doi.org/10.1111/j.1469-5812.2010.00672.x

Goffman, E. (1974). *Frame analysis: An essay on the Organization of Experience*. Harvard University Press.

Gottesman, A. J., & Hoskins, S. G. (2013). C.R.E.A.T.E. cornerstone: introduction to scientific thinking, a new course for STEM-interested freshmen, demystifies scientific thinking through analysis of scientific literature. *CBE Life Sciences Education, 12*(1), 59–72. https://doi.org/10.1187/cbe.12-11-0201

Gouvea, J., Sawtelle, V., & Nair, A. (2019). Epistemological progress in physics and its impact on biology. *Physical Review Physics Education Research, 15*, 010107.

Hall, K. L. (2013). *Examining the effects of students' classroom expectations on undergraduate biology course reform* (Doctoral dissertation). University of Maryland. Retrieved from http://drum.lib.umd.edu/bitstream/handle/1903/14080/Hall_umd_0117E_14168.pdf?sequence=1&isAllowed=y

Hall, K. L., Watkins, J. E., Coffey, J. E., Cooke, T. J., & Redish, E. F. (2011). Examining the impact of student expectations on undergraduate biology education reform. *AERA National Meeting*. https://arxiv.org/abs/1105.6349

Hammer, D. (1994). Epistemological beliefs in introductory physics. *Cognition and Instruction, 12*(2), 151–183. https://doi.org/10.2307/3233679

Hammer, D. (2000). Student resources for learning introductory physics. *American Journal of Physics, 68*, S52–S59. https://doi.org/10.1119/1.19520

Hammer, D., & Elby, A. (2002). On the form of personal epistemology. In B. K. Hofer & P. R. Pintrich (Eds.), *Personal epistemology: The psychology of beliefs about knowledge and knowing* (pp. 169–190). Lawrence Erlbaum Press.

Hammer, D., & Elby, A. (2003). Tapping students' epistemological resources. *Journal of the Learning Sciences, 12*(1), 53–91. https://doi.org/10.1207/S15327809JLS1201_3

Hammer, D., Elby, A., Scherr, R. E., & Redish, E. F. (2005). Resources, framing, and transfer. In J. Mestre & C. T. Greenwich (Eds.), *Transfer of learning from a modern multidisciplinary perspective* (pp. 89–120). Information Age Publishing.

Hayes-Klosteridis, J. (2019). *Exploring the dimensions of gender and student epistemologies in a reformed learner-centered organismal biology course: A mixed methods approach.* Dissertation. University of Maryland, College Park. Retrieved from https://drum.lib.umd.edu/bitstream/handle/1903/24949/Klosteridis_umd_0117E_20221.pdf?sequence=2

Hester, S. D., Nadler, M., Katcher, J., Elfring, L. K., Dykstra, E., Rezende, L. F., & Bolger, M. S. (2018). Authentic inquiry through modeling in biology (AIM-bio): An introductory laboratory curriculum that increases undergraduates' scientific agency and skills. *CBE Life Sciences Education, 17*(4), 1–23. https://doi.org/10.1187/cbe.18-06-0090

Hodson, D., & Wong, S. L. (2017). Going beyond the consensus view: Broadening and enriching the scope of NOS-oriented curricula. *Canadian Journal of Science, Mathematics and Technology Education, 17*(1), 3–17. https://doi.org/10.1080/14926156.2016.1271919

Hofer, B. K. (2000). Dimensionality and disciplinary differences in personal epistemology. *Contemporary Educational Psychology, 25*(4), 378–405. https://doi.org/10.1006/ceps.1999.1026

Hofer, B. K. (2001). Personal epistemology research: Implications for learning and teaching. *Educational Psychology Review, 13*(4), 353–383. https://doi.org/10.1023/A:1011965830686

Hofer, B. K. (2006). Domain specificity of personal epistemology: Resolved questions, persistent issues, new models. *International Journal of Educational Research, 45*(1–2), 85–95. https://doi.org/10.1016/j.ijer.2006.08.006

Hofer, B. K., & Bendixen, L. D. (2012). Personal epistemology: Theory, research, and future directions. In K. R. Harris, S. Graham, T. Urdan, C. B. McCormick, G. M. Sinatra, & J. Sweller (Eds.), *APA educational psychology handbook, Vol. 1. Theories, constructs, and critical issues* (pp. 227–256). American Psychological Association.

Hofer, B. K., & Pintrich, P. R. (1997). The development of epistemological theories: Beliefs about knowledge and knowing and their relation to learning. *Review of Educational Research, 67*, 88–140. https://doi.org/10.3102/00346543067001088

Kang, H., Thompson, J., & Windschitl, M. (2014). Creating opportunities for students to show what they know: The role of scaffolding in assessment tasks. *Science Education, 98*(4), 674–704. https://doi.org/10.1002/sce.21123

Kardash, C. M., & Howell, K. L. (2000). Effects of epistemological beliefs and topic-specific beliefs on undergraduates' cognitive and strategic processing of dual-positional text. *Journal of Educational Psychology, 92*(3), 524–535. https://doi.org/10.1037//0022-0663.92.3.524

Kelly, G. J., McDonald, S., & Wickman, P. O. (2012). Science learning and epistemology. In K. Tobin & B. J. Fraser (Eds.), *Second international handbook of science education* (pp. 281–291). Springer.

Lederman, N. G. (1992). Students' and teachers' conceptions of the nature of science: A review of the research. *Journal of Research in Science Teaching, 29*(4), 331–359. https://doi.org/10.1002/tea.3660290404

Lederman, N. G. (1999). Teachers' understanding of the nature of science and classroom practice: Factors that facilitate or impede the relationship. *Journal of Research in Science Teaching, 36*(8), 916–929. https://doi.org/10.1002/(SICI)1098-2736(199910)36:8<916::AID-TEA2>3.0.CO;2-A

Lederman, N. G., Abd-El-Khalick, F., Bell, R. L., & Schwartz, R. (2002). Views of nature of science questionnaire: Toward a valid and meaningful assessment of learners' conceptions of nature of science. *Journal of Research in Science Teaching, 39*(6), 497–521. https://doi.org/10.1002/tea.10034

Lising, L., & Elby, A. (2005). The impact of epistemology on learning: A case study from introductory physics. *American Journal of Physics, 73*(4), 372–383. https://doi.org/10.1119/1.1848115

Louca, L. T., Elby, A., Hammer, D., & Kagey, T. (2004). Epistemological resources: Applying a new epistemological framework to science instruction. *Educational Psychologist, 39*(1), 57–68. https://doi.org/10.1207/s15326985ep3901_6

Madsen, A., Mc Kagan, S. B., & Sayre, E. C. (2015). How physics instruction impacts students' beliefs about learning physics: A meta-analysis of 24 studies. *Physical Review Physics Education Research, 11*, 010115. https://doi.org/10.1103/PhysRevSTPER.11.010115

Momsen, J. L., Long, T. M., Wyse, S. A., & Ebert-May, D. (2010). Just the facts? Introductory undergraduate biology courses focus on low-level cognitive skills. *CBE Life Sciences Education, 9*(4), 435–440. https://doi.org/10.1187/cbe.10-01-0001

Moore, W. S. (2002). Understanding learning in a postmodern world: Reconsidering the Perry scheme of intellectual and ethical development. In B. K. Hofer & P. R. Pintrich (Eds.), *Personal epistemology: The psychology of beliefs about knowledge and knowing* (pp. 17–36). Lawrence Erlbaum.

Muis, K. R. (2004). Personal epistemology and mathematics: A critical review and synthesis of research. *Review of Educational Research, 74*(3), 317–377. https://doi.org/10.3102/00346543074003317

National Research Council. (2012). *A framework for K-12 science education: Practices, crosscutting concepts, and core ideas*. The National Academies Press.

Nehm, R. H. (2019). Biology education research: Building integrative frameworks for teaching and learning about living systems. *Disciplinary and Interdisciplinary Science Education Research, 1*, 15. https://doi.org/10.1186/s43031-019-0017-6

Pelaez, N., Anderson, T., Gardner, S.M., Yin, Y., Abraham, J.K., Bartlett, E., Gormally, C., Hill, J.P., Hoover, M., Hurney, C., Long, T., Newman, D.L., Sirum, K., & Stevens, M. (2017). The basic competencies of biological experimentation: Concept-skill statements. *PIBERG Instructional Innovation Materials*. Retrieved from http://docs.lib.purdue.edu/pibergiim/4/. Accessed 18 Oct 2021.

Pelaez, N., Anderson, T.R., Gardner, S.M., Yin, Y., Abraham, J.K., Bartlett, E.L., Gormally, C., Hurney, C.A., Long, T.M., Newman, D.L., Sirum, K., & Stevens, M.T. (2018). A community-building framework for collaborative research coordination across the education and biology research disciplines. *CBE Life Sciences Education, 17*(2), es2. Retrieved from https://doi.org/10.1187/cbe.17-04-0060

Perkins, K. K., Adams, W. K., Pollock, S. J., Finkelstein, N. D., & Wieman, C. E. (2005). Correlating student beliefs with student learning using the Colorado learning attitudes about science survey. *AIP Conference Proceedings, 790*, 61. https://doi.org/10.1063/1.2084701

Perry, J. W. (1970). *Forms of intellectual and ethical development in the college years: A scheme*. Jossey-Bass Publishers.

Prince, M. (2004). Does active learning work? A review of the research. *The Research Journal for Engineering Education, 93*(3), 223–231. https://doi.org/10.1002/j.2168-9830.2004.tb00809.x
Russ, R. S. (2014). Epistemology of science vs. epistemology for science. *Science Education, 98*(3), 388–396. https://doi.org/10.1002/sce.21106
Schommer, M. (1990). Effects of beliefs about the nature of knowledge on comprehension. *Journal of Educational Psychology, 82*(3), 498–504. https://doi.org/10.1037/0022-0663.82.3.498
Schommer-Aikins, M. (2004). Explaining the epistemological belief system: Introducing the embedded systemic model and coordinated research approach. *Educational Psychologist, 39*(1), 19–29. https://doi.org/10.1207/s15326985ep3901_3
Schommer-Aikins, M., & Duell, O. K. (2013). Domain specific and general epistemological beliefs. Their effects on mathematics. *Revistade Investigación Educativa, 31*(2), 317–330. https://doi.org/10.1187/cbe.10-10-0133
Schommer-Aikins, M., & Easter, M. (2006). Ways of knowing and epistemological beliefs: Combined effect on academic performance. *Educational Psychology, 26*(3), 411–423. https://doi.org/10.1080/01443410500341304
Schwartz, R., & Lederman, N. G. (2008). What scientists say: Scientists' views of nature of science and relation to science context. *International Journal of Science Education, 30*(6), 727–771. https://doi.org/10.1080/09500690701225801
Semsar, K., Knight, J. K., Birol, G., & Smith, M. K. (2011). The Colorado Learning Attitudes about Science Survey (CLASS) for use in biology. *CBE Life Sciences Education, 10*(3), 268–278. https://doi.org/10.1187/cbe.10-10-0133
Southard, K., Wince, T., Meddleton, S., & Bolger, M. S. (2016). Features of knowledge building in biology: Understanding undergraduate students' ideas about molecular mechanisms. *CBE Life Sciences Education, 15*(1), 1–16. https://doi.org/10.1187/cbe.15-05-0114
Stanger-Hall, K. F. (2012). Multiple-choice exams: An obstacle for higher-level thinking in introductory science classes. *CBE Life Sciences Education, 11*(3), 294–306. https://doi.org/10.1187/cbe.11-11-0100
Syed, M., & Nelson, S. C. (2015). Guidelines for establishing reliability when coding narrative data. *Emerging Adulthood, 3*(6), 375–387. https://doi.org/10.1177/2167696815587648
Tang, K. (2016). Constructing scientific explanations through premise–reasoning–outcome (PRO): An exploratory study to scaffold students in structuring written explanations. *International Journal of Science Education, 38*(9), 1415–1440. https://doi.org/10.1080/09500693.2016.1192309
Tannen, D. (1993). What's in a frame? Surface evidence for underlying expectations. In D. Tannen (Ed.), *Framing in discourse* (pp. 14–56). Oxford University Press.
Thompson, L., & Blankinship, L. A. (2015). Teaching information literacy skills to sophomore-level biology majors. *Journal of Microbiology & Biology Education, 16*(1), 29–33. https://doi.org/10.1128/jmbe.v16i1.818
Watkins, J., Hall, K. L., Redish, E. F., & Cooke, T. J. (2010). Understanding how students use physical ideas in introductory biology courses. *Proceedings of the Physics Education Research Conference, 1289*, 333–336. https://doi.org/10.1063/1.3515237
Wienhold, C., & Branchaw, J. (2018). Exploring biology: A vision and change disciplinary first-year seminar improves academic performance in introductory biology. *CBE-Life Sciences Education, 17*(2), 1–11. https://doi.org/10.1187/cbe.17-08-0158
Zangori, L., Forbes, C. T., & Schwarz, C. V. (2015). Exploring the effect of embedded scaffolding within curricular tasks on third-grade students' model-based explanations about hydrologic cycling. *Science & Education, 24*, 957–981. https://doi.org/10.1007/s11191-015-9771-9

Part II
Operationalizing and Planning: Designing Instruction to Promote Learning of Biological Experimentation

Chapter 5
Backward Designing a Lab Course to Promote Authentic Research Experience According to Students' Gains in Research Abilities

Zhiyong Cheng, Trevor R. Anderson, and Nancy J. Pelaez

5.1 Introduction

The importance of building students' critical thinking skills has been increasingly recognized by STEM (science, technology, engineering, and mathematics) educators, with the consensus that inclusion of authentic research practices in higher education is the key to achieving such a goal (Auchincloss et al., 2014; Irby et al., 2018a, b). In addition to fundamental basic knowledge, students should master relevant technical, scientific-inquiry, and problem-solving skills (Akuma & Callaghan, 2019; Lee & Songer, 2003; Manz et al., 2020; Novak & Treagust, 2018; Pellegrino, 2012; Wong & Hodson, 2009). To this end, Apprentice-Based Research Experiences (AREs) or Course-based Undergraduate Research Experiences (CUREs) stand out for the development of student competence to perform authentic research (Auchincloss et al., 2014; Shapiro et al., 2015). AREs include research internships, academic credit-based research (independent study), and volunteer research experience in a research lab, where students receive research training one on one with a researcher through a mentored independent research project. CUREs take place in a teaching lab where more students are involved in scientific discovery through

Z. Cheng (✉)
Food Science and Human Nutrition Department, Institute of Food and Agricultural Sciences, University of Florida, Gainesville, FL, USA
e-mail: z.cheng@ufl.edu

T. R. Anderson
Department of Chemistry, Purdue University, West Lafayette, IN, USA
e-mail: ander333@purdue.edu

N. J. Pelaez
Department of Biological Sciences, Purdue University, West Lafayette, IN, USA
e-mail: npelaez@purdue.edu

N. J. Pelaez et al. (eds.), *Trends in Teaching Experimentation in the Life Sciences*, Contributions from Biology Education Research,
https://doi.org/10.1007/978-3-030-98592-9_5

curriculum-based research projects. In contrast to a traditional recipe-like or cookbook lab that target technical skill training, AREs and CUREs increase the dimensions of laboratory learning such as "Use of science practices", "Discovery", "Broader relevance or importance", "Collaboration", and "Iteration" (Auchincloss et al., 2014; Shapiro et al., 2015).

In response to the calls for authentic research experiences implemented in undergraduate education, most colleges and universities have started to offer ARE and/or CURE training (Brownell & Kloser, 2015; Linn et al., 2015). However, attempts are needed to define optimal ways to orient and guide students and to measure benefits of such experiences such as gains in research abilities or conceptual understandings (Linn et al., 2015). Pelaez and colleagues suggest that the competencies in authentic research (or experimentation) can be classified and examined in seven areas (Irby et al., 2018a, Pelaez et al., 2017, 2018; Chap. 1 in this volume): (a) the ability to *Identify* gaps or limitations in current research knowledge through the review, filtering and synthesis of relevant literature; (b) the ability to generate a research *Question* and formulate hypotheses; (c) the ability to *Plan* feasible and ethical experiments to answer research questions or test hypotheses; (d) the ability to *Conduct* an investigation to achieve research goals; (e) the ability to *Analyze* and process data; (f) the ability to *Conclude* about data with inferences that are limited to the scope inherent in the experimental design; (g) the ability to *Communicate* research work in professionally appropriate modes, including visual, written, and oral formats. Accomplishment of the competencies requires well-designed learning objectives (or anticipated learning outcomes, ALOs) and assessments to track whether and how students are making progress (Irby et al., 2018a, b).

This chapter describes a backward design informed by Wiggins and McTighe (2005) to improve a lab course, in which learning objectives (or ALOs according to Irby et al., 2018a, b) are aligned to the seven areas of competencies and used to guide or navigate the designs of assessments and teaching/learning activities. In addition, students were engaged in literature research, gap analysis, identification of research questions, formulation of hypothesis and rationales, and project design and planning in addition to experimentation and data presentation. The question-driven research training pipeline served as a focused guide for students to conduct authentic research and accomplish the gains in research abilities. The study was approved by the University of Florida's Institutional Review Board (IRB) under protocol IRB201902746.

5.2 Backward Design of a Lab Course

The advancements in life sciences and biomedical research are so rapid that the traditional cookbook or recipe-like lab procedures cannot accommodate the needs of teaching students how to do science and research (Brownell & Kloser, 2015). In a "cookbook" lab setting, it provides detailed procedures for students to follow like a cooking recipe. Students are usually told of what to do step-wisely and what to expect at the end of the experiment. As such, finishing a procedure might become the goal or impression of a lab course to students who passively follow the "cookbook", and the opportunity for developing scientific research skills, such as identifying a research question, formulating a hypothesis with rationales, designing a study plan to address the questions, critical thinking for troubleshooting and alternative approaches, is quite limited (Auchincloss et al., 2014; Brownell & Kloser, 2015; Cheng, 2019). As such, there have been calls for shifting the focus of lab courses from cookbook labs to more authentic research experiences that target the learning goals of competencies, concepts, and skills associated with the planning, design, completion, and dissemination of experiments (Brownell & Kloser, 2015; Pelaez et al., 2017, 2018; Chap. 1 in this volume).

Here we describe a backward design of lab course to engage students in molecular nutrition research, a multi-disciplinary subject that ranges from nutritional biochemistry, molecular biology, cell biology, physiology, statistics, and bioinformatics. The backward design was manifested in two stages. First, the whole course was designed backwardly, using the learning objectives to navigate assessments and teaching/learning activities (Fig. 5.1). In addition to disciplinary science knowledge, the learning objectives aimed to build students' inquiry and critical thinking skills, particularly their competencies in authentic research, including the skills associated with gap/question identification, hypothesis/rational formulation, study planning, design, completion, and dissemination of experiments (Fig. 5.2). To this end, formative (e.g., trouble shooting, question/answer sections, and reflections) and summative (e.g., exams, protocol development, study report, and project presentation) assessments were customized to evaluate students' progress (Fig. 5.1). Teaching/learning activities were administrated in a hybrid manner that engaged students in research training both in and outside class though question-guided video watching, lecturing, in class exercise/practice, case studies, project design, and protocol development (Figs. 5.1 and 5.2).

The second stage of backward design began with the overarching goal of building students' critical thinking or problem-solving skills like a scientist (Fig. 5.2). To achieve this goal, students were guided through literature review to ***Identify*** gaps or questions of interest based on certain context and keywords (i.e., context and question). Once research questions were identified, students were asked to formulate (or reason out) hypothetic answers based on the known facts in the literature (i.e., hypothesis and rationales). To test the hypotheses, students were required to design a study, justifying sample size, the control of variables, and selection of measurements (or methods). Moreover, students needed to propose anticipated results, potential pitfalls, and alternative approaches (i.e., study plan and alternatives). These components are aligned with three areas of the ACE-Bio Competencies (Table 5.1), *Identify*, *Question*, and *Plan*, which are absent from a "cookbook" (or recipe-like) lab procedure that typically would be more focused on technical skill training (Fig. 5.2). In addition, students were not provided the stepwise procedure as in a traditional "cookbook" lab. Instead, students were engaged in development of their own experimental protocols based on materials they had been reading, question-guided video watching, observation, practice, and troubleshooting. It

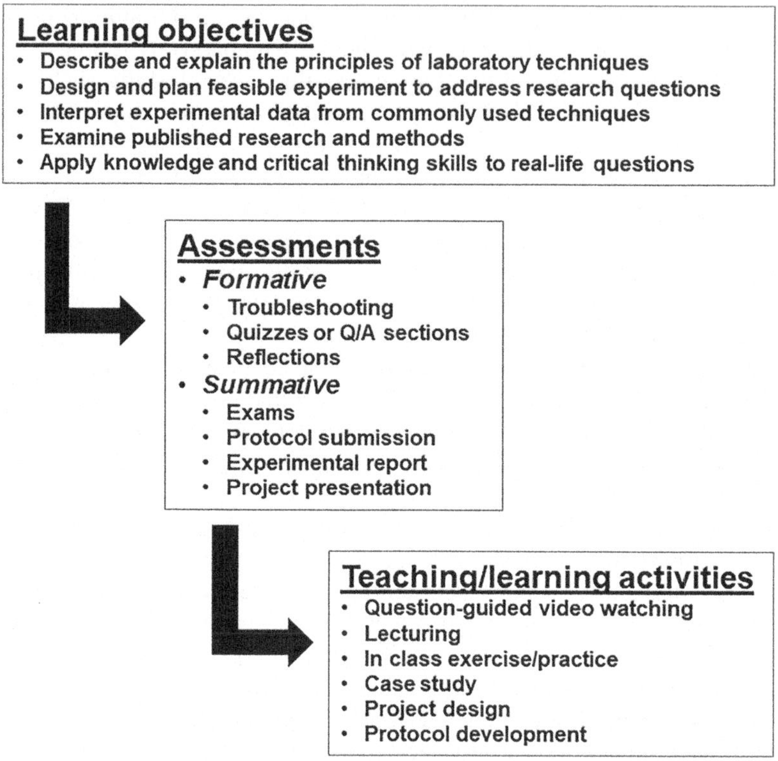

Fig. 5.1 Learning objective-driven backward design of laboratory courses. In contrast to traditional course design starting with teaching and learning activities that determine the ways of assessments and learning objectives, the learning objective-driven backward design starts with learning objectives (anticipated learning outcomes, or ALOs) that guide and navigate the design of assessments and teaching/learning activities

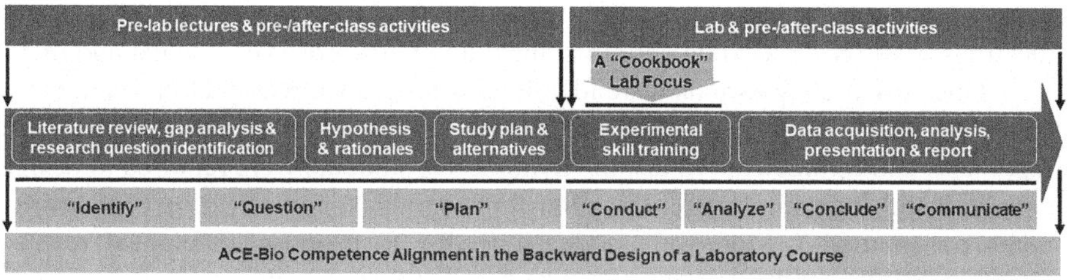

Fig. 5.2 Research question-driven backward design of a laboratory course. Rather than ask students to follow a recipe-like procedure to complete experiments, the research question-driven backward design engaged students in such active learning activities as (1) literature review, gap analysis, and research question identification (i.e., context & questions), (2) hypothesis and rationale formulation, (3) study plan and alternative approaches, (4) experimental skill training, and (5) data acquisition, analysis, presentation, and report. This design adopted a hybrid classroom setting where students were immersed in constant research and scientific inquiry activities in and outside class, which facilitated accomplishing seven areas of competencies in authentic research: *Identify*, *Question*, *Plan*, *Conduct*, *Analyze*, *Conclude*, and *Communicate*. A traditional "cookbook" (recipe-like) lab usually would have focused only on technical skill training

Table 5.1 Anticipated learning outcomes were aligned with the ACE-Bio Competencies (Pelaez et al., 2017; Chap. 1 in this volume)

Anticipated Learning outcomes (ALOs)	ACE-Bio Competences or Sub-Competences
Identify real-life questions concerning molecular nutrition via literature search and survey	Ability to *identify* gaps or limitations in current research knowledge Ability to filter and synthesize relevant literature Ability to generate a research *Question* and formulate hypotheses
Describe and explain the principles of laboratory techniques for molecular nutrition research Design and plan feasible experiments to address research questions Develop experimental protocol and conduct research Produce reproducible results	Ability to *Plan* feasible and ethical experiments Ability to *Conduct* investigation to achieve research goals
Interpret experimental data, reason, and draw conclusion	Ability to *Analyze*/process data and *Conclude* about data with inferences
Develop project reports for oral or poster presentation	Ability to professionally *Communicate* research work

facilitated maintaining high-level activities of inquiry, information synthesis, and critical thinking throughout the training process by taking advantage of the hybrid class administration, where students spent time outside class reading materials, watching videos, processing information, and drafting protocols. In-class practicing and troubleshooting provided the opportunities for student to find issues and make corrections, which resulted in validated or working protocols. Using the validated protocols students proceeded to project execution and problem solving through data acquisition, analysis, presentations, and reports (Fig. 5.2). Clearly, the identified research question played an important role in guiding the second stage of backward design from hypothesis development to study design and planning, to project execution and problem solving. Overall, the components of "experimental skill training" and "data acquisition, analysis, presentation, and report" are aligned with the other four areas of ACE-Bio Competencies, *Conduct*, *Analyze*, *Conclude*, and *Communicate*.

5.3 Assessment of Scientific Research Practices

Course designs serve for pedagogical goals, and assessment of whether and how much the pedagogical goals being achieved have been the theme of investigation (Brownell & Kloser, 2015; Dasgupta et al., 2016; Cooper et al., 2017; Hills et al., 2020). In particularly, it is challenging to know whether and what students actually learn from the instructions even though Likert-scale surveys are used to probe student confidence in research inquiry and problem-solving skills (Brownell & Kloser,

2015; Dasgupta et al., 2014, 2016). For instance, questions have been raised on whether students are truly skillful in scientific research practice or they merely think that they are (Brownell & Kloser, 2015, Linn et al., 2015). Recently, assessment instruments or rubrics have been developed to measure students' use of experimental design concepts and representations or diagnose students' experimental design knowledge and difficulties (Dasgupta et al., 2014, 2016). In addition, Anderson and colleagues suggested that to identify discovery-type research abilities that students actually develop, it is essential to first identify anticipated learning outcomes (ALOs) (Irby et al., 2018a, b, 2020). In such a scenario, assessments can be conducted according to the established and verified learning outcomes (or VLOs) versus the ALOs.

To assess students' progress and research abilities, we align the ALOs with the seven areas of ACE-Bio Competencies (Fig. 5.2, Table 5.1), and develop rubrics to evaluate the deliverables of each components, including gap/question identification, hypothesis and rational formulation, study design and completion, and dissemination of experiments. In the lab of studying glucose metabolic physiology, for instance, students are informed of the potential contributors to blood glucose, i.e., dietary carbohydrate, glycogenolysis (glycogen breakdown), and gluconeogenesis (de novo glucose production) in the liver.

Based on their contextual information, students were asked to review the literature on changes in postprandial glucose in healthy and diabetic individuals, and to formulate a hypothesis on whether and how hepatic gluconeogenesis contributes to postprandial blood glucose level (Fig. 5.3). The hypothesis must be supported by rationales derived from published research, and then be tested by student-proposed experimentation (i.e., project design and study planning) under the instructor guidance. Formulation of hypothesis/rationales and study design required students to do intensive information synthesis and critical thinking for problem solving. Students also needed to justify and make decisions on their research plan, e.g., feasibility, pros and cons, and alternative approaches. For instance, most students tended to propose employing the hyperinsulinemic-euglycemic clamp because the literature ranks it as a "gold standard" method to directly measure hepatic gluconeogenesis (Kim, 2009). However, the equipment is expensive and less accessible. Additionally, the clamp procedure demands several months of training for animal surgery and special care, thus raising questions about the feasibility of this proposed plan. As such, students then found alternative approaches to address their research questions. With the instructor's guidance, they adjusted their approaches with techniques more accessible like qPCR (quantitative polymerase chain reac(tion) and Western blotting to analyze gluconeogenesis (e.g., rate-limiting gluconeogenic enzyme PEPCK (Cheng, 2015; Cheng & White, 2012), see Fig. 5.3). Engaging students in research method evaluation and selection helped them navigate the problem-solving procedure, increased their motivation (or apparent eagerness) and dedication to learning new techniques and testing their hypotheses. Deliverables from students were evident, including (a) the identified gaps or research questions, (b) the formulated hypothesis and rationales with justification, (c) designed project/planned study justification, (d) developed protocols with troubleshooting notes, and (e) acquired data

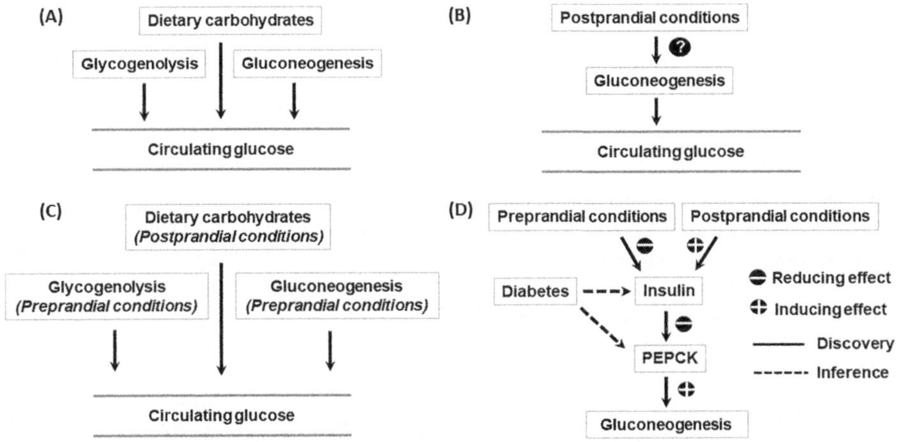

Fig. 5.3 Question-driven research and problem-solving activities. (**a–b**) Generic context about three contributors (dietary carbohydrates, glycogenolysis, and gluconeogenesis) to blood or circulating glucose in mammals was provided to students (panel **a**), and a question was identified as to whether and how hepatic gluconeogenesis may contribute to circulating glucose (panel **b**). The first set of inquiries for students to do through literature research was find out what conditions (preprandial or fasting state) account for gluconeogenesis and for glycogenolysis (preprandial state), dietary carbohydrates (postprandial or feeding state; panel **c**). The second set of inquiries for students to do through literature and discovery-type research was to delineate the regulatory relationship between postprandial conditions and gluconeogenesis (panel **d**). The discovery-type research was based on the hypothesis/rationales that postprandial state increases insulin secretion whereas preprandial state decreases it, and insulin inhibits rate-limiting gluconeogenic enzyme PEPCK to keep blood glucose on check. As a control experiment, parameters were measured under the preprandial conditions in parallel with the postprandial conditions. Based on the literature and experimental data (discovery) from healthy subjects, students were required to infer how gluconeogenesis responds to feeding (postprandial) conditions in diabetics. PEPCK, phosphoenolpyruvate carboxykinase

and resultant reports. With grading rubrics such as the example in Table 5.2 for project design, both the instructor and students knew where they were and what to improve.

The deliverables of project execution are shown in Fig. 5.4, which clearly demonstrates progress in student learning and mastery of Western blotting analysis of gluconeogenetic enzyme PEPCK (EC 4.1.1.32, PEPCK). Denatured protein samples (i.e., liver lysates from mice under fasting and feeding states, approved by IACUC) were provided for students to run Wester Blotting analysis, including sample loading, SDS-PAGE (sodium dodecyl sulfate polyacrylamide gel electrophoresis), protein transfer, antibody incubation and washing, enzymatic reactions and imaging. Figure 5.4a shows the results from students who completed a recipe-like procedure. Pronounced differences and progress was made when they developed and followed their own protocol based on materials from their readings (including the recipe-like procedures), video watching, observation, practicing, and note-taking (Fig. 5.4b). Troubleshooting required higher order critical thinking for

Table 5.2 Grading rubric for project design

Criteria	Ratings		Points
Background information was extensive	5 pts (full marks)	0 pts (no credit)	
Significance was evident	5 pts (full marks)	0 pts (no credit)	
Hypothesis was innovative and scientifically sound	5 pts (full marks)	0 pts (no credit)	
Rationales were logical and scientifically sound	5 pts (full marks)	0 pts (no credit)	
Methods were well-justified, feasible and provided sufficient details (e.g., power calculation and sample size, etc.)	5 pts (full marks)	0 pts (no credit)	
Expected results were clearly justified and in depth	5 pts (full marks)	0 pts (no credit)	
Potential pitfalls and alternatives were perceived and well-reasoned	5 pts (full marks)	0 pts (no credit)	
Conclusions clearly summarized how the project might advance our understanding of this topic	5 pts (full marks)	0 pts (no credit)	

problem solving (Table 5.3), which led to further progress in student learning and mastery of the technique as indicated by data quality in terms of signal, noise, and reproducibility (Fig. 5.4c).

Of note, among several potential causes proposed by students for the high background in Experiment A in Fig. 5.4, some were irrelevant (not logical or reasonable), and the proposed solutions had no chance to solve the problem (success chance marked as "–", Table 5.3); some that were pertinent, such as properly stripping undesired antibodies, washing the blots thoroughly, and preventing contamination of the gel or membrane, may have effectively reduced background noise (success chance marked as "+", Table 5.3). A similar scenario happened to Experiment B in Fig. 5.4, where student proposed to increase exposure time to visualize the missing areas of bands, but it did not address the transfer issue (the root issue). Therefore, it demanded disciplinary knowledge, intensive critical thinking and evidentiary reasoning (as discussed in Chap. 21 of this volume), and the instructor's close guidance to accomplish the goals.

5.4 Common Difficulties and Solutions

Several common difficulties were identified for both the students and the instructor during administering the class/lab course that followed a backward design. From the students' perspective, lack (or insufficiency) of disciplinary knowledge or prior experiences underlies the common difficulties (Table 5.4). For example, students had difficulties dealing with controversial topics or data in the literature because

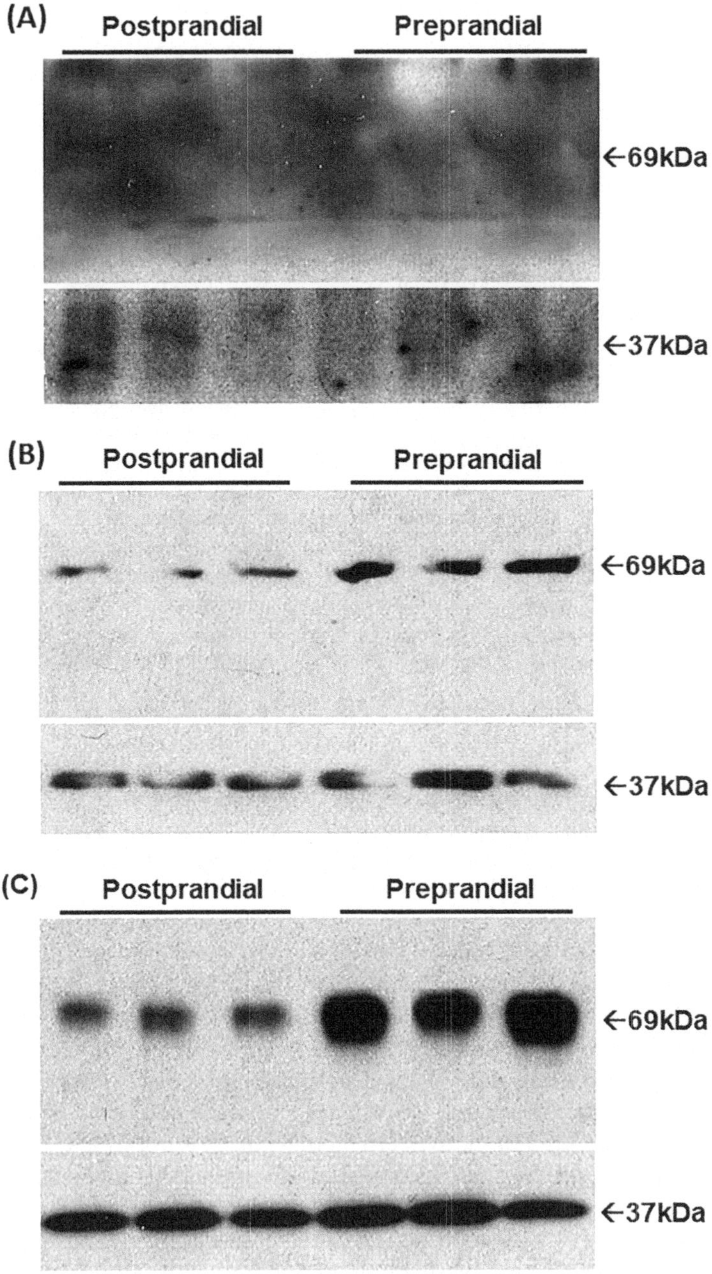

Fig. 5.4 Learning progress and deliverables (data quality) in Western blotting analysis of gluconeogenic marker. The expression of the gluconeogenic enzyme PEPCK (molecular weight: 69 kDa) in preprandial and postprandial states was analyzed by Western blot analysis, and glyceraldehyde 3-phosphate dehydrogenase (GAPDH, EC 1.2.1.12; MW: 37 kDa) was probed as the loading control. (**a**) The results of Western blotting analysis by the students who followed a recipe-like procedure. (**b**) The results of Western blotting analysis by the students who followed their own protocol developed based on materials they read, video protocols, observation, practicing, and notetaking. (**c**) The results of Western blotting analysis by students who did intensive troubleshooting and followed their own protocol developed based on materials they read, videos, observation, practicing, and notetaking

Table 5.3 Troubleshooting by students for the Western blotting analyses shown in Fig. 5.4

	Problems/issues	Proposed potential causes	Proposed solutions	Success chance
Experiment A	No clear bands	Low % of gel, pore size too large causing loss of proteins	Increasing % of gel to decrease the pore size and retain protein	–
	Bands undetectable or blurry	Incomplete stripping the blot results	Increasing stripping time without losing proteins	+
	High background	Exposure time too long	Reducing exposure time	–
	No clear bands, with blurry background	Insufficient washing	Washing thoroughly to remove non-specifically bound antibodies	+
	High background noise	Contamination from hands during the "sandwich" preparation	Being vigilant (e.g., with gloves) on preventing contaminations	+
Experiment B	Incomplete bands	Uneven sample loading	Using glycerol to ensure the sample sink to the bottom of the well	–
	Bands blurred out or partially visible	Insufficient exposure time	Increasing exposure time without background noise	+/–
	Random gaps in certain bands	Bad transfer buffer, or bubbles created during the "sandwich" preparation	Using quality buffer, or rolling out the "sandwich" properly to drive bubbles	+

Note: "+" refers to a high possibility to solve the problem successfully because the proposed causes and solutions are logical and reasonable; "–" refers to no chance to succeed in problem solving; "+/–" refers to a possibility to solve the problem but only partially

they didn't understand the content or know how to interpret the data in depth. During project design, some students tended to select one-sided data or claims to support their hypothesis and rationale or propose research approaches without being able to justify why. Calculation of sample size and proper control for variables appeared to be challenging for students due to lack of statistics knowledge and application experience. To address the issues, pre-lab lectures and activities should be designed to target bridging the gaps. For the Western blotting analysis of PEPCK, the disciplinary information needed by students was cross-disciplinary and included immunology (antigen-antibody biology), protein science, electrophoresis, and nutrition and metabolic physiology. Without disciplinary or interdisciplinary knowledge, it would have been difficult to do troubleshooting, design a scientifically sound project, or interpret the data correctly. However, pre-lab lectures on the principles and applications of the techniques, in parallel with paper discussion activities (examining published research), helped students significantly by bridging the knowledge gaps and building students' abilities to overcome the difficulties.

From the instructor's perspective, major challenges were associated with how to handle the increased logistics needs (e.g., time and resources). For a lab course with

Table 5.4 Common difficulties, underlying reasons, and ways to overcome these problems

Common difficulties	Causes or underlying reasons	Ways to fix or overcome
Controversies in scientific literature left behind	Lack of disciplinary knowledge	Examining published research
Biased evidence used in the development of hypothesis and rationales	Incomplete literature research Lack of prior experience of study design	Systematic literature research Designing a research project
Research methods proposed without balancing the pros and cons	Lack of in-depth understanding of methodology (principles, applications, costs, logistics)	Lecturing and demonstrating the basics of research methods Hands-on practicing lab skills
Sample sizes proposed based on similar studies, or reference values not representative for power calculation	Lack of prior experience of study design Lack of disciplinary knowledge	Lecturing the basics of statistics (e.g., α, β, SD) in power and sample size calculation Designing a research project
One control experiment used for multiple variables	Misunderstanding the functions of controls Lack of prior experience of study design	Lecturing the control of variables Lecturing the basics of study design
Data varying and inconclusive	Lack of in-depth understanding of experimental protocols	Developing a working protocol Conducting troubleshooting
Inability to infer or interpret data in depth	Lack of context/background information Lack of a whole picture of project design	Using project design to facilitate the proposal of expected results, potential pitfalls, and alternatives

pre-lab lectures, there are only 3–6 h per week to administer classes. To accomplish the ALOs and competencies in experimentation in the backward-designed lab course, a single experiment would take up to threefold more time.

To address the challenges, as advice to other educators, the following strategies can be considered:

- Implement a flipped or hybrid classroom model to increase students' pre- and after-class learning activities (Fig. 5.2);
- Strategically set the emphases on lab and lecture sections, e.g., focusing on the most accessible and frequently used techniques in lab sections and covering the advanced but less accessible techniques in lecture sections;
- Administer all sections in a lab setting, where, with all the lab resources accessible, the instructor and students have more flexibility to pracrise activities;
- Offer a "boot camp" option in the summer, when students have no less pressure from other classes but are able to dedicate more time to authentic research training, critical thinking, problem solving activities.

5.5 Conclusions

As STEM undergraduate education is shifting the focus of lab courses to authentic research experiences from traditional cookbook labs, it is critical to design effective teaching and assessment strategies or tools. This chapter describes a backward design of lab course in two stages: (1) using the learning objectives (or ALOs) to guide and navigate the design of assessments and teaching activities, and (Irby et al., 2018a) using research questions to guide and navigate the administration of class, where students are involved in literature research, gap analysis, questions identification, study design and planning, protocol development and troubleshooting, data acquisition, analysis, report, and presentation. This two-stage backward design enhances student engagement in scientific practices and discovery-type research, which facilitates building students' critical thinking skills, problem-solving ability, and creativity. Alignment of ALOs with the seven areas of competencies (*Identify*, *Question*, *Plan*, *Conduct*, *Analyze*, *Conclude*, and *Communicate*) clearly sets the milestones and clarifies a target for the assessments of gains in research abilities with grading rubrics. The instructor may encounter common difficulties or challenges that are associated with time commitment, resources, and prior disciplinary knowledge. To this end, use of flipped or hybrid classroom and summer bootcamp option may effectively address these challenges.

Acknowledgments This study was part of an Innovative Teaching Award (to Z.C.) project funded by the Academic Programs Section, Association of Public and Land-grant Universities. Collaborations and discussions that formulated ideas for this work were funded in part by NSF grant #1346567 for the Advancing Competencies in Experimentation–Biology (ACE-Bio) Network. Any opinions, findings, and conclusions or recommendations expressed are those of the authors and do not necessarily reflect the views of the NSF.

References

Akuma, F. V., & Callaghan, R. (2019). A systematic review characterizing and clarifying intrinsic teaching challenges linked to inquiry-based practical work. *Journal of Research in Science Teaching, 56*, 619–648. https://doi.org/10.1002/tea.21516

Auchincloss, L. C., Laursen, S. L., Branchaw, J. L., et al. (2014). Assessment of course-based undergraduate research experiences: A meeting report. *CBE Life Sciences Education, 13*(1), 29–40. https://doi.org/10.1187/cbe.14-01-0004

Brownell, S. E., & Kloser, M. J. (2015). Toward a conceptual framework for measuring the effectiveness of course-based undergraduate research experiences in undergraduate biology. *Studies in Higher Education, 40*(3), 525–544. https://doi.org/10.1080/03075079.2015.1004234

Cheng, Z. (2015). FoxO1: Mute for a tuned metabolism? *Trends in Endocrinology and Metabolism, 26*(8), 402–403. https://doi.org/10.1016/j.tem.2015.06.006

Cheng, Z. (2019). Backward planning of lab course to enhance students' critical thinking. *Life Science Teaching Resource Community – Physiology Education Communitry of Practice (PECOP) Blog*. https://blog.lifescitrc.org/pecop/2019/2012/2023/backward-planning-of-lab-course-to-enhance-students-critical-thinking/

Cheng, Z., & White, M. F. (2012). The AKTion in non-canonical insulin signaling. *Nature Medicine, 18*(3), 351–353. https://doi.org/10.1038/nm.2694

Cooper, K. M., Soneral, P. A. G., & Brownell, S. E. (2017). Define your goals before you design a CURE: A call to use backward design in planning course-based undergraduate research experiences. *The Journal of Microbiology & Biology Education, 18*(2). https://doi.org/10.1128/jmbe.v18i2.1287

Dasgupta, A. P., Anderson, T. R., & Pelaez, N. (2014). Development and validation of a rubric for diagnosing students' experimental design knowledge and difficulties. *CBE Life Sciences Education, 13*(2), 265–284. https://doi.org/10.1187/cbe.13-09-0192

Dasgupta, A. P., Anderson, T. R., & Pelaez, N. J. (2016). Development of the neuron assessment for measuring biology students' use of experimental design concepts and representations. *CBE Life Sciences Education, 15*(2), ARTN ar10. https://doi.org/10.1187/cbe.15-03-0077

Hills, M., Harcombe, K., & Bernstein, N. (2020). Using anticipated learning outcomes for backward design of a molecular cell biology course-based undergraduate research experience. *Biochemistry and Molecular Biology Education, 48*(4), 311–319. https://doi.org/10.1002/bmb.21350

Irby, S. M., Pelaez, N. J., & Anderson, T. R. (2018a). Anticipated learning outcomes for a biochemistry course-based undergraduate research experience aimed at predicting protein function from structure: Implications for assessment design. *Biochemistry and Molecular Biology Education, 46*(5), 478–492. https://doi.org/10.1002/bmb.21173

Irby, S. M., Pelaez, N. J., & Anderson, T. R. (2018b). How to identify the research abilities that instructors anticipate students will develop in a biochemistry course-based undergraduate research experience (CURE). *CBE Life Sciences Education, 17*(2), es4. https://doi.org/10.1187/cbe.17-12-0250

Irby, S. M., Pelaez, N. J., & Anderson, T. R. (2020). Student perceptions of their gains in course-based undergraduate research abilities identified as the anticipated learning outcomes for a biochemistry CURE. *Journal of Chemical Education, 97*(1), 56–65. https://doi.org/10.1021/acs.jchemed.9b00440

Kim, J. K. (2009). Hyperinsulinemic-euglycemic clamp to assess insulin sensitivity in vivo. *Methods in Molecular Biology, 560*, 221–238. https://doi.org/10.1007/978-1-59745-448-3_15

Lee, H. S., & Songer, N. B. (2003). Making authentic science accessible to students. *International Journal of Science Education, 25*, 923–948. https://doi.org/10.1080/09500690305023

Linn, M. C., Palmer, E., Baranger, A., Gerard, E., & Stone, E. (2015). Education. Undergraduate research experiences: Impacts and opportunities. *Science, 347*(6222), 1261757. https://doi.org/10.1126/science.1261757

Manz, E., Lehrer, R., & Schauble, L. (2020). Rethinking the classroom science investigation. *Journal of Research in Science Teaching, 57*, 1148–1174. https://doi.org/10.1002/tea.21625

Novak, A. M., & Treagust, D. F. (2018). Adjusting claims as new evidence emerges: Do students incorporate new evidence into their scientific explanations? *Journal of Research in Science Teaching, 55*, 526–549. https://doi.org/10.1002/tea.21429

Pelaez, N., Anderson, T. R., Gardner, S. M., Yin, Y., Abraham, J. K., Bartlett, E. L., Gormally, C., Hill, J. P., Hoover, M., Hurney, C. A., Long, T. M., Newman, D. L., Sirum, K., & Stevens, M. T. (2017). The basic competencies of biological experimentation: Concept-skill statements. *PIBERG Instructional Innovation Material Paper 4.* https://docs.lib.purdue.edu/pibergiim/4/. Accessed 8 Nov 2021

Pelaez, N., Anderson, T. R., Gardner, S. M., Yin, Y., Abraham, J. K., Bartlett, E. L., Gormally, C., Hurney, C. A., Long, T. M., Newman, D. L., Sirum, K., & Stevens, M. T. (2018). A community-building framework for collaborative research coordination across the education and biology research disciplines. *CBE Life Sciences Education, 17*(2), es2. https://doi.org/10.1187/cbe.17-04-0060

Pellegrino, J. W. (2012). Assessment of science learning: Living in interesting times. *Journal of Research in Science Teaching, 49*, 831–841. https://doi.org/10.1002/tea.21032

Shapiro, C., Moberg-Parker, J., Toma, S., et al. (2015). Comparing the impact of course-based and apprentice-based research experiences in a life science laboratory curriculum. *The Journal of Microbiology & Biology Education, 16*(2), 186–197. https://doi.org/10.1128/jmbe.v16i2.1045

Wiggins, G. P., & McTighe, J. (2005). *Understanding by design* (Expanded 2nd). Association for Supervision and Curriculum Development (ASCD).

Wong, S. L., & Hodson, D. (2009). From the horse's mouth: What scientists say about scientific investigation and scientific knowledge. *Science Education, 93*, 109–130. https://doi.org/10.1002/sce.20290

Chapter 6
Using the ACE-Bio Competencies Resource as a Course Planning Tool to Guide Students in Independent Research

Aeisha Thomas

6.1 Introduction

Research experience has been acknowledged as an essential part of an undergraduate biology curriculum as demonstrated in an action item in the seminal *Vision and Change* report (AAAS, 2011). Science students gain these skills in a variety of formats. Laboratory techniques, field methods, science writing and other research skills are of course taught in the typical science class. Students further experience undergraduate research in a range of modalities such as traditional apprentice type mentoring in research labs and course based undergraduate research experiences or CUREs (National Academies of Sciences, Engineering and Medicine (NASEM), 2017). In a study comparing research-based courses and undergraduate research experiences, both were found to help students albeit differently supporting the value of varied approaches in this area (Olivares-Donoso & Gonzalez, 2019). Brew (2013) presents the wheel framework for undergraduate research, which while not science specific, with its many layered contexts, learning outcomes and other parameters, is a picture of how truly diverse research instruction can and should be. Thus student research learning experiences differ and students often only encounter a subset or limited exposure to the necessary research skills in each experience (Auchincloss et al., 2014; NASEM, 2017). This is likely due to time or other practical constraints and specific aims (NASEM, 2017). One exception is the senior year capstone course in which students conduct independent research projects where they are responsible for all the steps involved. Biology capstone courses are often in a seminar format and can include research (Haave, 2015). Indeed, Olivares-Donoso and Gonzalez (2019) suggest theses as a way to get advantages of both course-based research and

A. Thomas (✉)
Crown College, St. Bonifacius, MN, USA
e-mail: thomasa@crown.edu

N. J. Pelaez et al. (eds.), *Trends in Teaching Experimentation in the Life Sciences*, Contributions from Biology Education Research,
https://doi.org/10.1007/978-3-030-98592-9_6

undergraduate research experiences. The class described herein has characteristics of both as a seminar course where the instructor guides multiple students as they perform their individual projects, potentially with a separate direct faculty researcher, that culminate in a final presentation and paper. In this unique learning context, appropriate pedagogical tools are of value. This chapter is a case report explaining an instructor's use of the resource developed by the ACE-Bio Network (Pelaez et al., 2017; Chap. 1 in this volume) to plan the research portion of an undergraduate senior seminar course.

The Advancing Competence in Experimentation-Biology Network published a list of Basic Competencies of Biological Experimentation with Concept-Skill statements for each competency (Pelaez et al., 2017; Chap. 1 in this volume). This resource, henceforth referred to as ACE-Bio Competencies, represents all the steps typically involved in doing biology research/experimentation. The seven Basic Competencies of Biological Experimentation of the ACE-Bio Network are ordered from *Identify*, *Question*, *Plan*, *Conduct*, *Analyze*, *Conclude*, to *Communicate*; which is how one would tend to conduct a research project (Pelaez et al., 2017; Chap. 1 in this volume). The ACE-Bio Competencies note that the steps are not always linear and so acknowledges the realistic need for adaptability as science research proceeds. Each competency is defined and fleshed out in a list of Concepts. An example is that the competency *Question* includes "The concepts of A. Observations, B. Research Questions, C. Models and D. Hypotheses." Each Concept is further delineated into a list of Skill Statements. The Skill Statements are action steps that one would take as a researcher. These detailed lists of skill statements are exhaustively inclusive and applicable to a range of Biology related fields. Hence ACE-Bio Competencies represent an exceptional practical tool for any biology researcher and consequently for pedagogy in this area.

Indeed ACE-Bio Competencies can be used as a framework for teaching and learning experimentation. While ACE-Bio Competencies have been useful for teaching a part of the experimentation process (e.g. Furrow et al., 2020), they are comprehensive, and as such, uniquely suited to settings where conducting all the competencies are required. ACE-Bio Competencies were chosen by the instructor in her first time teaching the course because the instructor was familiar with ACE-Bio Competencies (Furrow et al., 2020), and it seemed appropriate for the role of the main course planning tool for the research portion. Shulman et al. (2002) present important concepts for assessment somewhat derived from work by the California Assessment Collaborative (1993), such as setting goals, establishing assessments aligned with these goals, and assessment that is useful for teaching and student reflection. Further, Kruse (2018) notes that student use of rubrics can provide benefits such as reflection, terminology and direction, and ACE-Bio Competencies can be used similarly. These principles and others are easily implemented using ACE-Bio Competencies and may have influenced the course design process described in this chapter. This case report is a reflection highlighting that ACE-Bio Competencies have several characteristics amenable to course-based independent research since they are (a) inclusive and thus applicable to a range of biology-related areas, (b) comprehensive including all the steps required in a research project so that high standards are maintained, (c) concise so that time could be spent on the actual project in a one semester course, (d) composed of ordered higher-level competencies which can facilitate the framing and sequencing needed in a syllabus and to scaffold assignments, (e) hierarchical and actionable where each competency contains delineated Concept-Skill statements and so easily translatable to assignment development, and (f) realistic and reflective of the typical science research process. The resource was used to guide topic sequencing and timing and create scaffolded assignments that led to the final presentation and paper. The research component of this course was heavily based on ACE-Bio Competencies and thus course materials presented herein are extensions of this key resource modified for a specific course context. Permission has been obtained for the use of the ACE-Bio Competencies (Pelaez, et. al., 2017) and modifications in the publication of this chapter. It is hoped that instructors and research supervisors will find this helpful for directing students in independent research projects.

6.2 Course Context

This course occurred at a small faith-based private higher education institution. Students, normally in their senior year, from all the biology-related majors are required to take this capstone course. The course description outlines two main areas: (1) faith and science integration and (2) conducting and communicating independent research. A third minor component focused on preparing students for post-college life was included in this iteration. The course occurred over a 16-week semester. Classes were scheduled to meet twice a week and once during the last week (Finals Week). Usually Tuesdays were focused on faith and science integration while Thursdays were dedicated to the research project and post-college preparation. Students were expected to complete research work outside of class time. For the independent research project many variations are possible including the student continuing a project, starting a new project, working on a faculty-driven project, or coming up with their own project. Those students who worked on research with other faculty still needed to submit the work for the course. The instructor of this course, while monitoring the coursework and therefore research for all the projects, was not necessarily the immediate research supervisor of all the projects. Students can do a course that includes science research methods where students do a literature review, but this course is not required. Students who work on a research mentor's project typically are not involved in generating the research question (Seeling & Choudhary, 2016, Auchincloss et al., 2014), yet in this setting those students still had to navigate all research phases including in depth understanding of the research question. All students were expected to complete all the steps of a research project as a part of this course, regardless of previous work on the project or whether they were working on a faculty member's project.

6.3 Implementation

The ACE-Bio Competencies were the main tool for design of the research component of this course. It was used in planning the syllabus and assignments before the semester started. Simply put, the seven ACE-Bio Competencies were treated as the research steps and the concept skill statements used to create assignments and to direct students in those assignments. The pacing, sequence and details of these assignments were guided by the ACE-Bio Competencies. Students were therefore heavily exposed to ACE-Bio Competencies. Rubrics that contributed to the grading of the presentation and final paper were also partially based on ACE-Bio Competencies but are not discussed in this chapter.

This approach is in keeping with the assessment concepts already noted by Shulman et al. (2002) and Kruse (2018) since ACE-Bio Competencies facilitated identification of goals and associated assessment, as well as student reflection. Backward design is another highly regarded practice and involves the following sequence: (1) determining goals, (2) developing assessment and then (3) teaching strategies (Wiggins & McTighe, 2005). Although this course was not planned explicitly using this model, ACE-Bio Competencies as the source of both the overarching goals and action items for the assignments allowed easy alignment and the performance of the Concept Skills that students were learning during the assessment. While this was the case for most of the assignments, it was however not true of the final presentation and paper. To properly conduct backward design (Wiggins & McTighe, 2005), the grading criteria should have been created during the course planning phase.

The reader should note that the course occurred during the start of the COVID-19 pandemic and some changes were made to accommodate this. Other minor changes normal to course implementation also occurred. The course materials are presented in this case report as initially intended.

6.3.1 Syllabus

All seven competencies were used to provide the framework and guide the pacing for the research component of the course. The competencies were sequenced in the order presented in ACE-Bio Competencies as indicated in Table 6.1 to fit into the 16-week semester. The opportunity to present finished projects at venues beyond the classroom was also a consideration for timing. The order in ACE-Bio Competencies lent itself easily to the course setting since it established the same starting point for all students whether they were continuing a project or beginning a new one. The seven competencies were spread throughout the semester with appropriate time assigned to each competency. The most time was given to the *Conduct* competency so that students could do their research over a 5-week period. *Communicate* was given the second highest time-period as students prepared both a presentation and paper. *Identify* occurred over the first 2 weeks as students explored

Table 6.1 Sequence of competencies with relevant assignments (Modified from Pelaez et al., 2017 and Chap. 1 in this volume, used with permission)

Week	Competency	Concept & skills	Summary of assignment
1	*Identify*	A1, A2	Identify topic, key words, conduct literature search and refine reference list, and complete checklist
2	*Identify*	A3-B3	Identify gap and communicate as summary statement Write 3–4 paragraphs giving appropriate background information about current understanding of the topic to lead the reader to the gap identified Individual meeting with instructor Complete checklist
3	*Question*	All	Write a hypothesis statement Draw a diagram representing your idea, or equation or other model format Write 1–3 paragraphs commenting on your hypothesis Complete checklist
4	*Plan*	All	Generate a research plan to include the following: Draw a diagram showing the steps involved in conducting your research which should include a timeline with proposed dates of each step Create mock figures of predicted results Write 1–3 paragraphs of comments on your plan including addressing relevant skills from the checklist that were not completed in #1 and 2. E.g. identifying sources of error Reminder about IRB/Safety documentation Complete checklist
5–8	*Conduct*	All	Prepare powerpoint slides with data updates on work completed each week. Present lab book records or notes including where secure data is stored as appropriate
9	*Conduct*	All	Prepare powerpoint slides with data updates on work completed each week. Present lab book records or notes including where secure data is stored as appropriate Checklist was assigned for this last *Conclude* week
10	*Analyze*	All	Complete appropriate statistical analysis to generate final figures. Create powerpoint slides. Complete checklist
11	*Conclude*	All	Write the Results section of your paper using A1 and B1 from *Conclude* checklist and the analysis skills list for guidance Write the Discussion section of your paper using A2-3 and B1-9 from the *Conclude* checklist for guidance Complete checklist
12–13	*Communicate*	All	Prepare powerpoint presentation on your project and complete checklist Feedback on student work
14	*Communicate*	All	Paper draft and complete checklist
15	*Communicate*	All	Final thesis paper (Week 16 Reflection)

the literature and chose a specific direction. The remaining competencies were assigned 1 week each. The reader should note that ACE-Bio Competencies explicitly point out that science and thus the order of competencies is not always linear (Pelaez et al., 2017; Chap. 1 in this volume). Indeed, for practical purposes, Institutional Review Board application submission for relevant projects was scheduled early (weeks 1–2) even though it is a part of the *Plan* competency. Although the competencies were sequenced linearly, students could make changes as their projects progressed. The pacing of the course as shown in Table 6.1 worked well for this 16-week format to move students from the *Identify* competency to the *Communicate* competency. Week 16 occurred during Finals week and was used for a Reflection Assignment.

6.3.2 Assignments

The research component of the course represented 60.5% of the course grade. The course used a point grading system where the total points possible was 1000. There were 13 assignments each worth 10 points, the presentation was 80 points and the summative assessment final paper 375 points. Students also completed a reflection survey of all aspects of the course in week 16 (10 points), which will not be discussed in this essay. The syllabus included 10 points for human subject research training/application or safety documentation.

The assignments essentially walked students sequentially through completion of the competencies from *Identify* to *Communicate*. In other words, ACE-Bio Competencies were operationalized in the assignments. The assignments were tailored to the competencies as indicated in Table 6.1. These assignments scaffolded the research process so that students gradually built upon earlier work as the course and their projects progressed. Table 6.2 shows more details of assignments based on the *Identify* competency. As shown in these examples, each competency has a list of Concept Skill Statements. The action statements used in the Skills' Statements are particularly suited to the checklist approach used after competencies (or parts thereof) were completed. The reader should note that the checklist took a survey format and students answered other questions about the competencies that are beyond the scope of this essay. The checklist approach, while not a rubric, is still a valid instructional tool (Brookhart, 2013) and helpful for student reflection which is highly valued (Kruse, 2018; Shulman et al., 2002). In some assignments, students were directed to treat the concept skills statements as guides to help them complete the competencies in a proficient manner. This had similar advantages to a rubric since it would "keep students focused" as noted by Kruse (2018). Students were also expected to present updates to the class and this was built into relevant assignments. The instructor's goal for class discussions of these updates was to create an atmosphere simulating a traditional research group's lab meeting where students and the instructor gave feedback on projects. Students therefore received input and support from others as they worked on their individual projects.

Table 6.2 Examples of the *Identify* competency from the ACE-Bio Network aligned with assignments. Concepts and Skill statements were modified from Pelaez et al. (2017), used with permission

Assignment #1 The Concept:	#1 Skills: The ability to use the concept to:	Assignment #1: General topic, References and Checklist 1a
A. Relevant Background Knowledge	A1. ____Find appropriate sources of relevant scientific information (primary, secondary, etc.)	(a) Choose a topic. Bring PRINTED copies to class of (a) (i) What is the topic that you are interested in? (ii) List key words related to your topic of interest (b) Conduct a literature search using the key words. Please consult the instructor or the librarian if you need additional help with this. Complete any interlibrary loan requests as soon as possible (c) Choose a subset of references from your literature search as the core literature that you will use for your project. Look at your article abstracts and skim your references to help you with narrowing down the list (d) Checklist *The details of an additional question about previous coursework related to their project are not included since students were directed to still complete this assignment
	A2. ____Filter and evaluate the relevance of information from appropriate sources to the specific research focus	
Assignment #2 The Concept:	#2 Skills: The ability to use the concept to:	Assignment #2: Problem Identification, Checklist 1b
A. Relevant Background Knowledge	A3. ____Evaluate background information with critical scientific skepticism	After reviewing the articles that you chose, *identify* a gap in current understanding of your topic (a) Write a summary statement of the gap that you have identified (b) Write 3-4 paragraphs based on your references that should give appropriate background information about current understanding on your topic and lead the reader to the gap that you have identified. Treat the list of skills below as a guideline for what should be included (c) Checklist *The details of an additional question about previous coursework related to their project are not included since students were directed to still complete this assignment
	A4. ____Synthesize and apply current knowledge to generate a contextual foundation for the research problem.	
	A5. ____Reflect on the skills and knowledge needed in the relevant field before proceeding to do research	
B. A Gap in Current Knowledge	B1. ____Recognize a gap in current scientific knowledge that can be addressed with experimentation	
	B2. ____Reflect on limits of background knowledge related to the gap	
	B3. ____*Identify* a problem that is timely, relevant, and interesting, and, if addressed, could build on our foundational knowledge of science	

The basic strategy for designing the content of the assignments was to look at what was involved in the competency, generate activities to help students accomplish the task, and choose an appropriate time-period. As indicated in Table 6.1, students completed actions based on the competencies as they carried out the research process. The specifics of the assignment depended on the competency. For example, the *Identify* competency was broken into two assignments (see Table 6.2). The first was in essence a literature search using key words. The second assignment was for the remaining *Identify* competency Concept-Skills and involved the students completing a literature review. The students were to "*Identify* gaps" and summarize the current background of the field. Students who had done this in a previous course were still required to redo these assignments especially since new work could have been published. Students were also required to meet individually with the instructor following completion of these first two assignments. Although this was done for scheduling reasons, it allowed students a space to discuss their projects one on one early in the process. The instructor was of course available to meet with individuals any time during the semester.

The *Plan* competency is the most extensive and reflects the exhaustive nature of the ACE-Bio Competencies, which allow them to be applicable across a wide range of projects. Although concepts such as limitations from this competency are essential to all projects, ACE-Bio Competencies note that controls may not always be appropriate or "relevant." Additionally, the checklist part of the assignments acknowledged that all the concept skill statements were not applicable to all of the projects. Consequently, students were expected to complete the skills applicable to their research project thus creating research steps unique to their project. This exposed the students to both the essential science research steps while acknowledging that all research is not identical. These assignments moved them from the *Question* competency in a step-wise fashion to the *Conclude* competency even though it is recognized that research does not always move in a linear or stepwise manner (Pelaez et al., 2017; Chap. 1 in this volume).

The *Communicate* competency was approached differently. The course requirements of a final scientific presentation and paper are preset, and for some students they are used for institutional and departmental assessment. Students were given a rubric for their oral presentations. They completed their oral presentations, and then their peers completed feedback forms (Table 6.3) based on the *Communicate* competency list. Students then received the evaluations to help them revise and refine their papers.

Students were given the grading criteria that would be used to grade their senior thesis paper. The document containing the paper grading guidelines also directed students to two resources on paper writing (Frey, 2003; Hesselbach et al., 2012). A draft of their senior thesis paper was submitted to the instructor for initial review. The students then handed in the final paper. The ACE-Bio Competencies translated easily to familiar science research products such as literature search, research plan, record keeping and the formal oral presentation and science paper. For example, the sections characteristic of a science paper (Frey, 2003; Hesselbach et al., 2012; Timmerman et al., 2011) and ACE-Bio Competencies can be aligned:

Table 6.3 Feedback and scoring for presentations. Modified from Pelaez et al. (2017) and used with permission

The *Communicate* checklist from ACE-Bio Competencies has been provided below. To assist students, classmates provide feedback below on each presentation

(a) The skill was exhibited very well
(b) The skill was exhibited fairly well
(c) The student completed the skill but it needs improvement
(d) The student did not complete this skill

After the presentation, observations are recorded:

What did the student do exceptionally well?

For any responses where you checked c, please give an explanation of the problem and possible solutions

Please state any additional concerns about their presentation?

The concept:	Skill: The ability to use the concept to...	a	b	c	d
A. Representations	A1. ____Distill results into clear numeric and/or graphical forms that are aligned with the experimental objective/question/hypothesis				
	A2. ____Develop a predictive or explanatory model to summarize research finding				
B. Scientific Communication	B1. ____Construct scientific communications using standard conventions.				
	B2. ____Distinguish typical structure and detail of an oral versus a written presentation				
	B3. ____Tailor structure and content of a presentation to the probable audience (e.g., scientific vs. public)				
	B4. ____Construct a wide range of representations such as tables, graphs, slides, diagrams, animations and simulations to present main points clearly in written and oral presentations				
	B5. ____Select the representation that best depicts the data to allow for appropriate inferences				
C. Limitations	C1. ____Articulate limitations, unanswered questions, and the tentative nature of results (both positive and negative)				
	C2. ____Contrast results and findings with previously published scientific work				
	C3. ____Offer alternative hypotheses				
	C4. ____Construct a justification and counter-justification argument for each alternative, if possible				
D. Synthesis and Reflection	D1. ____Evaluate, analyze and explain the significance and implications of the research				
	D2. ____Revise an existing model based on observations or data				
	D3. ____Articulate how findings contribute to new knowledge that can drive further inquiry				
	D4. ____Propose follow up experiments based on inferences from predicted or actual results of experiments.				

Relevant Paper Section for the ACE-Bio Competencies (Pelaez, et al., 2017)

Identify	Introduction
Question	Introduction
Plan	Introduction, Methods, Results and Discussion
Conduct	Methods
Analyze	Results
Conclude	Results, Discussion
Communicate	All sections

Grading of the 10-point assignments was by completion and no rubric provided. The ACE-Bio Competencies and statement-skill lists were essentially treated as both learning goals/tasks and assignments simultaneously and thus proved valuable for teaching and learning in a similar manner to rubrics (Brookhart, 2013) and supported by Shulman et al. (2002). Final grading criteria for the presentation and paper were decided on during the course of the semester and given to the students for both the presentation and paper, but are not described in this essay. Although the students received the grading plans with enough time to direct their work, these decisions should have been made at the time of course planning similar to the other ACE-Bio Competencies based course materials. This would have better fulfilled backward design guidelines (Wiggins & McTighe, 2005) where the goals drive assessment. Readers should note that while students were given the grading approach ahead of time, some modification occurred during the grading process.

6.4 Comparison of ACE-Bio Competencies with Other Resources

Schaefer (2016) compared two resources for use in a specific course and this comparison example was followed for this study. Certainly, if ACE-Bio Competencies share many characteristics with other resources, that would confirm that it is indeed a comprehensive resource appropriate for thorough student exposure to research fundamentals. The final product of research projects, including for this course, is often the traditional science paper. The universal rubric (Timmerman et al., 2011) was developed for this type of paper and so represents commonly accepted standards. When compared to ACE-Bio Competencies at the general level, they contain all the concepts reviewed in the universal rubric except for writing focused assessment e.g. grammar (Timmerman et al., 2011). ACE-Bio Competencies therefore have similar standards even though some of the elaborations are different. The reader should note that the universal rubric (Timmerman et al., 2011) was a part of the grading criteria used in in this course. As expected, ACE-Bio Competencies contain several research steps beyond the communication focus of a paper and so had many more criteria. ACE-Bio Competencies are therefore a useful and valid approach to guide students to do research that results in key qualities of an important research end goal, the science paper.

The completed product is not the only evidence of quality research standards but the characteristics of the process are also important. BioSkills (Clemmons et al., 2020a, b) delineates details of the core competencies that are expected of undergraduate biology majors that resulted from the pivotal *Vision and Change* report (AAAS, 2011). When compared with ACE-Bio Competencies there is quite a bit in common with the "Process of Science," "Modelling," and "Quantitative Reasoning" competencies of BioSkills for undergraduate biology (Clemmons et al., 2020a, b). This is not surprising since these competencies are involved directly in the research process. However, the remaining BioSkills' competencies "Science & Society," "Interdisciplinary Nature of Science," and "Communication & Collaboration" as defined therein (Clemmons et al., 2020a, b) were absent or minimally covered in the ACE-Bio Competencies. These latter components, although important in science, are not necessarily explicitly involved in the research process but often still occur. For example, since these were independent projects in an area of the student's choosing, formal collaboration did not occur. Interestingly, since the students received regular feedback from their peers during class, there was still some level of collaboration as they went through the semester together. Thus, students guided through ACE-Bio Competencies are also completing research related work deemed as fundamental by BioSkills (Clemmons et al., 2020a, b) and *Vision and Change* (AAAS, 2011).

To further determine whether ACE-Bio Competencies are indeed thorough and encompass foundational research concepts, one final comparison was done. Six competencies were determined by Dirks and Knight (2016) when they merged Biology competencies from eight sources including *Vision and Change* (AAAS, 2011) from which BioSkills (Clemmons et al., 2020a, b) is based. These six competencies (Dirks & Knight, 2016) therefore represent significant core values since they reflect multiple sources. ACE-Bio Competencies include components of all but one which was "Appreciate and Apply the Interdisciplinary Nature of Science" (Dirks & Knight, 2016). Thus, not only do ACE-Bio Competencies contain the research relevant competencies (Clemmons et al., 2020a, b) and the traditional science paper components (Timmerman et al., 2011) articulated in specific resources, they also have almost all the competencies recognized in the more exhaustive inventory generated from this combinatorial approach (Dirks & Knight, 2016). ACE-Bio Competencies are therefore highly consistent with current expectations from various sources, highlighting the validity of its use in a teaching context.

6.5 Discussion

The perspective of the author, who was teaching the course for the first time, was that the research portion of this course ran smoothly because of the use of ACE-Bio Competencies. The instructor relied upon ACE-Bio Competencies to set the research steps, guide course pacing and develop assignments. Using one main resource facilitated consistency across the varied student projects in the course. Assignments were easily generated from this resource and provided a scaffolded approach. All the students therefore had similar research experiences despite the unique nature of their projects. Some insights from this course design process are highlighted as a bulleted list at the end of this chapter.

While ACE-Bio Competencies provided a useful framework, they were also adaptable. They were easily used to build the research component of this course with different types of biology research projects. Being able to fit a complete research project within time boundaries is difficult due to the unpredictable nature of research. ACE-Bio Competencies, while acknowledging the non-linearity of the research process (Pelaez et al., 2017; Chap. 1 in this volume), provided guidelines for the sequencing necessary for a syllabus. Moreover, students could select the steps applicable to them. The student process, therefore, remained similar yet was individualized. The plan described here could be easily modified to other course contexts. In a faculty survey, Coil et al. (2010) found that time is the most frequent limitation in teaching science process skills and the lack of a sense of "how to teach skills in a classroom format" was also a factor, albeit for fewer instructors. Although these authors were referring to overall instruction, they presented a freshman course as an example of a solution to provide the needed clear instruction in this area (Coil et al., 2010). The authors asserted that "these skills are rarely taught to undergraduates in an explicitly and scaffolded manner" (Coil et al., 2010). ACE-Bio Competencies provide straightforward Concept Skill statements that can be used to generate smaller assignments in a course as described in this case report.

Whereas ACE-Bio Competencies were helpful in this course context it should also be helpful for research supervisors. Seeling and Choudhary (2016), highlight similar activities to ACE-Bio Competencies as a part of undergraduate research experiences. The approach used in this course in some ways represents the apprentice style (NASEM, 2017). Thus, this resource could also be applied in that setting since it is short, yet comprehensive. A checklist-like approach for both the supervisor and students would be recommended. Brown et al. (2016) in an apprentice type setting, developed a syllabus and assessments for undergraduate researchers even though they were not in a course, and ACE-Bio Competencies could be the basis for something comparable.

Although the ACE-Bio Competencies are comprehensive, there are other components that could be added. Activities could be introduced alongside or integrated with the use of ACE-Bio Competencies. For example, an important learning outcome cited in BioSkills (Clemmons et al., 2020a, b) that was missing from the ACE-Bio Competencies is "Collaboration." In this course students interacted with each other in a similar manner to a research lab meeting since they were expected to give and receive feedback. ACE-Bio Competencies were the basis of the one instance of detailed peer feedback, which occurred for presentations. Similar review could have occurred for the other competencies to increase instances of cooperativity (Johnson et al., Chap. 22 in this volume). Furthermore, strategies that employ feedback have been used for undergraduate projects (Wieth et al., 2019; Reynolds & Thompson Jr, 2011; Chaps. 9, 12, and 20 in this volume).

The ACE-Bio Competencies provided an opportunity to expose students to science culture since it listed steps of a "competent biologist" (Pelaez et al., 2017)". Expectations were similar to those identified by other science pedagogy resources such as BioSkills (Clemmons et al., 2020a, b) and the universal rubric (Timmerman et al., 2011). Class meetings reflected a lab meeting type format typical of science research groups. The instructor regrets that this aspect of professional science culture may not have been highlighted as such to students. Competency sequencing in the course occurred linearly, yet this is not always the case in science (Chap. 1 in this volume). Perhaps the importance of this could have been reiterated by giving students an assignment where they are required to revisit an earlier step. Further, this would facilitate the practice of iteration which is typical of science research and recommended for students at the undergraduate and K-12 levels (Auchincloss et al., 2014; National Research Council, 1996, 2012). One possible implementation strategy could be an additional assignment requiring that students revise an earlier assignment by the time they get to the *Analyze* competency #5. This would leave some latitude for revision to occur when it was best for the individual student. Although this was a course context, the students did experience components of typical science culture.

Clearly there are books on research (e.g. Batavia, 2001; Patten & Bruce, 2014) and resources identified by Ottowitz and Halbreich (1993) in their annotated bibliography can be similarly used for course design. Research was an important component of this class. However, the focus was more on students doing science independently rather than giving them a formal research methods class, where a book would perhaps have been more appropriate. Online self-paced research courses (e.g. Behrman & Schnoes, n.d.) and *Entering research: A curriculum to support undergraduate and graduate research trainees* (Branchaw et al., 2020) also exist. While these resources include some of the research parameters relevant to this course, they are more appropriate for the apprentice model. Although the apprentice model can have "a structured approach" (Brown et al., 2016), the capstone course described herein, as a course required both substantial organization and flexibility since students conduct varied independent work. Course design approaches vary widely from more context-driven (e.g. Colabroy, 2011) to those where formal pedagogy is applied to a context (e.g. Hills et al., 2020). One goal of the wheel model presented by Brew (2013) is to acknowledge the "complexity of implementing undergraduate research and inquiry" which then drives another goal, which is to help with research instruction strategy selection. This instructor agrees since there is a wide range of science, technology engineering and math (STEM) research experiences (NASEM, 2017) and capstone courses even within a single discipline such as Biology (Haave, 2015) that indeed indicate the importance of identifying a resource that can be tailored to your situation. ACE-Bio Competencies were particularly useful to the author because they were concise, field-specific, familiar and easily used for the daunting task of teaching a course for the first time. This case report reflects a more practical approach to course design where ACE-Bio

Competencies more than amply met a need. Post-course assessment of ACE-Bio Competencies indicated that it is on par with established competencies (e.g. those described by Dirks & Knight, 2016) and thus valid for helping students reach current accepted standards. This is an important asset of a course planning tool since it means the derived course content is well-grounded. In this case report, ACE-Bio Competencies facilitated the timeliness and comprehensiveness needed when guiding students in the research process while they are actually conducting the research.

Finally, ACE-Bio Competencies are reflective of the scientific reality of the research process. Using ACE-Bio Competencies as a framework for this course provided students with a rich science research experience and the instructor with a powerful practical template. Naturally continued course improvement should occur, and some examples have already been presented. Undergraduate research pedagogy studies contain other possibilities. For example, Linn et al. (2015) in their review of undergraduate research identified significant areas including mentoring. The instructor in this instance essentially functioned as a research mentor in many of the ways highlighted by Linn et al. (2015), and literature in this area could highlight future considerations. Another important source for modification strategies is the student perspective on the use of ACE-Bio Competencies.

In summary, ACE-Bio Competencies were a significant factor in the success of this course and could be easily applied to other contexts. Others may consider the following recommendations for guiding students in independent research in a course context using ACE-Bio Competencies:

- Use a resource such as ACE-Bio Competencies that covers all the research components in a way that is inclusive enough to be applicable to all projects, but also adaptable to the individual project. Comprehensive resources are best suited to a course with diverse student projects.
- Start all students at the *Identify* competency and set identical requirements, even if they are continuing a project or are working on a faculty member's project to ensure that the students are engaged in all the research steps.
- Use backward design (Wiggins & McTighe, 2005) to create scaffolded assignments that sequentially walk students through the steps involved in a complete project. These assignments should represent familiar products of science research e.g., literature search and record keeping.
- Use the detailed Concept Skill Statements as steps involved in research methodology to give direction to students in assignments and as a basis for peer and instructor feedback. Have students treat the Concept Skill statements as checklists as they complete assignments.
- Include assignments that require students to revisit earlier steps, to program revision of earlier steps and demonstrate the nonlinear nature of scientific research (Pelaez et al., 2017; Chap. 1 in this volume).
- Align the parts of the research process (ACE-Bio Competencies) to the summative assessment, such as through writing or presenting the sections of a formal science paper or presentation.

Acknowledgments This work was partially funded by a stipend from the National Science Foundation (NSF) grant #1346567 for the Advancing Competencies in Experimentation–Biology (ACE-Bio) Network. Any opinions, findings, and conclusions or recommendations expressed are those of the author and do not necessarily reflect the views of the NSF. Pelaez, et al. (2017) was used to inform assignments and Tables 6.1, 6.2 and 6.3 and course assignments were modified from Pelaez et al. (2017) and are used in this publication with permission from the Rightsholder Nancy Pelaez. The full citation is Pelaez, Nancy et al. "The Basic Competencies of Biological Experimentation: Concept-Skill Statements" (2017). PIBERG Instructional Innovation Materials. Paper 4. http://docs.lib.purdue.edu/pibergiim/4. The original source is licensed under a Creative Commons Attribution-NonCommercial-ShareAlike 4.0 International License. The tables and assignments are available to anyone who contacts the author via email, Aeisha Thomas (thomasa@crown.edu), for a file that will be easy to modify and use in the classroom. Thanks go to Dr. Don Bouchard, Dr. Karen Gustafson, Mrs. De Anne Robb and Dr. Lynne Shenk for review, suggestions and helpful discussion; book editors and contributing authors for suggestions; Dr. Deanna Munson for reference suggestions; and Crown College for partial funding of publication costs.

References

American Association for the Advancement of Science (AAAS). (2011). In C. Brewer & D. Smith (Eds.), *Vision and change in undergraduate biology education: A call to action*. American Association for the Advancement of Science (AAAS). https://visionandchange.org/finalreport/. Accessed 18 Oct 2021.

Auchincloss, L. C., Laursen, S. L., Branchaw, J. L., Eagan, K., Graham, M., Hanauer, D. I., Lawrie, G., McLinn, C. M., Pelaez, N., Rowland, S., Towns, M., Trautmann, N. M., Varma-Nelson, P., Weston, T. J., & Dolan, E. L. (2014). Assessment of course-based undergraduate research experiences: A meeting report. *CBE Life Sciences Education, 13*(1), 29–40. https://doi.org/10.1187/cbe.14-01-0004

Batavia, M. (2001). *Clinical research for health professionals: A user-friendly guide*. Butterworth-Heinemann.

Behrman, S., & Schnoes, A. (n.d.). Let's experiment: A guide for scientists working at the bench (Self-Paced) *iBiology courses*. https://courses.ibiology.org/courses/course-v1:iBiology+LE+SP/about. Retrieved November 17, 2020.

Branchaw, J., Butz, A. R., & Smith, A. R. (2020). *Entering research: A curriculum to support undergraduate and graduate research trainees* (Navigating research and mentoring series). Macmillan.

Brew, A. (2013). Understanding the scope of undergraduate research: A framework for curricular and pedagogical decision-making. *Higher Education, 66*(5), 603–618.

Brookhart, S. M. (2013). *How to create and use rubrics for formative assessment and grading*. Association for Supervision and Curriculum Development.

Brown, A. M., Lewis, S. N., & Bevan, D. R. (2016). Development of a structured undergraduate research experience: Framework and implications. *Biochemistry and Molecular Biology Education, 44*(5), 463–474. https://doi.org/10.1002/bmb.20975

California Assessment Collaborative. (1993). *Charting the course toward instructionally sound assessment: A report of the alternative assessment pilot project*. Far West Laboratory, California Assessment Collaborative.

Clemmons, A., Timbrook, J., Herron, J., & Crowe, A. (2020a). BioSkills Guide. Core Competencies For Undergraduate Biology, (Version 5.0). *QUBES Educational Resources*. https://doi.org/10.25334/156H-T617

Clemmons, A., Herron, J., Crowe, A., & Timbrook, J. (2020b, January 16). BioSkills guide: Development and national validation of a tool for interpreting the vision and change core competencies. *Biorxiv*. https://doi.org/10.1101/2020.01.11.902882

Coil, D., Wenderoth, M. P., Cunningham, M., & Dirks, C. (2010). Teaching the process of science: Faculty perceptions and an effective methodology. *CBE-Life Sciences Education, 9*(4), 524–535. https://doi.org/10.1187/cbe.10-01-0005

Colabroy, K. L. (2011). A writing-intensive, methods-based laboratory course for undergraduates. *Biochemistry and Molecular Biology Education, 39*(3), 196–203. https://doi.org/10.1002/bmb.20496

Dirks, C., & Knight, J. K. (2016). Measuring college learning in biology. In R. Arum, J. Roksa, & A. Cook (Eds.), *Improving quality in American higher education: Learning outcomes and assessments for the 21 st century* (pp. 225–260). Jossey-Bass.

Frey, P. (2003). Guidelines for writing research papers. *Biochemistry and Molecular Biology Education, 31*(4), 237–241. https://doi.org/10.1002/bmb.2003.494031040241

Furrow, R. E., Price, R. M., Thomas, A., Yiny, Y., & Bernot, K. M. (2020). Vitamin C for colds? Writing LETTERS to synthesize and communicate results from multiple studies. *CourseSource.* https://doi.org/10.24918/cs.2020.8

Haave, N. C. (2015). Survey of biology capstone courses in American and Canadian higher education: Requirement, content, and skills. *Bioscene: Journal of College Biology Teaching, 41*(2), 19–26.

Hesselbach, R. A., Petering, D. H., Berg, C. A., Tomasiewicz, H., & Weber, D. (2012). A guide to writing a scientific paper: A focus on high school through graduate level student research. *Zebrafish, 9*(4), 246–249. https://doi.org/10.1089/zeb.2012.0743. Epub 2012 Oct 24. PubMed PMID: 23094692; PubMed Central PMCID: PMC3528086.

Hills, M., Harcombe, K., & Bernstein, N. (2020). Using anticipated learning outcomes for backward design of a molecular cell biology course-based undergraduate research experience. *Biochemistry and Molecular Biology Education, 48*(4), 311–319. https://doi.org/10.1002/bmb.21350

Kruse, M. (2018). 5 student-centered ways to use rubrics. *Reading and Writing Haven; A blog for Educators.* https://www.readingandwritinghaven.com/5-student-centered-ways-to-use-rubrics/

Linn, M. C., Palmer, E., Baranger, A., Gerard, E., & Stone, E. (2015). Undergraduate research experiences: Impacts and opportunities. *Science, 347*(6222), 1261757–1261757. https://doi.org/10.1126/science.1261757

National Academies of Sciences, Engineering, and Medicine (U.S.). Committee on Strengthening Research Experiences for Undergraduate STEM Students. (NASEM). (2017). In J. Gentile, K. A. Brenner, & A. Stephens (Eds.), *Undergraduate research experiences for stem students: Successes, challenges, and opportunities*. National Academies Press.

National Research Council. (1996). *National Science Education Standards*. National Academies Press. https://doi.org/10.17226/4962

National Research Council. (2012). *A framework for K-12 science education: Practices, crosscutting concepts, and Core ideas* (Committee on a Conceptual Framework for New K-12 Science Education Standards. Board on Science Education, Division of Behavioral and Social Sciences and Education). The National Academies Press.

Olivares-Donoso, R., & Gonzalez, C. (2019). Undergraduate research or research-based courses: Which is most beneficial for science students? *Research in Science Education, 49*(1), 91–107.

Ottowitz, W. E., 3rd, & Halbreich, U. (1993). Introducing the student and trainee to biomedical experimentation–a selective annotated bibliography. *Postgraduate Medical Journal, 69*(815), 728–732.

Patten, M. L., & Bruce, R. R. (2014). *Understanding research methods: An overview of the essentials* (9th ed.). Pyrczak Pub.

Pelaez, N., Anderson, T., Gardner, S. M., Yin, Y., Abraham, J. K., Bartlett, E., Gormally, C., Hill, J. P., Hoover, M., Hurney, C., Long, T., Newman, D. L., Sirum, K., & Stevens, M. (2017). The basic competencies of biological experimentation: Concept-skill statements. *PIBERG Instructional Innovation materials. paper 4.* http://docs.lib.purdue.edu/pibergiim/4/. Accessed 18 Oct 2021.

Reynolds, J. A., & Thompson, R. J., Jr. (2011). Want to improve undergraduate thesis writing? Engage students and their faculty readers in scientific peer review. *CBE-Life Sciences Education, 10*(2), 209–215. https://doi.org/10.1187/cbe.10-10-0127

Schaefer, J. E. (2016). The BRAIN initiative provides a unifying context for integrating Core STEM competencies into a neurobiology course. *Journal of Undergraduate Neuroscience Education, 14*(2), A97–A103.

Seeling, J. M., & Choudhary, M. (2016). Professional practices in undergraduate research programs. *Journal of Microbiology & Biology Education, 17*(2), 246–251. https://doi.org/10.1128/jmbe.v17i2.982

Shulman, J., Whittaker, A., & Lew, M. (2002). *Using assessments to teach for understanding: A casebook for educators*. Teachers College Press.

Timmerman, B. E. C., Strickland, D., Johnson, R., & Payne, J. (2011). Development of a "universal" rubric for assessing undergraduates' scientific reasoning skills using scientific writing. *Assessment & Evaluation in Higher Education, 36*(5), 509–547. https://doi.org/10.1080/02602930903540991

Wieth, M. B., Francis, A. P., & Christopher, A. N. (2019). Use of a creative problem solving (CPS) approach in a senior thesis course to advance undergraduate publications. *Frontiers in Psychology, 10*, 749. https://doi.org/10.3389/fpsyg.2019.00749

Wiggins, G. P., & McTighe, J. (2005). *Understanding by design*. Association for Supervision.

Chapter 7
Experiments in Data Mining: Using Digitized Natural History Collections to Introduce Biology Students to Data Science

Debra L. Linton, Elizabeth Ellwood, Lisa D. White, Natalie F. Douglas, and Anna K. Monfils

7.1 The Case for Integrating Data Mining into the Undergraduate Biology Curriculum

The post-secondary biology curriculum must be designed to both provide foundational biological knowledge and to train students in the processes of scientific research so that they are prepared to continue expanding our understanding of life on Earth as the next generation of biologists. Both of these tasks continue to become more complex as our knowledge increases and the methods that biologists use to study the world expand. In response to this challenge, biology education has moved away from an emphasis on transmission of knowledge to a focus on training students how to access, analyze, and evaluate information to construct their own understanding of new concepts (AAAS, 2011). This approach also promotes the learning of biological experimentation as students use the methods of biology to build their knowledge of biology.

Training future biologists in the skills they will need requires us to first identify what those skills are. Biologists use many different skills and sub-discipline specific

D. L. Linton (✉) · N. F. Douglas · A. K. Monfils
Departments of Biology and Communication Sciences & Disorders, Central Michigan University, Mt. Pleasant, MI, USA
e-mail: linto1dl@cmich.edu; dougl1nm@cmich.edu; monfi1ak@cmich.edu

E. Ellwood
iDigBio, Florida Museum of Natural History, University of Florida, Gainesville, FL, USA
e-mail: eellwood@floridamuseum.ufl.edu

L. D. White
University of California Museum of Paleontology, University of California at Berkeley, Berkeley, CA, USA
e-mail: ldwhite@berkeley.edu

N. J. Pelaez et al. (eds.), *Trends in Teaching Experimentation in the Life Sciences*, Contributions from Biology Education Research,
https://doi.org/10.1007/978-3-030-98592-9_7

practices to *Conduct* biological research. In this volume, we are examining trends in teaching "experimentation". A major difference between various methods of conducting scientific research is the distinction between what constitutes an "experiment" in different disciplines. For example, hypothesis-driven research focuses on testing a prescribed hypothesis through traditional experimentation. Hypothesis-driven experimentation requires an independent variable that can be manipulated and a dependent variable that can be measured. The researcher manipulates the independent variable and measures the response on the dependent variable. All other variables are maintained constant. There are multiple replicates. Data from this designed study are recorded and analyzed. Conclusions are made. This approach has been critical to advancing our understanding of many biological relationships, continues to be a fundamental method of biological research, and has traditionally been well represented in the undergraduate biology curriculum.

Hypothesis-driven experimentation aligns well with the traditional model of "The Scientific Method." This method has been reprinted in introductory level undergraduate Biology courses as a linear sequence through a series of steps. More realistic representations of science have come to include feedback loops between steps in the sequence to show how this process in reality is more dynamic and iterative. Even more authentic models of how scientific experimentation is done abandon any reference to a linear sequence and try to capture the complexity and variety of the processes we use to build scientific knowledge. For example, the ACE-Bio Network (Advancing Competencies in Experimentation - Biology) has defined competencies of Biological Experimentation that emphasize the equal weight and flexible order of seven major areas within experimentation. These include *Identify*, *Question*, *Plan*, *Conduct*, *Analyze*, *Conclude*, and *Communicate* (Pelaez et al., 2017; Chap. 1 in this volume). Similarly, UC Berkeley's Museum of Paleontology has developed a multi-faceted model of science, as well as a suite of resources (https://undsci.berkeley.edu) to help instructors design lessons that help students understand what science is and how it is done (Thanukos et al., 2010). This model of science focuses on the iterative nature of science and the need for science processes to be flexible, with many access points, evidence-based reasoning, an open science process with peer review, and linkages to the communities in which the science occurs.

Another approach to biological research and experimentation that must be introduced in the undergraduate biology curriculum is descriptive research. Descriptive research is based on observation, data collection, descriptive statistics, comparisons, modeling, and classification. This is a type of science commonly practiced in the "ologies" (e.g., botany, zoology, ecology, conservation, paleontology, taxonomy) and provides the foundation for communication about species and the documentation of observed changes over time, space, and scale. Descriptive research is a vital component of the process of science and the scientific method and requires a unique and specialized set of skills and knowledge, including but not limited to, an understanding of species limits, taxonomy, morphology, multivariate statistics, comparative morphology, coding, and specialized modeling. The critical importance of this type of science has been elevated by the volume of data emerging from the ecology and evolution communities, new technologies for imaging, data integration and analysis, and the integration of emerging computational and analytic skills relative to data science (NAS, 2018, 2020).

Advances in technology are allowing us to collect data about the world at an ever-increasing rate, resulting in an enormous volume of what we now call Big Data. In the life sciences, much of this data consists of information about the variation and distributions of organisms and the environments in which they live. This Biodiversity Data includes genetic data, environmental data, and organismal data. The enormity of the scale of these data that exist and the rate that they are being collected is difficult to even comprehend. For example, there are over 200 million DNA sequences in GenBank, with millions being added each year. The number of bases in GenBank has doubled every 18 months since 1982 (https://www.ncbi.nlm.nih.gov/genbank/statistics/). The monitoring of ambient weather conditions occurs continuously at tens of thousands of weather stations across the world. Millions of citizen scientists are contributing biodiversity data to publicly accessible databases. These data have the potential to revolutionize our study of life on Earth.

Science and scientists now have access to a volume and breadth of data that can be used to investigate important questions at a scale previously not possible (NAS, 2018). As the volume and accessibility of data about organisms and the environment continues to increase at an accelerating rate, the methods scientists are using to leverage these data to generate and test hypotheses are also undergoing a revolution. As educators we need to address the emerging, if not already emergent, need for the twenty-first century scientist to have new skills related to extracting useful information from those data. Data mining is this process of sorting through these "big" data sets to identify patterns, generate hypotheses, and perhaps establish causative relationships. These new data mining methodologies require new sets of knowledge and skills (Baker, 2017; Hampton et al., 2017; Nelson & Ellis, 2018).

However, training in these knowledge and skills has not kept up with the demand. This is true both for current scientists (Baker, 2017; Barone et al., 2017; Beardsley et al., 2018; Hampton et al., 2017) and for the training of future scientists (Hernandez et al., 2012, Strasser & Hampton, 2012). Biology educators have identified this gap in undergraduate training and have begun efforts to bring data science into biology classes (e.g., Feser et al., 2013; Gibson & Mourad, 2018; Kjelvik & Schultheis, 2019). Progress is being made. The bioinformatics community has identified bioinformatics core competencies (Wilson Sayres et al., 2018) for working with big genomic data and is developing strategies for integrating bioinformatics training into undergraduate biology courses (e.g., Honts, 2003; Madlung, 2018; Makarevitch et al., 2015). Similar work is needed in order to train students to use big biodiversity data to investigate some of the most important scientific issues and crises facing society (Ellwood et al., 2020). The availability of these data about the biotic and abiotic environment has increased our capacity to investigate large-scale issues of critical importance (e.g., climate change, resource management, zoonotic diseases). Biology programs need to produce biologists with the knowledge and skills to tackle these challenges. The competencies necessary for working with big biodiversity data consist of an intersection of basic data literacy, data-intensive research skills (Hampton et al., 2017), and an added set of knowledge and skills specific to these unique biodiversity data types. We need to identify these competencies and develop strategies to integrate them as a part of the undergraduate biology curriculum (Ellwood et al., 2020; Lacey et al., 2017; Michener & Jones, 2012). This integration will rely on a foundation of inquiry-based instruction, with students mining data to investigate and construct understanding of biological concepts.

7.2 Digitized Natural History Collections as a Gateway to Big Biodiversity Data

We suggest that digitized natural history collections present a set of data that is well-suited to introducing students to working with large, messy datasets while investigating foundational ecological principles. The data in natural history collections are based on species occurrence records, which are information on the presence of an individual from a defined species at a specific place and time. These occurrence data include both specimen-based and observation-based records of biodiversity over time and space. Information from occurrence records has proven to be highly valuable, allowing scientists to examine changes in distributions over time, perhaps in correlation with specific environmental factors, and to compare the distributions of different species (Nelson & Ellis, 2018; Soltis & Soltis, 2016). Occurrence data can be gathered by the collection of specimens, which are then preserved in a natural history collection, or may be based on records of observations.

Specimen-based data are the occurrence records that are based on archived biological specimens housed in a natural history collection. Scientists, naturalists, and explorers have been collecting and preserving specimens for hundreds of years. Preserved with the specimens is a variety of information about the organism, including collection date, location, habitat, images, morphological measurements, community assemblage, and phenology (Ellwood et al., 2020).

Observation-based data is information gathered by scientists, naturalists, and citizen scientists and represents an observation of an organism. Observation data is not linked directly to a physical specimen. While some data may be associated with just a location and date, other data types may be accompanied by a photo and detailed information (e.g., environmental conditions, geographic location, associated species, behavior, abundance, phenology). Similar to specimen-based data, these data can provide a wealth of information on biodiversity and are now included in many online databases accessible to researchers, educators, and the public.

Since 1999, museums and other collection holders have been engaged in a massive data mobilization effort to provide digital databases of archived specimens and observations (Nelson & Ellis, 2018). These data are aggregated in searchable portals that provide broad access to information on individual specimens, images of the specimens, and associated metadata. These efforts have vastly increased the accessibility and utility of biodiversity data. The online portals through which these data are accessible serve as aggregators, combining data from natural history collections around the world. Due to targeted funding, research scope, and other focused efforts, some of these data portals are taxon-specific. For example, VertNet (http://vertnet.org) specializes in occurrence records of vertebrates. Some portals are regional. SERNEC (http://sernecportal.org/portal/), the Southeast Regional Network of Expertise and Collections, has its own data portal covering the Southeastern United States. Other portals cover larger geographic regions. BISON (Biodiversity Information Serving our Nation – https://bison.usgs.gov) provides data from the United States that are held in U.S. Federal agencies. Some portals, like iDigBio (Integrated Digitized Biocollections – https://www.idigbio.org), aim to aggregate specimen-based occurrence data from existing U. S. collections. Other portals, such as GBIF (Global Biodiversity Information Facility – https://www.gbif.org) are working to aggregate all specimen-based and observation-based occurrence records worldwide.

Mining of these data is an emerging method of experimentation in biodiversity research (Bakker et al., 2020; Hedrick et al., 2020). Since 2011, over 250 publications acknowledged searching, accessing or downloading data from one the occurrence databases (Nelson & Ellis, 2018). Scientists use these existing data to look for patterns across large spatial and temporal scales. Beyond their explicit value for direct research using the collections data (e.g., Holmes et al., 2016; McLean et al., 2015; Meineke et al., 2018; Pyke & Ehrlich, 2010; Soltis & Soltis, 2016; Suarez & Tsutsui, 2004; Tewksbury et al., 2014) and the need to train future biologists to do such research, digitized natural history collections can serve as an accessible gateway for students into working with truly "big" data.

Digitized collections provide the opportunity for students at any institution to engage in structured, guided, or open-ended inquiry activities, while working with medium-sized data sets. A search for a species in one of these portals may return thousands of data records. This is likely a much larger data set than most students have worked with before, however it is orders of magnitude smaller than the datasets being collected by remote-sensing platforms and sensor networks deployed for some ecological research (Michener & Jones, 2012). Working with these smaller data sets can be a more accessible way to help students develop data mining skills that can be later be applied to bigger datasets.

The types of data available in digitized natural history collections are also appropriate to a variety of different types of analyses common in biological research (Soltis & Soltis, 2016). For example, students can look for changes in species distribution or morphology over time, explore relationships between two variables (e.g., specimen size vs. collection latitude), or make comparisons between different groups (e.g., # species collected in different habitat types). These types of investigations with biodiversity data provide opportunities for students to participate in all of the basic competencies of biological experimentation (Pelaez et al., 2017). Within the inquiry-based instruction framework, students can generate their own questions that can be investigated using collections data, plan their data mining and analysis procedures, conduct their investigation, analyze the results, make conclusions, and communicate their findings. Digitized natural history collections are also well-suited to introducing students to data science because they are inherently engaging and place-based (Cook et al., 2014; Ellwood et al., 2020; Lacey et al., 2017; Monfils et al., 2017; Powers et al., 2014). Students can select a favorite or local organism and explore the available data while learning and practicing the targeted skills.

7.3 Strategies for Integration of Data Mining into the Undergraduate Biology Curriculum

7.3.1 Data Mining Activities Targeting Specific Biology Concepts

The undergraduate biology curriculum is already packed with content and skills that students need to learn. Fortunately, the integration of data mining can be done in a way that supports the teaching of existing content rather than expanding it. National recommendations for best practices in biology education suggest that students should be learning concepts through discovery (AAAS, 2011). Existing biodiversity data, including digitized natural history collections, can be used to help students do just that.

Vision and Change in Undergraduate Biology Education (AAAS, 2011) identified five key content areas for undergraduate biology education. These five content areas have been expanded upon to provide a framework of overarching biological principles and specific concepts (BioCore) that should be included in an introductory biology curriculum (Brownell et al., 2014). Similarly, the Four-Dimensional Ecology Education (4DEE) Framework from the Ecological Society of America (Prevost et al., 2019) identifies Core Ecology Concepts (as well as Ecology Practices) to guide undergraduate ecology education. Many of the concepts identified in these frameworks could be investigated using biodiversity data (see Table 7.1 for examples) in an inquiry-based approach. Inquiry-based learning pedagogy has become common in undergraduate biology education, particularly for labs, but also in lecture courses that use an active learning approach (Sundberg et al., 2005; Ruiz-Primo et al., 2011; Makarevitch et al., 2015). In inquiry-based instruction, learners conduct experiments and other types of investigations to find answers to questions that have been posed and to construct knowledge that is new to them (Pedaste et al., 2015). Inquiry-based instruction has been shown to improve student learning of concepts in undergraduate science courses (e.g., Prince et al., 2009; Nybo & May, 2015).

Instructors could implement a single activity using biodiversity data to investigate one of these core concepts, or a full or partial sequence of biodiversity data activities could be built into an introductory Ecology course. These activities could be

Table 7.1 Example alignments of 4DEE Core Ecology Concepts and BioCore Concepts (Knowledge) with data mining activities using biodiversity data

4DEE Concept	BioCore Concept	Data mining activity
Autecology Resources and Regulators	A species' distribution is limited by available resources and by interactions between biotic and abiotic factors.	Examine potential correlation between some environmental variable of interest (e.g. precipitation) and the presence of different species.
Species Diversity	Biodiversity impacts many aspects of ecosystems.	Compare diversity of species occurrence records between different community types
Food Chains, Food Webs, Networks	Within an ecosystem, interactions among individuals form networks; changes in one node of a network can cause changes in other nodes directly or indirectly.	Use occurrence data to build a food web for a community
Competition, Predation, Mutualism	Competition, mutualism, and other interactions are mediated by each species' morphological, physiological, and behavioral traits.	Compare distributions or occurrence records for two competing species, predators and prey, plants and their pollinators
Global Climate Change	Ecosystems are not isolated and static--they respond to change, both as a result of intrinsic changes to networks of species and as a result of extrinsic environmental drivers.	Compare distributions of occurrence records for species over time Look for changes in phenology over time and potential disruption of interspecific interactions

designed as in-class group activities, homework assignments, or 2–3-h laboratory experiences, depending on the learning objectives and the time available to address them. Another design consideration for developing activities using biodiversity data is identifying what level of inquiry is appropriate (Bell et al., 2005) based on both the content and competency learning objectives. Different levels of inquiry can be achieved by varying the sources of the questions, the experimental design, and the conclusions (Bianchi & Bell, 2008). In a structured inquiry the instructor provides students with a question and the procedure for investigating the question. The student completes the procedure and learns the concept through discovery. In a guided inquiry the instructor provides the question, but students have to develop their procedure for investigating that question and derive their own conclusions. In an open inquiry, students develop their own questions, as well as the methods and conclusions.

When the learning objectives for an activity target a specific biological concept, the activity will likely be most effective at the structured inquiry level. In these cases, the instructor will need to have identified a dataset that they know will illustrate the target concept. Guided inquiry, in which the instructor provides the question, but students have to develop their procedure for investigating that question, may be appropriate for broader ecological questions. For example, if students are already familiar with the data portals, an instructor might pose a question like, “What is the relationship between species diversity and latitude?” Students could determine what data they could access to address this question and design their own data mining "experiment" to determine the answer and discover this ecological relationship.

7.3.2 Data Mining Activities Targeting Specific Biological Research Skills

Perhaps the approach where data mining of digitized natural history collections has the greatest potential is an open inquiry design, where students can develop their own questions. The approach allows students to participate authentically in the full scientific process, learning and practicing the skills they will need to be a biologist. *Vision and Change* identified "Core Competencies" which they define as "sets of skills linked to disciplinary practices" or the skills needed to be a biologist (AAAS, 2011). These competencies have been unpacked in the BioSkills Guide, a tool for interpreting the *Vision and Change* core competencies (Clemmons et al., 2020). The 4DEE identifies "Ecology Practices", which are the approaches and methods used in doing ecology (Prevost et al., 2019). Engaging in open inquiry experiences working with biodiversity data provides opportunities for students to learn and apply these processes of science (Table 7.2) while also learning additional data science skills. Using biodiversity data, students can practice tackling the complex problems that face society by mining databases and bridging multiple data types to explore challenging questions using an interdisciplinary approach (Hiong & Osman, 2013). Collections data can address this need as they have inherent interdisciplinary ties. The geography, geology, chemistry, and climate of habitats where species occur have crucial impacts on species' survival and resulting distribution.

7.3.3 Course-Based Undergraduate Research Experiences

To strengthen training in research skills, many undergraduate biology programs strive to give students more authentic research experiences. These experiences allow students to more closely model the skills of biology researchers and also increase students' science self-efficacy and bolster their identities as scientists (Olivares-Donoso & Gonzalez, 2019; Robnett et al., 2016). Undergraduate research experiences have also been shown to improve retention and enhance STEM-related career aspirations of students from underrepresented groups (Carpi et al., 2017; Chang et al., 2014).

Because logistics, time, and finite resources limit the ability for every undergraduate student to have a closely mentored individual experience in a scientists' research program, many innovative educators have developed course-based undergraduate research experiences (CURE) to provide an authentic research

Table 7.2 Example alignments of 4DEE Ecology Practices and BioSkills with data mining activities using biodiversity data skills

4DEE Practice	BioSkills	Data mining skills
Natural History	Scientific Thinking: Explain how science generates knowledge of the natural world.	Incorporate biodiversity data discovery into the scientific method Curate biodiversity data for reproducibility, preservation, and open access. Critically evaluate information derived from biodiversity datasets.
Quantitative reasoning and computational thinking	Quantitative and Qualitative Data Analysis: Apply the tools of graphing, statistics, and data science to analyze biological data.	Retrieve data from open sources for biodiversity data in a format that is usable for analysis. Use software tools to process and analyze biodiversity data.
Designing and critiquing investigations	Study Design Data Interpretation and Evaluation: Interpret, evaluate, and draw conclusions from data in order to make evidence-based arguments about the natural world.	Access appropriate data sources for specific types of biodiversity datasets. Critically evaluate information derived from biodiversity datasets.
Working collaboratively	Collaboration	Curate biodiversity data for reproducibility, preservation, and open access. Utilize collaborative workflows that facilitate real time information exchange and connectivity.
Communicating and applying ecology	Communication	Visualize biodiversity data using a variety of visualization tools and formats. Represent biodiversity data in a way that will clarify meaning, retain fidelity to the data source, and accurately represent the larger dataset.
Ethical dimensions	Ethics: Demonstrate the ability to critically analyze ethical issues in the conduct of science.	Practice ethical data gathering and usage

experience. CUREs occur within an undergraduate course, often a high-enrollment, introductory course. In a CURE, students work collaboratively on a research project that generates novel (potentially publishable) data or results. Course-based undergraduate research experiences have been shown to shift students' thinking about what it means to think like a scientist and develop specialized expertise, as well as improve students' ability to analyze and interpret data (Brownell et al., 2015; Olimpo et al., 2016). CUREs also improve student engagement and self-efficacy (Olimpo et al., 2016) and have the potential to make science more inclusive by providing research experiences for students with barriers to participating in traditional research opportunities (Bangera & Brownell, 2014).

The design of CUREs faces significant logistical constraints. In order to make the cost reasonable and the logistics manageable, most CUREs consist of students collaboratively collecting data to answer the same question, provided by the instructor. If each student, or even teams of students, were given the freedom to develop their own question and procedure, the cost of supplies and the logistics of acquiring them could be prohibitive. Providing a common question and procedure to the students keeps the cost as low as possible. Some post-secondary institutions lack the resources to implement even the lower cost designs of CUREs. However, a course-based undergraduate research experience based on data mining has the benefits of being low-cost (essentially free as long as you have access to computers and internet) and allowing students much greater freedom to develop and test their own questions.

A course-based undergraduate research experience based on data mining of digitized natural history collections can allow students to participate in all parts of the process of science and practice all of the competencies for biological experimentation (Pelaez et al., 2017; Chap. 1 in this volume). Students can *Identify* gaps in current knowledge, generate a research *Question* that can be investigated using the types of data found in digitized collections, formulate hypotheses, *Plan* and design data mining "experiments", *Conduct* the investigation by locating and downloading the appropriate data, process and *Analyze* the data, *Conclude* based on the evidence, and *Communicate* their results. Students undertaking a biodiversity data-based CURE can gain critical data skills, explore a diversity of questions, and get an authentic and publishable product.

Despite the fact that digitized natural history collections are significantly smaller than some of the truly "big" data sources, students will still need a good deal of scaffolding to help them gain the knowledge and skills necessary to participate in an open inquiry using these data. Undergraduate biology students, especially at the 100- and 200-level, are not likely to be familiar with the types of data available in digitized natural history collections. Students may also be coming in with limited quantitative skills and perhaps even math anxiety (Wachsmuth et al., 2017). While the stand-alone activities can be structured in such a way that students can follow detailed instructions without some of the higher-level data skills, open inquiry in a course-based undergraduate research experience will require students to have a deeper understanding of the sources, standards, and limitations of the different data types. Since biodiversity collections are so versatile, biodiversity data can be incorporated alongside standard lab content. Students build data acumen and statistical skills that can be applied to an open inquiry experience within a traditional lab experience.

As an example of what this scaffolding should include, consider the following sequence (Table 7.3) that could be implemented as a CURE in the laboratory portion of an undergraduate biology course. This research experience incorporates many of the BioSkills and 4DEE practices, while at the same time bringing in some foundational biodiversity data science knowledge and skills, identified below. This

Table 7.3 Example sequence and aligned data science knowledge and skills in a course-based undergraduate research experience using publicly available biodiversity data sets

Activity	Data science knowledge and skills
Students review a published research project that used existing digitized natural history collections data to answer a novel question about an endangered species. This activity provides context for the value of these data and the types of questions they can address.	Evaluate information derived from biodiversity datasets
Students are introduced to different data portals and the types of data they include. For example, this could include specimen-based and observation-based occurrence data, environmental data, and genetic data portals.	Identify portals, databases, and open data sources for biodiversity data.
Students discuss ethical considerations for using data collected by other scientists and attribution of data to the appropriate sources.	Practice ethical data gathering and usage
Students review the data standards used in aggregating large data sets. Aggregating data from potentially hundreds of sources requires coming to an agreement on the way the data will be described and recorded. These agreements are called data standards.	Use established data standards in biodiversity science
Students download data and learn how to "clean" the data. Data cleaning involves identifying errors or inconsistencies in the data and then correcting these issues.	Clean a dataset retrieved from aggregated biodiversity data sources Identify assumptions, gaps, and potential bias in a dataset
Students practice working with the data in discrete activities designed by the instructional team. These activities can align with content requirements in an introductory course. These activities should highlight the types of inferences that can, and cannot, be made using these data types. Students review appropriate statistical tests and visualizations for various question types.	Visualize biodiversity data using a variety of visualization tools and formats Perform basic statistical tests or large biodiversity datasets
Student teams define their own research questions, identify data needs, appropriate analyses, and visualizations, and review potential data sources.	Determine the nature and extent of biodiversity data needs relative to a specific question or problem
Students conduct research, consult with instructional team, trouble-shoot data problems, and iteratively refine questions and analyses. This portion typically occurs over multiple weeks, both synchronously during labs and asynchronously as students work on individual assignments they were given by the team.	Investigate publicly available biodiversity datasets when addressing a relevant scientific question or problem
Students report out the results of their research. Possible formats include oral presentations, journal-style write-ups, research poster, and infographics.	Represent biodiversity data in a way that will clarify meaning

sequence of activities provides a rich opportunity for students to participate in authentic research based on their own questions. Alternatively, a shorter sequence could be developed using subsets of the activities for a less intensive research experience.

7.4 Resources for Implementation

Despite the potential benefits of integrating data mining into the introductory biology curriculum, there are also challenges to overcome. Perhaps the largest challenge is that many, if not most, biology instructors at all levels were not themselves trained in data science (Barone et al., 2017). Instructors attempting to design their own data mining exercises may become frustrated with the enormity and messiness of the data. To address this challenge, biodiversity scientists and biology educators are working to develop and disseminate modules and training materials for integrating publicly-available biodiversity data into the biology curriculum (e.g., Lacey et al., 2017; Langen et al., 2014). The US National Science Foundation has funded initiatives to support these activities. One such initiative is Biodiversity Literacy in Undergraduate Education (BLUE), a research coordination network in undergraduate biology education (RCN-UBE). The goals of BLUE are to build community consensus on the knowledge and skills needed to work with these biodiversity data and bring together biodiversity researchers, data scientists, and biology educators to develop exemplar materials and facilitate the implementation of these materials and the mentored development of new materials by an extended network of biology instructors and mentors (Ellwood et al., 2020). BLUE works with the data aggregators (e.g., iDigBio, NEON) to develop educational materials using biodiversity data and run professional development workshops to train instructors in their use. These workshops allow for tailored coaching, problem solving, and follow-up to support implementation.

BLUE has identified a set of biodiversity literacy competencies, developed an extensive set of teaching modules to address those learning objectives, and developed and piloted an assessment instrument for evaluating student learning outcomes and module effectiveness. The curricular materials have been field tested at a variety of institution types and the modules include instructor materials to facilitate implementation. This set of resources includes stand-alone resources that align with specific ecological and evolutionary concepts (e.g., Linton et al., 2018) as well as a sequence of modules designed for use as a CURE, following the example outlined in Table 7.3.

BLUE educational materials are all published as Open Education Resources through Qubeshub.org and are also available at Biodiversityliteracy.com. To facilitate implementation, BLUE has accessory materials available to educators with answer keys, downloaded data sets, PowerPoints to introduce activities, and lab management timelines and best practices. BLUE regularly works with partner societies and BLUE collaborators to conduct webinars, run workshops, and facilitate faculty mentoring networks to aid in the adoption of BLUE modules.

7.5 Example Module

We have implemented both stand-alone activities and a version of this CURE in a 200-level Form and Function course at a large, midwestern, public institution every semester beginning with Spring 2018. In this course, students work in teams of four students on a research project that they design. Prior to the assignment of the team research project, students have completed several modules that introduce the types of natural history collections data, the data portals, data ethics, data standards, and data cleaning. Students have also used stand-alone modules to investigate several specific form and function concepts (e.g., water balance and temperature as potential limiting factors in determining species distribution). An example of such a module is included below.

This module contains sufficiently detailed instructions that it could be implemented as a stand-alone module or even as students' first introduction to digitized collections data. However, we recommend using it as part of a sequence, as described above. If students are already familiar with these types of data and the logistics of accessing and analyzing them, they will be in a better position to focus on the concepts being investigated, while at the same time practicing their data science skills.

This module, and others like it, are available through the BLUE website (www.biodiversityliteracy.com) or through the QUBESHub (https://qubeshub.org).

Investigating Changes in Species' Ranges Over Time Using Digitized Natural History Collections

Objectives

- Students completing this module will be able to:
- Predict how global change might influence species' distributions.
- *Analyze* changes in the distribution of species occurrence records over time.
- Use quantitative reasoning to collect, clean, and analyze data from a large, curated, aggregated dataset.

Introduction

- With global average temperatures rising since the 1940s, populations of species may no longer be able to successfully occupy the same geographic range as they once had. There is strong evidence that rising temperatures has already led to significant shifts in species' distributions. Additional human disturbances and invasive species have likewise altered the available habitat for species. In this module, you will review published research data describing this evidence and the predicted consequences of biodiversity redistribution. You will then investigate the distribution over time for a species of your choosing using data available from digitized Natural History Collections.

Activity 1: Review Primary Literature

1. Read the article provided (Pecl et al., 2017). This is a metanalysis co-authored by 41 biodiversity scientists.

2. Based on the metanalysis performed, how far, on average, have terrestrial taxa moved poleward per decade? How far for marine taxa?
3. How have the ranges of terrestrial organisms living on mountainsides and fish in the oceans changed as a result of warming?
4. How can changes in species' distributions lead to changes in biotic interactions?
5. Give an example of how biodiversity redistribution can impact human well-being.

Activity 2: Research a Species

1. Identify a species you are interested in investigating. Each member of your team should investigate a different species, but work together and compare your progress as you complete each step of the procedure. What species will you be investigating?
2. For the species you select, do some web research about its habitat and life history. Briefly summarize this information below.
3. Make a prediction. Do you think that the range of this species has shifted over the last century? What features of your species' biology led you to make this prediction.

Activity 3: Generate Distribution Maps

- You will be using the GBIF biodiversity data portal. The Global Biodiversity Information Facility includes both specimen-based and observation-based natural history collection data. We will be using GBIF records to estimate species ranges and range changes over time.

Please note that the absence of a species record from a specific place at a specific time cannot be considered proof that it wasn't present in that location in that year, so this is only an estimate of species ranges that we can use to generate hypotheses. However, since we cannot go back in time and carry out a systematic sampling protocol, for some species these are the only data available to make these range change estimates.

Also note that while we are looking for possible range shifts correlated with increased global temperatures, there are other factors that might lead to changes in species' ranges over time. What might some of these factors be?

Procedure 1

1. Navigate to the GBIF website @ https://www.gbif.org
2. Create a user account if you don't already have one.
3. Click on "Species" above the search box.
4. Enter the name of your species in the search box. Watch out for autocorrect!
5. Search results will appear to the right. Click on the one that best matches what you are looking for.
6. Download. Directly below the map will be a slider for time of collection ranging anywhere between 1600 and the current date.
7. Take screen shots of the map in 20-year increments beginning with 1900–1920 and ending with 2000–current year. Note: Some organisms do not have data on

specific years, use the closest number to the target data (e.g., 1941 instead of 1940 or 1999 instead of 2000).

8. Paste your six maps into the appropriate boxes below.
9. Do the maps provide evidence of a potential shift in the range of your species? If so, describe the shift for each data range.

   - 1900–1920
   - 1920–1940
   - 1940–1960
   - 1960–1980
   - 1980–2000
   - 2000–2020

Procedure 2

1. Return to your GBIF search results and click on the "Explore" button below the map.
2. In the menu section to the left, open the "year" tab and set the range to 1900–present.
3. Select "Download" from the menu above the specimen records, then "CSV" to download your data set.
4. Open your dataset in Excel. You will need to import the CSV file to split the data into separate columns.
5. Open a new spreadsheet tab and paste a copy of your data in the new tab. Always keep a copy of the raw data to document your procedure. Clean your data by removing unnecessary columns. The only essential columns are "decimalLatitude" and "year", but consider keeping at least "decimalLongitude" and "basisOfRecord" to provide some context for your data and always keep the unique record identifier column, in this case "gbifID".
6. Sort by "Year" and remove any rows that do not have the year recorded.
7. Sort by "Latitude" and remove any rows that do not have the latitude recorded.
8. Remove any obvious outliers.
9. Create a scatterplot of latitude vs year (with axes labeled). Copy and paste your graph into the space below.

Activity 4: Interpret Your Data

1. What might be some predicted impacts on ecosystems (including humans) if your species did have a change in range over the last century?
2. Is there evidence of a latitudinal change in range for your species? Summarize the data.
3. What other factors (besides temperature) might influence the range of your species

7.6 Summary

The integration of data mining of digitized natural history collections into undergraduate biology courses provides several major advantages. First, lessons designed following this approach support the use of inquiry-based instruction, which has been shown to improve student learning over traditional lecture-based teaching, as discussed above. The strategies and resources presented here can allow students to learn biological concepts and traditional experimentation skills through inquiry-based discovery and participation in authentic research experiences. Second, but just as important, the approach we have outlined introduces students to data science, a powerful and essential set of techniques for the future of biological research. Most undergraduate biology programs do not currently include a strong foundation in data science, yet we have an obligation to train students in this emerging field. To not do so would limit students career options and leave the discipline lacking the next generation of scientists trained to harness these powerful data methodologies to answer large-scale biological questions of critical importance. Finally, digitized natural history collections data have the advantage of providing a complex, but not overwhelming, dataset that can serve as an entry point for students into the field of data science.

To implement this approach, we recommend the following for instructors:

- Identify core biological concepts that can be addressed in activities using biodiversity data science.
- Identify data sets and analyses that align with these core concepts.
- Design inquiry-based lessons where students learn and practice biodiversity data skills while constructing understanding of the targeted core concepts.
- Review the resources described in this chapter as sources of exemplar lessons already developed using this approach.

Acknowledgments The BLUE (Biodiversity Literacy in Undergraduate Education) RCN-UBE was funded by a U.S. National Science Foundation (NSF) Grant DBI-1730526. Any opinions, findings, and conclusions or recommendations expressed in this article are those of the authors and do not necessarily reflect the views of the NSF.

References

American Association for the Advancement of Science (AAAS). (2011). *Vision and change in undergraduate biology education: A call to action* (C. Brewer & D. Smith (Eds.)). https://visionandchange.org/finalreport/. Accessed 18 Oct 2021.

Baker, B. (2017). Big data opens promising career paths for biologists. *BioScience, 67*, 100–122.

Bakker, F. T., Antonelli, A., Clarke, J. A., Cook, J. A., Edwards, S. V., Ericson, P. G. P., Faurby, S., Ferrand, N., Gelang, M., Gillespie, R. G., Irestedt, M., Lundin, K., Larsson, E., Matos-Maraví, P., Müller, J., von Proschwitz, T., Roderick, G. K., Schliep, A., Wahlberg, N., … Källersjö, M. (2020). The Global Museum: Natural history collections and the future of evolutionary science and public education. *PeerJ, 8*, e8225. https://doi.org/10.7717/peerj.8225

Bangera, G., & Brownell, S. E. (2014). Course-based undergraduate research experiences can make scientific research more inclusive. *CBE Life Sciences Education, 13*, 602–606.

Barone, L., Williams, J., & Micklos, D. (2017). Unmet needs for analyzing biological big data: A survey of 704 NSF principal investigators. *PLoS Computational Biology, 13*, art. e1005755.

Beardsley, T. M., Gropp, R. E., & Verdier, J. M. (2018). Addressing biological informatics workforce needs: A report from the AIBS Council. *BioScience, 68*, 847–853.

Bell, R. L., Smetana, L., & Binns, I. (2005). Simplifying inquiry instruction. *The Science Teacher, 72*(7), 30–33.

Bianchi, H., & Bell, R. (2008). The many levels of inquiry. *Science and Children, 46*(2), 26–29.

Brownell, S. E., Freeman, S., Wenderoth, M. P., & Crowe, A. J. (2014). BioCore Guide: A tool for interpreting the core concepts of vision and change for biology majors. *CBE Life Sciences Education, 13*, 200–211.

Brownell, S. E., Hekmat-Scafe, D. S., Singla, V., Chandler Seawell, P., Conklin Imam, J. F., Eddy, S. L., Stearns, T., & Cyert, M. S. (2015). A high-enrollment course-based undergraduate research experience improves student conceptions of scientific thinking and ability to interpret data. *CBE Life Sciences Education, 14*, 1–14.

Carpi, A., Ronan, D., Falconer, H., & Lents, N. (2017). Cultivating minority students: Undergraduate research increases self-efficacy and career ambitions for underrepresented students in STEM. *Journal of Research in Science Teaching, 54*(2), 169–194. https://doi.org/10.1002/tea.21341

Chang, M., Sharkness, J., Hurtado, S., & Newman, C. (2014). What matters in college for retaining scientists and engineers from underrepresented racial groups. *Journal of Research in Science Teaching, 51*(5), 555–580. https://doi.org/10.1002/tea.21146

Clemmons, A., Timbrook, J., Herron, J., & Crowe, A. (2020). BioSkills guide: Development and national validation of a tool for interpreting the *vision and change* core competencies. *CBE Life Sciences Education, 19*, 4. https://doi.org/10.1187/cbe.19-11-0259

Cook, J. A., Edwards, S. V., Lacey, E. A., Guralnick, R. P., Soltis, P. S., Soltis, D. E., Welch, C. K., Bell, K. C., Galbreath, K. E., Himes, C., Allen, J. M., Heath, T. A., Carnaval, A. C., Cooper, K. L., Liu, M., Hanken, J., & Ickert-Bond, S. (2014). Natural history collections as emerging resources for innovative education. *BioScience, 64*, 725–734.

Ellwood, E. R., Sessa, J. A., Abraham, J. K., Budden, A. E., Douglas, N., Guralnick, R., Krimmel, E., Langen, T., Linton, D., Phillips, M., Soltis, P. S., Studer, M., White, L. D., Williams, J., & Monfils, A. K. (2020). Biodiversity science and the twenty-first century workforce. *BioScience, 70*, 119–121.

Feser, J., Vasaly, H., & Herrera, J. (2013). On the edge of mathematics and biology integration, improving quantitative skills in undergraduate biology education. *CBE Life Sciences Education, 12*, 124–128.

Gibson, J. P., & Mourad, T. (2018). The growing importance of data literacy in life science education. *American Journal of Botany, 105*, 1953–1956.

Hampton, S. E., Jones, M. B., Wasser, L. A., Schildhauer, M. P., Supp, S. R., Brun, J., Hernandez, R. R., Boettiger, C., Collins, S. L., Gross, L. J., Fernandez, D. S., Budden, A., White, E. P., Teal, T. K., Labou, S. G., & Aukema, J. E. (2017). Skills and knowledge for data-intensive environmental research. *BioScience, 67*, 546–557.

Hedrick, B. P., Heberling, J. M., Meineke, E. K., Turner, K. G., Grass, C. J., Park, D. S., Kennedy, J., Clarke, J. A., Cook, J. A., Blackburn, D. C., Edwards, S. V., & Davis, C. C. (2020). Digitization and the future of natural history collections. *BioScience, 70*, 243–251.

Hernandez, R. R., Mayernik, M. S., Murphy-Mariscal, M. L., & Allen, M. F. (2012). Advanced technologies and data management practices in environmental science: Lessons from academia. *BioScience, 62*, 1067–1076.

Hiong, L. C., & Osman, K. (2013). A conceptual framework for the integration of 21st century skills in biology education. *Research Journal of Applied Sciences, Engineering and Technology, 6*, 2976–2983.

Holmes, M. W., Hammond, T. T., Wogan, G. O. U., Walsh, R. E., LaBarbera, K., Wommack, E. A., Martins, F. M., Crawford, J. C., Mack, K. L., Bloch, L. M., & Nachman, M. W. (2016). Natural history collections as windows on evolutionary processes. *Molecular Ecology, 25*, 864–881. https://doi.org/10.1111/mec.13529

Honts, J. E. (2003). Evolving strategies for the incorporation of bioinformatics within the undergraduate cell biology curriculum. *Cell Biology Education, 2*, 233–247.

Kjelvik, M. K., & Schultheis, E. H. (2019). Getting messy with authentic data: Exploring the potential of using data from scientific research to support student data literacy. *CBE Life Sciences Education, 18*(es2), 1–18.

Lacey, E. A., Hammond, T. T., Walsh, R. E., Bell, K. C., Edwards, S. V., Ellwood, E. R., Guralnick, R., Ickert-Bond, S. M., Mast, A. R., McCormack, J. E., Monfils, A. K., Soltis, P. S., Soltis, D. E., & Cook, J. A. (2017). Climate change, collections and the classroom: Using big data to tackle big problems. *Evolution: Education and Outreach, 10*, 2. https://doi.org/10.1186/s12052-017-0065-3

Langen, T. A., Mourad, T., Grant, B. W., Gram, W. K., Abraham, B. J., Fernandez, D. S., Carroll, M., Nuding, A., Balch, J. K., Rodriguez, J., & Hampton, S. E. (2014). Using large public datasets in the undergraduate ecology classroom. *Frontiers in Ecology and the Environment, 12*, 362–363.

Linton, D. L., Monfils, A. K., Phillips, M., & Ellwood, E. (2018). The effect of climate change on butterfly phenology. *Teaching Issues and Experiments in Ecology, 13*(7). https://doi.org/10.25334/Q46Q5B

Madlung, A. (2018). Assessing an effective undergraduate module teaching applied bioinformatics to biology students. *PLoS Computational Biology, 14*, e1005872.

Makarevitch, I., Frechette, C., & Wiatros, N. (2015). Authentic research experience and "big data" analysis in the classroom: Maize response to abiotic stress. *CBE Life Sciences Education, 14*, 1–12.

McLean, B. S., Bell, K. C., Dunnum, J. L., Abrahamson, B., Colella, J. P., Deardorff, E. R., Weber, J. A., Jones, A. K., Salazar-Miralles, F., & Cook, J. A. (2015). Natural history collections-based research: Progress, promise, and best practices. *Journal of Mammalogy, 97*, 287–297.

Meineke, E. K., Davies, T. J., Daru, B. H., & Davis, C. C. (2018). Biological collections for understanding biodiversity in the Anthropocene. *Philosophical Transactions of the Royal Society B: Biological Sciences, 374*. https://doi.org/10.1098/rstb.2017.0386

Michener, W. K., & Jones, M. B. (2012). Supporting ecology as a data-intensive science. *Trends in Ecology & Evolution, 27*, 85–93.

Monfils, A. K., Powers, K. E., Marshall, C. J., Martine, C. T., Smith, J. F., & Prather, L. A. (2017). Natural history collections: Teaching about biodiversity across time, space, and digital platforms. *Southeastern Naturalist, 16*, 47–57.

National Academies of Sciences, Engineering, and Medicine. (2018). *Envisioning the data science discipline: The undergraduate perspective* (Interim report). The National Academies Press.

National Academies of Sciences, Engineering, and Medicine. (2020). *Biological collections: Ensuring critical research and education for the 21st century*. The National Academies Press. https://doi.org/10.17226/25592

Nelson, G., & Ellis, S. (2018). The history and impact of digitization and digital data mobilization on biodiversity research. *Philosophical Transactions of the Royal Society B, 374*, 20170391. https://doi.org/10.1098/rstb.2017.0391

Nybo, L., & May, M. (2015). Effectiveness of inquiry-based learning in an undergraduate exercise physiology course. *Advances in Physiology Education, 39*, 76–80.

Olimpo, J. T., Fisher, G. R., & DeChenne-Peters, S. E. (2016). Development and evaluation of the *Tigriopus* course-based undergraduate research experience: Impacts on students' content knowledge, attitudes, and motivation in a majors introductory biology course. *CBE Life Sciences Education, 15*, 1–15.

Olivares-Donoso, R., & González, C. (2019). Undergraduate research or research-based courses: Which is most beneficial for science students? *Research in Science Education, 49*(1), 91–107. https://doi.org/10.1007/s11165-017-9616-4

Pecl, G. T., Araújo, M. B., Bell, J. D., Blanchard, J., Bonebrake, T. C., Chen, I. C., Clark, T. D., Colwell, R. K., Danielsen, F., Evengård, B., Falconi, L., Ferrier, S., Frusher, S., Garcia, R. A., Griffis, R. B., Hobday, A. J., Janion-Scheepers, C., Jarzyna, M. A., Jennings, S., … Williams, S. E. (2017). Biodiversity redistribution under climate change: Impacts on ecosystems and human well-being. *Science, 355*(6332), eaai9214. https://doi.org/10.1126/science.aai9214

Pedaste, M., Mäeots, M., Siiman, L. A., de Jong, T., van Riesen, S. A. N., Kamp, E. T., Manoli, C. C., Zacharia, Z. C., & Tsourlidaki, E. (2015). Phases of inquiry-based learning: Definitions and the inquiry cycle. *Educational Research Review, 14*, 47–61.

Pelaez, N. J., Anderson, T. R., Gardner, S. M., Yin, Y., Abraham, J. K., Bartlett, E., Gormally, C., Hill, J. P., Hoover, M., Hurney, C., Long, T., Newman, D. L., Sirum, K., & Stevens, M. (2017). The basic competencies of biological experimentation: Concept-skill statements. *PIBERG Instructional Innovation Material Paper 4.* Retrieved from http://docs.lib.purdue.edu/pibergiim/4. Accessed 18 Oct 2021.

Powers, K. E., Prather, A., Cook, J. A., Woolley, J., Bart, H. L., Jr., Monfils, A. K., & Sierwald, P. (2014). Revolutionizing the use of natural history collections in education. *Science Education Review, 13*, 24–33.

Prevost, L., Sorensen, A. E., Doherty, J. H., Ebert-May, D., & Pohlad, B. (2019). 4DEE – What's next? Designing instruction and assessing student learning. *Bulletin of the Ecological Society of America, 100*, 1–6.

Prince, M. J., Vigeant, M. A. S., & Nottis, K. (2009). A preliminary study on the effectiveness of inquiry-based activities for addressing misconceptions of undergraduate engineering students. *Education for Chemical Engineers, 4*, 29–41.

Pyke, G. H., & Ehrlich, P. R. (2010). Biological collections and ecological/environmental research: A review, some observations and a look to the future. *Biological Reviews, 85*, 247–266.

Robnett, R., Chemers, M., & Zurbriggen, E. (2016). Longitudinal associations among undergraduates' research experience, self-efficacy, and identity. *Journal of Research in Science Teaching, 52*(6), 847–867. https://doi.org/10.1002/tea.21221

Ruiz-Primo, M. A., Briggs, D., Iverson, H., Talbot, R., & Shepard, L. A. (2011). Impact of undergraduate science course innovations on learning. *Science, 331*, 1269–1270.

Soltis, D. E., & Soltis, P. S. (2016). Mobilizing and integrating big data in studies of spatial and phylogenetic patterns of biodiversity. *Plant Diversity, 38*, 264–270.

Strasser, C. A., & Hampton, S. E. (2012). The fractured lab notebook: Undergraduates and ecological data management training in the United States. *Ecosphere, 3*, 1–18. Art116.

Suarez, A. V., & Tsutsui, N. D. (2004). The value of museum collections for research and society. *BioScience, 54*, 66.

Sundberg, M. D., Armstrong, J. E., & Wischusen, E. W. (2005). A reappraisal of the status of introductory biology laboratory education in U.S. colleges and universities. *American Biology Teacher, 67*, 525–529.

Tewksbury, J. J., Anderson, J. G. T., Bakker, J. D., Billo, T. J., Dunwiddie, P. W., Groom, M. J., Hampton, S. E., Herman, S. G., Levey, D. J., Machnicki, N. J., Martinez del Rio, C., Power, M. E., Rowell, K., Salomon, A. K., Stacey, L., Trombulak, S. C., & Wheeler, T. A. (2014). Natural history's place in science and society. *BioScience, 64*, 300–311.

Thanukos, A., Scotchmoor, J. G., Caldwell, R., & Lindberg, D. R. (2010). The real process of science-understanding science at http://www.understandingscience.org and Science 101: Building the foundations for real understanding. *Science, 330*(6012), 1764–1765. https://doi.org/10.1126/science.1186994

Wachsmuth, L. P., Runyon, C. R., Drake, J. M., & Dolan, E. L. (2017). Do biology students really hate math? Empirical insights into undergraduate life science majors' emotions about mathematics. *CBE Life Sciences Education, 16*(ar49), 1–10.

Wilson Sayres, M. A., Hauser, C., Sierk, M., Robic, S., Rosenwald, A. G., Smith, T. M., Triplett, E. W., Williams, J. J., Dinsdale, E., Morgan, W. R., Burnette, J. M., III, Donovan, S. S., Drew, J. C., Elgin, S. C. R., Fowlks, E. R., Galindo-Gonzalez, S., Goodman, A. L., Grandgenett, N. F., Goller, C. C., … Pauley, M. A. (2018). Bioinformatics core competencies for undergraduate life sciences education. *PLoS ONE, 13*(6), e0196878.

Chapter 8
A Framework for Teaching and Learning Graphing in Undergraduate Biology

Stephanie M. Gardner, Aakanksha Angra, and Joseph A. Harsh

8.1 Introduction

Anyone who has spent even a few moments considering how to use a graph to *Communicate* information or to make a decision will notice that success in doing so is influenced by multiple, interacting factors. Graphing is a learned practice, which involves the application of a collection of concepts and more specific practices to create and decode visual representations of data to make and discern patterns, ask questions, and describe phenomena (Bowen et al., 1999; Roth & Bowen, 2001; Hegarty, 2011; Padilla et al., 2018). Reflecting the ubiquity and importance of graphs across disciplines and our increasingly data-driven society (e.g., NRC, 2003; NSF, 2016), emphasis in K-16 science education over the last three decades has been placed on developing student competence with constructing and analyzing these common data displays for functional literacy (Padilla et al., 1986; George et al., 1996; AAMC-HHMI Committee, 2009; AAAS, 2011; NRC, 2015). Because of the challenges in graphing that learners of all ages, and even experts continue to demonstrate in STEM and medical fields (Roth & Bowen, 2001; Schriger & Cooper, 2001; Bowen & Roth, 2005; Schriger et al., 2006; Roth, 2013; Rougier et al., 2014;

S. M. Gardner (✉)
Department of Biological Sciences, Purdue University, West Lafayette, IN, USA
e-mail: sgardne@purdue.edu

A. Angra
Department of Biology, Georgia State University, Atlanta, GA, USA

Office of Medical Education, University of Minnesota, Minneapolis, MN, USA
e-mail: angra002@umn.edu

J. A. Harsh
Department of Biology, James Madison University, Harrisonburg, VA, USA
e-mail: harshja@jmu.edu

N. J. Pelaez et al. (eds.), *Trends in Teaching Experimentation in the Life Sciences*, Contributions from Biology Education Research,
https://doi.org/10.1007/978-3-030-98592-9_8

Weissgerber et al., 2015, 2019), it is important to consider how instructional enhancements to develop graphing competence can be guided by our understanding of the underlying processes involved in the reading, interpretation, and transformation of graph data.

While the activities of graph construction, reading, and interpretation involve diverse cognitive processes, they are, however, inextricably linked (e.g., Lehrer & Schauble, 2000; Shah & Freedman, 2011). The purpose of a graph display is to communicate data to a viewer, whether to the graph creator themself or others, in order to make meaning of data and draw inferences. In addition, in reading and interpreting a graph, one needs to deconstruct its components and evaluate them, imagining the natural measurements and observations, which the displayed data represent (Bowen et al., 1999). Therefore, while graphing is a common practice across disciplines and often taught as a generalizable skill, domain-specific knowledge is essential to effective design and interpretation of graph data in a field of study (Roth & Bowen, 2001; Shah & Freedman, 2011).

In this chapter we argue that teaching graphing should be done in the context of engaging in scientific practices instead of using isolated, decontextualized activities and instruction. We present a literature-supported practical framework for teaching graphing in undergraduate biology and use evidence from two case studies to illustrate its implementation to provide instructors with potential models to enact to improve student competence. The framework represents the synthesis of our own work seeking to understand (Maltese et al., 2015; Angra & Gardner, 2017; Harsh et al., 2019) and advance (Angra & Gardner, 2016; Harsh & Schmitt-Harsh, 2016) the development of graphing skills as well as draws from relevant research in the learning sciences, statistics education, biology education, and cognitive sciences. This framework is situated in the specific context of biology which speaks to the integration of concepts and practices from statistics and perceptual cognition tied to the inquiry practices and embodied understanding of the field during graphing. We do, however, acknowledge that even within biology there is diversity in the concepts, norms, inquiry methods, and data across subdisciplines in this broad field that can affect graph understanding and use (Bowen et al., 1999; Roth & Bowen, 2001). The first study will highlight the impact of a semester-long teaching intervention in an upper division biology course on graph construction and the second study will highlight the impact of a shorter-term instructional unit on graph construction and graph reading in an introductory non-majors course.

8.2 Essential Features of Instructional Design and Their Application to Teaching Graphing

Regardless of the concepts and practices that are the target of instruction, there are essential features of instructional design, which promote student learning. Therefore, before we present our framework for teaching for graph competence in the college biology classroom, we will highlight these design features, with some examples from the context of graphing. We advocate for a backward design approach to instructional design by initially setting *learning objectives* (specific, measurable competencies) prior to considering *assessments that demonstrate evidence of student understanding and learning*, and then subsequently develop teaching activities to support students in their learning (Wiggins & McTighe, 1998; Allen & Tanner, 2007). In this way, there is alignment between what students do (the activities), the evidence of their learning (performance on assessments), and the target for learning (the learning objectives) (NRC, 2001; Martone & Sireci, 2009).

8.2.1 Learning Objectives

While individual instructors should articulate their own graphing-oriented learning objectives specific to their student population and instructional context (e.g., lab or lecture setting, subdiscipline of biology, etc.), there are existing resources available to use or adapt from curricular frameworks and reform documents in secondary (NGSS, 2013; College Board, 2019) and undergraduate science education (AAAS, 2011). The landmark consensus document for undergraduate life science education, *Vision and Change* (AAAS, 2011) includes a broad competency and a suggested demonstration of competence related to graphing. The core competency of 'ability to use quantitative reasoning' is demonstrated by applying quantitative analysis to interpret biological data, for example, by developing and interpreting graphs. While the inclusion of graphing in the recommendations for undergraduate life science education is important, graphing is a complex practice comprising multidisciplinary concepts and practices. A finer grain size of the *Vision and Change* skills was developed by Clemmons et al. (2020). They articulated learning objectives at both the program and course level, with two areas being relevant for graphing: Process of Science and Quantitative Reasoning. Finally, the ACE-Bio Competencies for Experimentation in the Life Sciences (Pelaez et al., 2017; Chap. 1 in this volume) provides the finest grain size statements of concept-skill pairs which can be crafted into learning objectives. For graphs, the competency areas of *Plan* (the ability to plan feasible and ethical experiments to answer research questions or test hypotheses), *Analyze* (the ability to analyze and process data), and *Conclude* (the ability to draw conclusions about data with inference that are limited to the scope inherent in the experimental design) are relevant. While many of the concept-skill pairs in the ACE-Bio Competencies resemble practices and objectives from the other sources summarized here, they are framed specifically within the context of biological experimentation based on what a competent biologist would know and know how to do (Table 8.1).

Table 8.1 Select ACE-Bio Concept-Skill pairs in areas related to graphing (Pelaez et al., 2017)

Competency area	Concept	Skill
<u>Plan</u>	Representations	Construct a visual representation of predicted results
<u>Analyze</u>	Data curation Data analysis Statistics Data summary	Construct appropriate ways to organize data (e.g. tables, figures) Explore and reduce raw data to discern trend and summarize relationships among variables Identify outliers and/or errant data by generating criteria for inclusion or rejection of data Display appropriate comparisons (i.e. detect natural groupings) Conduct transformations that facilitate statistical or other analytic tests Conduct computations for summariz case study ing/interpreting findings <u>Analyze</u> clean data using discipline-appropriate methods based on the measurements collected and the experimental questions. Choose and conduct statistical tests that are appropriate for the type/nature of data Choose and conduct statistical tests that are aligned with hypotheses and experimental research methods Generate statistics for a sample to summarize and/or describe parameters for a whole population (e.g., mean, median, measures of variance). Appropriately identify a legend, label axes, and select appropriate scale to graph findings Considering the variables intended for comparisons, select an appropriate graphical type for the particular data type (e.g. contingency tables, bar graphs, histograms, scatterplots, etc.) Display findings with a representation that is effective in summarizing trends or major findings, including illustrating contrasts among categorical groups where relevant
<u>Conclude</u>	Patterns and relationships Inferences and conclusions	Describe trends in numeric and visual representations of data Interpret whether the results suggest a causal mechanism beyond simple correlation Distinguish biologically-meaningful trends from expected natural biological variability Generalize results to an appropriate level (more than single experiment, less than universal) Connect analysis of results with valid claims or conclusion in a logical way Evaluate limitations of the findings and limitations that determine scope of inference (experimental and practical limitations) Compare results to other previously reported results and reconcile differences Align conclusion with analyses, hypotheses, research question(s), and existing knowledge Determine and articulate whether data support or refute hypotheses and predictions Express uncertainty by discussing limitations of data analysis (sources of error, inaccurate measurement, and sample bias, statistical significance vs. biological relevance)

8.2.2 Assessment to Reveal Student Knowledge and Competence

Assessments are tools that instructors can use to monitor student knowledge and learning to adjust their teaching in-process and give students feedback on their progress (formative assessments) or evaluate student learning, competence, or knowledge as compared against a benchmark at the conclusion of an instructional unit (summative assessments). Regardless of purpose (summative versus formative), assessment activities that provide coherent insight to student reasoning are guided by three key design characteristics, including: the definition of competence or mastery of particular knowledge and skills in the domain (*cognition*), appropriate tasks or performances that will yield evidence to student proficiency (*observation*), and the drawing of inferences (*interpretation*) on the basis of the evidence collected from the observations (NRC, 2001; Mislevy & Haertel, 2006). Represented as the "assessment triangle" (NRC, 2001), it is essential for effective assessment that each of the three design aspects – or vertices – are well considered individually, and connect to each other in a meaningful way. In classroom assessment, it is also of critical importance to consider the *model of cognition*, which places focus on howstudents progress towards higher proficiency levels as well as the evaluation of changes in knowledge and competencies over time by using tasks along a continuum of cognitive complexity (Brown & Wilson, 2011). This approach allows instructors to make decisions in teaching activities that can support student learning progressions.

In biology, graphing is a practice that is woven throughout the process of inquiry and experimentation. It is a practice that is often implicitly learned to varying degrees of success prior to (e.g. Padilla et al., 1986; Tairab & Khalaf Al-Naqbi, 2004) and in the college science classroom (e.g. Picone et al., 2007; Bray Speth et al., 2010), but explicitly used to explore patterns, draw conclusions, and make predictions (Table 8.1). Therefore, we advocate for the use of authentic assessments which aim to allow students to demonstrate their existing graphing knowledge and competence (diSessa et al., 1991; Smith et al., 1993; Hammer, 1996; NRC, 2000) in response to carefully designed tasks that replicate real world practices in the discipline (Roth & Roychoudhury, 1993; Wiggins, 1998). In this frame, to aptly assess student strengths and difficulties in graphing as well as to positively impact learning, biology students should be engaged in realistic inquiry practices that permits the opportunity to grapple with the complexities of taking data from its collection through to its representation and interpretation.

Beyond one's prior knowledge and familiarity with the guidelines or conventions about how to "best" display data, the decision-making process in graphing is often guided by tacit understanding or important considerations of the data itself. As an example, it is inappropriate to put a regression line in a graph of quantitative and categorical data; however, it is common to connect two quantitative variables from different categories with a line to facilitate the ability to discern trends (e.g. blood glucose levels with and without a therapeutic drug treatment). At the same time, data representations should make the same meaning to viewers (the designer and others), but this is complicated as learners often view data through alternative perspectives which subjectively influences how they prefer to engage with, use, or depict data (Konold et al., 2014).

Taking these considerations into account, it can then be argued that taking a single artifact (i.e. a constructed graph) or claim (i.e. an interpretation of a graph) is insufficient to gain meaningful insight into a student's graphing competence as well as hamper teaching practices to address challenges and support students in development of their graph-related skills. As has been noted by others when assessing scientific reasoning (e.g., Berg & Smith, 1994; NRC, 2007), we have also found that asking students to provide statements describing their decision making processes in addition to the graph they constructed (Harsh et al., 2013; Angra, 2016; Angra & Gardner, 2018) or the claim that they made (Harsh et al., 2019) provides a more complete picture of their ability to understand, describe, and explain graph data. For example, in a pilot study which laid the foundation for the teaching intervention described in Case Study 1, we asked students to reflect on the advantages and disadvantages of the graph their team chose to summarize the findings from their experiment that week in an undergraduate inquiry physiology laboratory class (Fig. 8.1). These reflections allowed us to better characterize student understanding of graph design and their reasoning during data analysis and display in a way that would not be possible by simply looking at the final graphs they created. Here, we have also found that one's ability to sufficiently create a graph may not represent their understanding of the data at hand and/or graph theory as students often rely on intuition and prior experience in graph construction (Harsh et al., 2013; Angra & Gardner, 2018).

8.3 Framework for Teaching Graphing in Undergraduate Biology

Teaching graphing in the undergraduate classroom to engage diverse learners while promoting critical thinking and quantitative literacy can be a challenging task for instructors. Graphs are visual representations meant to facilitate exploration of data and communicate meaning to the person creating a graph and for others who view a graph. In Table 8.2 below, we describe a framework of five instructional design features for teaching graphing in biology, along with definitions and examples from our own work and additional examples from fields of learning sciences, statistics education, biology education and cognitive science. We have divided the instructional design features into three areas: activity design, instructor roles and interactions, and student behaviors. The outlined design features can be implemented separately or in combination depending on the instructional goals and context of the learning activity and/or course curriculum. Each of the detailed design features have been reported to contribute to students' graphing skills, with the potential for additive effects when used jointly given the diverse cognitive processes involved in making and using graphs (i.e. "some is good, more is better").

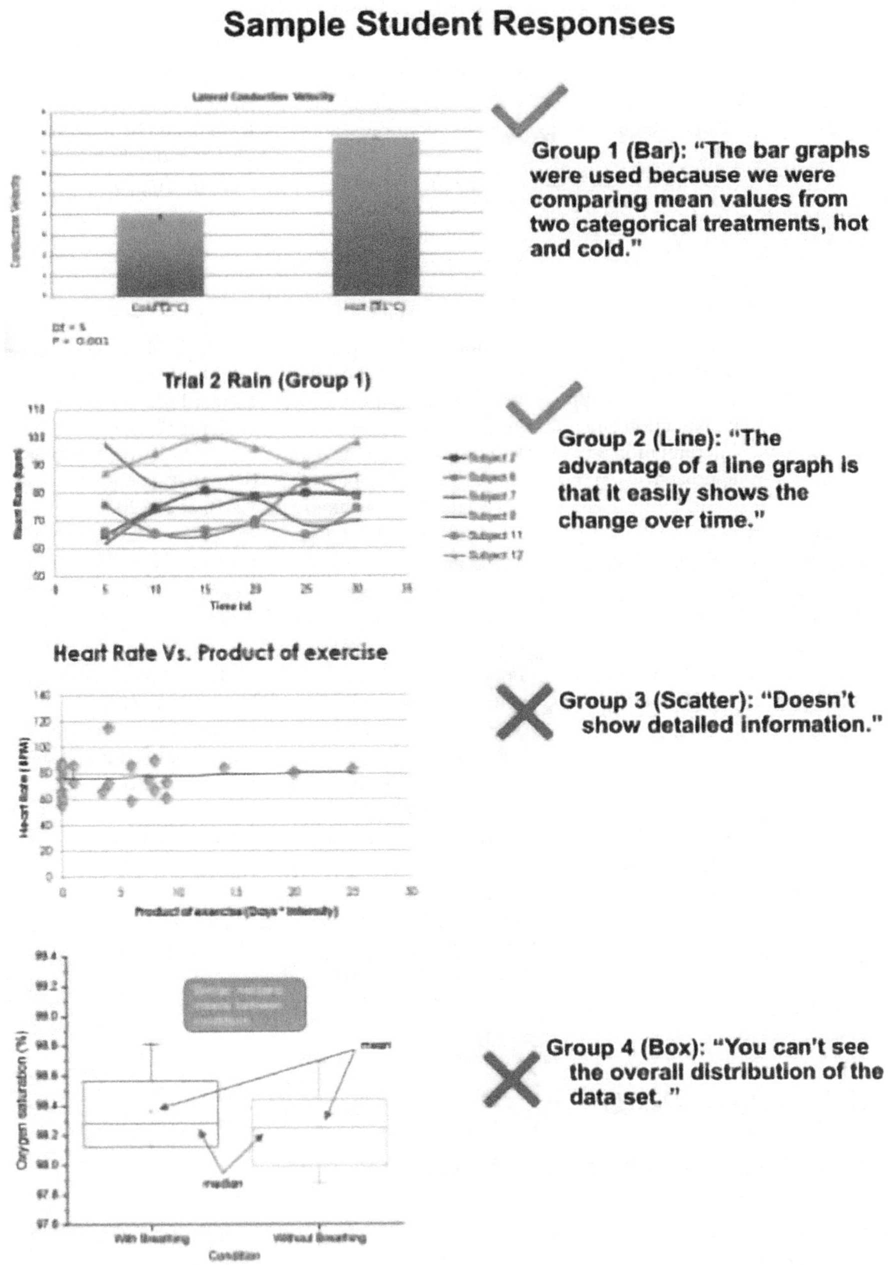

Fig. 8.1 Examples of student graphs with reasoning for advantages and disadvantages: Figure shows students' correct responses to the advantages of their graph (top) and incorrect disadvantages (below). This figure is representative of the types of responses students gave, i.e. students more often and correctly provided the advantages of bar and line graphs but more often provided incorrect disadvantages of other graph types such as scatter plots and box and whisker plots

Table 8.2 Framework for designing instruction for graphing in undergraduate biology

	Instructional Design Features	Definition/example	Supporting References (examples)
Activity design	Engage with real-world, messy data	Gather and evaluate data from inquiry or research investigations Unfiltered, raw data Biological variation Variation due to measurement precision and error, human error, sampling	Kastens et al. (2015), Kjelvik and Schultheis (2019) and Schultheis and Kjelvik (2020)
	Encourage a multi-step data analysis and graph construction and interpretation approach	Reading the data (planning, making sense of the data, work by hand, use graphing software) Reading between the data (making comparisons, observing relationships) Reading beyond the data	Curcio (1989), Kellman (2000), Patterson and Leonard (2005), Franconeri et al. (2012), Angra and Gardner (2016), Harsh and Schmitt-Harsh (2016) and Harsh et al. (2019)
Instructor roles and interactions	Intentional and explicit instruction	Modeling the process of data analysis, graph construction, and decoding and interpreting graphs	Collins et al. (1987) and Dennen (2004)
Student behaviors	Collaborative and social practice	Follow norms of the scientific community for analysis, representation, and interpretation claims embedded in the discipline Science is collaborative	Roth and McGinn (1997), Bowen et al. (1999), Schultheis and Kjelvik (2015) and Schultheis and Kjelvik (2020)
	Engage students in evaluation and reflection	Critique representations affordances and limitations of analysis and representational choices) Evaluate alignment between representation and purpose in the context of inquiry	diSessa (2004) and Angra and Gardner (2016)

8.3.1 Activity Design

8.3.1.1 Engage with Real-Word, Messy Data

Historically, undergraduate biology textbooks represent data in graphs in a schematized manner in which trends and differences are artificially clear and obvious (Hoskins et al., 2007; Rybarczyk, 2011; Angra & Gardner, 2018). This oversimplified model communicates to students what their data and graphs should look like when they create or interpret graphs. Students should have first-hand experience collecting, working with, and making sense of raw, unfiltered data with either biological or human-error variation (Kjelvik & Schultheis, 2019; Schultheis & Kjelvik, 2020). This challenges them to think critically of the relevant variables and the form of data needed to visualize relationships in a graph as well as formulate explanations on trends and outliers for themselves or for others viewing their graphs (Wild & Pfannkuch, 1999; Lehrer & Schauble, 2007; Pelaez et al., 2017). Previous literature has shown a positive association between transforming raw data and interpreting graphs (Picone et al., 2007; Morrison & McDuffie, 2009; Bray Speth et al., 2010). Instructors can also use this as an opportunity to connect raw data to the biological concepts covered in the classroom and also the importance of consistent data collection in the laboratory (e.g. Pelaez et al., 2017). While experiences for data collection can be easily accomplished in laboratory courses, it is also possible for lecturers to utilize case studies (e.g. Data Nuggets, HHMI Biointeractive) or publicly available data sets (e.g. data.gov; usafacts.org) in large lecture classrooms.

8.3.1.2 Encourage a Multi-step Data Construction and Interpretation Approach

Students should be provided step-by-step instruction on graph construction, reading, and interpretation. In relation to graph construction, it is first important to decouple the practices of graph design (e.g., display type selection, variable manipulation) and computer-aided data visualization as two constituent activities for graphing that reflect different cognitive processes (Patterson & Leonard, 2005; Roth & McGinn, 1997). Further, it has been argued that the benefits of computer tools in graph generation (i.e. high degree of accuracy and ease in generating alternative displays for examination) may subsequently "release the learner from thinking about how to identify and plot variables (which) has greatly contributed to such inability to construct graphs" (Tairab & Khalaf Al-Naqbi, 2004, p. 130). As such, a three-step graph construction approach drawing on separate, but linked, cognitive processes is recommended (Tairab & Khalaf Al-Naqbi, 2004; Patterson & Leonard, 2005). The first step is to guide students to explore the data and then formulate a *plan* to "best" represent the data to communicate the findings in a manner that permits comparable meaning making for all viewers and allows arguments to be drawn. As the goals of the graph should align with the question at hand, students should be prompted to make decisions on the type of graph they wish to construct based on the data type, the relationships between the plotted variables they wish to display, and the form that the data will take in their graph (i.e. raw, summarized, transformed). In the next step, students should sketch out or *draw* by hand what they think the graph representation should look like by manipulating and conceptualizing how the data are plotted. Finally, students should use technological tools to *visualize* data and variable relationships for interpretive purposes.

A step-wise approach is also recommended in teaching students how to read and interpret a graph through instructional scaffolds that target student difficulties in making sense of and using graphs (Angra & Gardner, 2016, 2017). First, in a manner comparable to the general process by which science experts direct their attention when *reading* a new graph image (Harsh et al., 2019), students should be trained to initially focus on the graph framework (e.g. axes, scale) and then contextual features (e.g. variable labels, title, caption) before moving on to the provided graph data. By deconstructing the graph in such a way, learned emphasis is placed on the information relevant to effectively *interpreting* the presented data relationship. In the next steps, students should be guided through practice to interpret graph data at the three levels of comprehension as outlined by Curcio (1989), including: reading the data (i.e. drawing facts explicitly stated in the graph), reading between the data (i.e. making comparisons between variables, trends) and reading beyond the data (i.e. drawing inferences and making predictions from the data). Examples of graphing instructional tools that can help instructors and students with these steps are published (See step-by-step guide and guide to tables and figures in Angra & Gardner, 2016; Harsh & Schmitt-Harsh, 2016; Harsh et al., 2019).

8.3.2 *Instructor Roles and Interactions*

8.3.2.1 Intentional and Explicit Instruction

When teaching it is important to provide intentional and explicit instruction to students on transforming data into graphs and interpreting the take-home message to make instructors' hidden, tacit knowledge visible to the students. This is a crucial element in teaching graphing because an instructor can model their own graphing practices for students and provide them the opportunity to gain insight on not only the how but also the why rationale on data transformation, construction, and interpretation. A more formal and elaborated means of intentional and explicit instruction is to adopt the instructional approach called the cognitive apprenticeship model (CAM), which is a social learning method that helps novices transition towards expertise (Dennen, 2004). Unlike the earlier apprenticeship models which were used for trade purposes that stressed the importance of physical skills, the CAM focuses on developing cognitive and metacognitive skills by participating in authentic learning experiences (Collins et al., 1987; Dennen, 2004). The CAM consists of 6 components that draw on the social constructivist learning theory, developed by Lev Vygostsky (1962), which emphasizes the collaborative nature of learning, reiterating interaction between the instructor and learner: (1) modeling of the instructional materials and processes of thinking by the instructor, (2) coaching the students through observations and offering suggestions, where necessary (3) scaffolding the learning by providing guidance to the students, (4) allowing students to articulate their knowledge and thinking, (5) engaging the students in reflective tasks that allow them to compare and modify their knowledge with that of the instructor and/or other students, and (6) encouraging students to formulate and test new research questions and hypotheses, in order to apply their knowledge. Steps in the CAM model such as articulation, reflection, and exploration are habitual to, generally (Hogan &

Maglienti, 2001) and graphing (diSessa, 2004; Maltese et al., 2015; Angra & Gardner, 2017; Harsh et al., 2019).

8.3.3 Student Behaviors

8.3.3.1 Collaborative and Social Practice

As disciplinary (biology) and sub disciplinary-specific (e.g. ecology) knowledge (e.g. theories, concepts, norms, and heuristics) influences one's ability to problem solve and learn content, students should be given the opportunity to develop and apply graphing skills through collaborative, inquiry-based activities within context of the field of study. Such contextualized exposure to graphing through disciplinary inquiry and community norms is essential for students to derive deep insights and conceptual understanding of the biological systems under examination (Roth & McGinn, 1997; Bowen et al., 1999). Disciplinary knowledge of data, which is grounded in the theories, concepts, measurements, and biological concepts of the discipline will allow students to communicate their findings effectively. As well, the acquisition of "disciplinary habits of mind", as it relates to graph construction and interpretation, by providing students experience in the decision-making processes typical of and expected within the discipline benefits their graphing competence as well as their ability to use and refine evidence-based claims for scientific argumentation and general communication (Bowen et al., 1999).

8.3.3.2 Evaluation and Reflection

The final element in our framework is to give students the opportunity to evaluate and reflect on graphs that they have made or graphs that they are reading and interpreting from external sources (e.g. primary literature, media, textbooks, peers). Self-reflection and critical thinking are some of the best ways for students to evaluate and think about their learning (Tynjälä, 1999) and are indicative of expert graphing practices (diSessa, 2004; Angra & Gardner, 2017). Therefore, graphing competence is not merely the ability to generate and make meaning from representations such as graphs (representational competence). Rather, full competence involves what diSessa and colleagues have called meta-representational competence (MRC) (diSessa et al., 1991; diSessa, 2004). MRC in graphing includes the abilities to: (1) design new graphs, (2) explain graphs, (3) understand the role a graph plays, and (4) critique and evaluate the affordances and limitations of a given graph over other possible alternatives. As mentioned previously helping our students understand the role of a graphs within inquiry and experimentation and how graphing choices can affect what is conveyed is essential to helping our students develop graphing competence. Although oral reflections via interviews are a rich source of data to convey to the researcher what the student is thinking (Angra &

Gardner, 2017), they are oftentimes not practical in a classroom setting. Written reflections are useful when they probe students to think deeply. In order to increase students' confidence and refine their critical thinking skills, and the learning or reflection component of the MRC, reflections should be performed and encouraged numerous times throughout a class.

8.4 Student Learning Graphing as Part of Inquiry and Experimentation

We advocate that instruction and student activity to learn and practice graphing be integrated within the practice in inquiry and experimentation and not as isolated, decontextualized activities (Roth, 2013). As is highlighted in the ACE-Bio Competencies, graphing runs through many parts of experimentation (Table 8.1) and is not merely a final end-product from an inquiry. In addition, by learning graphing as a natural part of inquiry, when students make and/or read and interpret graphs, they can do so in a deeper manner. The students are not only connected to the data because they were involved in its collection, but they will have a better understanding of the nature of and variation within the biological system under study, the affordances and limitations to the instrumentation and measurement systems, and know more of the biological significance (See Table 2 in Auchincloss et al., 2014). This will allow them to make claims that are not merely a superficial read of the data, but are claims that can reach beyond the data.

Here we present two case studies from our own work in undergraduate biology which exemplify this authentic practice approach to teaching graphing and highlight the features of our framework (Table 8.2). The case studies represent different implementations of the design framework as the first study details an upper-level laboratory experience with graphing interventions woven throughout the semester while the second study reports on the impact of a short-term ecology-focused graphing unit in an introductory course for non-science majors.

8.4.1 Case Study 1: Inquiry-Embedded Sustained Graphing Intervention in an Upper-Division Biology Course

The first case study summarizes data collected over four years from an upper-level physiology laboratory course at a large public R1 Midwestern University. Broadly, we characterize the graph construction practices and describe the impact of repeated use of evidence- based graphing materials: step-by-step guide (Angra & Gardner, 2016), guide to data displays (Angra & Gardner, 2016), and the graph rubric (Angra & Gardner, 2018) on students' graph construction skills.

8.4.1.1 Participants and Laboratory Context

Under approved an approved IRB protocol (#1210012775), we compare 139 students from the non-intervention semesters (Spring 2013 and 2014; these students did not learn with the evidence-based graphing materials) with 123 students from the intervention semesters, (Spring 2015 and 2016; these students had formal instruction on and unlimited usage of the evidence-based graphing materials). Across all semesters, student populations did not vary across several demographic dimensions (e.g. race/ethnicity, binary gender, sub-majors within biology, number of credit hours) and performance in the course as measured by exam performance. The structure, context and physiology content of the laboratory curriculum, and laboratory instructors also remained the same across all four semesters. The labs were designed to be ***inquiry-based*** and open-ended, to allow students to ***collaborate*** in small teams to design experiments. Four times over the semester students engaged in inquiry and could ***collect real-world and messy data***, and represent in their graphs and verbally ***reflect on the variation in their data*** during and after their short group PowerPoint presentations of their data. The lab experiments consisted of topics in physiology including: neuroscience, endocrine, cardiovascular, and a topic of student's choice in the final lab. All labs required students to actively engage with science practices such as: formulating a research question and hypothesis, planning and designing an experiment for that weeks' lab topic, identifying and defining key variables, and ***transforming, analyzing, and interpreting data*** (Padilla, 1990; Roth & Roychoudhury, 1993; Pelaez et al., 2017).

8.4.1.2 Teaching Intervention Design

This teaching intervention study was guided by the cognitive apprenticeship model (CAM; see description under"***intentional and explicit instruction***" and Table 8.2). Active engagement with experimental design, data collection, graph construction at multiple time points throughout the semester, and repeated practice and reflection are ideal practices since the instructional model is based on the premise that involvement in the subject matter in an authentic situated learning environment encourages the novice to learn and aids in accelerating them towards expertise (Dennen, 2004; Roth & Bowen, 2001; McFarland, 2010). In addition to pursuing their roles as instructors, the nature of this lab allowed the instructors to serve as coaches and mentors to the students. Instructors in this laboratory setting usually engage in conversations with students, coach students, and have them articulate their thinking, in order to better guide them towards expertise (Gormally et al., 2009).

When the graphing materials were modelled to the students, we discussed the importance of reflection and alignment. To guide students with creating appropriate graphs, we asked them to reiterate their research question, hypothesis, and the variables they were manipulating and those that they were measuring. Student groups in both the non-intervention and intervention semesters were encouraged to make exploratory graphs of their data because it helps to visualize the patterns and trends showcased by the data. However, students were informed that at least one graph that they used in their PowerPoint presentation should be aligned with their research question and hypothesis.

Students were given multiple opportunities throughout the semester to ***reflect*** on graphs and graph construction choices in both intervention and non-intervention semesters. All students reflected during the oral group PPT presentation. Students in the intervention semesters were additionally prompted to reflect as they created the graphs using the Step-by-Step Guide (Angra & Gardner, 2016).

8.4.1.3 Data Collection and Analysis

To analyze the quality of the graphs constructed by student groups over the non-intervention and intervention semesters, graphs were removed from their PowerPoint presentations, along with the research question and hypothesis, and individually scored using the graph rubric (Angra & Gardner, 2018). The last subcategory featured in the graph rubric, alignment, refers to choosing and constructing a graph that displays the data in a manner that facilitates the answering of the research question or evaluation of a hypothesis. Because some graphs created by students might be used for data exploration or designed to highlight additional data trends, the evaluation of alignment was not conducted on all graphs. Below we report and discuss the results of the graph alignment separately along with scores from graph mechanics, communication, and choice, which are other areas in our graph rubric (Angra & Gardner, 2018). A Kruskal-Wallis ANOVA was used to compare the non-intervention semester with the intervention semester in the graph rubric categories of mechanics, communication, and choice.

8.4.1.4 Findings

Across the four laboratory topics, in both the non-intervention and intervention semesters, bar graphs were the most common type of graph constructed. In Fig. 8.2 below, we show representative examples of bar graphs from the first and last labs in the intervention and non-intervention semesters. Noticeable differences in graphs produced between the two semesters are the lack of proper graph mechanics (non-intervention semester is missing descriptive labels and units), type of data plotted (non-intervention semester more often plotted data in a way that hinders quick understanding such as paired bars for all experimental trials), and alignment between the graph and the research question and hypothesis (lack of alignment in the non-intervention semester).

The overall scores for bar graphs did not change very much over the course of the non-intervention semester, with the average score of 59% of the 15 bar graphs constructed for the first lab and 40 bar graphs constructed in the last lab. However, an increase was noticed in the intervention semesters where the bar graph average began at 61% (15 bar graphs constructed) in the first lab and 74% in the last lab

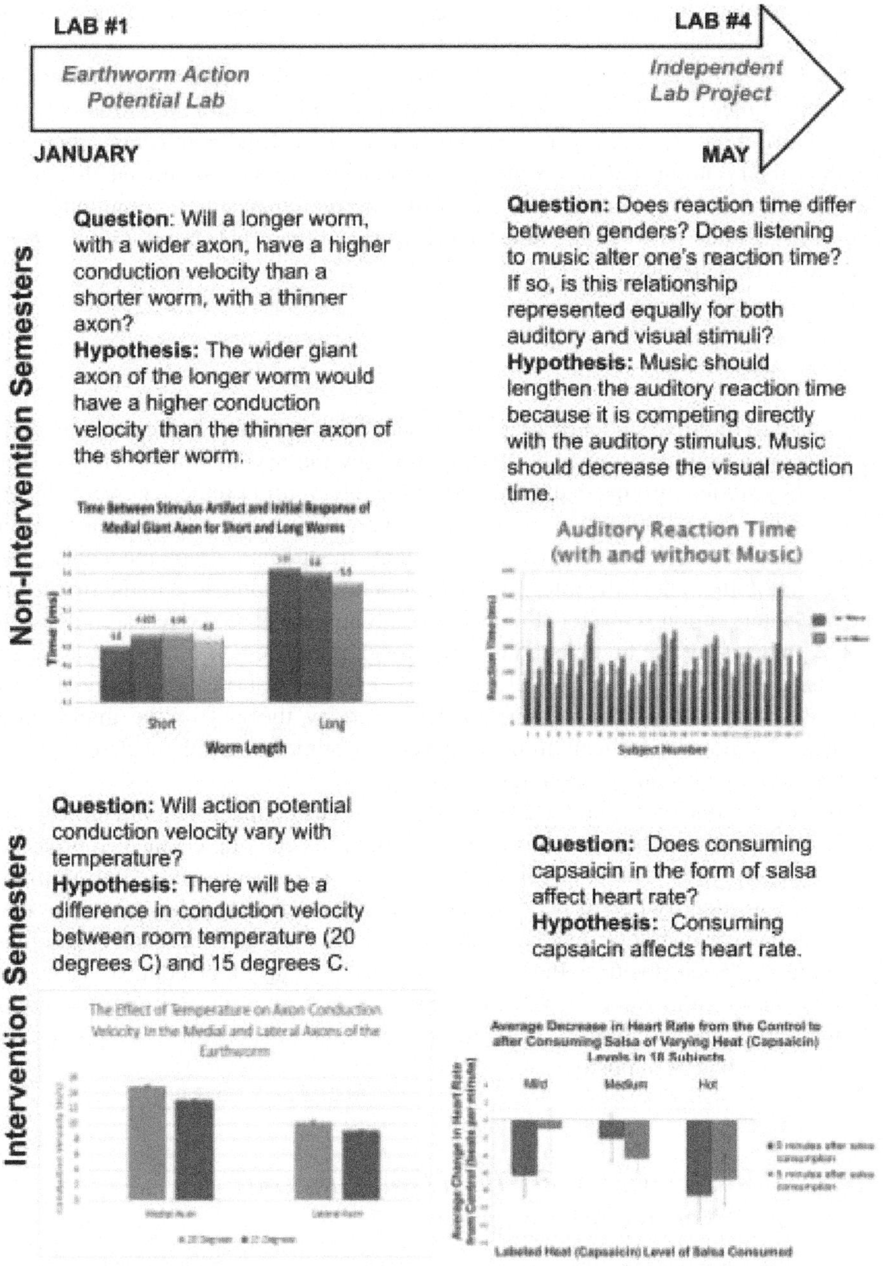

Fig. 8.2 Illustration of the varieties of graphs produced by student groups at the start and end of the non-intervention (top) and intervention semesters (bottom). Student group research questions and hypotheses are provided for context. The independent lab project was a topic of the student groups' choosing using the instruments and methods used during the semester. Scores for student graphs for the non-intervention semesters from lab #1 to lab #4 were 61% and 67% respectively and scores for the graphs constructed in the intervention semesters for lab #1 and 4 were 78% and 100%, respectively

(39 bar graphs constructed), with the majority of bar graphs scoring close to the maximum possible.

Looking at the data with the graph rubric framework (mechanics, communication, and choice) for all types of graphs made for the first lab in the semester a significant difference ($p < .05$) was seen in graph mechanics and graph choice between the non-intervention and intervention semesters. A highly significant difference ($p < .0005$) was noticed in all three categories of graph mechanics, communication, and graph choice between the non-intervention and intervention semesters for the last lab in the semester.

8.4.1.5 Discussion

We aimed to improve students' graphing practices through a multifaceted, authentic practice approach in an upper-division physiology inquiry laboratory. Graphs constructed by student groups in the final lab in the intervention semesters were significantly better ($p < .0005$) than the graphs constructed during the non-intervention semesters. We noticed immediate improvements in students' graph mechanics, and choice in the first lab of the intervention semester, when compared to the first lab in the non-intervention semester (Fig. 8.2). Graph alignment to the research question and hypothesis for the first lab did not vary much from the non-intervention semester. This could be because it is a difficult skill to master that even professionals have difficulty with (Rougier et al., 2014). However, by the end of the intervention semester, we noted significant improvements with students' graph mechanics, communication, choice, and graph alignment.

We believe that evidence-based graphing materials coupled with intentional and explicit instruction is beneficial to student learning instead of piecemeal approaches used to teach skills. The comprehensive nature of the teaching intervention limits the ability to identify one or two critical elements which lead to better student outcomes, however instructor feedback on the ease of using this approach to teaching graphing has been positive. Indeed, subsequent work in other contexts (data not shown) shows similar findings in the areas of graph mechanics, communication, choice, and alignment. Finally, students' perception on the usefulness of the graphing materials has been positive.

8.4.2 Case Study 2: Short-Term Graphing Unit in a Non-majors Course

The results of Case Study 1 highlight the positive impact of multiple evidence-based practices and the CAM instructional framework on laboratory students' graph interpretation and construction skills over the course of a term. The second case study we present summarizes the design, implementation, and testing of a short-term unit to foster students' graphing skills. Motivated by our graphing research with participants across the expert-novice continuum (Harsh et al., 2013, 2019; Maltese et al., 2015), this ecology-focused unit deliberately incorporated the design features discussed above (and see Table 8.2) to emphasize the elements and practice of graphing to address common challenges that undergraduates face when making and using graphs.

8.4.2.1 Participants and Context

The study was conducted at a second large Midwestern public research university in an undergraduate non-science majors' course intended to develop student competencies necessary for scientific literacy. This course was offered each semester and was populated by non-science majors, the majority of whom were pre-service K-8 teachers in their first or second year of study. The integrated lecture-laboratory course met for 115 minutes twice a week as students engaged in activities within the context of environmental science topics (e.g., climate change, soil and water quality) to practice process, quantitative, and communication skills as well as extend their understanding of science and its role in society.

8.4.2.2 Unit Design

The graphing unit was situated within a water quality module, where students assessed the quality of an on-campus stream using collected benthic macroinvertebrate data. Here, as learning is most successful in context (Brown et al., 1989), students were able to practice graphing skills while engaged in the hands-on collection and analysis of stream quality data. This topic is relatively approachable across levels of science background expertise as most students are familiar with streams and/or environmental issues often associated with water quality (e.g., pollution). It was also anticipated that student interest – which has been repeatedly shown to benefit engagement and motivation (Hidi & Renninger, 2006) – would be triggered and maintained in sampling an on-campus stream given task novelty and potential personal relevance.

The stream ecology graphing unit (or SEGU) was developed based on a multipoint design framework drawn from the literature that reflects the key instructional features described above for advancing students' competency in graphing (Table 8.2). Throughout the four-part unit, students worked ***collaboratively*** to make sense of and communicate data and the teaching activities were guided by frequent ***assessment*** of student skills and understanding. Of particular value was the collection of pre-unit performance tasks to establish a baseline for student abilities in graph construction and interpretation.

At the onset of the unit, students were introduced to the topic of water quality and how data derived from biosurveys of aquatic organisms provide valuable information on the biological and physical condition of streams. In small groups, students collected and identified benthic invertebrates using sampling practices comparable to those employed by ecologists in monitoring stream communities (***authentic experience***). In the second part of the unit, student teams compiled their sampling data within and across sections to calculate metrics of system perturbation, and practiced presenting this data in a graphical form via a ***two-step approach*** by first transforming their data by hand before using statistical software (i.e. Microsoft Excel) for data visualization and interpretation. Bookending these learning activities, ***intentional and explicit instruction*** initially exposed students to graph design topics (Kosslyn, 2006) and included instructor modeling of graph display, reading, and interpretation as well as self-reflection of the employed practices. After presenting the information, students were asked to ***evaluate and reflect*** on their decisions in graph display as well as their understanding of the data at the levels outlined by Curcio (1989; i.e. reading the data, reading between the data, and reading beyond the data). The latter conversations often highlighted data variability (***complex/messy data***) as many students lacked prior experience in the collection and analysis of raw data. In the third part of the unit, students constructed graphs – again using a two-step approach – to compare their measurements of on-campus stream health to conditions at over 250 sites collected across the state as part of a formalized biomonitoring program of water quality as classified by resident biota. Given the variation present in data reported across sites, students gained experience in the handling and analysis of authentic large, complex data sets in which they had to identify relevant information for site comparisons. Throughout their graphing activities, students were again encouraged to ***reflect***, in small and large group discussions, on the suitability of their data display and their understanding of the demonstrated trends as well as peer-review other's graphs. In the final part of the unit, students were tasked with writing a hypothetical article for a university or local news story regarding the health of the on-campus stream with emphasis placed emphasis on the students' use of the constructed graphs as evidential support for their general conclusions (***communication***).

8.4.2.3 Data Collection and Analysis

At the beginning of the unit, the study was introduced, and volunteers for study participants were recruited from two classes taught by one of the chapter authors (JH). A total of 39 students (out of 40) volunteered and provided data used in the IRB approved study, of which all were non-science majors in their first or second academic year. To characterize students' graph construction and interpretation skills, performance tasks were adopted from existing instruments designed to measure college students' abilities to draw and interpret graphs (Picone et al., 2007; Bray Speth et al., 2010; Maltese et al., 2015). In accompaniment, open-response prompts were used to document student reflections on their decision-making in graph design. Two items were adopted from Picone et al. (2007) that asked participants about their affect toward graphing (i.e. frustration, anxiety when faced with graph data) on a 5-point Likert-scale were also included in the test. Students were given this short 10-item test at the onset and completion of the unit, with comparable pre- and posttest items. A vetted scoring rubric (Harsh et al., 2013) based on Kosslyn's essential display components (2006) was used to rate students' graph drawings, with responses to open-ended graph interpretation and data display rationalization prompts scored via rubric for appropriateness. Test data were statistically examined using paired t-tests for evidence of changes in student graphing skills and affect in response to unit participation. In addition, at the end of the unit, we administered a questionnaire that included closed- and open-ended prompts about their perceptions of the lesson activities (e.g., what they like, what could be improved) as well as demographic information. Quantitative questionnaire data were analyzed using descriptive statistics while qualitative data were coded and analyzed to identify major themes.

8.4.2.4 Findings

We first describe changes in students' graph competencies and then discuss what features of the unit were identified to support learning. With respect to graph creation, we found that students' ability to display data graphically improved as a result of participation in the short-term graphing unit. More specifically, students made substantial improvements in rationalizing their decision-making process in graph design (i.e. variable placement and selection of graph type), which shifted from personal preference and visualization (i.e. what looked "best") to a focus on the nature of the data being displayed. In addition, students scored significantly higher in the areas of content representation (i.e. data accuracy) and labelling (e.g., scaling) on the post-test graph construction tasks compared with their pre-test responses (Table 8.3). Likewise, while there was minimal change in the basic interpretative activities of identifying variables and individual data points, student efforts in the more advanced graph reading skill of elaborating on trends observed between variables (or "seeing between or beyond the data" as described by Curcio, 1989, Friel et al., 2001) were found to increase significantly over the unit. Concomitant with these cognitive skill-based gains, most students reported lower levels of anxiety (56%) and frustration (52%) when faced with graph data at the end of the unit in comparison to the start.

Eighty-six percent of students indicated the unit helped develop their graph construction and/or interpretation skills. Consistent with the unit framework, the students spoke positively about the various ways in which the activity contributed to their graphing abilities. Most students (>85%) appreciated the problem context and unit features of collecting and analyzing their own data, working collaboratively, reflecting on their graph decision-making, engaging with complex data sets, and the explicit instruction of graphing practices. Here, several students remarked in open-response that this was the first time learning how to make sense and use graphs. Students, to a lesser degree (70%), identified the contributions of the two-step approach to their learning; however, it is noteworthy, that negative feedback

Table 8.3 Pretest and posttest scores of students' (n = 37) graph data transformation and comprehension skills, and affective responses to graph data (paired t-tests; 95% confidence intervals shown in parentheses). When multiple test items were used to measure comparable skills, those items were grouped for analysis (as noted in parentheses)

Item	Point value possible[a]	Pretest mean (± SD)	Posttest mean (± SD)	P
Graph drawing skills				
Reasoning for variable positioning in graph construction	2	0.64 (0.5)	1.15 (0.5)	0.000
Reasoning for using the employed graph type to represent the data (n = 2 questions)	4	2.01 (0.5)	2.58 (0.8)	0.000
Transformation of data into graphs (n = 2 questions)				
Framework items (i.e., graph type, axial layout, variable positioning)	10	9.12 (1.0)	8.74 (0.9)	0.109
Content items (i.e., accuracy of data, effective data representation)	8	6.61 (1.2)	7.47 (0.7)	0.000
Label items (i.e., proper identification of variables, labels, scale)	13	7.04 (1.7)	10.90 (1.5)	0.000
Graph Reading skills				
Identification of independent and dependent variables	2	1.49 (0.1)	1.65 (0.8)	0.384
Interpretation of graphs demonstrating single-variable relationships (n = 2 questions)	2	1.54 (0.5)	1.88 (0.3)	0.000
Interpretation of a graph demonstrating mathematical concepts (i.e., slope)	1	0.11 (0.3)	0.11 (0.3)	1.0
Affect				
Level of anxiety when faced with graph data[b]		2.32 (0.8)	2.05 (0.6)	0.010
Level of frustration when faced with graph data[b]		2.19 (0.8)	1.95 (0.7)	0.010

Table reprinted from Harsh and Schmitt-Harsh (2016) with permission of University of California Press

[a]Maximum potential score for the given skill. Graph-drawing skills were evaluated from the pretests and posttests using an existing scoring rubric (Harsh et al., 2013), and graph-reading skills were evaluated on the basis of response correctness

[b]Items measured on a 4-point scale: 0 = lowest, 4 = highest

regarding this practice centered on the perceived repetitive nature of drawing the graphs by hand prior to visualizing electronically.

8.4.2.5 Discussion

The results of Case Study 2 demonstrate the effectiveness of short-term graphing interventions on improving college biology students' graphing competence. Although the unit only spanned a few class meetings, our results suggest the learning experience contributed to the development of students' ability to create and interpret graph data as well as their attitudes toward graphing. These gains can be attributed to the multi-point design framework – situated in an authentic context – focused on improving graph competency, which the students largely favored as part of the unit activities. This was further reflected during data-based lecture and lab activities later in the term as students were anecdotally observed employing the introduced graph display and interpretation strategies as well as references made to what they did during the SEGU. Our findings support the value of the graph-focused design features, which have the potential to be transferred to other problem contexts (e.g., plant physiology, urban ecology) as well as the lecture setting due to their generalizability.

8.5 Conclusions and Implications for Instructors

In this chapter, we discussed the design and use of a practical framework to answer the broad question "How can we as instructors effectively teach graphing in undergraduate biology?" In two different studies we illustrated the implementation of a common set of design features in context for the development of students' graph interpretation and construction skills (summarized in Table 8.4). Taken together, the results of both studies indicate that short- and long-term interventions based on the cognitive apprenticeship model (CAM) have the potential to improve student performance in making and using graphs as well as their affect toward graphing. Our results are in line with that of prior K-16 studies showing that the general evidence-based teaching activities of providing students the opportunity to participate in authentic inquiry (data collection and analysis), frequent assessment, and collaboration positively contributes to students' graphing skills (Roth & McGinn, 1997; Tairab & Khalaf Al-Naqbi, 2004; Bowen & Roth, 2005; Picone et al., 2007; Morrison & McDuffie, 2009; Bray Speth et al., 2010; Grumbine, 2010; McFarland, 2010). At the same time, our findings align with prior work that highlights the impact of graph-specific teaching practices (Roth & McGinn, 1997; Bowen & Roth, 1998; Bray Speth et al., 2010; Schultheis & Kjelvik, 2015) which needs to go beyond the simple integration of graphs into coursework and presented materials. As Lehrer and Romberg (1996) pointed out, it is unreasonable for educators to expect students' "data modeling to spring forth, like Athena from the head of Zeus, in the form practiced by scientists and mathematicians" (p. 70). We suggest that interventions to support the development of students' graph competencies should expose students to explicit instruction on the use and design of graphs (e.g., modeling, scaffolding) and provide them opportunities to engage with real-world "messy" data, multiple-step approaches for graph display and interpretation, and opportunities to reflect on their graph decision-making processes.

While our data suggest a fairly robust model of instructional features for graph learning, there remain questions for future research. For example, while the findings of our work and that of other teaching activities are promising (e.g. Bray Speth

Table 8.4 Summary of the application of the framework for teaching graphing in undergraduate biology

Instructional Design Features	Case Study 1, Inquiry-embedded sustained graphing intervention in an upper-division biology course	Case Study 2, Short-term graphing unit in a non-majors general biology course
Engage with real-world, messy data	Over the semester, students: Worked in small groups to design original experiments, collect data,interpret their messy data, construct graphs, and present their findings in a PPT presentation. Repeated this for four labs over the semester.	Over multiple class meetings students: Collected and analyzed aquatic macroinvertebrate data for assessing stream quality, Handled and analyzed a large "messy" public data set
Data analysis and graph construction and interpretation approach	Students utilized the step-by-step guide to help them think of the best way to present their data in a graph that aligns with their research question, hypothesis. Students also used the guide to data displays and the graph Rubric to make sure they had appropriate graph mechanics, communication, and choice.	Students utilized a two-step approach for data display in which they first drew the graph by hand prior to generating a visualization using statistical software. Students were prompted to consider and discuss graph data in small and large group settings at multiple levels
Intentional and explicit instruction	The cognitive apprenticeship model and its practices were used in the course. Formal instruction on the graphing materials began in the second week of the semester (for the intervention group), with modeling of the graphing materials and having students practice using these materials. Throughout the intervention semesters, instructors utilized the published graphing materials to coach and encourage students to become independent thinkers and graph makers.	At the onset of the unit, students were introduced to graph design and use through explicit instruction and published materials. The instructor modeled a step-wise approach in how to (a) draw, (b) visualize, (c) read, and (d) interpret graph data in support of student efforts. Regular small and large group discussions were held in which students were to reflect on their decision-making, critique others work, and to debrief to address problems.
Collaborative and social practice	Students worked in small groups to design experiments, collect data, construct graphs, and present their findings in a PowerPoint presentation. After the oral presentations, students were instructed to use the graph rubric to critique peer graphs and reflect on their own knowledge of the advantages and disadvantages of graphs.	Students collaboratively worked in small teams through the activity to collect, analyze, interpret, and display data as well as reflect on their decision-making processes. Small group activities were followed with large-group discussions. Students communicated their findings in an article format

(continued)

Table 8.4 (continued)

Instructional Design Features	Case Study 1, Inquiry-embedded sustained graphing intervention in an upper-division biology course	Case Study 2, Short-term graphing unit in a non-majors general biology course
Evaluation and reflection	Students were encouraged to reflect on graphs at three different points in each lab. First, immediately after constructing the graph in the third phase of the step-by-step guide. Second, during the oral group PPT presentation. Third, at the end of the lab presentations, during peer graph critique.	Students reflected on their decision-making in graph construction and interpretation at multiple points throughout the activity. These activities included small team and large group discussion as well peer-reviews.

et al., 2010; Schultheis & Kjelvik, 2015; DeBoy, 2017; Kirby et al., 2019), further research is needed to understand the longitudinal impacts of these "one-time" interventions as well as how graphing instruction can be vertically integrated throughout the curriculum as a cross-cutting theme as one's graph comprehension is grounded in their embodied knowledge of the domain and its representational practices (Roth & Bowen, 2001). In respect to the latter, prime areas to focus such training is through research activity (in and out of the classroom) and exploring primary literature, which are regularly identified by science practitioners as means they learned graphing in the field (Bowen et al., 1999; Harsh et al., 2019). Next steps in research also should focus on the connection between one's abilities in graph interpretation and construction.

In summary we offer the following implications for instructors:

- Engage students in scientific inquiry and experimentation as it is practiced by scientists, including collaboration and enmeshed in the concepts, methods, measurements, and features of the biological subdiscipline
- Create learning opportunities for students to engage in the practice of graphing embedded within authentic inquiry contexts which generate with messy data
- Provide explicit instruction on graphing which involves deconstructing the process, decision making, and reflective evaluation that are practiced by experts
- Evaluate student competence with graphing using authentic assessments that are integrated as part of their investigations
- Encourage students to evaluate and reflect on their graphing choices and rationale to promote deeper practice and make their thoughts explicit

Acknowledgements This work benefited from the research on graph construction in undergraduate biology conducted under NSF # 1726180 (SMG) and the collaborations and opportunities within the ACE-Bio Network (NSF # 1346567). Any opinions, findings, and conclusions or recommendations expressed are those of the authors and do not necessarily reflect the views of the NSF. We are grateful to Julia Gouvea for her helpful feedback on this chapter. Many thanks to the members of the Purdue International Biology Education Research Group (PIBERG) and the James Madison University Biology Education Research Group (jmuBERG) for their valuable feedback and assistance over the years. In addition, we would like to acknowledge Adam Maltese and Mikaela Schmitt-Harsh for their significant contributions to multiple projects included here. We are also grateful to the students and faculty who contributed their time to our respective work.

CRediT Author Contribution Statement *All authors contributed equally to:* Conceptualization, Methodology, Writing, Visualization. *SMG contributed to*: Supervision.

References

AAMC-HHMI Committee. (2009). *Scientific foundations for future physicians* (pp. 26–29). Association of American Medical Colleges.

Allen, D., & Tanner, K. (2007). Putting the horse back in front of the cart: Using visions and decisions about high-quality learning experiences to drive course design. *CBE—Life Sciences Education, 6*(2), 85–89.

American Association for the Advancement of Science (AAAS). (2011). *Vision and change in undergraduate biology education: A call to action* (C. Brewer & D. Smith, Eds.). Washington, DC. https://visionandchange.org/finalreport/. Accessed 18 Oct 2021.

Angra, A. (2016). *Understanding, evaluating, and diagnosing undergraduate student difficulties with graph choice and construction* (Doctoral dissertation, Purdue University).

Angra, A., & Gardner, S. M. (2016). Development of a framework for graph choice and construction. *Advances in Physiology Education, 40*(1), 123–128.

Angra, A., & Gardner, S. M. (2017). Reflecting on graphs: Attributes of graph choice and construction practices in biology. *CBE—Life Sciences Education, 16*(3), ar53.

Angra, A., & Gardner, S. M. (2018). The graph rubric: Development of a teaching, learning, and research tool. *CBE—Life Sciences Education, 17*(4), ar65.

Auchincloss, L. C., Laursen, S. L., Branchaw, J. L., Eagan, K., Graham, M., Hanauer, D. I., Lawrie, G., McLinn, C. M., Pelaez, N., Rowland, S., Towns, M., Trautmann, N. M., Varma-Nelson, P., Weston, T. J., & Dolan, E. L. (2014). Assessment of course-based undergraduate research experiences: A meeting report. *CBE—Life Sciences Education, 13*(1), 29–40. https://doi.org/10.1187/cbe.14-01-0004

Berg, C. A., & Smith, P. (1994). Assessing students' abilities to construct and interpret line graphs: Disparities between multiple-choice and free-response instruments. *Science Education, 78*, 527–554.

Bowen, G. M., & Roth, W. M. (1998). Lecturing graphing: What features of lectures contribute to student difficulties in learning to interpret graphs? *Research in Science Education, 28*, 77–90.

Bowen, G. M., & Roth, W. M. (2005). Data and graph interpretation practices among preservice science educators. *Journal of Research in Science Teaching, 42*(10), 1063–1088.

Bowen, G. M., Roth, W. M., & McGinn, M. K. (1999). Interpretations of graphs by university biology students and practicing scientists: Toward a social practice view of scientific presentation practices. *Journal of Research in Science Teaching, 36*, 1020–1043.

Bray Speth, E., Momsen, J. L., Moyerbrailean, G. A., Ebert-May, D., Long, T. M., Wyse, S., & Linton, D. (2010). 1, 2, 3, 4: Infusing quantitative literacy into introductory biology. *CBE—Life Sciences Education, 9*(3), 323–332.

Brown, N. J., & Wilson, M. (2011). A model of cognition: The missing cornerstone of assessment. *Educational Psychology Review, 23*(2), 221.

Brown, J. S., Collins, A., & Duguid, P. (1989). Situated cognition and the culture of learning. *Educational Researcher, 18*(1), 32–42.

Clemmons, A., Timbrook, J., Herron, J., & Crowe, A. (2020). BioSkills Guide. Core Competencies for Undergraduate Biology, (version 5.0). *QUBES Educational Resources*. https://doi.org/10.25334/156H-T617

College Board. (2019). *AP biology course and exam description*. https://apcentral.collegeboard.org/pdf/ap-biology-course-and-exam-description-0.pdf?course=ap-biology

Collins, A., Brown, J. S., & Newman, S. E. (1987, January). *Cognitive apprenticeship: Teaching the craft of reading, writing and mathematics* (Technical Report No. 403). BBN Laboratories, Centre for the Study of Reading, University of Illinois.

Curcio, F. R. (1989). Developing graph comprehension. Elementary and middle school activities. National Council of Teachers of Mathematics, Inc., 1906 Association Drive, Reston, VA 22091.

DeBoy, C. A. (2017). Student use of self-data for out-of-class graphing activities increases student engagement and learning outcomes. *Journal of Microbiology & Biology Education, 18*(3), 1–9.

Dennen, V. P. (2004). Cognitive apprenticeship in educational practice: Research on scaffolding, modeling, mentoring, and coaching as instructional strategies. *Handbook of Research on Educational Communications and Technology, 2*(2004), 813–828.

diSessa, A. A. (2004). Metarepresentation: Native competence and targets for instruction. *Cognition and Instruction, 22*(3), 293–331.

diSessa, A. A., Hammer, D., Sherin, B., & Kolpakowski, T. (1991). Inventing graphing: Metarepresentational expertise in children. *Journal of Mathematical Behavior, 10*, 117–160.

Franconeri, S. L., Scimeca, J. M., Roth, J. C., Helseth, S. A., & Kahn, L. E. (2012). Flexible visual processing of spatial relationships. *Cognition, 122*(2), 210–227. https://doi.org/10.1016/j.cognition.2011.11.002

Friel, S. N., Curcio, F. R., & Bright, G. W. (2001). Making sense of graphs: Critical factors influencing comprehension and instructional implications *Journal for Research in Mathematics Education, 32*(2), 124–158.

George, M., Bragg, S., de los Santos, A. G., Jr., Denton, D. D., Gerber, P., Lindquist, M. M., Rosser, J. M., Sanchez, D. A., & Meyer, C. (1996). *Shaping the future: New expectations for undergraduate education in science* (Mathematics, engineering and technology). National Science Foundation.

Gormally, C., Brickman, P., Hallar, B., & Armstrong, N. (2009). Effects of inquiry-based learning on students' science literacy skills and confidence. *International Journal for the Scholarship of Teaching and Learning, 3*(2), n2.

Grumbine, R. (2010). Using data-collection activities to enrich science courses. *The American Biology Teacher, 72*(6), 369–372.

Hammer, D. (1996). More than misconceptions: Multiple perspectives on student knowledge and reasoning, and an appropriate role for education research. *American Journal of Physics, 64*, 1316–1325.

Harsh, J. A., & Schmitt-Harsh, M. (2016). Instructional strategies to develop graphing skills in the college science classroom. *The American Biology Teacher, 78*(1), 49–56.

Harsh, J. A., Maltese, A. V., & Warner, J. M. (2013). The development of expertise in data analysis skills: An exploration of the cognitive and metacognitive processes by which scientists and students construct graphs. In *American Association for the Advancement of Science (AAAS) national meeting*, Boston, MA

Harsh, J. A., Campillo, M., Murray, C., Myers, C., Nguyen, J., & Maltese, A. V. (2019). "Seeing" data like an expert: An eye-tracking study using graphical data representations. *CBE—Life Sciences Education, 18*(3), ar32.

Hegarty, M. (2011). The cognitive science of visual-spatial displays: Implications for design. *Topics in Cognitive Science, 3*(2011), 446–474. https://doi.org/10.1111/j.1756-8765.2011.01150.x

Hidi, S., & Renninger, K. A. (2006). The four-phase model of interest development. *Educational Psychologist, 41*(2), 111–127.

Hogan, K., & Maglienti, M. (2001). Comparing the epistemological underpinnings of students' and scientists' reasoning about conclusions. *Journal of Research in Science Teaching: The Official Journal of the National Association for Research in Science Teaching, 38*(6), 663–687.

Hoskins, S. G., Stevens, L. M., & Nehm, R. H. (2007). Selective use of the primary literature transforms the classroom into a virtual laboratory. *Genetics, 176*(3), 1381–1389.

Kastens, K. A., Krumhansel, R., & Baker, I. (2015). Thinking big—Transitioning your students from working with small, student-collected data sets towards "big data". *Science Teacher, 82*(5), 25–31.

Kellman, P. J. (2000). An update on gestalt psychology. In B. Landau, J. Sabini, J. Jonides, & E. Newport (Eds.), *Perception, cognition, and language: Essays in honor of Henry and Lila Gleitman*. MIT Press.

Kirby, C. K., Fleming-Davies, A., & White, P. J. (2019). The figure of the day: A classroom activity to improve Students' figure creation skills in biology. *The American Biology Teacher, 81*(5), 317–325.

Kjelvik, M. K., & Schultheis, E. H. (2019). Getting messy with authentic data: Exploring the potential of using data from scientific research to support student data literacy. *CBE—Life Sciences Education, 18*(2), es2.

Konold, C., Higgins, T., Russell, S. J., & Khalil, K. (2014). Data seen through different lenses. *Educational Studies in Mathematics, 88*(3), 305–325.

Kosslyn, S. M. (2006). *Graph design for the eye and mind*. Oxford University Press.

Lehrer, R., & Romberg, T. (1996). *Exploring children's data modeling. Cognition and Instruction, 14*(1), 69–108.

Lehrer, R., & Schauble, L. (2000). Developing model-based reasoning in mathematics and science. *Journal of Applied Developmental Psychology, 21*(1), 39–48.

Lehrer, R., & Schauble. (2007). Contrasting emerging conceptions of distribution in contexts of error and natural variation. In M. Lovett & P. Shah (Eds.), *Thinking with data*. New York. https://doi.org/10.4324/9780203810057

Maltese, A. V., Harsh, J. A., & Svetina, D. (2015). Data visualization literacy: Investigating data interpretation along the novice—Expert continuum. *Journal of College Science Teaching, 45*(1), 84–90.

Martone, A., & Sireci, S. G. (2009). Evaluating alignment between curriculum, assessment, and instruction. *Review of Educational Research, 79*(4), 1332–1361.

McFarland, J. (2010). Teaching and assessing graphing using active learning. *MathAMATYC Educator, 1*(2), 32–39.

Mislevy, R. J., & Haertel, G. D. (2006). Implications of evidence-centered design for educational testing. *Educational Measurement: Issues and Practice, 25*(4), 6–20.

Morrison, J., & McDuffie, A. R. (2009). Connecting science and mathematics: Using inquiry investigations to learn about data collection, analysis, and display. *School Science and Mathematics, 109*(1), 31–44.

National Research Council. (2000). *How people learn: Brain, mind, experience, and school: Expanded edition*. National Academies Press. https://doi.org/10.17226/9853

National Research Council. (2001). *Knowing what students know: The science and Design of Educational Assessment*. National Academies Press.

National Research Council. (2003). *BIO2010: Transforming undergraduate education for future research biologists*. National Academies Press.

National Research Council. (2007). *Taking science to school: Learning and teaching science in grades K-8*. The National Academies Press. https://doi.org/10.17226/11625

National Research Council. (2015). *Guide to implementing the next generation science standards*. The National Academies Press. https://doi.org/10.17226/18802

National Science Foundation. (2016). *Ten big ideas*. https://www.nsf.gov/news/special_reports/big_ideas/

NGSS Lead States. (2013). *Next generation science standards: For States, by States*. National Academies Press.

Padilla, M. J. (1990). The science process skills. *Research Matters-to the science Teacher, 9004*. https://narst.org/researchmatters/science-process-skills

Padilla, M. J., McKenzie, D. L., & Shaw, E. L. (1986). An examination of the line graphing ability of students in grades seven through twelve. *School Science and Mathematics, 86*, 20–26.
Padilla, L. M., Creem-Regehr, S. H., Hegarty, M., et al. (2018). Decision making with visualizations: A cognitive framework across disciplines. *Cognitive Research, 3*, 29. https://doi.org/10.1186/s41235-018-0120-9
Patterson, T. F., & Leonard, J. G. (2005). Turning spreadsheets into graphs: An information technology lesson in whole brain thinking. *Journal of Computing in Higher Education, 17*(1), 95–115.
Pelaez, N.J., Anderson, T.R., Gardner, S. M., Yin, Y., Abraham, J. K., Bartlett, E., Gormally, C., Hill, J. P., Hoover, M., Hurney, C., Long, T., Newman, D.L., Sirum, K., & Stevens, M. (2017). *The basic competencies of biological experimentation: Concept-skill statements (PIBERG Instructional Innovation Material Paper 4)*. Retrieved from http://docs.lib.purdue.edu/pibergiim/4. Accessed 18 Oct 2021.
Picone, C., Rhode, J., Hyatt, L., & Parshall, T. (2007). Assessing gains in undergraduate students' abilities to analyze graphical data. *Teaching Issues and Experiments in Ecology, 5*(July), 1–54.
Roth, W. M. (2013). Undoing decontextualization or how scientists come to understand their own data/graphs. *Science Education, 97*(1), 80–112.
Roth, W. M., & Bowen, G. M. (2001). Professionals read graphs: A semiotic analysis. *Journal for Research in Mathematics Education, 32*(2), 159–194.
Roth, W. M., & McGinn, M. K. (1997). Graphing: Cognitive ability or practice? *Science Education, 81*, 91–106.
Roth, W. M., & Roychoudhury, A. (1993). The development of science process skills in authentic contexts. *Journal of Research in Science Teaching, 30*(2), 127–152.
Rougier, N. P., Droettboom, M., & Bourne, P. E. (2014). Ten simple rules for better figures. *PLoS Computational Biology, 10*(9), e1003833.
Rybarczyk, B. (2011). Visual literacy in biology: A comparison of visual representations in textbooks and journal articles. *Journal of College Science Teaching, 41*(1), 106.
Schriger, D. L., & Cooper, R. J. (2001). Achieving graphical excellence: Suggestions and methods for creating high-quality visual displays of experimental data. *Annals of Emergency Medicine, 37*(1), 75–87.
Schriger, D. L., Sinha, R., Schroter, S., Liu, P. Y., & Altman, D. G. (2006). From submission to publication: A retrospective review of the tables and figures in a cohort of randomized controlled trials submitted to the British Medical Journal. *Annals of Emergency Medicine, 48*(6), 750-6, 756.e1-21. https://doi.org/10.1016/j.annemergmed.2006.06.017. Epub 2006 Sep 15.
Schultheis, E. H., & Kjelvik, M. K. (2015). Data nuggets: Bringing real data into the classroom to unearth students' quantitative & inquiry skills. *The American Biology Teacher, 77*(1), 19–29.
Schultheis, E. H., & Kjelvik, M. K. (2020). Using messy, authentic data to promote data literacy and reveal the nature of science. *The American Biology Teacher, 82*(7), 439–446.
Shah, P., & Freedman, E. G. (2011). Bar and line graph comprehension: An interaction of top-down and bottom-up processes. *Topics in Cognitive Science, 3*(3), 560–578.
Smith, J., diSessa, A., & Rochelle, J. (1993). Misconceptions reconceived: A constructivist analysis of knowledge in transition. *The Journal of the Learning Sciences, 3*, 115–163.
Tairab, H. H., & Khalaf Al-Naqbi, A. K. (2004). How do secondary school science students interpret and construct scientific graphs? *Journal of Biological Education, 38*(3), 127–132.
Tynjälä, P. (1999). Towards expert knowledge? A comparison between a constructivist and a traditional learning environment in the university. *International Journal of Educational Research, 31*(5), 357–442.
Vygotsky, L. (1962). Thought and Word. In L. Vygotsky, E. Hanfmann, & G. Vakar (Eds.), *Thought and language* (Studies in Communication) (pp. 119–153). MIT Press. https://doi.org/10.1037/11193-007
Weissgerber, T. L., Milic, N. M., Winham, S. J., & Garovic, V. D. (2015). Beyond bar and line graphs: Time for a new data presentation paradigm. *PLoS Biology, 13*(4), e1002128.
Weissgerber, T. L., Winham, S. J., Heinzen, E. P., Milin-Lazovic, J. S., Garcia-Valencia, O., Bukumiric, Z., Savic, M. D., Garovic, V. D., & Milic, N. M. (2019). Reveal, Don't conceal:

Transforming data visualization to improve transparency. *Circulation, 140*, 1506–1518. https://doi.org/10.1161/CIRCULATIONAHA.118.037777

Wiggins, G. (1998). Ensuring authentic performance. Chapter 2. In *Educative assessment: Designing assessments to inform and improve student performance* (pp. 21–42). Jossey-Bass.

Wiggins, G., & McTighe, J. (1998). What is backward design? In *Understanding by design* (1st ed., pp. 7–19). Merrill Prentice Hall. Retrieved from https://web.archive.org/web/20160721163755/http://www.fitnyc.edu/files/pdfs/Backward_design.pdf

Wild, C. J., & Pfannkuch, M. (1999). Statistical thinking in empirical enquiry. *Statistical Review, 67*, 223–265.

Part III
Implementation and Student Engagement: Guiding Learners to Do Experiments and Use Representations in Biological Research

Chapter 9
Teaching Undergraduate Students How to *Identify* a Gap in the Literature: Design of a Visual Map Assignment to Develop a Grant Proposal Research Question

Anne E. Kruchten and Jenean H. O'Brien

9.1 Background

How do we teach students to become competent or even expert researchers in biology? Unpacking the process of research reveals that there are multiple aspects of developing experimentation skills. To be a successful researcher, students must understand the field, learn to frame a research question, identify available and relevant methodologies to answer the question, set up the experiments, collect and analyze data, revise the question and methods based on the results, repeat new experiments, analyze new data, and more. Figure 9.1 depicts these steps in the process of scientific experimentation. The field of biology education and the broader scientific discipline both recognize the need for our students to develop competency in scientific experimentation. One of the core competencies of the *Vision and Change: A Call to Action* report (AAAS, 2011) is the "Ability to Apply the Process of Science," emphasizing the need for biology undergraduates to develop and practice experimentation skills. A collaborative group of educational specialists and biology researchers in the ACE-Bio Network (NSF# 1346567) drafted a set of seven competencies that scientists draw from to perform experimentation (Pelaez et al., 2017; Chap. 1 in this volume). These seven competencies – *Identify*, *Analyze*, *Communicate*, *Question*, *Conduct*, *Plan*, *Conclude* – are at the core of our efforts to teach students to become competent scientific researchers.

Within the time constraints of a single undergraduate course, it is often difficult to effectively engage students in all the aspects of experimentation. Figure 9.1 depicts pedagogical approaches to teaching scientific experimentation. Wilson and colleagues (Wilson & Rigakos, 2016) developed a method for evaluating students'

A. E. Kruchten · J. H. O'Brien (✉)
Biology Department, The College of St. Scholastica, Duluth, MN, USA
e-mail: akruchten@css.edu; jobrien@css.edu

N. J. Pelaez et al. (eds.), *Trends in Teaching Experimentation in the Life Sciences*, Contributions from Biology Education Research,
https://doi.org/10.1007/978-3-030-98592-9_9

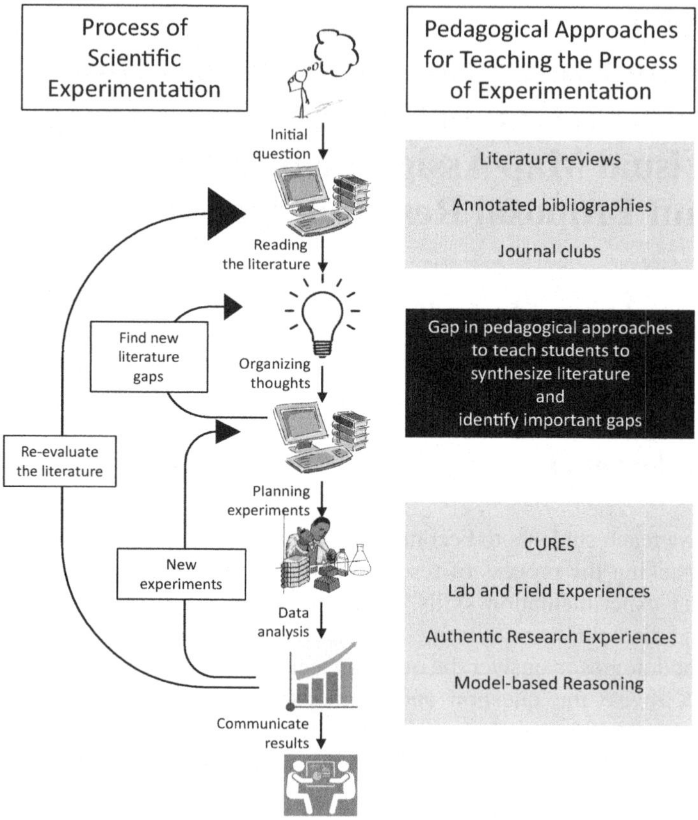

Fig. 9.1 Pedagogical approaches for teaching scientific experimentation. The process of scientific experimentation is represented from top to bottom, from the initial idea to the communication of results. At right, example approaches of teaching experimentation are shown. Literature reviews (Cole et al., 2013), annotated bibliographies (Soneral & Wyse, 2015), journal clubs (Wiegant et al., 2011), and many others can be used to teach students to read the literature. CUREs (Course-based Undergraduate Research Experiences) (CUREnet, 2020), laboratory and field experiences (Hole, 2018), authentic research experiences (Indorf et al., 2019), model-based reasoning (Hester et al., 2018), and many others can teach the skills of experimental planning, data analysis, and communicating results. The black box depicts what we believe to be a gap in pedagogical approaches: teaching students to synthesize the literature and identify meaningful scientific questions from gaps in the literature. This chapter provides a pedagogical approach to fill that gap

changes in understanding the process of scientific experimentation that demonstrated students' understanding of the scientific process improved after teaching interventions. Often, laboratory or field courses focus on aspects of experimentation including methodologies, experimental set ups, and collecting and analyzing data. In these courses, several studies and approaches focus on student engagement in authentic research experiences (CUREnet, 2020; Hester et al., 2018) and interpreting data using model-based reasoning (Zagallo et al., 2016). For example, using an Authentic Inquiry through Modeling (AIM-Bio) approach, Hester and colleagues help students avoid experiments that simply confirm expected results and instead focus on using models to iteratively arrive at explanations for their results (Hester et al., 2018). These practices often model how we as scientists perform science in order to expand our understanding rather than confirm our hypotheses. These experiences have been shown to help students better understand the scientific process and develop more confidence in their abilities as scientists.

Experimentation includes laboratory and field methodology, but it also includes *Identifying* a gap in the literature and developing an experimental question. In many biology courses, a laboratory or field component for experimentation or modeling is not available or is separate from the lecture component of the course. These courses are good scenarios for teaching experimentation skills such as developing experimental questions that precede the hands-on laboratory or field component. Science writing is an important component of an undergraduate science curriculum (Libarkin & Ording, 2012; Woodin et al., 2010) that can be used to develop these experimentation skills. A research question developed for a written grant proposal is significantly different from a question a student might pose in a lab for a preliminary experiment. The question a scientist poses for a laboratory experiment is typically a testable question in which one variable is changed during the experiment (Sirum & Humburg, 2011). A research question, in contrast, is more comprehensive and often includes multiple experimental methodologies to test several hypotheses (Herek, 2011).

To develop a strong research question, the student must have a significant understanding of the existing scientific literature in the area and an ability to *Identify* the gaps in the literature. There are few examples in the literature that address how to teach undergraduates to develop a synthetic view of the literature and *identify* gaps in it. While many undergraduate science courses implement review articles and research papers as science writing assignments (Cole et al., 2013; Jude, 2017), another option is a research proposal writing project (Evans et al., 2016). Research proposal assignments develop critical thinking skills in the form of evaluation and application of information in addition to relevant writing skills (NRC, 2003). Many courses that incorporate grant writing require a literature review (Cole et al., 2013), annotated bibliography (Soneral & Wyse, 2015), or even an informal “chalk talk” (Wiegant et al., 2011) as part of the proposal project.

For an undergraduate writing a research proposal for the first time, the literature review often starts with a very broad question, such as “how does drug X affect cells?” Clearly this type of question is too broad for a research question, but it provides an entry point for the literature searches and the process of developing keywords for the search. The Writing Center at George Mason University has produced a number of guides for writers, including two on writing a literature review (The_Writing_Center, 2018c) and organizing a literature review (The_Writing_Center, 2018b, c). Similar guides can be found around the internet, and many have similar pointers. Guidance includes reading multiple studies, looking for patterns that emerge on your topic, and determining what evidence supports your research and how best to organize that evidence. The George Mason Center (The_Writing_Center, 2018b, c) suggests a number of structural patterns for organization, including topical, debate, chronological, distant to close (in topic relevance), and seminal studies. The literature review often seems easy for students, as they have spent their academic careers being taught to conduct searches, cite their sources, and fill a certain number of pages with a prescribed number of words.

In experimentation, the purpose of the literature review is to understand the field well enough to develop a good research question. As a student progresses further into their literature review, it becomes apparent that their initial question is probably too broad. Ratan and colleagues (Ratan et al., 2019) argue that a good scientific research question follows the FINERMAPS acronym: **F**easible, **I**nteresting, **N**ovel, **E**thical, **R**elevant, **M**anageable, **A**ppropriate, **P**otential value and publishability, and **S**ystematic. Additionally, a research question should be clear, concise, focused, and arguable (The_Writing_Center, 2018a). For undergraduates, however, this often becomes a stumbling block. While the students may have completed a literature review, the review often lacks a synthetic component that allows a student to understand a whole field. While advanced scientists may dedicate significant time to understanding the state of the discipline, undergraduates are typically completing a literature review while balancing multiple other classes and responsibilities and are wading into a literature that includes both unrecognized words and methodologies.

In addition to the issues students may have in creating a synthetic view of the literature, little guidance is available on how to move from the broad literature review to the development of a detailed research question that fills a gap in knowledge. In the scientific literature, instructions for developing a research question are often aimed at experienced scientists writing grants who already have a grasp of the scientific literature. When searching for general guidance about developing research questions, the examples often do not provide suggestions for an approach to synthesize and understand the detailed scientific knowledge found in primary scientific literature. Phillips and colleagues (Phillips et al., 2017) describe this process of identifying a gap in scientific knowledge as "problematizing". Drawn from both K-12 and undergraduate studies in physics education, observations of student discussions demonstrated that during conversations students work to articulate what they do not know or understand. They argue that "problematizing" is a common trait in all scientific endeavors, and that faculty can specifically support this process in the classroom to promote the skills of scientific experimentation. Writing a literature review, often as a summary of papers, does not always help students develop a problematizing approach to reading the literature. Further, a literature review can act as a static summary of the field, not demonstrating the dynamic nature of experimentation. Novak and colleagues (Novak & Treagust, 2017) found that after students make claims based on evidence they have reviewed, it can be challenging for students to incorporate new evidence into their summary explanation. We want our students to recognize the iterative, continuously testing nature of research by developing a research question from their literature review.

While the FINERMAPS acronym and other descriptions of research questions are useful for judging the quality of a research question, they do not help students move from writing a literature review as a simple summary of articles to problematizing from the complex material. Here we suggest a new teaching tool for helping students move from the literature review to identifying the gap in knowledge on which they want to base their research question. In the cell biology literature, each publication often describes one very small portion of a complex set of activities occurring in a cellular system. Suggested mechanisms for organizing a literature review (topical, debate, chronological, distant to close (in topic relevance), and seminal studies) may not help students to synthesize material and *Identify* a gap in knowledge. We propose the use of a signaling pathway map to visually synthesize the research collected during a literature review and *Identify* areas that need to be elucidated with further experimental research.

Signaling pathway maps are a commonly used visual representation of the actions occurring in a cell, particularly actions such as cellular signaling cascades. Many biotechnology companies use signaling pathway maps to illustrate the molecular actions of their products (for an example, see https://www.bio-rad.com/en-us/prime-pcr-assays/pathway/signal-transduction-pka-signaling). In publications, many researchers use signaling pathway maps to diagram a model of their research results, making this a common mechanism for cell biologists, biochemists, and many other scientists to effectively *Communicate* information. To be clear, signaling pathway maps should not be confused with concept maps or mind maps in which relationships between concepts are depicted.

In the signaling pathway map assignment we have designed, students collect pieces of data from published works about their area of interest and incorporate each piece of data (with citations) onto a visual map of a cell. As the map is built, students begin to recognize what contextual components need to be added, such as the type of cell harboring the activities, the timing of events, cellular interactions, etc. These visual cues direct students to continue their literature review in specific directions until they have exhausted the literature available in an area. When the student cannot connect the dots between parts of the map, this emerges visually as a gap in knowledge and can be marked with a question mark. These question marks become the basis for specific research questions for a grant proposal. Of note, while there are several published papers about how to help students read and interpret signaling pathway maps (Emtage et al., 2016; Kramer & Thomas, 2006), none that incorporated creation of a signaling pathway map as an assignment were found. The production of a signaling pathway map summarizing current literature aids students in the development of a research question within our semester-long grant proposal project. This scaffolded assignment has benefited our students and is designed to serve other students similarly.

9.2 Methods

9.2.1 Educational Setting

This semester-long scaffolded research proposal assignment has been implemented in full for four semesters as part of a junior/senior undergraduate level cell biology course that serves as an elective for biology majors and is required for biochemistry majors at a comprehensive Masters level institution. Typically, 20–25 students are enrolled in each section of the course, split 50:50 between these two majors. Many but not all these students are interested in and successfully matriculate into graduate and medical degree programs. Overall, student buy-in and commitment to this course is rather high. At the end of this course, a successful student should be able to: understand fundamental and advanced concepts in cell biology, effectively develop and analyze experiments in cell biology and effectively communicate in the verbal, written, and visual forms common to cell biologists. The project described here is one of the major assessments for the course; other major assessments are two exams and critical thinking essay assignments for each unit. The course is taught in a classroom designed for active learning, with six tables positioned around the outside of the room, each of which seats six students facing each other. Each table has its own wall-mounted whiteboard and a digital screen that can be connected to the main classroom computer or to individual devices within each group. The course meets three times a week for 65 min each session, for 15 weeks followed by one final exam session.

9.2.2 The Assignment

The complete assignment described here consists of two parts. First students individually research a drug of interest and incorporate each piece of data they find into a visual map of a cell. Second, students use the map they have developed to ***Identify*** a research ***Question*** and write a research grant proposal that is reviewed by their peers, approximating the process of an NIH study section. The assignment is fully scaffolded during the semester (Fig. 9.2), both with instructor and peer feedback and with the course resources of the cell biology content and skills. The intent is for each sub-assignment to build and provide support for the next sub-assignment. The creation of the map drives the formation of a research question, the research question drives the writing of the proposal, and the finished proposal provides the understanding necessary for a funding review meeting. The sections that follow provide detail on the implementation of each part of the assignment.

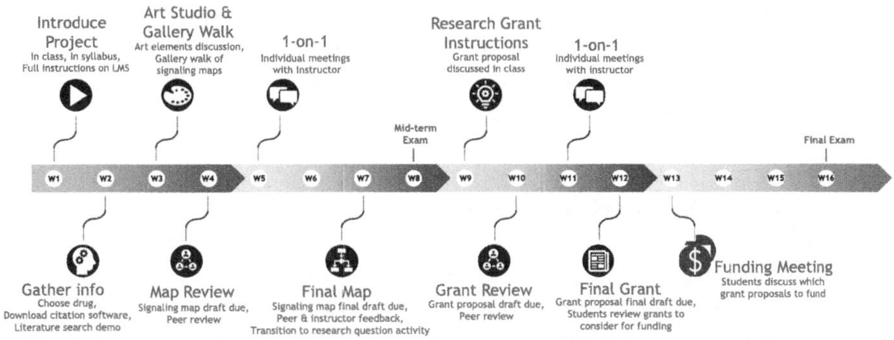

Fig. 9.2 Project timeline locating each major activity across the 16 weeks (W) of the semester. Activities listed above the timeline are instructor driven, while activities listed below the timeline are primarily completed by the students

9.2.2.1 Day One

Students are introduced to the project on the first day of the course. This introduction includes a verbal description from the instructor of the two major components (the pathway map and the grant proposal), presentation of a visual example in the form of a previous student's pathway map, and reference to the syllabus. The syllabus contains brief descriptions of the two major components, referral to full instructions including grading rubrics (Appendix) on the course learning management system (LMS) for these two major components, and a schedule of due dates for all major and minor components. Further, students are encouraged to download citation management software (EndNote is free for our students, Zotero also has a free version available) by day three of the course to begin organizing their sources. Students are tasked with identifying a drug of interest (or several) by day four of the course, with the term drug loosely defined as a chemical that interacts with a cell and initiates a signaling reaction in the cell. Students often choose drugs that they are personally interested in, ranging from therapeutics (antidepressants, cancer chemotherapies) to endogenous ligands (insulin, epinephrine) to illicit drugs (marijuana).

After students have chosen a drug (within week 2), they are reminded in class that they have access to the pathway map project instructions on the LMS and the instructor leads a brief discussion about these. Specifically, in class students are instructed to do the following:

> Research your drug in the primary (and secondary) literature. As you read each article, add a visual summary of the article to your pathway map. Every item you add to your map should be properly referenced using a numbering system that corresponds to Council of Science Editors (CSE, the citation style adopted for use by the biology department) style references in a numbered bibliography. Focus on the direct effects of your drug on signal transduction and the initiation of a signaling cascade. When you cannot find information about a step on your map, such as how the pathway moves from point A to point B, label this step with a question mark.

Following this discussion, the instructor walks through how to perform a literature search, utilizing library resources and biology-specific search engines. Several examples of pathway maps from previous students are also shown, and the details of the assignment are reviewed in the grading rubric.

9.2.2.2 Art Studio and Gallery Walk

During the third week of the course, we meet for a full class period in an art studio, where one of our art faculty members takes the instructional lead. This lesson is a combination of a two-part activity. First, the instructor provides a brief lecture on common art elements including line, shape, value, texture, and color. Part two of the activity involves students working in pairs with a blank sheet of poster-sized paper and several markers. The instructor provides instructions on how to draw a line (from corner to corner, jagged, short, etc), and partners take turns quickly following the instructions until a random collection of lines is created. Each group's drawing is hung on the wall next to each other facing the whole class. Quickly by show of hands, we all vote on our favorite. Surprisingly, but consistently, there is always one clear favorite. We then discuss the favorite random drawing in terms of the art concepts discussed earlier. The student pairs then take a different group's drawing and are tasked to improve it based on the principles they just learned in the lecture. Again, all drawings are posted on the wall for the class to review together at the end of the course period. We vote again for our favorite, with the votes this time usually falling more evenly across all of the drawings – indicating effective use of the recently discussed art elements. Students are encouraged to keep this art experience in mind as they design their own pathway map and these art elements are included in the pathway map grading rubric.

In one of the class periods (week 3) soon after this art studio date, we incorporate a short activity involving professionally designed pathway maps. Posters of signaling pathway maps (acquired from various molecular biology reagent vendors) are placed around the active learning classroom. Students are directed to take a gallery walk together in small groups around the room and discuss what they like and do not like about each poster. Often the art elements learned in the studio lesson are part of this discussion as is the clarity (or lack thereof) presented by the layout and/or detail of each poster. Again, students are encouraged to keep these conversations in mind as they design their own pathway map. One student anonymously responded to the prompt "in what ways was the pathway map project beneficial to you" with the comment,

> I also liked how you showed all of the examples of good or bad pathway maps. That helped me with what I wanted to do for my own pathway map.

This short gallery walk exercise had a lasting impact as this feedback was not solicited until the end of the course, many weeks and several stages past this phase of the project. We also continue to discuss how the art elements are used (well and not so well) when we look at various graphics throughout the course, from textbook figures to data from primary research articles we read and discuss together.

At the beginning of almost every class period that follows, students are reminded about the due date of the first draft of their pathway map and the instructor provides time to respond to any questions. Building on the instructions described above, the instructor reminds students that this pathway map is a way to visually summarize what they find as they are reviewing the literature. Students should be focused on *trying* to find a place (or places) on their map that are unknown, which they represent with question marks. Further, students are tasked to write a one sentence question on their pathway map that represents a research question option for them to test through the experiments they will propose in their grant proposal. Therefore, because students understand that this pathway map acts as an assignment to help them prepare for their grant proposal, they can recognize the scaffold design of these project assignments. This understanding is represented in the following student quote:

> I found developing the pathway map made it much easier to develop a research question than creating an annotated bibliography because I could actually visualize the concept as a whole and assess my learning through what I physically and artistically created. For me, looking at a piece of art and visualizing it that way (makes it) much easier to digest the information than looking at more words. I believe that's why figures are so helpful in textbooks.

9.2.2.3 Draft Pathway Map and Peer Review

A draft of each student's individual pathway map is due at the end of the fourth week of the course. Students turn in an electronic copy to the course LMS for minor points and bring two hard copies to class. Each student receives two peers' reviews of their map and reviews two of their peers' maps. Students are provided a hard copy of the peer review sheet to write their feedback on, which consists of four simple questions and a copy of the grading rubric to complete (Appendix). This feedback is not collected by the instructor but given back to each student along with the hard copy of their pathway map, which may also be directly written upon by peer reviewers.

9.2.2.4 Individual Student Meetings with Instructor

In the week following peer review (week 5), students are invited (not required – although 90–100% participate every semester) to meet for 20 min in-person individually with the course instructor. The instructor does not review the pathway maps at all before this meeting and students are instructed to bring a hard copy to this meeting. There seem to be many advantages to this meeting for all involved. Time grading drafts disappears and instead is spent giving oral feedback to students directly. Because this feedback is part of a conversation, this often results in more understandable feedback as students ask clarifying questions immediately. This also ensures that the students 'hear' the feedback, as there is no guarantee that written feedback is read by all students (Glover & Brown, 2006). The student and instructor can physically point to different parts of the map during this discussion and hand-write notes directly on the map.

Perhaps literally outside the scope of this project, the student and instructor get a chance to interact outside the classroom – which allows students to learn where to physically find the instructor if needed in the future. We have noticed that office hours are used at a much higher rate after this activity. This experience also contributes to building relationships between the students and instructor, adding to the community aspect of the class. This experience also helps the instructor explain to students about the process of science, as sometimes students find conflicting information during their literature search. Students leave with a better understanding of how research occurs and that conflicts are a natural part of the scientific process that can be resolved through further experimentation.

9.2.2.5 Final Pathway Map, Feedback and Grant Proposal Instructions

Final drafts of each student's pathway map are due at the beginning of week seven, leaving students space to shift focus to the mid-term exam during week eight. Students again upload an electronic copy of their pathway map to the LMS and bring two hard copies to class. A brief second round of peer review occurs in class, where students respond to two questions (Appendix) that provide feedback for students to reflect on as they start working on the next major component: their grant proposal. During this class period, the instructor also discusses how to develop a research question and students use a worksheet to aid with this process, transitioning from their map to their proposal (Appendix).

The instructor aims to return feedback with plenty of time for students to consider as they prepare their grant proposal draft. Although this "final" draft of the pathway map turned in during week seven is the version graded for course points, students are told the pathway map can be revised for two additional uses in the course. First, the pathway map is required to be included as a figure in their grant proposal. Further, near the end of the course, each student's pathway map is printed as a two foot by two foot poster that they orally present in a campus-wide research symposium. Revised pathway maps are welcomed for both of these applications.

After the mid-term exam (when we hope attention can be achieved at a higher rate again), at the end of week eight, the instructor discusses the instructions for the grant proposal in class. Briefly, this grant proposal is composed of three sections, an Introduction, Background and Proposed Research, in addition to the Literature Cited list. The Introduction is a one-page summary of background and work proposed, very similar to a Specific Aims Page. Students are required to develop a single overarching research question and one specific aim to address this question, and describe these in the Introduction. In the Proposed Research section, students must describe one experimental approach addressing their specific aim and one (briefer) alternative experimental approach in the event the primary approach does not work for technical reasons. The primary research approach should include a conceptual framework (why each piece is included), a design (major differences between experimental and control groups and/or treatment types), methods (including how results are analyzed) and an interpretation of expected outcomes (with both what the data would look like if the hypothesis is correct AND what the data would look like if the hypothesis is not supported). Diagrams explaining the overall experimental process and demonstrating theoretical results (for both outcomes) are encouraged with inclusion of their pathway map as a figure required. Students are referred to Chap. 10 – Writing Research Proposals from Pechenik's *Short Guide to Writing About Biology* (Pechenik, 2012) and the assignment instructions posted on our course LMS for further details.

9.2.2.6 Primary Literature Discussions

Throughout the course, students are assigned primary research articles to read before class, present in small groups and then discuss with the whole class. For each of these articles, specific questions are provided for the students to focus on during their reading and discussions that cover a variety of topics. Several of these topics were designed to give students practice that would directly aid with their grant proposal preparation. Specifically, the focus for one primary research article is to define the following for *specific* data figures: the question the researchers were asking, the method they used to test this question, and an explanation of how that method works. For another primary research article, the discussion focused more on how the data in one figure answered one research question, and what the data would look like if it did not support the hypothesis. Finally, another primary research discussion focused on the overall paper, defining what the overarching question was, what the researchers' hypothesis was, and what specific aims they set out to test. These discussions serve as scaffolded activities that give students hands-on opportunities, with instructor feedback, on skills they will utilize when writing their proposals.

9.2.2.7 Grant Proposal Draft, Peer Review and Individual Student Meetings with Instructor

Drafts of the grant proposal are due during the tenth week, and again peer review is performed in class. Students are provided with a peer review form (Appendix) that includes several questions and directed tasks. We find that having two peers review each draft controls for less detailed/helpful reviewers and provides more valuable feedback for each student.

In the week following peer review (week 11), students are invited to meet for 20 min in-person individually with the course instructor again to review their grant proposal draft. At this meeting, we often begin by having the student read their first paragraph of the Introduction aloud. Then the instructor asks them if what they just read would convince the funder to give them a million dollars? If they hesitate – we ask them to show us where their argument for that is located? If they can find it, we discuss moving the argument earlier in the proposal. If they cannot find such an argument in their draft, we discuss what that could look like. For the Background section, we have the student write the purpose of each paragraph in the margins. This helps to lead a discussion on what is missing or what is unnecessary. We have also noted that pointing out to students that this is often the section where the figure of the pathway map fits best, and that they then need to explain the map in the text, is a helpful way for students to recognize the topics that need to be discussed in this section. Finally, in the Proposed Research section, we focus on having the students point out controls for each experiment, which can lead to conversations about choosing appropriate methods.

9.2.2.8 Final Grant Proposal and Funding Meeting

The final draft of the grant proposal is due in week 12, 3 weeks before the end of the semester. Proposals are assigned to funding groups (4–5 proposals/group, each group has 4–5 students) based on topic, and students are assigned to review topics that they did not write about, to ensure that they do not review their own proposal. Students are provided a funding review form (Appendix) to complete for two proposals, one acting as a primary reviewer and one acting as a secondary reviewer. Each proposal in a funding group only gets read by two reviewers, who then lead the discussion about it on the funding meeting day (week 13). At this meeting, student reviewers are instructed to summarize the proposal's ideas, use the review form to highlight the strengths and weaknesses of the proposal, and decide as a group on an overall score, where 1 = highly recommended for funding and 5 = not recommended for funding. These scores are shared with the instructor so that the funding group gets credit for their work, but not with the student authors of the proposals to try to keep the process emotion-free. After reviewing all the proposals, each funding group makes a consensus recommendation on the top two proposals (ranked one and two) within their section, and this information is shared with the whole class. The process is anonymous and the student reviews do not count toward the writer's grade. Review scores range quite a bit and we find that students are honest and critical with their reviews. The instructor usually provides some sort of small prize (such as a $5 coffee card) and the students often take quite a bit of pride in learning that their proposal was funded by their peers.

9.3 Results

To assess the effectiveness of this project on gap identification and research question development, we designed an assessment rubric (Table 9.1) to assess the following criteria. This rubric was used for our research process, not for grading

Table 9.1 Assessment rubric criteria (Central Map Question, Novel, Scope, Testable, Methods description) defined and associated scores (0, 1, 2) from student cohort presented as frequency (out of n = 43 students) and percentage (%). The Central Map Question was assessed using students' signaling pathway maps, and all other criteria were assessed from students' grant proposals

Assessment Rubric	0	1	2
Central map question	The central question is undefined/out of context AND the unknown element is not highlighted on map.	The central question is undefined/out of context OR the unknown element is not highlighted on map.	The focus of the pathway map highlights the central question while effectively framing it in the context of the cell. The unknown element is highlighted on the map
Frequency of scores	4/43 (9.3%)	9/43 (20.9%)	30/43 (69.8%)
Novel	The research question has no relation to map or the gap targeted by the research question is already known/not a gap	The research question does not reflect a full understanding of the map	The research question clearly targets the gap in the literature
Frequency of scores	3/43 (7.0%)	3/43 (7.0%)	37/43 (86%)
Scope	The research question is too broad – Covers the whole map	The research question tries to accomplish too much – Covers a significant portion of the map	The research question is specific to the gap on the map and asks only one question
Frequency of scores	1/43 (2.3%)	7/43 (16.3%)	35/43 (81.4%)
Testable	Totally unrelated method identified (sometimes because no technique exists currently to test this question)	Indirect method identified that will not clearly support or refute the hypothesis	Direct method identified that provides evidence toward hypothesis
Frequency of scores	1/43 (2.3%)	11/43 (25.6%)	31/43 (72.1%)
Methods description	Techniques are not visually described AND are not clearly referenced.	Techniques are not visually described OR are not clearly referenced.	Techniques are clearly visually depicted or appropriately referenced from the literature.
Frequency of scores	1/43 (2.3%)	10/43 (23.3%)	32/43 (74.4%)

student work. *Central Map Question* refers to whether the gap (unknown element) is clear on the pathway map and highlights the main associated question. *Novel* refers to whether the research question in the grant proposal targets the gap identified on the pathway map. *Scope* refers to the specificity of the research question in the grant proposal. *Testable* refers to whether research techniques (direct or indirect) are available to answer the question and are identified correctly by the student. This criterion assesses the research question and method choice simultaneously. Finally, the *Methods description* refers to whether the student can explain associated research methods appropriately, as a demonstration that they understand the methods. This criterion assesses the methods section of the grant proposal as opposed to the research question itself.

Scores from our 43 student cohort (Table 9.1) demonstrate that overall, students are meeting expectations, as the most frequent scores are 2 (out of 2 points total) for all criteria. Students tend to do better on the *Novel* (37/43 scored a 2) and *Scope* (35/43 scored a 2) criteria than the other criteria, which supports the improved research question development goal of this scaffolded project.

We noted a slight, statistically non-significant trend for students that do not do as well on the *Central Map Question* criterion to also score lower on the other criteria (data not shown). This indicates it may be helpful to increase student awareness of how well their map demonstrates a gap and focuses around an associated research question. For this, we have considered adding a specific question on the peer review form asking the peer reviewer to list the identified gap(s) and re-state the associated question in their own words. Of note, in the current design, gap identification and its connection to research question development is emphasized during the individual student meetings with the instructor.

Speaking to the effectiveness of the peer review and individual student meetings with the instructor, among other aspects of this project, we often observe much improvement from the pathway map draft (Fig. 9.3 A,C,E) to the final version (Fig. 9.3B,D,F). Specifically, we often see an improved use of art elements (including color representations, arrows, and use of space) and level of detail. For example, in the student exemplars in Fig. 9.3, all three students use the same color in their final maps to indicate precursor molecules and their final products (CYP384 becomes methylated (Fig. 9.3B), PIP_2 becomes DAG and IP_3 (Fig. 9.3D) and GDP becomes GTP (Fig. 9.3F)), while this level of detail and color representation was missing from their original drafts. Additionally, all three students modify the arrows to differentiate time/relationship aspects in their final drafts compared to more simplistic arrow use in their first drafts. Further, both Student 1 and Student 3 better utilize space to indicate details relevant to one cell type versus another in their final maps (untreated vs treated cancer cells by Student 1 (Fig. 9.3B), liver vs heart vs lung cells by Student 3 (Fig. 9.3fF). All three students also indicate increased awareness/understanding of spatial location within the cells in their final maps, as organelles including the cell membrane, nucleus, and sarcoplasmic reticulum are more clearly indicated. Altogether, these changes demonstrate an increased understanding of the material and of how to best visually represent these concepts.

In addition to the changes observed visually in the pathway map draft to final versions, the research questions also evolve during the individual steps of this scaffolded project (Table 9.2).

Commonly, there are several gaps identified in the draft pathway map and the students begin to focus their research questions onto one of these gaps by the final pathway map step, as observed for Student 3. Also, students often think they have

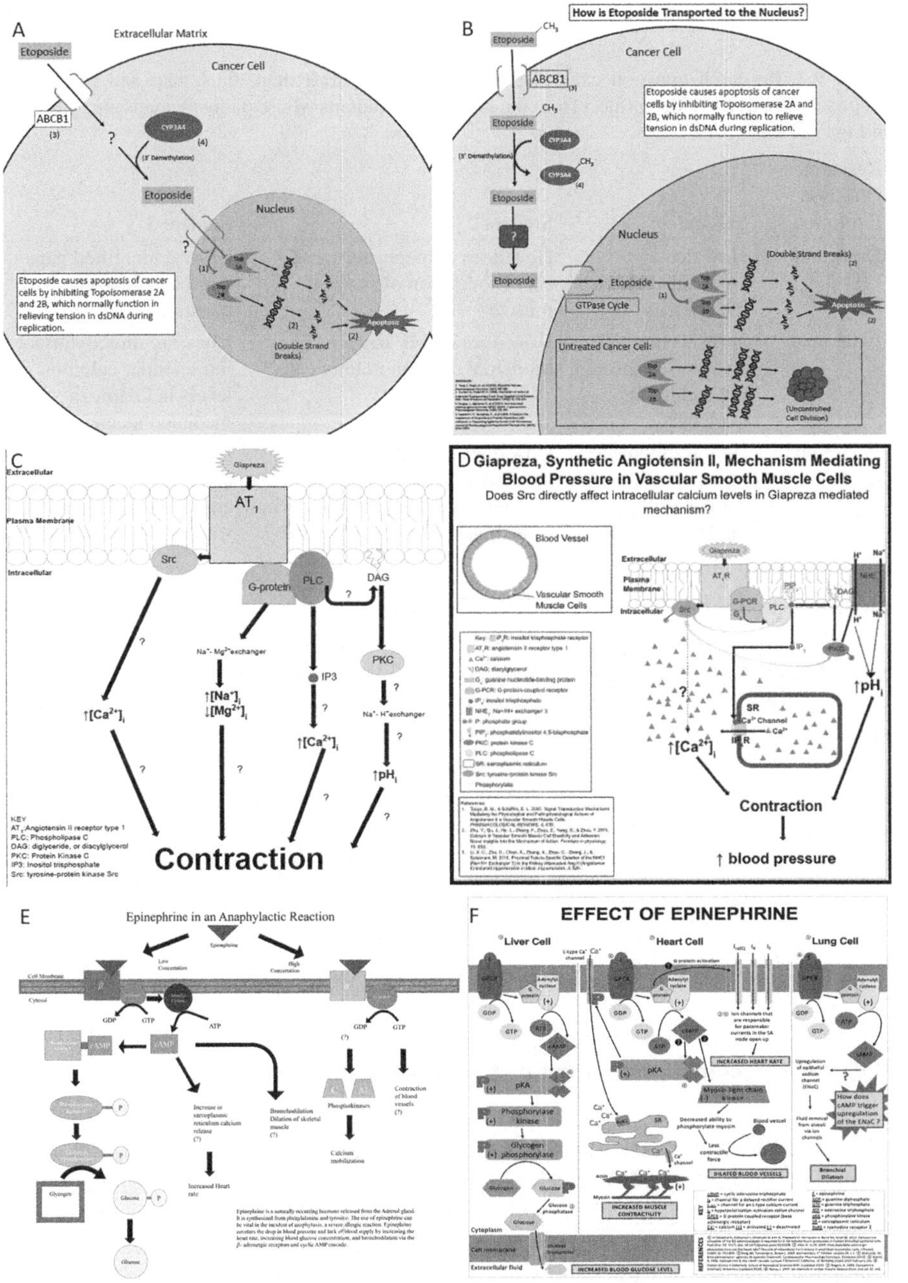

Fig. 9.3 Draft pathway maps (A, C, E) and final pathway maps (B, D, F) created by three students (Student 1 = A, B; Student 2 = C, D; Student 3 = E, F). In A, note there are two question marks, and in the final draft (B) there is a central question mark located between the vertical and horizontal pathways. In C, the student's draft contains multiple question marks, but the final draft (D) includes a single question mark between Src and (Ca^{2+}), denoting the student's improved understanding of the field. The student in E also included multiple question marks, but in F demonstrated a significantly altered understanding of the types of cells involved in anaphylaxis and included a question mark only in the lung cell

Table 9.2 Research question examples at three stages (draft map, final map and final grant proposal) of the scaffolded project from the same three students whose pathway maps are presented in Fig. 9.3

Research Question Evolution	Student 1	Student 2	Student 3
Draft map	How does etoposide get into the nucleus?	How does epinephrine cause bronchodilation of skeletal muscle?	Several identified gaps without a central question presented.
Final map	How is etoposide transported through the cell to the nucleus?	How does cAMP trigger bronchial dilation in lung cells?	Does Src directly affect intracellular calcium levels in Giapreza mediated mechanism?
Final grant proposal	How is etoposide transported through the cell to the nucleus?	How does cAMP trigger upregulation of the epithelial sodium channels (ENaC) (during bronchial dilation in lung cells?	How does Src increase calcium to stimulate contraction in vascular smooth muscle cells?

identified an unknown gap at the draft map stage, but proceed to find published data that fills this gap during their continued research to prepare the final map. For example, from draft to final map, Student 1 realized the etoposide nuclear transport process is known, but identified that how etoposide gets transported to the nucleus is still unknown. Sometimes students have incorrect information in their draft maps that also gets corrected with further research, observed when Student 2's question changed from draft to final map to correct the cell type. When progressing from the pathway map to the grant proposal, the research questions often become even more specific (Students 2 and 3). We have observed that this specification or narrowing of scope commonly occurs as students grapple with method selection. If their research question was too broad for one method to answer, students realize they need to modify their research question.

9.4 Discussion and Implications for Instructors

The ACE-Bio Network (Advancing Competencies in Experimentation) has highlighted seven biological research competencies: *Identify*, *Analyze*, *Communicate*, *Question*, *Conduct*, *Plan*, *Conclude*. This signaling pathway map approach focuses on the competencies of *Identify*, *Question*, and *Communicate*. We have found that the visual mapping process helps students to better understand their literature review and *Identify* key gaps in the current scientific understanding in an area. This understanding helps students to develop better research questions that are targeted at the gap in knowledge (novel) and are focused at the right scope and level for a research project.

Through direct instruction and their personal experience in the process, students recognized how the visual maps helped them to synthesize the field and develop questions. In the course evaluation, we asked two open ended questions. The first was, "In what ways was the pathway map project beneficial to you?" Students responded with a variety of answers about their experiences:

> It was actually kind of fun to see all my research come together and create a signaling map. I think it was beneficial to me because it gave me practice researching and coming up with an unknown question.
>
> I greatly like expanding my comfort zone and working with different search engines and Powerpoint more.
>
> The pathway map project was super cool. I really enjoyed it and I think it was very helpful for understanding how to read a pathway map.
>
> It drove me to really dig deep on a topic and build the skills to develop a complete understanding on my own as well as helping me practice effectively communicating that understanding.

We asked the students a second open-ended question: "In what ways was the grant proposal useful?"

> I liked the new style of writing that could actually be applied to my life in the future.
>
> Better my scientific writing skills.
>
> Since I'm going to grad school, this was a very valuable experience to have since I will no doubt be writing these in the very near future. It was good to see the basics and know how to construct those and what to look for in good grant proposals. I will also be looking over grant proposals for the science community so the peer review was a fantastic exercise as well.
>
> The grant proposal was very helpful with my scientific writing and with writing grant proposals in the future.
>
> It helped me learn how to search for gaps in the understanding of a topic and it helped me practice my scientific problem solving skills to develop a way to fill that understanding gap.

From these responses and in-class discussions, it was clear that students understood how this work could be applied to other areas of their lives, including job applications, funding opportunities, graduate school applications, and more. In addition, during the study section activity, we emphasized that this type of group deliberation is also used by search and hiring committees for a variety of professions. This helped them to understand how a review group can be influenced by the clarity of writing (or lack thereof) in professional documents such as applications.

This assignment has been particularly effective as a teaching tool because this visual summary of research itself facilitates a conversation between faculty and student or between a student and their peers. The visual mapping process helps make it clear if a student has (1) performed a significant amount of literature searching and (2) synthesized the material into a deep understanding of the cellular system. The instructor can use these visual cues to quickly help a faltering student develop a more effective search approach early in the process. The instructor can also provide one on one instruction in difficult areas, or potentially redirect a student towards an area better aligned with their interests or abilities.

9.4.1 Implications for Instructors

To assist instructors who are new to this area, we offer the following guidelines for teaching:

- Identify a colleague to help you with the artistic elements of a signaling map. Elements of art including line, shape, value, texture, and color are essential for clear visual communication of scientific ideas. This engages students to utilize art elements when designing their visual signaling pathway map.
- Include structured peer review exercises that direct students to comment on specific aspects of the signaling pathway maps and proposal drafts to produce useful feedback. Peer review forms and guidance can be found in the Appendix.
- Take the time for one-on-one feedback sessions with students. This process takes less time than grading, produces more constructive feedback, and strengthens the mentoring relationship between instructor and student.
- Encourage a growth mindset in your students by encouraging the process of drafts, feedback, and learning.
- Whenever possible, relate the activities of the project to real world experiences, such as how the funding session conversations are similar to how admissions committees review applications, search committees review candidates, etc.
- This experience, where students visually synthesize published research in order to identify a gap in the research literature, prepares students for professional work in addition to teaching cell biology and experimentation.

Any field in which experimentation is taught could potentially modify this approach. For cell biology, the map of a cell serves as a template on which students map out data in a geographical and temporal context. Within the larger field of biology, an ecologist might use a food web as a template instead of a cell. In chemistry, the template for mapping might be the context of a long string of reactions between a substrate and a final product. A physicist might develop a sketch of an experimental set up with many different materials upon which the student maps information discovered about each of the materials. The important aspect of the assignment is to provide a template for visually synthesizing the information gained through the literature search in order to find the gap in knowledge.

In the future, there are opportunities to link this research proposal project more closely with students' future careers as scientific researchers. Several students each year produce work that is a suitable first draft for federal research funding such as the NSF Graduate Research Fellowship Program. About a quarter to a third of students in this class have plans for graduate research training post graduation. We are considering offering opportunities for students to develop their proposal into a full grant proposal or mentor them to use the signaling pathway map approach to develop a new topic for an NSF proposal. This approach could encourage more students to learn about graduate research programs, the funding available, and career possibilities in the experimental sciences.

In order to help undergraduate students develop strong experimentation skills, we need to provide opportunities for them to engage with all the competencies that lead to strong scientific skills. *Identifying* gaps in knowledge and then designing research questions to address these gaps is a difficult task to master. We believe the visual signaling pathway map approach provides resources for instructors to use to support their students in this area. Together with additional experiences in methodology design, data collection, and data analysis, these approaches can help our students be prepared for careers in scientific experimentation.

Appendix: https://quarry.css.edu/islandora/object/CSSrepository%3A41471

CrediT Author Statement

Anne Kruchten: Conceptualization, Methodology, Writing- Original Draft, Writing- Review & Editing **Jenean O'Brien**: Methodology, Writing- Original Draft, Writing- Review & Editing

Acknowledgments Sincere thanks to Paul LaJeunesse, Assistant Professor of Art, for collaborating on the artistic instruction for our students. We extend our appreciation to all of our BIO 3600 students for participating in this project, and special thanks to students Matthew Hinaus, Samuel Pitzen, and Mary Vue for approving inclusion of their signaling pathway maps as examples in this article. We also extend our appreciation to colleagues Melanie Talaga and Jennifer Maki for critical reading of our manuscript drafts.

References

American Association for the Advancement of Science (AAAS). (2011). *Vision and change in undergraduate biology education: A call to action*. Washington, DC. Retrieved December 23, 2020 from https://live-visionandchange.pantheonsite.io/wp-content/uploads/2013/11/aaas-VISchange-web1113.pdf

Cole, K. E., Inada, M., Smith, A. M., & Haaf, M. P. (2013). Implementing a Grant proposal writing exercise in undergraduate science courses to incorporate real-world applications and critical analysis of current literature. *Journal of Chemical Education, 90*(10), 1316–1319. https://doi.org/10.1021/ed400130s

CUREnet. (2020). *What is a CURE?* Retrieved from https://serc.carleton.edu/curenet/whatis.html

Emtage, L., Bradbury, L., Coleman, N., Devenport, D., Nietzel, A., & Grew, J. C. (2016). Cell signaling pathways – A case study approach. *CourseSource* Retrieved from https://www.coursesource.org/sites/default/files/downloads/Emtage-Cell%20Signaling%20Pathways-A%20Case%20Study%20Approach%20.pdf

Evans, H. G., Heyl, D. L., & Liggit, P. (2016). Team-based learning, faculty research, and Grant writing bring significant learning experiences to an undergraduate biochemistry laboratory course. *Journal of Chemical Education, 93*(6), 1027–1033. https://doi.org/10.1021/acs.jchemed.5b00854

Glover, C., & Brown, E. (2006). Written feedback for students: Too much, too detailed or too incomprehensible to be effective? *Bioscience Education, 7*(1), 1–16. https://doi.org/10.3108/beej.2006.07000004

Herek, G. M. (2011). Developing a theoretical framework and rationale for a research proposal. In W. Pequegnat, E. Stover, & C. A. Boyce (Eds.), *How to write a successful research Grant application: A guide for social and behavioral scientists* (pp. 137–145). Springer US.

Hester, S. D., Nadler, M., Katcher, J., Elfring, L. K., Dykstra, E., Rezende, L. F., & Bolger, M. S. (2018). Authentic inquiry through modeling in biology (AIM-bio): An introductory laboratory curriculum that increases undergraduates' scientific agency and skills. *CBE: Life Sciences Education, 17*(4), ar63. https://doi.org/10.1187/cbe.18-06-0090

Hole, T. N. (2018). Working and learning in a field excursion. *CBE: Life Sciences Education, 17*(2), ar24. https://doi.org/10.1187/cbe.17-08-0185

Indorf, J. L., Weremijewicz, J., Janos, D. P., & Gaines, M. S. (2019). Adding authenticity to inquiry in a first-year, research-based, biology laboratory course. *CBE: Life Sciences Education, 18*(3), ar38. https://doi.org/10.1187/cbe.18-07-0126

Jude, B. A. (2017). Strategies for developing a science communicator through authentic science writing assignments. *CourseSource*. https://doi.org/10.24918/cs.2017.5

Kramer, I., & Thomas, G. (2006). Meeting report: Teaching signal transduction. *CBE: Life Sciences Education, 5*(1), 19–26. https://doi.org/10.1187/cbe.05-11-0127

Libarkin, J., & Ording, G. (2012). The utility of writing assignments in undergraduate bioscience. *CBE: Life Sciences Education, 11*(1), 39–46.

National Research Council. (2003). *BIO2010: Transforming undergraduate education for future research biologists*. The National Academies Press.

Novak, A. M., & Treagust, D. F. (2017). Adjusting claims as new evidence emerges: Do students incorporate new evidence into their scientific explanations? *Journal of Research in Science Teaching, 55*(4), 526–549. https://doi.org/10.1002/tea.21429

Pechenik, J. (2012). *A short guide to writing about biology* (8th ed.). Pearson.

Pelaez, N.J., Anderson, T.R., Gardner, S. M., Yin, Y., Abraham, J. K., Bartlett, E., Gormally, C., Hill, J. P., Hoover, M., Hurney, C., Long, T., Newman, D.L., Sirum, K., & Stevens, M. (2017). The basic competencies of biological experimentation: Concept-skill statements. *PIBERG Instructional Innovation Material Paper 4*. Retrieved from http://docs.lib.purdue.edu/pibergiim/4. Accessed 18 Oct 2021.

Phillips, A. M., Watkins, J., & Hammer, D. (2017). Problematizing as a scientific endeavor. *Physical Review Physics Education Research, 13*(2), 020107. https://doi.org/10.1103/PhysRevPhysEducRes.13.020107

Ratan, S. K., Anand, T., & Ratan, J. (2019). Formulation of research question – Stepwise approach. *Journal of Indian Association of Pediatric Surgeons, 24*(1), 15–20. https://doi.org/10.4103/jiaps.JIAPS_76_18

Sirum, K., & Humburg, J. (2011). The experimental design ability test (EDAT). *Bioscene: The Journal of College Biology Teaching, 37*, 8–16.

Soneral, P. A., & Wyse, S. (2015). Your tax dollars at work: A mock grant writing experience centered on scientific process skills. *CourseSource*. https://doi.org/10.24918/cs.2015.16

The_Writing_Center. (2018a). *How to write a research question*. George Mason University.

The_Writing_Center. (2018b). *Organizing literature reviews*. George Mason University.

The_Writing_Center. (2018c). *Writing a literature review*. George Mason University.

Wiegant, F., Scager, K., & Boonstra, J. (2011). An undergraduate course to bridge the gap between textbooks and scientific research. *CBE: Life Sciences Education, 10*(1), 83–94. https://doi.org/10.1187/cbe.10-08-0100

Wilson, K. J., & Rigakos, B. (2016). Scientific process flowchart assessment (SPFA): A method for evaluating changes in understanding and visualization of the scientific process in a multidisciplinary student population. *CBE: Life Sciences Education, 15*(4), ar63. https://doi.org/10.1187/cbe.15-10-0212

Woodin, T., Carter, V. C., & Fletcher, L. (2010). Vision and change in biology undergraduate education, a call for action—Initial responses. *CBE: Life Sciences Education, 9*(2), 71–73.

Zagallo, P., Meddleton, S., & Bolger, M. S. (2016). Teaching real data interpretation with models (TRIM): Analysis of student dialogue in a large-enrollment cell and developmental biology course. *CBE: Life Sciences Education, 15*(2). https://doi.org/10.1187/cbe.15-11-0239

Chapter 10
Virtual Microscope: Using Simulated Equipment to Teach Experimental Techniques and Processes

Cecilia I. Casali, Rocio A. Moreira Szokalo, Bruno J. Santacreu, Lucila G. Pescio, Laura Bonofiglio, Daniela J. Romero, and Nicolás O. Favale

10.1 Introduction

10.1.1 Simulation-Based Education

In recent years, undergraduate science education has been exploring new teaching and learning practices to develop the basic competences in biological experimentation (Pelaez et al., 2017; Chap. 1 in this volume). The design of effective strategies thus includes platforms for cognitive thinking and spaces for knowledge transfer from professors and researchers, all based on high flexibility and limited time. Simulation has been used as a training technique in the aeronautical industry and military fields since the early 1900s. The first flight simulator was developed in 1929, and its complexity and sophistication improved progressively with the integration of computer-based systems. The translation of simulation into health

C. I. Casali · B. J. Santacreu · L. G. Pescio · D. J. Romero · N. O. Favale (✉)
Departamento de Ciencias Biológicas. Cátedra de Biología Celular y Molecular, Universidad de Buenos Aires. Facultad de Farmacia y Bioquímica, Buenos Aires, Argentina
e-mail: ccasali@ffyb.uba.ar; bsantacreu@ffyb.uba.ar; lucilagpescio@ffyb.uba.ar; djromero@docente.ffyb.uba.ar; nofaval@ffyb.uba.ar

R. A. Moreira Szokalo
Departamento de Bioquímica Clínica. Cátedra de Bioquímica Clínica II. Citogenética Humana y Genética Toxicológica, Universidad de Buenos Aires. Facultad de Farmacia y Bioquímica, Buenos Aires, Argentina
e-mail: rmoreira@docente.ffyb.uba.ar

L. Bonofiglio
Departamento de Microbiología, Biotecnología, Inmunología y Genética. Cátedra de Microbiología, Universidad de Buenos Aires. Facultad de Farmacia y Bioquímica, Buenos Aires, Argentina
e-mail: lbonofi@ffyb.uba.ar

N. J. Pelaez et al. (eds.), *Trends in Teaching Experimentation in the Life Sciences*, Contributions from Biology Education Research,
https://doi.org/10.1007/978-3-030-98592-9_10

education has undergone almost exponential growth (Rosen, 2008), and simulation-based education has emerged as a solid strategy offering students a bridge between the classroom and professional reality in different scenarios (Salas Perea & Ardanza Zulueta, 1995). Indeed, this training method can be used not only to replace or expand the real experience through guided practice (Bernardo, 2017), but also to enable students to build up knowledge based on personal decision-making about real biology-related problems (Beux & Fieschi, 2007), while reflecting on errors as a learning method (Porath, 2000).

Simulation-based education has been used in different areas of sciences, with examples including training without patients and inter-professional communication to share pathology interpreted from images in microscopy slides (Lindgren & Schwartz, 2009). In this way, students as active participants have safely practiced and developed research abilities to achieve their learning goals (Brydges et al., 2015). In addition, Shegog et al. (2012) showed that the use of simulation is a positive complement to learn in the biology classroom because it helps to promote both procedural and declarative knowledge of molecular biology procedures. On the other hand, the rapid progress of mobile and wireless communication has allowed the development of ubiquitous learning (u-learning) environments, where students learn from the real world with access to digital resources (Hwang, 2014).

Since modern devices can offer virtual data and authentic, simulated experiences, their integration into curricula can bring about a structural shift away from classrooms as the primary learning space (Dede, 2011). Integrating technologies such as simulations can help teachers engage their students in observational, correlational, and experimental inquiry investigations; and may ease the necessary transition from traditional teacher-centered, lecture-style teaching to more learner-centered, reform-based teaching (Maeng et al., 2013). Considering the changes thus promoted in teaching and learning as a more continuous process, faculty can take advantage of u-learning and articulate simulation-based education to create active-learning environments for students anywhere and anytime using computer or mobile devices (Burbules, 2012).

10.1.2 Virtual Microscope as a Simulation Tool

Microscopy uses a fundamental tool for health professions including bio-chemistry, pharmacy, medicine, dentistry, and veterinary sciences. Microscope manipulation and precise observation are thus essential skills that require sustained and assisted training.

Microscopy learning in massive courses requires a large number of microscopes and samples, long observation times, and the supervision of qualified staff for students to acquire expertise and develop critical thinking skills. This is particularly relevant in undergraduate courses, where the number of students is often above 100 per class. Therefore, the problem is easy to recognize but difficult to solve in practical terms.

For these reasons, an ideal training program should meet at least the following requirements:

- Sufficient number of microscopes for the number of students (ideally 1:1).
- At least one representative sample per student.
- Time to practice according to students' needs.
- Permanent access to instructional assistance.

However, undergraduate courses generally have a large number of students with limited teaching staff and practice time (Saco et al., 2016). Due to the difficulty in teaching microscopy effectively to large undergraduate classes, simulated microscopy training proves a valuable tool to be incorporated into e-learning approaches. A growing trend across science education is to use computer-assisted instruction to enhance traditional teaching strategies (Cook et al., 2008; Paulsen et al., 2010), as these digital platforms can significantly help to solve the problems enumerated above. The benefits include (a) an unlimited number of samples, (b) the fact that each sample may be observed by several students at the same time, (c) out-of-classroom access to the microscope, and (d) collaborative learning experience (Triola & Holloway, 2011).

Several projects are currently addressing the use of digital platforms for microscopy teaching. Most of these platforms emerged from histology and pathology teaching (Kuo & Leo, 2019) and are referred to as "virtual microscopes" (VM). However, this term has no generally accepted definition (Glatz-Krieger et al., 2006), as it may refer to a wide range of platforms with different tools and even diametrically opposed functions. VM comprise mostly two types: those focusing on training in precise observation and sample analysis and those focusing on training in optical and mechanical use of functioning parts and their adjustment. However, some VM could combine both characteristics.

One of the most widespread platform types today is classified as whole slide imaging (WSI). WSI technology involves digitizing a complete specimen and making it accessible via the internet to computers or mobile devices, thus al-lowing unlimited and simultaneous remote access by different users to the same specimen (Saco et al., 2016). In addition, digital samples do not degrade over time and can then be distributed instantly to multiple students with no risk of breakage, loss, or contamination. WSI has been mainly aimed at training students and professionals in recognizing structures. This technology is generally presented as a panoramic image where users can apply the zoom tool to magnify a specific image area much as they do with an online map, with some projects even using Google map API for visualization (Triola & Holloway, 2011). Although zooming on the panoramic image may or may not be consistent with objective magnification (Foad, 2017; Tian et al., 2014), these platforms allow students to collaboratively create content by annotation sharing (Triola & Holloway, 2011). WSI also enables users to simultaneously visualize several specimens and staining techniques, thus facilitating observation and comparison (Paulsen et al., 2010). As an additional advantage, the developers of this type of platform have pointed out that users find WSI images to be always in focus (Ordi et al., 2015; Saco et al., 2016). Therefore, students become easily acquainted with the specimen and can immediately focus on histological features. This characteristic, which can be regarded as an advantage depending on the context, can also be considered a disadvantage, as it does not reflect real microscope use.

In contrast, other VM types are designed for student training on how to use and adjust the optical and mechanical parts of a microscope for a detailed observation of the specimens (BioNetwork's Virtual Microscope, n.d.; UD Virtual Compound Microscope – University of Delaware, n.d.). In this case, the main criticism focuses on the small number of samples, with the sample sometimes representing a single field plane.

VM – especially those based on WSI – have been extensively used in microscopy teaching. Several studies have even addressed the impact of these platforms on teaching success compared to the use of the conventional microscope, with encouraging results (Kuo & Leo, 2019). These tools combined with a team-based learning approach have been found to improve learning efficiency (Farah & Maybury, 2009). Kuo and Leo have collected these and other studies in an interesting systematic review (Kuo & Leo, 2019), which reinforces the promising prospects of the use of the VM in microscopy teaching. An increasing number of institutions are incorporating VM as a regular teaching tool (Hanna et al., 2015; McBride & Drake, 2018); however, a shift from the conventional microscope to the exclusive use of the VM should be further evaluated as there may be a need to develop better applications (Pantanowitz et al., 2012).

10.2 Our Aim to Improve Science Education with Virtual Microscopy

Although the advent of new information technologies may certainly generate a wide range of interesting tools, their application in teaching environments should be in keeping with pedagogical decisions. In this context, when we considered our own VM, our design was based on reproducing what students observe with the conventional microscope. For this purpose, we designed our VM (Facultad de Farmacia y Bioquímica- Virtual Microscope – FFyB-VM) not only focusing on slice recognition and identification, but also on the different parts of the microscope, which could be adjusted to improve specimen observation. For this reason, our design aimed to integrate both types of VM mentioned above. Our VM may hence be thought of as an ideal on-demand microscope for each student, which may not only help overcome microscopy-training difficulties, but also provide an innovative strategy for distance education and student-centered learning about experimentation with the microscope as a research tool.

Our platform has virtual knobs to place the sample into focus and adjust the diaphragm aperture and light intensity for the clearest image. Also, users can change the objectives and virtually move the specimen to visualize specific sample features. The use of FFyB-VM entails active student participation and thus provides training opportunities in focus adjustment, a key research skill. Students can also learn to better formulate hypotheses and design experiments, as well as interpret results and draw conclusions through specifically designed tasks. In addition, FFyB-VM offers the possibility of deferred feedback, thus allowing students to later receive the corrections made by the teachers. Students may take photographs of the experiments, edit and send them to teachers with comments, thus tracking their own learning. Since the goal of FFyB-VM is to provide a powerful tool through information and communication technology (ICT), the simulation provides students access from any mobile device or computer so they can practice with no space or time constraints.

FFyB-MV is versatile and can be adapted to specific learning goals. For instance, in a class aiming at specimen focus training, exercises are designed to guide students in the use of the coarse and fine adjustment knobs and the adjustment of light level and diaphragm aperture when changing objectives; in this case, students may then send the teacher a photo of the focused specimen for evaluation. In turn, if the goal is specific cell type recognition, exercises are planned to help students recognize, for example, lymphocytes in a blood smear. Moreover, if the goal is the development of more complex skills such as hypothesis-driven research, exercises are designed to guide students in sample comparison, e.g., control and drug-treated, through specific instructions and open-ended questions. Parts of the ACE-Bio Competencies framework was useful in guiding student engagement with experimentation and in persuading learners to reflect on their experiences and findings. In sum, FFyB–VM provides a panel with exercises and instructions that are associated with the supporting slides and can be tailored to relevant learning objectives. The exercises in FFyB-VM are meant to boost students' ability to organize content into a coherent cognitive structure, combine new and relevant prior knowledge, and apply information to new situations or problems.

FFyB-VM has been designed not only to overcome microscopy-training difficulties, but also to enhance virtual education and offer hands-on research experience. In view of this, we wondered whether our VM could be a suitable tool to teach basic research competencies in our field. As exercises can be designed for students to develop different levels of technical and critical thinking skills, the essential use of FFyB-VM will depend on the defined anticipated learning outcomes (ALOs) (Irby et al., 2018).

Here we present two exercises using the platform for a Cellular and Molecular Biology class and we analyze two aspects: a) FFyB-VM as a platform (What was users' opinion about the platform?) and b) what FFyB-VM contributed to student learning (Did students acquire observation/research abilities?). The two exercises developed in the FFyB-VM platform were designed to be used in the Cellular and Molecular Biology course for biochemistry and pharmacy students at Facultad de Farmacia y Bioquímica – Universidad de Buenos Aires. However, one of the exercises was done by students while they were taking the Cellular and Molecular Biology class, while the other one was completed after the course had finished. The first exercise allowed students to improve their ability in microscope handling and to develop technical skills that are essential in basic research. Additionally, students were able to generate questions and communicate results on the basis of sample observations (*Identify*, a gap in current knowledge: recognize a gap in current scientific knowledge that can be addressed with experimentation) (Pelaez, et al., 2017;

Chap. 1 in this volume). The second exercise involved the observation and analysis of treated and control samples to evaluate the effect of a potentially therapeutic drug. The purpose was to promote students' critical thinking and research skills which involve recognizing drug effects, formulating hypotheses, drawing conclusions in scientific reports, and proposing new experiments to test their hypothesis. The design of this exercise was influenced by the ACE-Bio Competencies framework (*Question*, hypothesis: generate multiple explanations of the natural world that are testable and potentially falsifiable; *Analyze*, data analysis: analyze clean data using discipline-appropriate methods based on the measurements collected and the experimental questions; *Communicate*, synthesis and reflection: evaluate, analyze and explain the significance and implications of the research; also propose follow up experiments based on inferences from predicted or actual results of experiments) (Pelaez, et al., 2017; Chap. 1 in this volume).

With a view to further developing microscopy training conducted in previous Cellular and Molecular Biology courses and optimizing technical skill learning, we analyzed students' experience and aimed to determine whether learning with FFyB-VM represented an authentic professional-like research simulation scenario for undergraduate students.

10.2.1 FFyB-VM in a Cellular and Molecular Biology Class

The first exercise involved cell counting in the Neubauer chamber. A viable cell count is essential for the study of eukaryotic cells for different purposes, such as cell culture management in biological research to track treatment effectiveness and quality control in industrial processes. Total and viable cells can be determined by a Trypan blue exclusion test, which is based on the principle that live cells' intact membranes exclude Trypan blue, whereas non-viable cells' compromised membranes do not. The stained suspension of cells was placed in the Neubauer chamber and counted under the brightfield microscope. In this exercise, the students ($n = 19$) observed the chamber with the stained cell suspension and were asked to count the cells (Fig. 10.1). FFyB-VM allowed students to place the sample into focus with the coarse focus and fine focus knobs, adjust light intensity and diaphragm aperture, select the appropriate objectives, and move the specimen to analyze the sample. As part of the exercise, students were asked to do a series of tasks through the platform such as taking a picture and cropping it and answering open-ended and multiple-choice questions. Students submitted their answers through the platform, and then teachers graded answers and made suggestions for each activity, thus generating feedback and enabling students to repeat exercises as many times as necessary to ensure learning.

The implementation of this exercise and the platform performance were evaluated through a survey. To analyze data, the questions were classified into three main categories: FFyB-VM vs reality, FFyB-VM as a platform (How user-friendly was FFyB-VM?) and the relevance of the proposed exercise (Table 10.1). In addition,

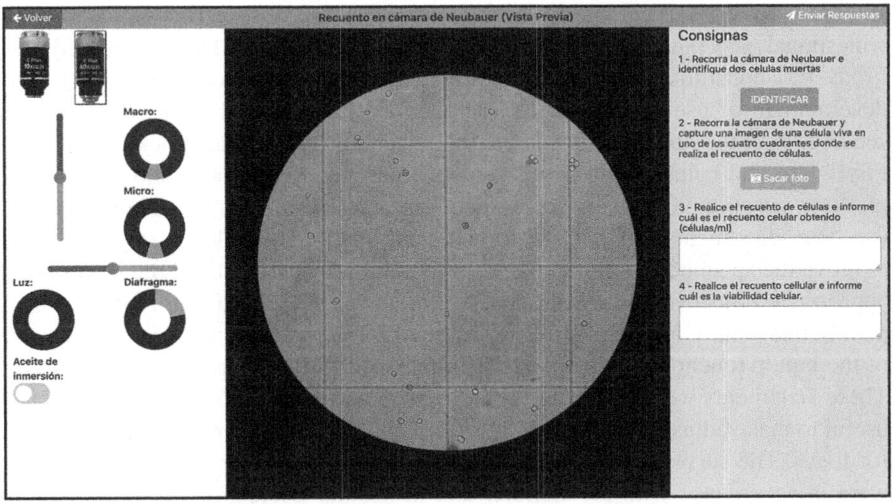

Fig. 10.1 First exercise: cell counting in the Neubauer chamber. The left panel shows the optical and mechanical parts of a microscope, such as objectives, coarse focus (Macro) and fine focus (Micro) knobs, light (Luz) and diaphragm (Diafragma). The blue bars allow to move the specimen along the x and y axis. Users can also add immersion oil (aceite de inmersión). The center panel shows the ocular view for students to visualize the Neubauer chamber with the stained cells. The right panel shows the exercise instructions (consignas), such taking and editing photos and answering open-ended questions. At the end of activities, students are asked to submit their answers through the platform using the button at the top of the panel (Enviar respuestas)

Table 10.1 Students' experience in the FFyB-VM cell counting exercise according to a survey

Survey statement	% Agree/strongly agree (n = 19)
FFyB-VM vs reality	
The VM was a good simulator of a real microscope	89.5
The VM allowed you to acquire technical abilities that are useful for the use of a real microscope	53
FFyB-VM as a platform	
The VM parts (focus knobs, light intensity and diaphragm aperture, objectives, etc) were easy to adjust	36
The VM platform was friendly to navigate	100
The VM was a positive experience	74
Relevance of the proposed exercise	
The proposed exercise was in accordance with the Cellular and Molecular Biology class curriculum	100
The type of instructions in the exercise were adequate	95
The proposed exercise was a positive experience	79

student opinions and comments were collected in the survey. The results showed widespread acceptance in all the categories evaluated. Students reported that FFyB-VM constituted a good simulator of the real microscope and allowed the development of useful abilities for the use of a conventional microscope. In this sense, 53% of the students agreed with this statement and 21% did not know whether FFyB-VM had helped them learn technical abilities. Results also showed that the platform constituted a positive experience and, in general, was friendly and easy to use. The adjustment of FFyB-VM parts (coarse and fine focus knobs, stage controls to move the specimen, light intensity and diaphragm aperture) and the selection of appropriate objectives were sometimes difficult for students using a laptop touchpad. In this regard, 52.6% of students reported focus problems associated to the use of the laptop touchpad, which made them regard FFyB-VM as not easy to handle. These statements were collected in the opinion and comments section and proved useful to make adjustments to the platform such as setting the mouse scroll function for focus. The survey also assessed whether the topics and the type of instructions were in accordance with the Cellular and Molecular Biology class. In this sense, results showed that the exercise constituted a positive experience and that it was in alignment with the topics studied. The opinion and comments surveyed were later used to improve the platform and design future exercises.

Furthermore, we evaluated what students had learned with FFyB-VM (Fig. 10.2). Some students (34%) were initially unable to distinguish viable from non-viable cells. Surprisingly, 58% students were unable to perform cell counting correctly on the first attempt. However, when students were given feedback and then allowed to do the exercise again, the percentages improved, with 82% identifying viable and non-viable cells and 75% carrying out cell counting correctly. FFyB-VM users also showed a tendency to improve cell viability calculation, as 68% students effectively assessed cell viability on the first attempt, but 76% did it properly after teachers' feedback. This exercise and the information obtained in the survey allowed us to

Fig. 10.2 FFyB-VM platform contribution to student learning. The bar graph represents the percentage of students who successfully completed the exercise on the first attempt (lower case a, b and c) and on the second attempt after feedback (upper case A, B and C). The letters represent exercise instructions: a-take a picture of viable cells, b-count cells/ml, and c-calculate cell viability. *, $p < 0.05$; t test

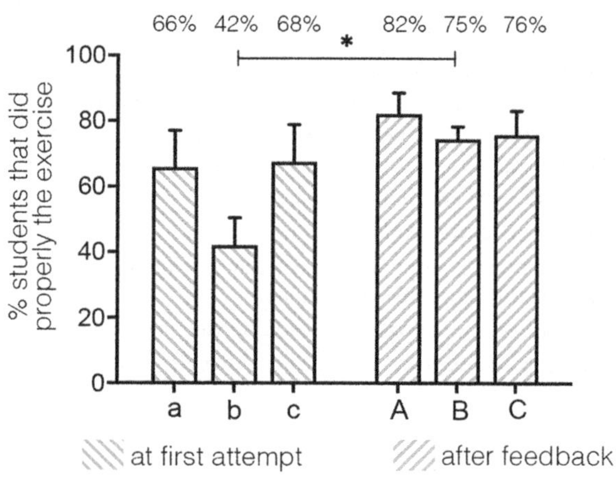

consider FFyB-VM a helpful tool for students' learning and a good simulator of the real microscope.

The second exercise was designed to evaluate undergraduate students' research abilities in the observation and analysis of fluorescence microscopy samples. In this case, we evaluated whether students who had completed the Cellular and Molecular Biology class were able to detect cell alterations (cytoskeleton changes, apoptosis, cell division, nuclear fragmentation, etc.) caused by a new drug, and then hypothesize its mechanism of action. In this exercise, students had two images taken with fluorescence microscopy. One photo corresponded to cells treated with a drug that causes cytoskeleton depolymerization and the other one corresponded to cells without treatment (control). Both cell samples had been processed through fluorescence microscopy using phalloidin-FITC and actin-mCherry to label the actin cytoskeleton, and Hoechst 33258 to label nuclei. FFyB-VM allowed students to select the specimen, control or drug-treated cells, place each sample into focus, adjust diaphragm aperture and light intensity, select the objectives and add oil immersion for the 100x objective. Students analyzed the two samples (Fig. 10.3), did the tasks included in the exercise, such taking and cropping a representative picture of both conditions, and answered open-ended and multiple-choice questions.

One group, defined as Group A (n = 9 students), was asked to observe complete specimens in FFyB-VM and analyze drug effects, while the other group, Group B (n = 12 students), was asked to observe/*Analyze* two photos (control and drug-treated cells) that were representative of the specimen shown in FFyB-VM without using FFyB-VM. In both cases, students were asked to write a brief report in which they proposed, based on their observations, a possible mechanism for the drug and the experiments needed to test their hypothesis. The results obtained by the two groups of students were analyzed through a rubric, and students' experience was evaluated through a survey. At the end of these exercises, students in Group B were given access to FFyB-VM to compare the two methods and answer the survey.

The survey results (Table 10.2) indicated that FFyB-VM was an effective tool for simulating the use of a real microscope and that the platform was easy to use and navigate. Interestingly, about 95% students reported that the adjustment of FFyB-VM parts was easier in this opportunity. For instance, setting the mouse scroll for focus based on students' comments in the first exercise proved a real improvement in VM part handling. Survey results also led us to add a cutting tool to adjust photo capture, a visual reference of stage movement according to objectives, and the possibility to send comments and photos to the teacher. In this way, the feedback obtained from the comments section was used as a guide in decision-making about platform improvements.

A general analysis of the survey showed that students considered FFyB-VM a useful tool to tackle the exercise, but only 48% students considered that FFyB-VM facilitated the formulation of a hypothesis related to drug effects on the actin cytoskeleton. However, as shown in Fig. 10.4, 78% of Group A students preferred FFyB-VM over the photos to formulate hypotheses, while 67% of Group B students had no preferences for FFyB-VM or the photos, and 25% preferred FFyB-VM over

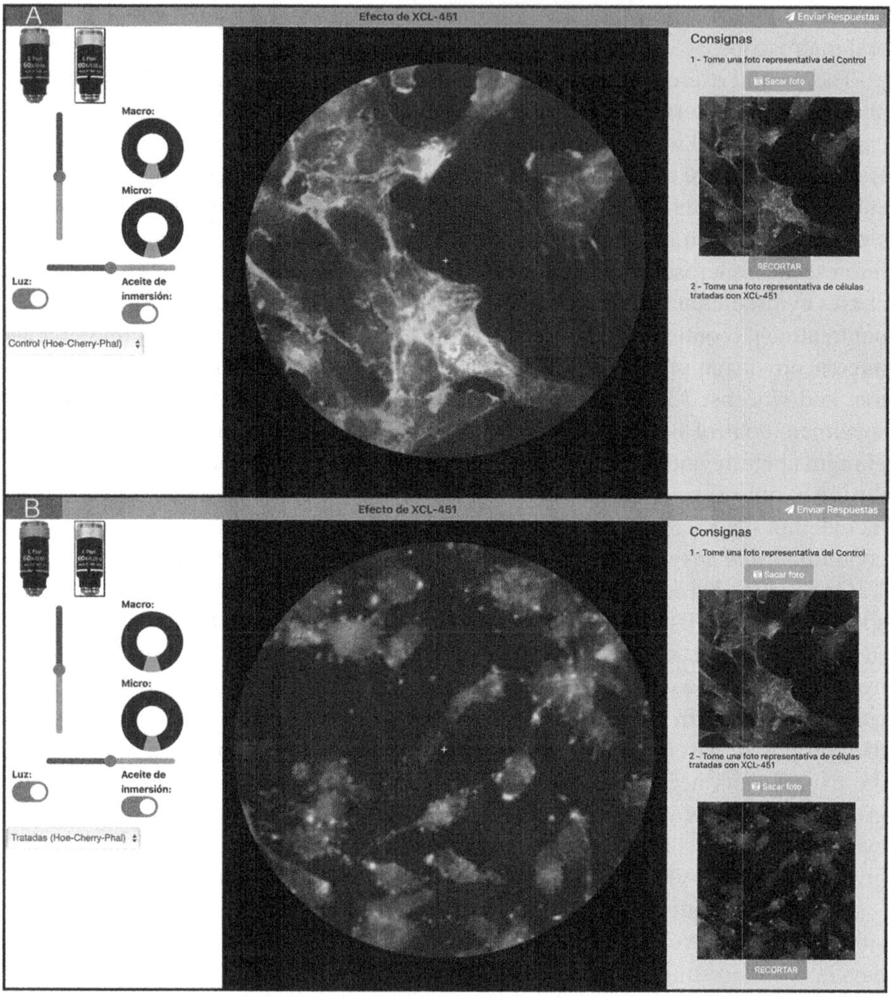

Fig. 10.3 Second exercise: effect of actin depolymerization drug on cell cultures. (**a**) Control cells; (**b**) Drug-treated cells. In both cases, the left panel shows the optical and mechanical parts of a fluorescence microscope, such as objectives, coarse focus (Macro) and fine focus (Micro) knobs and light. The blue bars allow to move the specimen along the x and y axis. Users can also add immersion oil (aceite de inmersión) and, in this exercise, they are asked to select the sample to observe (control or drug-treated cells) in the selection window at the bottom of the panel. The center panel shows the ocular view for students to visualize the sample selected. The right panel shows the exercise instructions, such as taking and editing photos. At the end of activities, students are asked to submit their answers through the platform using the button at the top of the panel (Enviar respuestas)

Table 10.2 Students' FFyB-VM research experience survey

Survey statement	% Agree/strongly agree (n = 21)
FFyB-VM vs reality	
The VM was a good simulator of a real microscope.	86
The VM allowed you to acquire technical abilities that are useful for the use of a real microscope.	62
FFyB-VM as a platform	
The VM platform was friendly to navigate.	100
The VM parts (focus knobs, light intensity and diaphragm aperture, objectives, etc.) were easy to adjust.	95
The use of the VM was a good experience.	100
FFyB-VM as a tool to acquire biology research competences	
The VM provided an advantage to analyze and solve the exercise over the two photos.	67
The use of the VM facilitated hypothesis formulation.	48

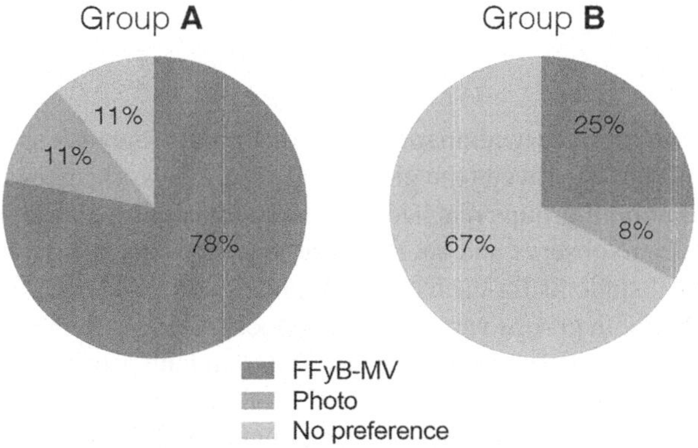

Fig. 10.4 Preferences for the use of FFyB-VM or photos to formulate hypotheses. Group A (n = 9 students) was asked to observe and *Analyze* drug effects on FFyB-VM, while Group B (n = 12 students) was asked to observe/*Analyze* two photos (control and drug-treated cells) which were representative of the specimen shown on FFyB-VM

the photos. These results suggest that FFyB-VM could be applied to learning about research abilities.

The reports were analyzed and scored through an analytic rubric (Mertler, 2001) to evaluate the following aspects: observation competence, recognition of actin-based structures, formulation of hypotheses based on the observations, coherence in experimental design and consistency among hypotheses, research questions, existing knowledge, analysis and conclusion (Table 10.3). Student responses were scored in each category as accomplished (2 points), partially accomplished (1 point), or not accomplished (0 points), and then added up to obtain a total score.

Table 10.3 Scoring rubric

	Not accomplished	Partially accomplished	Accomplished
Observation competence	No differences observed between drug-treated and control samples	Observation of actin depolymerization in treated cells	Observation of actin depolymerization, cell-cell junction loss and/or nuclear changes in drug-treated cells
Recognition of actin-based structures	No differences identified between G-actin and F-actin	Recognition of differences between G-actin and F-actin	Recognition of differences between G-actin and F-actin and elaboration of hypothesis on the basis of such differences
Hypothesis formulation	No hypothesis formulated	Make a minimum hypothesis about cytoskeleton	Observations and previous knowledge used to elaborate hypothesis
Relationship hypothesis-experimental design	Experimental design inconsistent with the hypothesis.	Experimental design consistent with the hypothesis but inconclusive	Experimental design consistent with the hypothesis and conclusive
Consistency	No integration	Integration of 2 or 3 elements	Integration of the 4 elements
Points	**0**	**1**	**2**

Both groups achieved a median total score of 7 points (data not shown). Although this lack of differences between the groups may seem to work against FFyB-VM, in fact it shows that both groups reached similar conclusions because they had representative samples to observe and analyze. For these reasons, we decided to evaluate the percentage of students for each category and aspect (Table 10.4).

We found that both groups were capable of recognizing cell alterations produced by the drug. All students from Group A and 75% students from Group B were able to recognize actin-based structures. Although all students gave multiple explanations and predicted associations between treatment conditions (*Question,* Research questions: develop novel, relevant, and testable research questions based on patterns or properties of components observed in biological systems or described in primary literature) (Pelaez et al., 2017), more students from Group A (89%) were able to generate hypotheses related to their observations than students from Group B (66.7%). This result is in agreement with the survey's answers by Group A students, who preferred the FFyB-VM modality to elaborate hypotheses.

Despite the fact that a large number of students from Group A (55.5%) failed to meet the goal of designing an experiment consistent with their hypothesis, we believe that this aspect may be more closely related to their previous knowledge –i.e. know-how of laboratory techniques, such as western blot and polymerase chain reaction– than to the use of FFyB-VM. Additionally, Group A students (77.7%) outperformed Group B students (66.6%) in integrating all aspects in the report.

Even if FFyB-VM users greater ability to formulate hypotheses may need further confirmation in future studies, these encouraging results and students' surveys and comments indicate that FFyB-VM may be regarded as a useful tool for research

Table 10.4 Students' achievement for each aspect, expressed as a percentage score

	Group A (FFyB-VM) (n = 9)			Group B (photos) (n = 12)		
	% Not accomplished	% Partially accomplished	% Accomplished	% Not accomplished	% Partially accomplished	% Accomplished
Observation competence	0.0	11.1	88.9	0.0	25.0	75.0
Actin recognition	0.0	33.3	66.7	25.0	25.0	50.0
Hypothesis	0.0	11.1	88.9	8.3	25.0	66.7
Relationship hypothesis/ experimental design	55.5	22.2	22.2	33.3	25.0	41.7
Consistency	22.2	55.5	22.2	33.3	33.3	33.3

simulation outside the wet laboratory and student training in sample analysis and hypothesis generation.

10.3 Discussion

Several studies have shown that simulation improves learning (Lateef, 2010; Shapiro et al., 2004). The use of simulation models of real situations and problems gives students the opportunity to become involved in decision-making and develop skills along the learning process. In this sense, the use of FFyB-VM enriched students' learning by providing a near-real microscope available on demand. We observed that students used the tool at different times according to their lifestyles, which shows that FFyB-VM provided flexibility to accommodate different learners' paces and to promote self-directed learning (Kuo & Leo, 2019). Feedback between students and teachers also offered an advantage, especially for students who continued failing to reach the goal.

A common aspiration of teachers is to introduce research experiences and to help students develop competencies that will allow them to tackle real research problems after graduation (Narang et al., 2018; Tekkol & Demirel, 2018). This can certainly be attained through direct student involvement in scientific research, although the associated costs seem rather high.

Our data show that untrained students using FFyB-VM learned to recognize and *Analyze* different microscopy image types. These findings allow us to consider FFyB-VM to be a helpful tool for students' training in essential technical skills for basic research. Additionally, FFyB-VM shows potential in the design of activities focusing on critical thinking skills such as formulating research questions, elaborating hypotheses and drawing conclusions, all of which can make students more familiar with small research projects. In challenging students with professional-like research questions, these projects may also boost students' interest and motivation in science. For these reasons, the VM emerges as a source of opportunity to apply the scientific method and, moreover, it opens the door to the knowledge of multiple intelligences required for investigation (Shearer, 2018). In our experience, the ACE-Bio Competencies framework was suitable for guiding the design of the FFyB-VM platform exercises. This framework helped us to provide support for students to engage in designing their research investigation and reflect on their learning. Additionally, the ACE-Bio Competencies guided us to improve the exercises and to create new ones focused on engaging students with research behaviors in addition to technical microscopy skills, work which will continue with future implementation of the MV-platform for engaging students with biological experimentation.

In sum, FFyB-VM presents versatility in allowing design and constraint adjustments according to students' and teachers' expectations and demands.

A successful platform requires not only good design but also regular updates. Whereas design is based on the views of a reduced group of teachers, updates rely on both students' and teachers' experience. For this reason, successful updating requires the design of comprehensive surveys on platform strengths, weaknesses and suggestions. FFyB-VM student surveys unveiled users' issues and allowed improvements to our VM functions for focusing and handling. Also, teachers' feedback allowed us to develop new tools for different learning activities (e.g. a cropping tool to adjust photo capture). In addition, these changes compelled teachers to adapt exercises to better match their learning objectives. In other words, digital platforms may be thought of as dynamic ever-evolving tools whose initial design should leave room for future updates.

10.4 How to Design a Good Simulator

As the development of equipment simulators including optical and mechanical parts, real samples, and strategies for virtual undergraduate education may present a major challenge, this section provides insights and suggestions based on our experience with the FFyB-VM.

First, design should be user-centered to clearly articulate the learning outcomes expected. Competencies and concepts to be taught should be identified to allow the incorporation of suitable exercises into the platform. Equally important, optical and mechanical parts should be designed to faithfully represent the equipment in question. In this sense, design should involve cell biology teachers, researchers, designers, developers, and other stakeholders. Given the differences in terminology among disciplines or in experiences in the use of simulators, the design process requires a dynamic and collaborative environment based on effective communication.

Second, simulator fidelity is also essential. Among several virtual tools available, software design should aim at a realistic use of the tool and its features. For instance, in the case of microscopes, manipulation of brightness and contrast could be added to the platform. However, both parameters should be available only at a later digitization stage, not on the virtual microscope itself. The more realistic the features recreated in the platform, the more challenging the experience for students, as it will help them learn to tackle scientific obstacles much like what they will face with real equipment. Exercises for engaging students with simulated tool might be informed by the ACE-Bio Competencies framework (*Question*, hypothesis: generate multiple explanations of the natural world that are testable and potentially falsifiable, for example) to persuade learners to reflect on their research experiences and findings with the simulator to support student engagement with experimentation. Last, but not least, a feedback space should be built into the platform for support and performance tracking, so that students can make suggestions that may influence learning decisions to meet goals.

The following aspects should ideally be taken into account in equipment design:

- Learning goals and equipment features to be reproduced should be clearly established. A short abstract may be written to keep the aim of the project in mind.

- Regular multidisciplinary meetings should be held to achieve effective communication, train developers in handling the equipment and, if possible, let them use the equipment.
- Some digital tools and shortcuts –though tempting– should only be incorporated if they are part of the original equipment for simulator fidelity.
- Other digital tools may indeed prove useful to strengthen the learning experience, even if not part of the original equipment; examples include photo capture and cropping and sharing information to improve collaborative education.
- The platform should allow for constraint adjustments according to class demands and different student proficiency levels.
- As students are meant to learn how to use real equipment and not only digital tools, ongoing systematic surveys should be used to obtain feedback on the use of the platform in comparison real equipment in order to make the necessary adjustments.

Third, a good science simulator should offer the challenges that a researcher experiences when doing experiments in a real laboratory. Thus, it is important to offer these challenges and motivations in the simulator's experience. The following aspects should ideally be taken into account in engaging students who are learning both technical skills and higher-order skills such the experimental presearch process online:

- Design exercises to combine new and relevant prior knowledge and apply information to new situations or problems. Exercises should be sufficiently cognitively demanding to challenge students but still have sufficient prior knowledge, so they do not feel frustrated.
- Design exercises that mirror the real science-world needs, challenging students with professional-like research questions.
- Focus the activities and feedback to students on skills informed by the ACE-Bio Competencies framework such as formulating research questions, elaborating hypotheses and drawing conclusions, with exercises to guide students to reflect on these aspects of their simulated research experiences.
- Introduce research experiences in the simulator to develop competencies that allow students to tackle real research problems they will face after graduation.

A final recommendation to ensure authentic integration of simulation-based teaching into the curricula is to provide training and support on virtual simulator design and/or implementation for all members of the faculty who are involved.

10.5 Conclusion

Findings reported here support the idea that FFyB-VM was effective for two educational purposes: on the one hand, it allowed the diagnosis, follow-up and improvement of students' performance and, on the other hand, it boosted undergraduate students' research abilities. We hope this chapter helps other groups to design similar activities and to test their platforms to potentiate learning about experimentation with simulation of authentic research tools online.

Acknowledgments The authors want to thank Universidad de Buenos Aires, Facultad de Farmacia y Bioquímica: Dean Prof. Dr. Cristina Arranz, Prof. Dr. Laura E. Schreier, Prof. Dr. Marcela Radice, Prof. Dr. Elida Alvarez, Prof. Dr. Marta Carballo, Prof Dr. María del Carmen Fernández, Dr. Marcela M. López Nigro. Also, we want to thank Asesoría Pedagógica: Mg. Maria Paz Florio and Mg. Fernando Salvatierra; and Centro de Servicios Informáticos (CESIN): Ana Villafañe and Alberto Ferreiro from Facultad de Farmacia y Bioquímica, Universidad de Buenos Aires.

This work was supported by grants: Programa UBATIC II: Proyecto Institucional (EXP-UBA: 25.1001/2011 Director: Prof. Dr. Laura E. Schreier) from Universidad de Buenos Aires and Programa UBATIC III (EXP-UBA: 57.6661/2017 Director: Prof. Dr. Marcela Radice) from Universidad de Buenos Aires.

References

Bernardo, A. (2017). Virtual reality and simulation in neurosurgical training. *World Neurosurgery, 106*, 1015–1029. https://doi.org/10.1016/j.wneu.2017.06.140

BioNetwork's Virtual Microscope. (n.d.). https://www.ncbionetwork.org/iet/microscope/

Brydges, R., Manzone, J., Shanks, D., Hatala, R., Hamstra, S. J., Zendejas, B., & Cook, D. A. (2015). Self-regulated learning in simulation-based training: A systematic review and meta-analysis. *Medical Education, 49*(4), 368–378. https://doi.org/10.1111/medu.12649

Burbules, N. (2012). Ubiquitous learning and the future of teaching. *Encounters in Theory and History of Education, 13*, 3–14. https://doi.org/10.24908/eoe-ese-rse.v13i0.4472

Cook, D. A., Levinson, A. J., Garside, S., Dupras, D. M., Erwin, P. J., & Montori, V. M. (2008). Internet-based learning in the health professions: A meta-analysis. *JAMA, 300*(10), 1181–1196. https://doi.org/10.1001/jama.300.10.1181

Dede, C. (2011). *Emerging technologies, ubiquitous learning, and educational transformation BT – Towards ubiquitous learning* (C. D. Kloos, D. Gillet, R. M. Crespo García, F. Wild, & M. Wolpers, Eds., pp. 1–8). Springer.

Farah, C. S., & Maybury, T. (2009). Implementing digital technology to enhance student learning of pathology. *European Journal of Dental Education: Official Journal of the Association for Dental Education in Europe, 13*(3), 172–178. https://doi.org/10.1111/j.1600-0579.2009.00570.x

Foad, A. F. A. (2017). Comparing the use of virtual and conventional light microscopy in practical sessions: Virtual reality in Tabuk University. *Journal of Taibah University Medical Sciences, 12*(2), 183–186. https://doi.org/10.1016/j.jtumed.2016.10.015

Glatz-Krieger, K., Spornitz, U., Spatz, A., Mihatsch, M. J., & Glatz, D. (2006). Factors to keep in mind when introducing virtual microscopy. *Virchows Archiv, 448*(3), 248–255. https://doi.org/10.1007/s00428-005-0112-2

Hanna, M. G., Pantanowitz, L., & Evans, A. J. (2015). Overview of contemporary guidelines in digital pathology: what is available in 2015 and what still needs to be addressed? *Journal of Clinical Pathology, 68*(7), 499 LP–505. https://doi.org/10.1136/jclinpath-2015-202914

Hwang, G.-J. (2014). Definition, framework and research issues of smart learning environments – A context-aware ubiquitous learning perspective. *Smart Learning Environments, 1*(1), 4. https://doi.org/10.1186/s40561-014-0004-5

Irby, S. M., Pelaez, N. J., & Anderson, T. R. (2018). Anticipated learning outcomes for a biochemistry course-based undergraduate research experience aimed at predicting protein function from structure: Implications for assessment design. *Biochemistry and Molecular Biology Education, 46*(5), 478–492. https://doi.org/10.1002/bmb.21173

Kuo, K.-H., & Leo, J. M. (2019). Optical versus virtual microscope for medical education: A systematic review. *Anatomical Sciences Education, 12*(6), 678–685. https://doi.org/10.1002/ase.1844

Lateef, F. (2010). Simulation-based learning: Just like the real thing. *Journal of Emergencies, Trauma, and Shock, 3*(4), 348–352. https://doi.org/10.4103/0974-2700.70743

Le Beux, P., & Fieschi, M. (2007). Virtual biomedical universities and e-learning. *International Journal of Medical Informatics, 76*(5), 331–335. https://doi.org/10.1016/S1386-5056(07)00060-3

Lindgren, R., & Schwartz, D. L. (2009). Spatial learning and computer simulations in science. *International Journal of Science Education, 31*(3), 419–438. https://doi.org/10.1080/09500690802595813

Maeng, J. L., Mulvey, B. K., Smetana, L. K., & Bell, R. L. (2013). Preservice teachers' TPACK: Using technology to support inquiry instruction. *Journal of Science Education and Technology, 22*(6), 838–857. https://doi.org/10.1007/s10956-013-9434-z

McBride, J. M., & Drake, R. L. (2018). National survey on anatomical sciences in medical education. *Anatomical Sciences Education, 11*(1), 7–14. https://doi.org/10.1002/ase.1760

Mertler, C. (2001). Designing scoring rubrics for your classroom. *Research & Evaluation, 7*.

Narang, A., Velagapudi, P., Rajagopalan, B., LeBude, B., Kithcart, A. P., Snipelisky, D., & Sinha, S. S. (2018). A new educational framework to improve lifelong learning for cardiologists. *Journal of the American College of Cardiology, 71*(4), 454–462. https://doi.org/10.1016/j.jacc.2017.11.045

Ordi, O., Bombí, J. A., Martínez, A., Ramírez, J., Alòs, L., Saco, A., Ribalta, T., Fernández, P. L., Campo, E., & Ordi, J. (2015). Virtual microscopy in the undergraduate teaching of pathology. *Journal of Pathology Informatics, 6*, 1. https://doi.org/10.4103/2153-3539.150246

Pantanowitz, L., Szymas, J., Yagi, Y., & Wilbur, D. (2012). Whole slide imaging for educational purposes. *Journal of Pathology Informatics, 3*(1), 46. https://doi.org/10.4103/2153-3539.104908

Paulsen, F. P., Eichhorn, M., & Bräuer, L. (2010). Virtual microscopy-the future of teaching histology in the medical curriculum? *Annals of Anatomy / Anatomischer Anzeiger : Official Organ of the Anatomische Gesellschaft, 192*(6), 378–382. https://doi.org/10.1016/j.aanat.2010.09.008

Pelaez, N., Anderson, T. R., Gardner, S. M., Yin, Y., Abraham, J. K., Bartlett, E., Gormally, C., Hill, J. P., Hoover, M., Hurney, C., Long, T., Newman, D. L., Sirum, K., & Stevens, M. (2017). The basic competencies of biological experimentation: Concept-skill statements. *PIBERG Instructional Innovation Material*, Paper 4. Retrieved October 18, 2021 from https://docs.lib.purdue.edu/pibergiim/4/

Porath, M. (2000). The culture of education by Jerome Bruner. *Canadian Journal of Education / Revue Canadienne de l'éducation, 25*, 236–238. https://doi.org/10.2307/1585957

Rosen, K. R. (2008). The history of medical simulation. *Journal of Critical Care, 23*(2), 157–166. https://doi.org/10.1016/j.jcrc.2007.12.004

Saco, A., Bombi, J. A., Garcia, A., Ramírez, J., & Ordi, J. (2016). Current status of whole-slide imaging in education. *Pathobiology, 83*(2–3), 79–88. https://doi.org/10.1159/000442391

Salas Perea, R. S., & Ardanza Zulueta, P. (1995). La simulación como método de enseñanza y aprendizaje. *Educación Médica Superior, 9*, 3–4.

Shapiro, M. J., Morey, J. C., Small, S. D., Langford, V., Kaylor, C. J., Jagminas, L., Suner, S., Salisbury, M. L., Simon, R., & Jay, G. D. (2004). Simulation based teamwork training for emergency department staff: Does it improve clinical team performance when added to an existing didactic teamwork curriculum? *Quality & Safety in Health Care, 13*(6), 417–421. https://doi.org/10.1136/qhc.13.6.417

Shearer, B. (2018). Multiple intelligences in teaching and education: Lessons learned from neuroscience. *Journal of Intelligence, 6*(3), 38. https://doi.org/10.3390/jintelligence6030038

Shegog, R., Lazarus, M. M., Murray, N. G., Diamond, P. M., Sessions, N., & Zsigmond, E. (2012). Virtual transgenics: Using a molecular biology simulation to impact student academic achievement and attitudes. *Research in Science Education, 42*(5), 875–890. https://doi.org/10.1007/s11165-011-9216-7

Tekkol, İ. A., & Demirel, M. (2018). An investigation of self-directed learning skills of undergraduate students. *Frontiers in Psychology, 9*, 2324. https://doi.org/10.3389/fpsyg.2018.02324

Tian, Y., Xiao, W., Li, C., Liu, Y., Qin, M., Wu, Y., Xiao, L., & Li, H. (2014). Virtual microscopy system at Chinese medical university: An assisted teaching platform for promoting active learning and problem-solving skills. *BMC Medical Education, 14*(1), 74. https://doi.org/10.1186/1472-6920-14-74

Triola, M. M., & Holloway, W. J. (2011). Enhanced virtual microscopy for collaborative education. *BMC Medical Education, 11*(1), 4. https://doi.org/10.1186/1472-6920-11-4

UD Virtual Compound Microscope – University of Delaware. (n.d.).

Chapter 11
Introductory Biology Students Engage in Guided Inquiry: Professional Practice Experiences Develop Their Scientific Process and Experimentation Competencies

Porché L. Spence

11.1 Introduction

The *Vision and Change in Undergraduate Biology Education: A Call to Action* provides recommendations for integrating inquiry-based research experiences into undergraduate biology curriculum (AAAS, 2011). Undergraduate instructors have been tasked to facilitate and motivate learning by creating a student-centered, interactive environment that prepares students to tackle "real world" issues. Helping students recognize connections between what they learn in their science courses and their daily decisions is essential for producing a citizenry capable of making informed decisions about our natural world. Adopting a competency-based teaching approach emphasizes the demonstration of applying the process of science, utilizing quantitative reasoning; recognizing the interdisciplinary nature of science; communicating and collaborating with their peers; and understanding the interconnections between science and society (AAAS, 2011). Introductory biology courses provide the foundation for conceptual understanding and core competencies for biology majors and non-biology majors. For non-science majors, an introductory biology course is one of the few science courses required to attain their undergraduate degree. Engaging students in the process of science in introductory courses foster an awareness and appreciation for science while learning core competencies (AAAS, 2011) and skills. Planning and conducting experiments, data analysis and

P. L. Spence (✉)
Department of Biological and Biomedical Sciences, North Carolina Central University, Durham, NC, USA

Department of Natural Resources and Environmental Design, North Carolina Agricultural and Technical State University, Greensboro, NC, USA
e-mail: plspence@nccu.edu

N. J. Pelaez et al. (eds.), *Trends in Teaching Experimentation in the Life Sciences*, Contributions from Biology Education Research,
https://doi.org/10.1007/978-3-030-98592-9_11

interpretation, communication and accurate documentation (Pelaez et al., 2017; Chap. 1 in this volume) are core skills for students aspiring to enter into science and non-science fields.

Many students, regardless of major, enroll in introductory biology courses with limited science reasoning and process skills. Students with limited prior knowledge who engage in weekly disconnected laboratory experiences consisting of step-by-step instructions and predetermined results often develop misconceptions about the scientific process (Coker, 2017). Participating in laboratory activities designed with predetermined results prevents students from learning problem solving strategies and targeting higher-level cognitive skills required to solve "real world' problems (McLoughlin & Hollingworth, 2005; Zoller, 2000). Instead of introducing a new topic each week in lab, it has been recommended to teach fewer concepts at a greater depth over a multi-week period to allow students the opportunity to perform the scientific process (AAAS, 2011). Structuring the laboratory sessions into multi-week modules focusing on major course topics allows the instructors to scaffold the laboratory curriculum and emphasizes scientific reasoning and process skills (Coker, 2017).

Implementing inquiry-based curricula in biological laboratory courses has positive impacts on student learning gains (Beck et al., 2014), especially for underprepared undergraduate students (Blumer & Beck, 2019). Buck et al. (2008) has created a rubric to assist instructors with determining the level of inquiry fostered by undergraduate laboratory curriculum. According to the rubric, guided inquiry occurs when the instructor provides the research question and the data collection methods. The students are responsible for conducting data analysis and communicating the results (Buck et al., 2008). Guided inquiry provides a hands-on student-centered and engaging learning experience for students and refreshing teaching experience for instructors. Lord and Orkwiszewski (2006) report that students enjoy inquiry-based laboratory curricula more than the traditional "cookbook" laboratory exercises. Therefore, undergraduate educators should consider teaching science using an inquiry-based instruction, because students retain the information longer and they learn how to apply the knowledge and skills (Lord & Orkwiszewski, 2006). A meta-analysis conducted by Beck et al. (2014) provides evidence that supports student enjoyment with engaging in inquiry-based experiments. Guided inquiry modules improve scientific reasoning and experimental design skills for underprepared undergraduate students across diverse institutions (Blumer & Beck, 2019).

Scaffolding inquiry curricula is recommended, because it creates opportunities for repeat performance and increases student confidence in scientific reasoning and process skills (D'Costa & Schlueter, 2013). The learning objectives should be defined to teach students the course concepts while emphasizing scientific process (AAAS, 2011), because students learn science by practicing and emulating professional practices performed by scientists (Bell, 2011). Spence et al. (2020) describe a team-taught and scaffolded "student-scientist" curriculum to educate students in an introductory biology course focusing on molecules and metabolic processes. The curriculum used the entire scientific process integrated with professional practices commonly performed by scientists. With this curriculum, students practiced preparing and peer-reviewing research proposals, conducting experiments, documenting experiments in laboratory journals, and creating and presenting a poster to communicate their scientific findings to their peers, instructors, and science faculty (Spence et al., 2020). Crafting the curriculum using a modular format allowed the instructors to design activities to allow students to revisit concepts while practicing scientific skills. Moreover, a scaffolded student-centered curriculum with hands-on research experiences provided an effective, engaging, and rewarding educational experience for undergraduate students (Spence et al., 2020).

Several action items and core skills proposed in the Vision and Change report (AAAS, 2011) can be met by engaging students in introductory biology courses with guided-inquiry and professional practices that promote ACE Bio experimentation competency skills (Pelaez et al., 2017). For this report, inquiry-based instruction in introductory biology laboratory courses for first year biology majors, non-biology majors, and nonscience majors was designed with scaffolding according to the ACE Bio experimentation competency skills framework using the process of Understanding by Design (UbD) (Wiggins & McTighe, 2011). The implementation was focused on engaging students in the process of science via collaborating as a research team, writing research proposals, independently documenting research experiments in laboratory journals and completing self-reflection worksheets to evaluate their own learning and to foster the following ACE-Bio experimentation competencies: *Plan, Conduct, Analyze, Conclude* and *Communicate.* We illustrate use of some tools that were used to support meaningful engagement of students with experimentation. Findings reveal benefits and challenges students experienced as they engaged with the scaffolded modules of guided investigation that were implemented.

11.2 Using Understanding by Design Framework to Scaffold Inquiry-Based Laboratory Curriculum

The Understanding by Design (UbD) framework centers around curriculum planning and assessment design with an emphasis on student achievement and clear learning goals. The UbD framework predicates on a three-stage "backward-design" process that emphasizes the application of content knowledge through performing "real-world" tasks to achieve enduring understanding (Wiggins & McTighe, 2011). This framework has been recommended for developing effective learning modules to cultivate scientific inquiry and thinking (AAAS, 2011; Cooper et al., 2017; Minbiole, 2016). The UbD is employed to assist with scaffolding effective high-quality modules to enhance student learning gains (Cooper et al., 2017) and to engage students, especially students who lack interest in learning science.

An effective curriculum incorporates design standards such as expectations, effective instruction, learning activities, assessment of performance goals, sequence and coherence (Wiggins & McTighe, 2011). The desired outcome (develop skills in

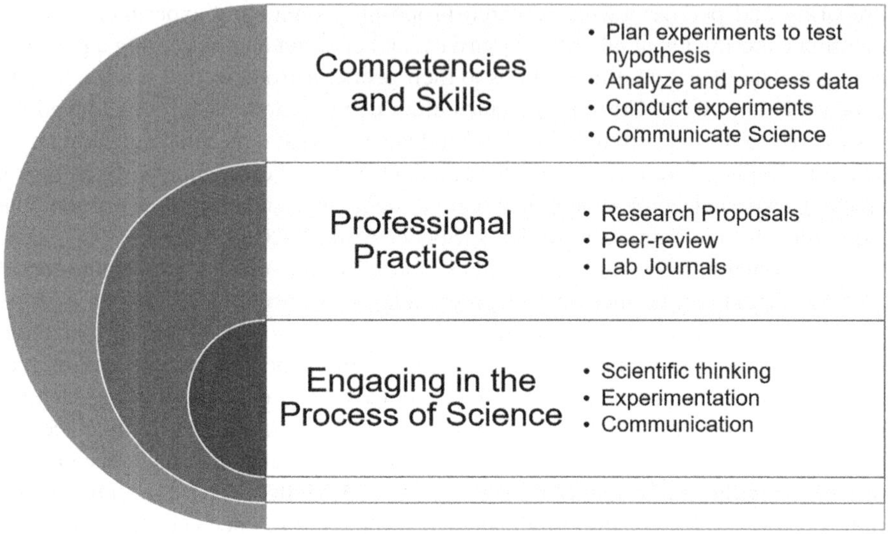

Fig. 11.1 Desired results for the inquiry-based curriculum

scientific thinking, experimentation, and professional communication by engaging in the process of science) is determined before creating assessments (laboratory journals, research proposals and self-reflection worksheets) and learning activities (inquiry-based experiments and peer-review) (Fig. 11.1). Evidence of learning is gathered from using grading rubrics based on a standardized-scale to measure student performance of professional practices and communicating science via research proposal and lab journals. Self-reflection worksheets provide students the opportunities to evaluate and reflect on their own learning gains. When students engage authentically in the scientific process and practice the application of conceptual knowledge, they are capable of explaining, interpreting, applying, and empathizing with content provided in the traditional lecture and scaffold laboratory sessions as well as self-assess their own learning (Wiggins & McTighe, 2011). Students engage in experimentation to practice the skills and apply the knowledge to different situations. Through the UbD framework, the instructors coach students through the process of science and provide an effective curriculum to measure their students' knowledge and skills while performing "real world" tasks.

11.3 Implementing the Guided-Inquiry Integrated with Professional Practices

The guided-inquiry curriculum described here was scaffolded over a full-semester (15 weeks) in the required laboratory session of introductory biology courses designed for first-year biology majors, non-biology majors, non-science majors, and early college high school students from diverse socio-demographic and ethnic backgrounds. The student population entered these introductory biology courses with various levels of prior knowledge and scientific reasoning and process skills. Introductory biology courses designed for biology majors and students who intend to take upper-level biology lecture topics focus on evolution, kingdom of living organisms and fundamental principles of ecology. Biodiversity and seed germination were the focal points for the laboratory performance activities and assessments. Introductory biology courses designed for undergraduate students (i.e., freshmen, sophomore, juniors) majoring in nursing, psychology, exercise sport science, criminal justice, environmental science, physical education, sports medicine, and early childhood development lecture focus on a basic introduction to physical and chemical properties of biological molecules and their interactions with the function and organization of cells. Laboratory sessions focused on metabolic processes associated with lactose intolerance and alcohol fermentation emphasizing biofuel production. The student-scientist laboratory curriculum (Spence et al., 2020) was adapted to meet the learning outcomes for the introductory biology course for non-biology majors. Both introductory biology courses met weekly for a two-hour laboratory session and two 75-min lectures over a full semester (15-weeks). Even though the learning activities and core concepts were different, the expectations, instruction, assessment, sequence, and coherence were the same for both introductory biology courses with an emphasis on the scientific process.

11.3.1 Learning Plan

During the entire semester, a research team of four students worked together to complete collaborative assessments and performance tasks. After each performance task, one student from each research team input their data into a Microsoft Excel spreadsheet to create a class dataset. Each class dataset was shared with students online via Blackboard so they could independently calculate the basic statistics and create graphs using Microsoft Excel, which students then documented in their lab journals. The scientific process (module 2) was taught following data analysis (module 1) because how data is collected and analyzed is essential planning when designing experiments (Fig. 11.2). Each component of the scientific process was discussed in detail so students could understand why scientists need a universal process to study the natural world. It was emphasized that the scientific process is non-linear with many components (Thanukos et al., 2010). PowerPoint presentations with embedded think-pair-share and clicker questions were used to facilitate the information in each module.

The semester began with students completing a "Student introduction" worksheet to provide information about their educational background and feedback on their prior experience with the scientific process. The laboratory curriculum began with students engaging in independent and collaborative performance tasks and assessments designed to teach the basic descriptive statistics (i.e. sample size, mean,

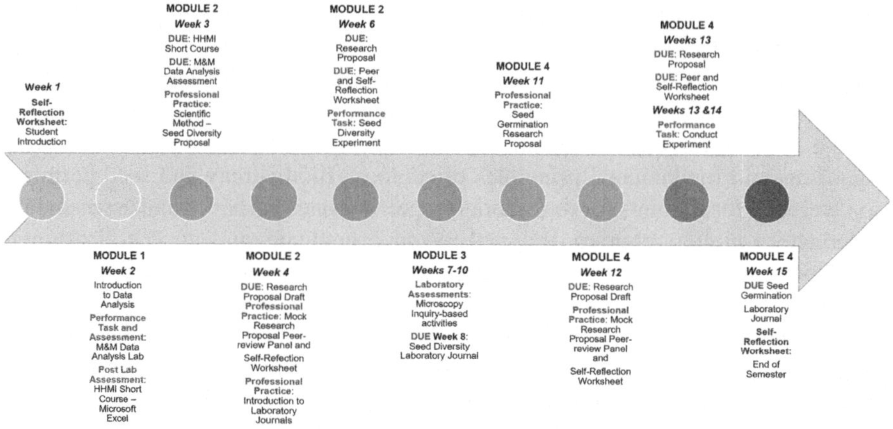

Fig. 11.2 Full-semester (15-week) inquiry-based curriculum timeline

standard deviation, and standard error), sample replication, and graph interpretation (Table 11.1). During the first module, students spent 2 weeks learning the importance of data collection, analysis, and interpretation. Students were introduced to Microsoft Excel as a tool to calculate basic statistics and create graphs. The second module consisted of students spending the next 4 weeks learning about the importance of biodiversity and performing the scientific process by collaborating as research teams to prepare a research proposal and participate in a mock peer-review panel. Furthermore, students were introduced to documenting experiments and communicating science in laboratory journals. While performing the scientific process, students practiced collecting data, calculating basic statistics using Microsoft Excel and interpreting results. During the last module, students spent a total of 6 weeks practicing the scientific process by engaging in both collaborative and independent assessments while learning about the impact of salt on seed germination. Students collaborated to prepare a research proposal describing a "hypothesis-driven" experiment. Additionally, students learned how to use GraphPad QuickCalcs t-test calculator to conduct t-test analysis (GraphPad Software, 2018). Figure 11.2 provides a timeline for instructors to navigate the curriculum and illustrate the amount of time allowed for students to engage in professional practices and complete the performance assessments in each module.

11.3.2 Desired Outcome

Module 1 was a two-week unit designed to help students practice quantitative reasoning skills necessary for the remaining modules. The desired student learning outcomes for module 1 (Table 11.1) were therefore different from modules 2 and 4 (Tables 11.2 and 11.3). After completing module 1, students were capable of using

Table 11.1 UbD framework for module 1

Module 1 topic: Data analysis Time frame: 2 weeks		
Stage 1 Desired results	**Teaching goals**: This module introduces students to general statistics, data replication, graphs and Microsoft Excel. **Learning outcomes**: Students will be skilled at using Microsoft Excel to calculate sample size, mean, median, standard deviation, standard error and to create graphs.	
Stage 2 Evidence	**Beginning assessment** Individual assessment	**Data analysis, interpretation, and basic statistic with Microsoft Excel** ***(Biological competencies: <u>Analyze</u> C3) (Blooms taxonomy: applying)*** Students are tasked with a HHMI Short Course Tutorial assessment and Practicing Data Analysis Using m&m's ® inquiry-based laboratory activity to assess their knowledge and skill to calculate the mean, sample size, standard deviation, standard error using Microsoft Excel.
	Performance task and assessment Collaborative and individual	**Data analysis and interpretation using m&m's** ***(Biological competencies: <u>Conduct</u> A1 and <u>Analyze</u> A1, C3) (Blooms taxonomy: Applying)*** Students are tasked with a Practicing Data Analysis Using m&m's ® inquiry-based laboratory activity to assess their knowledge and skill to calculate the mean, sample size, standard deviation, standard error as well as create graphs using Microsoft Excel. Additionally, students are evaluated on their ability to interpret their results.
	Other evidence Individual	**Self-reflection worksheet** ***(Blooms taxonomy: Evaluating)*** At the beginning of the semester, students complete a self-reflection to introduction themselves and to self-report their science course taken in high school and experience with the process of science.
Stage 3 Learning plan	Major learning activities include the following: Complete a Student Introduction self-reflection worksheet Participate in a class discussion – Data Analysis PowerPoint with embedded think-pair-share and clicker questions Learn the difference between qualitative and quantitative data Learn the importance of sample replication and the importance of using class datasets to conduct data analysis Read Basic Statistics Handout Complete HHMI Short Course – Microsoft Excel as a pre-lab assessment: Teaching Statistics and Math Using Spreadsheet Tutorials and Galapágos Finches https://www.hhmi.org/biointeractive/spreadsheet-data-analysis-tutorials Participate in a collaborative inquiry-based exercise: Practice Data Analysis Using m&m's ®	

Biological Competencies = ACE-Bio Experimentation Competencies (Pelaez et al., 2017)

Table 11.2 UbD framework for module 2

Module 2 topic: Scientific process emphasizing biodiversity Time frame: 4 weeks		
Stage 1 Desired results	**Teaching goals**: This module introduces the scientific process and concepts associated with biodiversity. **Learning outcomes:** Students will be able to explain the process of science and the difference between "discovery science" and "hypothesis-driven science." Students will collaboratively and independently demonstrate their knowledge and scientific inquiry skills. Each research team will communicate science by writing a collaborative research proposal detailing a plan to conduct a seed diversity project. Each student will independently document the details of their experiment and if an action plan is necessary to preserve the biodiversity for these island communities in their laboratory journal.	
Stage 2 Evidence	**Assessment** Collaborative	**Research proposal** ***(Biological competencies: <u>Conduct</u> A1; <u>Plan</u> B3, B6, I4)*** ***(Blooms taxonomy: Applying)*** Research proposal is a collaborative assessment designed to measure students' knowledge about food insecurity, biodiversity and ecological concepts, scientific process, basic statistics, graphs, and interpretation of potential findings. Each research team will write a research proposal to describe in detail how they plan to carry out a "discovery-based" experiment.
	Performance task and assessment Collaborative	**Mock peer-review panel** ***(Blooms taxonomy: Evaluating)*** Prior to students submitting their research proposal for a grade, they will undergo a mock peer-review process (scientific community). Students collaborate as a committee to read and critique the research proposal prepared by another research team. Student will use the research proposal grading rubric to provide written feedback to assist with improving their research proposal.

(continued)

Table 11.2 (continued)

Module 2 topic: Scientific process emphasizing biodiversity Time frame: 4 weeks		
	Performance task Collaborative	**Seed diversity experiment** ***(Biological competencies: Plan I4; Conduct A1 and Analyze C3) (Blooms taxonomy: Analyzing)*** Students will act as research scientists for USDA to measure and compare the species richness, relative abundance, and biodiversity index of food options for two island communities using seeds a model. Each research team is responsible for identifying and counting the seed "collected" from each island community to determine which one is more diverse. One person from each research team will enter their species richness and biodiversity index values into a Microsoft Excel spreadsheet to create a class dataset. The class dataset will be used to conduct the basic statistics and t-test analysis.
	Assessment individual	**Laboratory journal** ***(Biological competencies: Conduct A1, C1, C4; Analyze A1, C3, D1, D2; Conclude B6; Communicate D1)*** ***(Blooms taxonomy: Analyzing and evaluating)*** The laboratory journal is an assessment that measures students' skills to document an experiment, calculate basic statistics, conduct t-tests, interpret results, create graphs and self-reflect on their learning gains. Students will use Microsoft Excel to calculate basic statistics and GraphPad QuickCalcs: t-test calculator (https://www.graphpad.com/quickcalcs/ttest1.cfm) to compare species richness and biodiversity index from an island located in the Caribbean Island and an island in the South Pacific Ocean. Students will use their evidence to determine if an action plan is necessary to preserve the biodiversity for these island communities and provide recommendations for further research if needed.
	Other evidence Individual	**Self-reflection worksheets** ***(Blooms taxonomy: Evaluating)*** Students will complete self-reflection worksheets before and after serving on the mock peer-review panel and after submitting their final version for a grade.

(continued)

Table 11.2 (continued)

Module 2 topic: Scientific process emphasizing biodiversity Time frame: 4 weeks	
Stage 3 Learning plan	Major learning activities include the following: Learn the scientific process by participating in a class discussion – Scientific Process PowerPoint with clicker questions Learn brief overview of biodiversity and ecological concepts by participating in a class discussion Use a research proposal template to write a detailed experimental design Complete research proposal preparation self-reflection worksheet Serve on a Mock Research Proposal Peer-Review Panel Complete research proposal follow-up self-reflection worksheet Learn to use a laboratory journal as a tool to document the details of experiments – Lab Journal PowerPoint Lecture Identify, sort and count seeds Complete a peer and self-reflection worksheet

Table 11.3 UbD framework for module 4

Module 4 topic: Effect of salt concentrations on seed germination Time frame: 6 weeks		
Stage 1 Desired results	**Teaching goals:** Support students as they practice the scientific process while applying the knowledge and skills learned in units 1 and 2. **Learning outcomes:** Students will be able to explain the effect impacts of increasing salt concentrations on the rate of seed germination of two angiosperms (corn and black beans). Students will revisit seed diversity and learn about plant diversity. Additionally, students will be able to discuss the impact of these results on ecosystem services, food supply, and sustainability. Students will collaboratively and independently demonstrate their knowledge and scientific inquiry skills. Each research team will communicate science by writing a collaborative research proposal detailing a plan to conduct a seed germination project. Each student independently documents the details of their experiment in a laboratory journal.	
Stage 2 Evidence	**Assessment** Collaborative	**Research proposal** ***(Biological competencies: <u>Conduct</u> A1 and <u>Plan</u> B3, B6, C1, C2, D1) (Blooms taxonomy: Applying)*** Research proposal is a collaborative assessment designed to measure knowledge and skills for planning an experiment focusing on salt concentrations and seed germination. Each research team will learn how to plan a "hypothesis-driven" experiment by developing a hypothesis explain with scientific reasoning, identify each variable (independent, dependent, experimental and standardized) and explain the purpose for including the variable in the study, materials and techniques for collecting and analyzing data, and anticipated outcomes. Each research team is required to provide a timeline for completing the study.

(continued)

Table 11.3 (continued)

Module 4 topic: Effect of salt concentrations on seed germination Time frame: 6 weeks		
	Performance task and assessment Collaborative	**Mock peer-review panel** ***(Blooms taxonomy: Evaluating)*** Prior to students submitting their research proposal for a grade, they will undergo a mock peer-review process (scientific community). Students collaborate as a committee to read and critique the research proposal of prepared by another research team. Student will use the research proposal grading rubric to provide written feedback to assist with improving their research proposal.
	Performance task Collaborative	**Seed germination experiment** ***(Biological competencies: <u>Plan</u> I4; <u>Conduct</u> A1; <u>Analyze</u> A1, C3) (Blooms taxonomy: Analyzing)*** Students will act as research scientists for USDA to study the impact of salt concentrations (0%, 1.0% and 3.5%) on the seed germination a monocot (corn) and eudicot seed (black bean) using pocket seed viewers*. Each research team is responsible for measuring the shoot and root growth over a 2 week period. One person from each research team will enter shoot and root data into a Microsoft Excel spreadsheet to create a class dataset. The class dataset will be used to conduct the basic statistics and t-test analysis. **Dr. Biology. "Dr. Biology's Virtual Pocket Seed Experiment." ASU – Ask A Biologist. 15 Dec 2009. ASU – Ask A Biologist, Web. 1 Apr 2017. http://askabiologist.asu.edu/experiments/vpocketseeds*
	Assessment Individual	**Laboratory journal** ***(Biological competencies: <u>Plan</u> E3, I4; <u>Conduct</u> A1, A2, C1, C4; <u>Analyze</u> A1, C3, D1, D2; <u>Conclude</u> A1, B6, B8; <u>Communicate</u> C1, D1) (Blooms taxonomy: Analyzing and evaluating)*** The laboratory journal is an assessment that measures students' skills to document an experiment, calculate basic statistics, conduct t-tests, interpret results, create graphs, and self-reflect on their learning gains. Students will use Microsoft Excel to calculate basic statistics and GraphPad QuickCalcs: t-test calculator to compare the shoot and root growth of the different seeds for each salt concentration.

(continued)

Table 11.3 (continued)

Module 4 topic: Effect of salt concentrations on seed germination Time frame: 6 weeks		
	Other evidence Individual	**Self-reflection worksheets** ***(Blooms taxonomy: Evaluating)*** Students will complete self-reflection worksheets before and after serving on the mock peer-review panel and after submitting their final version for a grade. Students will complete two additional self-reflection worksheets: Peer and self-reflection and End of Semester self-reflection.
Stage 3 Learning Plan	Major learning activities include the following: Learn concepts associated with plants diversity and angiosperm seed germination Learn about the environmental issues associated with salinization in water and soil Use a research proposal template to write a detailed experimental design Complete research proposal preparation self-reflection worksheet Serve on a Mock Research Proposal Peer-Review Panel Complete research proposal follow-up self-reflection worksheet Document the details of seed germination experiment in a lab journal Measure shoot and root growth Complete a peer and self-reflection worksheet and end of Semester self-refection worksheet	

Microsoft Excel to calculate basic statistics and create graphs to illustrate their data. Students transferred knowledge and skills from module 1 to analyze and illustrate the data collected in modules 2 and 4.

Module 2 was a 4-week unit which focuses on students learning professional skills such as working collaboratively to prepare a detailed research proposal, participate in a mock peer-review panel, and independently documenting their experiments in their lab journals (communicating science). During the 6-week module 4, students used the skills gained in Modules 1 and 2 to demonstrate their knowledge and skills for performing professional practices, experimentation, scientific thinking, and communicating science.

11.3.3 Evidence from Learning Assessments

The modules described in this book chapter include both formative assessments (mock peer-review panel, performance tasks, and self-reflection worksheets) and summative assessments (research proposals and laboratory journals). The learning goals were transparent for each assessment and performance task so students understood the requirements and time commitment to complete each learning activity. Grading rubrics designed for each assessment used the same standardized assessment scale for each criterion (Assessment Scale: 4 = Excellent; 3 = Good, 2 = Fair;

1 = Needs Improvement). Prior to beginning each summative assessment, students were provided grading rubrics tailored specifically for each summative assessment. The instructor discussed each grading rubric with the class to clarify the expectations to successfully complete the assessment. During the mock peer-review panel, students used the research proposal grading rubric to evaluate the research proposal drafts.

Performance tasks were framed around authentic environmental issues to show students the interrelationships between human well-being and healthy ecosystems. Integrating guided-inquiry with professional practices was meant to help students achieve the basic competencies of biological experimentation (Pelaez et al., 2017; Chap. 1 in this volume) and to stimulate their curiosity about the environmental issues impacting our natural world. Performance tasks were developed from adapting pre-existing structured inquiry activities to guided-inquiry experiments (Buck et al., 2008). Students *Communicate* science while collaborating to write a research proposal and independently documenting their experiments in their laboratory journals. Serving on mock panels to peer-review research proposals taught the importance of effectively communicating science.

Students were encouraged to use their textbook and reliable scientific literature as resources for completing their assessments. Adequate time was given during and outside of the laboratory session to complete each assessment and receive clarification from the instructor. Mock peer-review panels provided students with the opportunity to evaluate a research proposal prepared by their peers before improving and submitting their proposal for a grade. The instructor reviewed the comments and feedback from each group prior to returning the research proposal with feedback to the original research team. Instructors guided each research team via answering questions during and outside the of the laboratory sessions. Written feedback from the instructor was provided to each research team when they receive their graded research proposals. Feedback from the instructor in module 2 was intended to enhance the quality of the research proposal draft prepared during module 4. Students self-assessed their learning experiences throughout the semester by completing worksheets with open-ended questions. Self-reflection worksheets were assigned as formative assessments. At the end of the semester, students completed an "End of Semester" self-reflection worksheet to assess their learning gains and perception about the course.

11.4 Evidence for Learning Outcome Achievement

11.4.1 Research Proposals

Chung and Behan (2010) report that students working collaboratively on research proposals become motivated to learn and hone their ability to apply and *Communicate* science. During Modules 2 and 4, students created research proposals to outline their experiment, which gave students the opportunity to prepare to *Conduct* the experiment and learn about the peer-review process. The instructor lead a class discussion about the importance of scientists writing research proposals to receive funding from agencies to pay for their research projects. It was disclosed to students that other scientists with similar expertise are recruited to evaluate the research proposals and assist the funding agency with deciding which research proposal should receive funding. After clarifying the purpose of this professional practice, students were tasked with serving on a mock double-blind peer-review panel. Each research team was required to prepare a research proposal draft in order to participate in the mock peer-review panel. As an added incentive, the research proposal with the highest score from the instructor was "funded" with five extra credit points (Spence et al., 2020).

Students completed a title page with their names, institution and department, majors, emails, proposed project title, proposed duration, project start date, and end date. The research proposal template worksheets consisted of a series of open-ended questions to assist the student groups with planning and preparing for their experiment. To make sure students understood the assigned task, research teams provided a research question and study objectives. Students used their textbook and reliable scientific literature to provide background information about the topic. Students were encouraged to use the scientific literature provided to them via Blackboard as well as from their own literature searches for reliable resources to find evidence to justify their answers. Using these resources helped students think more deeply about the content. In-text citations and references in APA format were required.

During module 2, students collaborated to perform a discovery-science project (Table 11.2). The research proposal template was tailored around students providing a predication with evidence to support it. The required research proposal criteria consisted of background information about the topic, an evidence-based prediction, materials and methods, anticipated data analysis and outcomes, responsibilities for each team member and a list of references in APA format (Table 11.4). Research teams were given the option to properly complete their cover page. The instructor was lenient on writing mechanics and correct reference style with in-text citations.

Module 4 provided a hypothesis-driven experience (Table 11.3). The research proposal template was designed for students to provide a hypothesis with evidence to justify it. Students were expected transfer their knowledge and skill for collaborating and preparing research proposals while demonstrating their knowledge about the impacts of salt on seed germination. Therefore, the expectations were higher for the seed germination research proposal in module 4 than the seed diversity experiment in module 2. These expectations were communicated to the class while identifying the assessment requirements and clarifying the criteria on the grading rubric. During module 4 the required research proposal criteria included a completed cover page with proposed title, background information, evidence-based hypothesis, variables, methods, and techniques for collecting data and analyzing the data, anticipated outcomes, timeline illustrating the duration of the project, responsibilities for each team member, writing mechanics and references with in-text citations in APA format (Table 11.5). There were additional criteria for students to include such as cover sheet, variables, methods, and techniques for collecting data, scope of project, and writing mechanics.

Table 11.4 Module 4 research paper grading rubric

Research paper grading rubric and criteria **Assessment scale:** 4 = Excellent; 3 = Good; 2 = Fair; 1 = Needs Improvement	**Points possible**	**Points earned**
Background information In your words what is biodiversity Explain the importance of preserving the biodiversity for island communities. What type of seeds does your research team expect to find on islands located in the Caribbean Sea and South Pacific Islands? Explain how a decline in seed diversity can impact food security. What is your research question? Explain the objectives for your study. Discuss how scientists can us this study to develop a hypothesis-driven research project and further explore the biodiversity of tropical islands located in the Caribbean Sea and South Pacific Ocean.	**28**	
Prediction: That matches your experimental design. Write a prediction for comparing the biodiversity of seeds collected from the two island communities. Discuss the scientific reasoning behind your prediction.	**8**	
Materials & methods In your own words, summarize how your research team plans to conduct this study. – Include all the materials required to complete the assignment.	**4**	
Data analysis: Explain how you will record your data. What will be measured? Explain which formulas/equations will be used in your study? Explain which basic statistical methods will be used to analyze the data? Explain which websites and/or computer programs will be used to analyze the data? Explain which websites and/or computer programs will be used to create a visual illustration of the data?	**20**	
Anticipated outcomes In order to REJECT your prediction, describe the data your research team will need to measure during the experiment. Explain your reasoning. In order to ACCEPT your prediction, describe the data your research team will need to measure during the experiment. Explain your reasoning.	**8**	
References: List of the scientific literature cited this document – At least 4 references should be provided	**4**	
Total points	**76**	

Adapted from: Spence et al. (2020)

Students were given one in-lab session and 1 week outside of the laboratory session to draft their written research proposal paper. The instructor did not provide suggestions for collaborative group work. However, after observing peer interactions during the laboratory sessions, the instructor reminded each research team about the importance of providing their contribution and sharing workload. Students were required to explain how each team member will contribute to the project in their research proposal. Research teams were not assessed based on their ability to

Table 11.5 Module 4 written research proposal paper grading rubric

Written research proposal paper grading rubric and criteria **Assessment scale:** 4 = Excellent; 3 = Good; 2 = Fair; 1 = Needs Improvement	**Points possible**	**Points earned**
Cover page Proposed title accurately describes the project All information on cover page provided.	**8**	
Background information Clearly stated the research question Explained seed germination Explained the purpose for conducting the study Explained the importance for studying this research problem* Demonstrated a mastery of the literature on the topic	**20**	
Hypothesis Provided a testable "if.., then.." statement Explained the scientific reasoning that supports the hypothesis	**8**	
Variables Explained the purpose for all the variables in the study.	**4**	
Materials and techniques for collecting data Summarized, in your own words, how your research team conducted this study. Included a budget for all the materials required to complete the experiment. Limitations of the study are identified and discussed	**12**	
Methods and techniques for analyzing data Explained what was measured and the equipment used to measure it Explained the plan for conducting data analysis and the website or computer programs used to create graphs	**8**	
Anticipated outcomes In order to <u>REJECT</u> your prediction, describe the data your research team will need to measure during the experiment. Explain your reasoning. In order to <u>ACCEPT</u> your prediction, describe the data your research team will need to measure during the experiment. Explain your reasoning.	**8**	
Scope of project Provided a timeline illustrating the sequence and duration of your project Described how each team member will contribute to the project	**8**	
Writing mechanics Correct Reference Style – APA format Many sources are used and all places need citation are documented Extensive reference list of scholarly literature that was integrated in the proposal	**16**	

(continued)

Table 11.5 (continued)

Written research proposal paper grading rubric and criteria **Assessment scale:** 4 = Excellent; 3 = Good; 2 = Fair; 1 = Needs Improvement	**Points possible**	**Points earned**
Documentation Correct Reference Style – APA format Many sources are used and all places need citation are documented Extensive reference list of scholarly literature that was integrated in the proposal	**12**	
Total points	**104**	
Optional* (up to 4 pts) Discussed the implications on ecosystem services	**4**	
OVERALL RESEARCH PROPOSAL SCORE		

Adapted from: Spence et al. (2020)

collaborate while writing their research proposals and serving on the mock peer-review panel. After each research team submitted their rough draft, students independently completed a Research Proposal Preparation self-reflection worksheet, which consist of six open-ended questions (Spence et al., 2020). The final research proposal was due to the instructor 1 week after students participate in the mock review panel. One student from each research team submitted their research proposal to the instructor via email, thus allowing students to include color figures and illustrations without worrying about printing costs. Students copied their "collaborators" on the email so everyone was aware of the proposal actually submitted for a grade. After submitting their research proposal, each student completed a peer and self-evaluation worksheet to provide information about their group collaboration (Spence et al., 2020).

11.4.2 Mock Peer-Review Panels

Peer-review is a useful active-learning strategy for undergraduate research courses as well as a skill needed for the workforce (Odom et al., 2009). Peer-review can also serve as a tool for instructors to identify gaps in knowledge (Halim et al., 2018). The mock research proposal review panels engaged students in the critique process and educated them about the importance of the peer-review process as a component of the scientific process. Students were required to bring three copies of their drafted research proposal to the laboratory session to participate in the mock review panel. One copy with the collaborators' names was given to the instructor. The other two copies were given to another research team during the mock peer-review panel.

Students worked as small research groups to evaluate and critique a research proposal from another group using a grading rubric (Odom et al., 2009). The research proposal grading rubric revealed the evaluation criteria during the peer-review process. The grading rubric was an adaptation of the research proposal grading rubric used by Spence et al. (2020). Modifications consisted of including a standardized assessment scale for each criterion, anticipated outcomes, responsibilities for each team member and at least four references listed in the reference section. Students provided written feedback to assist the other group with improving their work as well as a recommendation with justification for funding. After participating in the mock peer-review panel, students completed a mock research proposal review committee follow up worksheet.

During this activity, students demonstrated they were capable of detecting grammatical and spelling errors, inconsistent information, and lack of scientific evidence to justify their overall score and recommendation for funding. Students deemed the peer-review process as a beneficial experience for improving their research proposal and preparation for conducting their experiment. Serving on the mock research proposal committee allowed students to work together to help each other understand the concepts and the expectations for completing the research proposal. For example, students commented:

> I did find it as a valuable experience. Taking the time to dissect each section of the proposal helped me to further my analysis into each portion. I learned how I should be answering and organizing.

> It was a learning experience. Yes! In college it really was something new to me because I never did anything like it in high school. It definitely is continuing to teach me how to work as a group.

> It is useful because it allows us to get a feel for the process of conducting research and being a part of a group/committee, which is valuable as a biology major.

It was easy for some groups to work together to evaluate the research proposal whereas other groups found the experience to be challenging because it was difficult to agree on the feedback or evaluate a poorly written rough draft. Comments from two students illustrated this:

> It was challenging for me because not only did you have to read their information, you also had to determine if their work should be funded. Some of the group members might not feel the same as you do. You have to talk and come up with a solution.

> I think it can be challenging in certain aspects as you are all different people attempting to work as a hive mind. Some people may be more critical while others may be more forgiving.

Students' viewpoints about providing another peer feedback on their work varied. Some students felt pressure "grading" their peers' work and uneasy because they believed they did not have the knowledge to give feedback. Students perceived the feedback as an additional help from their peers and appreciated their assistance with pointing out items and information that was incorrect or missing from their proposal. Moreover, this experience provided students with the opportunity to reflect on the quality of their proposal. These benefits were captured in the following student reflections:

> I feel like giving constructive feedback was a good way for me to reflect on my own work, but I am not use to criticizing other's work in the science field. It was a new experience but an enlightening one.

> At first I felt as if I had to go easy on my classmates because I don't want them to grade mine poorly, but then I thought its best to be honest so they are able to learn from their mistakes.

> After serving on a mock research proposal panel, I felt providing constructive feedback to another student scientist about their study design was a lot of pressure because we had to judge another group's work and analyze what they needed to improve if they needed to.

> I felt a little bad about giving feedback. It wasn't that I was rude about it. I just felt like me saying that their idea shouldn't be funded was a little wrong.

> I felt good about it, because learning that scientists do critique each other's work to make the results better.

Improvements on summative assessments combined with student self-reflection responses indicate that participating in mock peer-review panels increases students' confidence in improving their research proposal and fostered additional knowledge gains such as explaining science, formatting references, and funding research projects. This experience promoted the importance of clearly explaining science with sufficient details and learning science can be explained in many different ways. Two students noted:

> I learned how in answering a question sometime you can partially provide answers while not fully giving all the details, thus being wrong. I learned the value of understanding the question being asked before answering.

> Learning that just because you are right doesn't mean someone else is wrong.

Prior to this experience, many of the students did not know how research projects were funded and the level of competition associated with receiving funding. Since the research team with the highest score would receive "funding," they were motivated to work hard to prepare a quality proposal. Students realized the challenges with providing a recommendation for funding. One student commented:

> Deciding whether or not the project deserved funding was new to me, but it made me think about all the other crazy science projects that receive loads of funding.

Many of the students entered college with limited knowledge about properly formatting references. The rough draft demonstrated their lack of knowledge. Partaking in the mock peer-review panels highlighted the importance of conducting research and properly citing resources using APA style. Students recognized that references needed to be in APA format and they learned not to overthink their work.

11.4.3 Experimentation

The laboratory exercises were simple and cost-effective experiments with a "real world" environmental focus. Students performed the inquiry-based activities in groups of four. Data collection for the seed diversity experiment occurred over one laboratory session (Table 11.2). Seed germination measurements occurred over a two-week period (Table 11.3). During both experiments, the data collected from each research team was entered into a class dataset (Microsoft Excel spreadsheet) to increase the sample size and so that student could learn about data replication. Students were expected to include the data collected by their research team and class data in their lab journal.

11.4.4 **Laboratory** ***Journals***

Maintaining an accurate laboratory journal is an essential skill for practicing scientists and a key component of the scientific process (Roberson & Lankford, 2010; Schreier et al., 2006). Schriever et al. (2006) provides a snapshot of the standards of maintaining research records. In addition to containing data, laboratory journals are used to record detailed descriptions of laboratory protocols, data calculations, interpretations of results, and all communication associated with the project. While learning how to keep a laboratory journal, students were introduced to the importance of good record-keeping for replicating studies and publishing research (Schreier et al., 2006).

The students were taught how to properly document their experiments in their own laboratory journal, which was a composition notebook. They were expected to write in permanent ink only, and they used transparent type to adhere documents into their journal, placing their initials partly on the tape and on the journal page. They numbered, signed, and dated each journal page. The students were encouraged to list their collaborators and they included pictures of their experiments. A table of contents was recorded on page one of their journals. Students were given the option to choose the order of information written in their lab journal as long as it matched the table of contents. Each student worked independently to document the details of their experiment in their lab journal and they each did calculations of basic statistics using the class excel spreadsheet. The grading rubric was an adaptation of the lab journal Grading Rubric used by Spence et al. (2020). Modifications consisted of including a standardized assessment scale for each criterion, a list of collaborators, a research proposal, and the addition self-reflection questions. Lab journals were always due at the end of each experiment. A new composition notebook was provided at the beginning of each new experiment. The laboratory journal grading criteria is provided in Table 11.6.

About half of the students self-reported that they had prior experience with keeping a laboratory journal. Most students demonstrated their ability to properly record

Table 11.6 Lab journal grading rubric

Lab journal items scoring criteria **Assessment scale:** 4 = Excellent; 3 = Good; 2 = Fair; 1 = Needs Improvement	**Points possible**	**Points earned**
Appearance Table of contents is present and accurate Neat look, legible, written in ink and pleasant to behold Pages are numbered Any charts, graphs, etc. are affixed inside the notebook (not sticking out) Signature on each page.	20	
Collaboration – List of research team members, if applicable	4	
Date for each entry – on each page	4	
Goal: What do you plan to accomplish and learn during this study?	4	
Research proposal – Should be taped inside the laboratory journal	4	
Notes Write detailed notes on the related to the research topic during the lecture and from scientific literature Record data collected during the study Make note of any problems or limitations during the study	16	
Results & conclusion Was the goal of this experiment accomplished? Interpret the data collected during the study Data tables (2) – (sample size, mean, standard deviation, std. error & t-test) Provide graphs/figures with detailed descriptions Write a detailed summary of your study using the evidence collected: Use the species richness, relative abundance, and biodiversity index to discuss which island community (Maroon or Gray) is more diverse? Which seed has the greatest relative abundance on each Island? Is the species richness and biodiversity significantly different on the two islands? Is an action plan necessary to preserve the biodiversity for these island communities? What is your recommendation future research?	40	
Reflection What did you learn from conducting this study? As a result of this on experiment I learned... Because of this experiment, I would like to learn more about... This experience relates to my career by...	16	
Total points	**108**	
Bonus points* (up to 4 pts) Diagrams related to the study Pictures of your experiment or background information Concept maps created for the experiment by research team	**12**	
OVERALL LAB JOURNAL SCORE		

Adapted from Spence et al. (2020)

the details of their experiment and calculate basic statistical skills using Microsoft Excel. A lack of effort and procrastination prevented some students from accurately documenting their experiments in the lab journals. Students self-reported learning several basic biological competency skills, such as how to create graphs and calculate basic statistics. They recognized that they were capable of effectively explaining their results and the implications of their experiments. A few students acknowledged that they were learning biological concepts while improving their scientific writing skills. As students learned the science concepts, they appreciated the importance of accurately interpreting figures and citing evidence from reliable scientific literature to support their conclusions. Furthermore, documenting information in their laboratory journal taught them how to collect and record data, look deeper into the information, and to use Microsoft Excel.

11.5 Teachable Moments

Allowing students to authentically engage in the process of science can bring about unanticipated teachable moments for students (Haug, 2014). Undergraduate educators should capitalize on these moments because they can enhance the student learning experience. There were incidences in these courses associated with the mock peer-review panel and experimental design that resulted in teachable moments.

The instructor prepared a control and various salt solutions using deionized water. Each research team built their pock-seed viewers (Ortiz, 2009) and poured the respective solution in each bag. After a few days, several students recognized seeds in the control seed viewers were not germinating. When several research teams observed what looked like microbial contamination in their seed pocket viewers, they were encouraged to write detailed notes about the microbial contamination and to take pictures to include in their lab journals. During a class discussion, students considered the possibility that they were observing effects of contaminated solutions. The contamination may have occurred from the deionized water or unclean glassware. These were unintended issues such as those a scientists could experience while conducting a research experiment. Although that normally scientists would restart their experiment and use non-contaminated solutions, it was close to the end of the semester, so there was not enough time to restart the experiment to generate more reliable data. The students were encouraged to think about the environmental problems associated with seed germination (growing crops) exposed to both salt and microbial contamination (polluted water). Although the research teams were disappointed, because they were looking forward to measuring the shoot and root growth and conducting the data analysis but they did not have data to analyze, they still learned a valuable lesson about how salt and microbial contamination could inhibit seed germination. Later it was determined that the deionized water was the source of contamination.

Research teams participated in double-blind mock peer-review panels to critique research proposals and provide written feedback. The instructor reviewed the comments and feedback from groups prior to returning the research proposal with feedback to the original research team. One semester, some disturbing remarks were found among comments and feedback that was discouraging, demeaning and disrespectful. This disappointment was discussed along with the difference between providing "constructive feedback" and "hurting someone's feelings." During this discussion, neither group was identified. There was a shockwave of reactions followed by many side-bar conservations. After the initial reactions subsided, a few brief personal stories about how receiving rejections after submitting research proposals and manuscripts for publication had impacted the instructor's confidence as a scientist. Students considered that "it is not what you say, but how you say it" and to take setbacks as motivation to "prove you are capable and can do the work." Several students asked questions throughout the discussion. After concluding the class discussion, the research proposals were returned with feedback and some encouraging words to each team. A week later, they submitted their revised research proposal for a grade. The team that had received the inappropriate peer-feedback earned the highest research proposal grade in the class. The following week, the graded research proposals were passed back along with the announcement of the team who received the "funding." The look on their faces was priceless. We were so proud of how hard they worked to improve their research proposal. This experience taught students about the power in the words we choose to use and the impact of providing criticism in a sensitive manner.

11.6 Student Opinions About the Curriculum

Student self-reflection responses indicated that scaffolded modules infused with guided-inquiry and professional practices promoted understanding and performance of the scientific process. The majority of students began the introductory biology courses with limited knowledge about the process of science. Students' end-of-semester self-reflections provided encouraging evidence that incorporating guided-inquiry in introductory biology courses was an effective approach to teach students several basic competencies of biological experimentation. Engaging in this inquiry-based curriculum fostered students' knowledge about the process of science, as demonstrated by these student comments:

> I know about how scientist analyze and conclude data. I know more about how to design and experiment and carry out data analyses.

> I know the scientific process is the process scientists use to answer any question. Ask a question, hypothesis, operational, analyze data, conclusion.

> Research, hypothesis, experiment, data. You have to do all of these things to figure something out in a science world.

> I know that the process of science is not just a one and done thing. Science never stops and in everything you are trying to figure out or study takes time and is a process to get to the most correct solution.

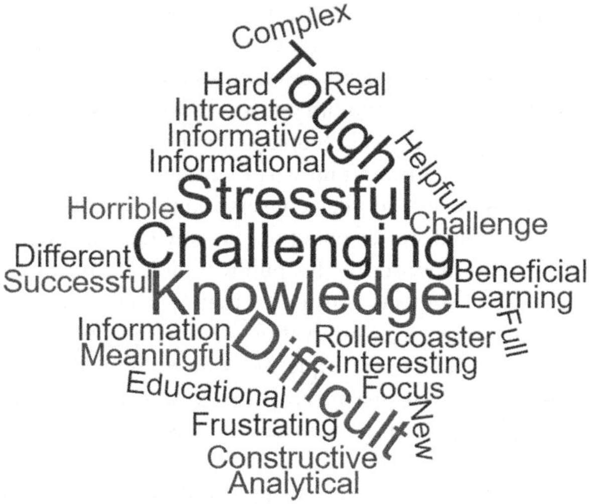

Fig. 11.3 Word cloud illustrating the frequency of words students used to describe the course. Larger words have greater frequencies

Students were asked to describe this curriculum in one word, which resulted in an extensive word list. The word cloud frequency analysis (Zygomatic, 2003) illustrated the most recurrent words were "knowledge," "difficult," "stressful" and "challenge" (Fig. 11.3).

Students explained their views associated with their chosen word:

> Challenging. The course really makes you think especially when writing proposals and doing a lab journal.

> Challenging. It's not "hard" it just requires work, time and focus.

> Educational. I've learned so many things in just 3 months in this class. I didn't learn this much in this amount of detail my whole high school science career.

> Constructive. To me this course was a loaded course where every day there was always something to do and learn. Either we were going over PowerPoints or in the lab doing observation/projects and sometimes it was both, but there was always something we had to do.

> Knowledge. I say that because it is impossible to not learn one new thing in a science class because it is constantly teaching you new stuff.

> Knowledge. This class really allows me to gain knowledge about science and myself because of just how different it is from high school and how much I need to really step it up.

Despite students' beliefs about the laboratory curriculum in these courses, most of the students would recommend this lab to their peers as demonstrated by the following remarks:

> Yes, because it would probably make them more aware and treat the Earth better. Also it is fun and easy, they would have a good time performing it.

> I would recommend this lab to my friends, because this lab was very fun but you also learn while having fun.

> Yes I would it was very fun and make you think like a scientist.

Teaching the scientific process integrated with professional practices yielded results comparable those reported from other undergraduate research studies (Cooper et al., 2017; D'Costa & Schlueter, 2013; Wiggins & McTighe, 2011). Some students initially struggled and became frustrated with inquiry-based research, because it was their first experience performing the whole scientific process. Coker (2017) reported that his students experienced anxiety, lack of confidence, and struggled to complete research using the entire scientific process during the first experiment. However after receiving feedback, his students performed better and begin to enjoy their research experience. Other studies described similar findings of their students enjoying and preferring inquiry-based instruction (Cooper et al., 2017; D'Costa & Schlueter, 2013). Inquiry-based instruction stimulates "stituational interest" when students actively engage in new learning experiences while collaborating with their peers (Palmer 2009). Overall students who are willing to invest the time and effort in their learning are capable of achieving the desired outcome. An inquiry-based course at the introductory biology level is a good starting point for students to understand scientific reasoning and process skills. Findings agree with others who report that additional inquiry-based laboratory courses across the undergraduate science curriculum are needed to close the achievement gap (Blumer & Beck, 2019).

11.7 Benefits and Challenges for Students

Implemention of this curriculum encouraged students to think beyond "rote performance" of the scientific process and students were engaged in "practice-based" instruction (Berland et al., 2016). Integrating guided-inquiry and professional practices in introductory biology courses yielded benefits and challenges for students. In terms of benefits, this curriculum promoted the application of biology concepts, scientific process, planning experiments and recording-keeping skills. Students learned to explain science in greater detail and to connect conceptual knowledge. Students demonstrated skills in collaboration, critical thinking, time management, and communication. They also recognized the importance of self-assessment and taking ownership of their learning.

Students also faced several challenges. A substantial amount of time and hard work was required to successfully complete the collaborative and independent assessments. The knowledge and skills gained in this course was student-dependent because students took ownership of their own learning. Many students had expressed limited exposure to the scientific process and thought of science as finding quick predetermined answers and documenting those answers on a worksheet. Students with limited prior knowledge of the scientific process and who lacked confidence in their ability to learn the concepts and skills often became intimidated and disinterested in the course. These students required additional motivation and encouragement to achieve the desired outcomes. Students were not notified about the inquiry-based laboratory instruction prior to enrolling in the course. Since introductory biology courses are graduation requirements, some students felt pressured and stressed to complete the course. However, scaffolding guided inquiry instruction coupled with motivational strategies has been reported to ease students' apprehension and allow them to focus on learning science (Palmer, 2009).

11.8 Considerations for Single-Instructor Introductory Biology Courses

11.8.1 Diverse Student Populations

Students from diverse backgrounds should have the opportunity to learn about the natural world and scientific process through inquiry based experimentation. Instructors need to design and create curriculum for introductory science courses with the assumption that students have limited experience with performing the scientific process (Coker, 2017; D'Costa & Schlueter, 2013). When inquiry-based assignments are introduced at the beginning of the course, students become intimidated and overwhelmed with the thought of the amount of work required to complete the tasks. To ease these fears, instructors need to be transparent with their students about the benefits for participating in the performance tasks, the course expectations and grading criteria. Additionally, instructors have to constantly motivate their students to keep them encouraged and focused on completing their assignments and achieving the desired outcome. Students need to be reminded that regardless of their major that learning the scientific process can be applied to their prospective professions and everyday life. For example, when students learn how to do good record-keeping through properly maintaining a laboratory journal, they are learning data documentation and science communication. So for students aspiring to be medical doctors and nurses, keeping a detailed lab journal provides practice for meticulously documenting information in a patient's medical chart.

11.8.2 Implementing the Curriculum

Teaching science is most effective in a laboratory setting because students engage in hands-on experiments that allow them to practice skills and apply knowledge to different situations. Guided-inquiry requires students to apply, analyze, and evaluate information. Guided-inquiry combined with professional practices (i.e. research proposals, peer-review, and laboratory journals) is an effective strategy for teaching the core research competencies (AAAS, 2011) and basic concept skills for biological experimentation (Pelaez et al., 2017; Chap. 1 in this volume) in introductory biology courses. This curriculum provided an exciting, fun, and rewarding teaching experience for instructors. Witnessing students blossom into scientists (or discover science is not their cup of tea) is profound.

Instructors can use class datasets when implementing guided inquiry-based curriculum. Class datasets are Microsoft Excel spreadsheets created by combining the data from each student research team currently enrolled in the course. Utilizing a class dataset is essential, and it serves many key purposes. First, the class dataset provides a method for teaching students the importance of sample replication, variation, and size. The students can observe differences in the basic statistics based on the data collected by their research team in comparison to values in the class dataset. Secondly, it encourages students to practice using Microsoft Excel. It is important for students to utilize this tool so they can conduct their own data analysis in this course and in any subsequent course with a data analysis component. Thirdly, the class dataset provides an answer key for assessing whether students have learned the targeted competencies skills. With guided inquiry, the data values vary amongst the research teams. Therefore the class dataset serves as an answer key, thus making it easier and time efficient for grading. The answer key will change each time the class is taught, because the measured values will change with each new research task. During unprecedented periods when instructors are unable to provide students with a hands-on research experience, merging the student data collected over several semesters into one large class dataset can provide an inquiry-based research experience. Students can use these merged class datasets to conduct statistical analysis while learning biological concepts. This experience can open a dialogue for students to learn the value of secondary data and how to use secondary data to conduct research.

Using the UbD framework to scaffold the curriculum requires time to plan and create an effective curriculum for students to achieve "desired outcomes." Once the course is designed, implementing the course is simple. However, grading, especially for courses with single instructors, can be time consuming because most assessments consist of open-ended responses. For instructors who teach courses consecutively during the academic year, planning and modifying the curriculum becomes less demanding. Student self-reflections are recommended to be included as a formative assessment after each unit and professional practice activity. Self-reflection worksheets are valuable tools for students to reflect on their ability to learn the course material as well as the instructor's ability to effectively teach them. Completion of this curriculum will not influence every student to aspire to become a research scientist, but it puts them on the path of becoming informed citizens.

11.8.3 Recommendations for Engaging Students in Experimentation

In summary, consider the following suggestions when engaging students in inquiry-based experimentation and professional practices:

- Students from diverse backgrounds can actively engage in the scientific process, new learning experiences, and professional practices performed in the scientific community.
- Scaffolding multi-week guided inquiry-based learning experiences or lab activities provides students with opportunities to practice and apply scientific techniques and concepts as part of the course curricula as well as foster skills in time management, critical thinking, and ACE-Bio experimentation competencies: *Plan, Conduct, Analyze, Conclude* and *Communicate*.
- Grading rubrics serve as useful tools to help guide and list the expectations to successfully complete the task, especially during peer-review activities.
- Collaborative inquiry-based activities allow students to learn from each other and enhance their communication skills.
- For students to achieve the desired outcomes, they need supportive and motivational instructors to guide and mentor them.

Acknowledgments A portion of this research is supported by the National Science Foundation grant #1346567 for the Advancing Competencies in Experimentation – Biology (ACE-Bio) Network. Any opinions, findings, and conclusions or recommendations expressed in this article are those of the author and do not necessarily reflect the views of the NSF. A special thanks goes to Nancy Pelaez, Trevor Anderson, and Stephanie Gardner for providing an opportunity to contribute a book chapter. Thanks also go to Nancy Peleaz (monitoring editor) and Michelle Harris for their invaluable feedback. Furthermore, I thank Paula H. Spence for transcribing open-ended student responses and reviewing this book chapter.

References

American Association for the Advancement of Science (AAAS). (2011). *Vision and change in undergraduate biology education: A call to action.* Washington, DC. Retrieved December 23, 2020 from https://live-visionandchange.pantheonsite.io/wp-content/uploads/2013/11/aaas-VISchange-web1113.pdf

Beck, C., Butler, A., & Burke da Silva, K. (2014). Promoting inquiry-based teaching in laboratory courses: Are we meeting the grade? *CBE-Life Sciences Education, 13*(3), 444–452. https://doi.org/10.1187/cbe.13-12-0245

Bell, E. (2011). Using research to teach an "troduction to biological thinking". *Biochemistry and Molecular Biology Education, 39*(1), 10–16.

Berland, L. K., Schwarz, C. V., Krist, C., Kenyon, L., Lo, A. S., & Reiser, B. J. (2016). Epistemologies in practice: Making scientific practices meaningful for students. *Journal of Research in Science Teaching, 53*, 1082–1112. https://doi.org/10.1002/tea.21257

Blumer, L. S., & Beck, C. W. (2019). Laboratory courses with guided-inquiry modules improve scientific reasoning and experimental design skills for the least-prepared undergraduate students. *CBE Life Sciences Education, 18*(1), ar2-ar2. https://doi.org/10.1187/cbe.18-08-0152

Buck, L. B., Bretz, S. L., & Towns, M. H. (2008). Characterizing the level of inquiry in the undergraduate laboratory. *Journal of College Science Teaching, 38*(1), 52–58.

Chung, H.-M., & Behan, K. J. (2010). Peer sharing facilitates the effect of inquiry-based projects on science learning. *The American Biology Teacher, 72*(1), 24–29.

Coker, J. S. (2017). Student-designed experiments: A pedagogical design for introductory science labs. *Journal of College Science Teaching, 46*(5), 14–19.

Cooper, K. M., Soneral, P. A. G., & Brownell, S. E. (2017). Define your goals before you design a CURE: A call to use backward design in planning course-based undergraduate research experiences. *Journal of Microbiology & Biology Education, 18*(2). https://doi.org/10.1128/jmbe.v18i2.1287

D'Costa, A. R., & Schlueter, M. A. (2013). Scaffolded instruction improves student understanding of the scientific method & experimental design. *BioOne, 75*(1), 18–28.

GraphPad Software. (2018). *QuickCalcs: T-test calculator.* GraphPad Software https://www.graphpad.com/quickcalcs/ttest1/?Format=SD

Halim, A. S., Finkenstaedt-Quinn, S. A., Olsen, L. J., Gere, A. R., & Shultz, G. V. (2018). Identifying and remediating student misconceptions in introductory biology via writing-to-learn assignments and peer review. *CBE—Life Sciences Education, 17*(2), ar28. https://doi.org/10.1187/cbe.17-10-0212

Haug, B. (2014). Inquiry-based science: Turning teachable moments into learnable moments. *Journal of Science Teacher Education, 25*, 79–96. https://doi.org/10.1007/s10972-013-9375-7

Lord, T., & Orkwiszewski, T. (2006). Moving from didactic to inquiry-based instruction in a science laboratory. *The American Biology Teacher, 68*(6), 342–345. https://doi.org/10.1662/0002-7685(2006)68(342:DTIIIA]2.0.CO;2

McLoughlin, C., & Hollingworth, R. W. (2005). Problem solving in the sciences: Sharing expertise with students. In C. McLoughlin & A. Taji (Eds.), *Teaching the sciences: Learner-centered approaches* (pp. 105–121). Food Product Press, Inc.

Minbiole, J. (2016). Improving course coherence & assessment rigor: "Understanding by design" in a nonmajors biology course. *The American Biology Teacher, 78*(6), 463–470. https://doi.org/10.1525/abt.2016.78.6.463

Odom, S., Glenn, B., Sanner, S., & Cannella, K. A. S. (2009). Group peer review as an active learning strategy in a research course. *International Journal of Teaching and Learning in Higher Education, 21*(1), 108–117.

Ortiz, E. (2009). *Pocket seed viewer*. ASU – Ask A Biologist. Retrieved from https://askabiologist.asu.edu/content/pocket-seed-viewer

Palmer, D. H. (2009). Student interest generated during an inquiry skills lesson. *Journal of Research in Science Teaching, 46*, 147–165. https://doi.org/10.1002/tea.20263

Pelaez, N., Anderson, T. R., Gardner, S. M., Yin, Y., Abraham, J. K., Bartlett, E, Gormally, C, Hill, J. P., Hoover, M., Hurney, C., & Long, T., Newman, D. L., Sirum, K., & Stevens, M. (2017). *The basic competencies of bio-logical experimentation: Concept-skill statements*. PIBERG Instructional Innovation Materials. Paper 4. Retrieved October 18, 2021 from https://docs.lib.purdue.edu/pibergiim/4/

Roberson, C., & Lankford, D. (2010). Laboratory notebooks in the science classroom. *The Science Teacher, 77*(1), 38–42.

Schreier, A. A., Wilson, K., & Resnik, D. (2006). Academic research record-keeping: Best practices for individuals, group leaders, and institutions. *Academic medicine: Journal of the Association of American Medical Colleges, 81*(1), 42–47. https://doi.org/10.1097/00001888-200601000-00010

Spence, P. L., Phillips, R. S., McAllister, A. R., White, S. L., & Hollowell, G. P. (2020). Student-scientist curriculum: Integrating inquiry-based research experiences and professional development activities into an introductory biology laboratory course. *Journal of Microbiology & Biology Education, 21*(3). https://doi.org/10.1128/jmbe.v21i3.2225

Thanukos, A, Scotchmoor, J. G., Caldwell, R., & Lindberg, D. R. (2010). The real process of science-understanding science at http://www.understandingscience.org and Science 101: Building the Foundations for Real Understanding. *Science, 330*(6012), 1764–1765. https://doi.org/10.1126/science.1186994.

Wiggins, G. W., & McTighe, J. (2011). *The understanding by design guide to creating high-quality units*. Association for Supervision and Curriculum Development.

Zoller, U. (2000). Teaching tomorrow's college science courses-are we getting it right? *Journal of College Science Teaching, 29*(6), 409–414.

Zygomatic. (2003). *Wordcloud*. Retrieved from https://www.wordclouds.com/

Chapter 12
Feedback and Discourse as a Critical Skill for the Development of Experimentation Competencies

Janet M. Batzli, Michelle A. Harris, Dennis Lee, and Heidi A. Horn

12.1 Introduction

Scientific discourse is the collaborative and reciprocal exchange of ideas, arguments, and feedback that happens among practicing scientists in research labs everyday. Scientists brainstorm, debate, troubleshoot, argue, and teach/learn from each other in formal and informal contexts; in lab meetings, bench conversations, hallway talk, Q&A presentations, and through the process of scientific peer review. Scientific discourse integrates scientific language within the norms, values, and behaviors of scientists, and within their scientific community of practice at large (Lave & Wenger, 1991). Research on how scientists construct knowledge points to the importance of delivery and uptake of feedback in scientific communities (Walsh & McGowan, 2017). Furthermore, scientific discourse through frequent and varied forms of feedback is central to the development of student-centered classrooms and scientific competencies described in the AAAS (2011) *Vision and Change in Undergraduate Education* report. This national call to action appeals to instructors to provide learning environments where students practice "ongoing, frequent, and multiple forms of feedback" as part of a student-centered approach to teaching (AAAS, 2011). If scientific discourse is so important to how we think, reason, and

J. M. Batzli (✉) · M. A. Harris
Biology Core Curriculum (Biocore), University of Wisconsin-Madison, Madison, WI, USA
e-mail: jcbatzli@wisc.edu; maharris@wisc.edu

D. Lee
BSCS Science Learning, Colorado Springs, CO, USA
e-mail: dlee@bscs.org

H. A. Horn
Department of Biology, Edgewood College, Madison, WI, USA
e-mail: hhorn@edgewood.edu

N. J. Pelaez et al. (eds.), *Trends in Teaching Experimentation in the Life Sciences*, Contributions from Biology Education Research,
https://doi.org/10.1007/978-3-030-98592-9_12

behave as scientists, how do we integrate scientific discourse into our learning outcomes and teach it as an essential skill for a practicing scientist?

Despite evidence that instructor feedback influences achievement of student learning outcomes (Baranczyk & Best, 2020), there are few studies connecting faculty and peer feedback to the development of experimentation competencies. This chapter introduces the idea of teaching the skill of scientific discourse through feedback in courses where students do and communicate authentic science, consistent with course-based undergraduate research experiences or CUREs (Corwin et al., 2014). Philosopher of science, Helen Longino, has long argued that scientific knowledge is socially legitimized when there are public forums where scientific information can be presented and critiqued, public standards can be discussed, participants hold intellectual equality that is tempered by expertise, and participants incorporate critique into their work when appropriate (Longino, 2002). Scientific discourse aids students in developing the norms for explanation, argumentation, and justification in the classroom community (Russ et al., 2016). Indeed, when instructors probe students for evidence, it enhances explanatory rigor (Grinath & Southerland, 2019), and when students probe their peers for explanations, they elicit disciplinary reasoning (Leupen et al., 2020) that informs their science and their development as scientists.

As we practice science in our classrooms, we consider consistent, high quality feedback and the scientific discourse that grows from it as the underlying 'glue' that binds together the experimentation competencies that students are learning. We contend that reciprocal feedback (written and oral) among students and between students and their instructors is an integral part of teaching experimentation. As students learn to give and receive feedback through iterative trials (and errors) in their own process of experimentation, they learn a fundamental principle about scientific knowledge itself – that it is based in evidence, born collaboratively, subject to skepticism and critique, and meant to be reflected on and revised as we create new knowledge. Learning how to solicit, receive, and give meaningful feedback in the context of experimentation offers a way to engage in scientific discourse.

In this chapter we introduce feedback as both a form of assessment and as a mode of entry for students into a disciplinary scientific community.

Our goal is twofold:

1. To demonstrate why feedback is important to the scientific enterprise, and how to support students as they learn to give and receive feedback as a scientific competency.
2. To describe how feedback is strategically scaffolded within a multi-week laboratory curriculum to help students achieve different levels of experimentation competencies.

The ACE-Bio competencies were an important affirmation of our instructional sequence where students do and communicate authentic research. In this chapter we describe an integrative lab sequence in which students repeatedly learn to use feedback as discourse as they engage in a sequence of independent research projects. We have found that scientific discourse and feedback are key components that serve as the 'glue' between each of the ACE-Bio experimentation competencies.

12.2 Background

The scientific competencies identified by the ACE-Bio network, collectively referred to as 'experimentation' competencies, are aligned with and implemented through a three-semester laboratory course sequence called Biology Core Curriculum (Biocore) at University of Wisconsin- Madison. This CURE sequence consists of 8–9 process of science or experimentation units, 2–3 units per semester over three semesters. Classes meet for four hours each week, consisting of a one-hour discussion section that precedes a three-hour laboratory class meeting. Each unit asks students to form *Question*s from their observations, *Identify* hypotheses, *Plan* and *Conduct* experiments, *Analyze* and interpret data and make *Conclusions*, and finally *Communicate* their science in the form of papers, posters, and presentations similar to professional scientists (Pelaez et al., 2017 and Chap. 1 in this volume; underlined/italicized basic competencies of biological experimentation from the ACE-Bio Network are consistent with Batzli et al., 2018). In Biocore, giving and receiving feedback through student-student and student-instructor dialog is a primary learning goal that is integrated into the process of science and experimentation competencies along with social skills of collaboration and interpersonal communication. With that, our teaching goals in Biocore are to foster students' practice of experimentation competencies with rich opportunities for feedback to prepare students to construct knowledge like professional biologists (Magnus, 2000), and ready our students for employers that value both technical and social skills of practicing scientists (Hora et al., 2016).

Scientific discourse, through the process of giving, receiving, and incorporating feedback at each step and upon each iteration of a cycle of experimentation, is an important skill for students to learn as they develop as scientists. Yet this skill is often overlooked as a learning outcome in science classrooms or is considered solely the domain of instructors providing unidirectional feedback to students. In Biocore, we establish a classroom culture that focuses on feedback as a learning goal. Establishing norms for science dialog and discourse promotes students' comfort as they solicit, give, and receive feedback from their fellow students and instructors.

Figure 12.1 illustrates how we consider the process of science as cyclic and iterative. Although the experimentation competencies (highlighted in white petals) are diagrammed as discrete phases in this simplistic model, in practice, they are skills and intellectual processes that are continuously referred to, reflected on, and revised throughout each process of science cycle or experimentation unit. Overlaid on the model is feedback (gray "propeller" petals); student-student or peer feedback and student-instructor feedback. In our classroom, feedback is integrated into all parts of the process, but is particularly important and intentional at two critical checkpoints: (1) the proposal phase: after developing a preliminary plan, but before making the investment to conduct an experiment and (2) the analysis phase: after making preliminary conclusions about the data and hypothesis, but before communicating the conclusions more broadly (i.e. publishing results). In both of these phases, we

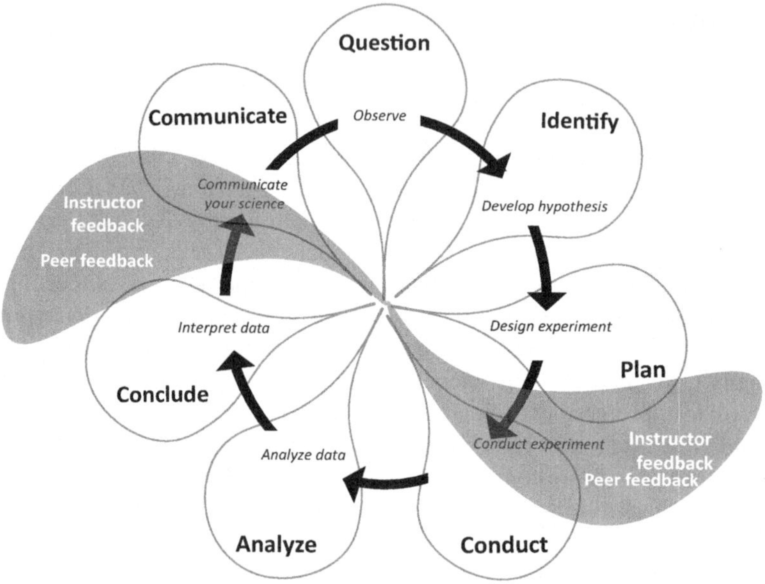

Fig. 12.1 Process of science cycle model overlaid with ACE-Bio experimentation competencies (white petals) and critical opportunities for instructor and peer feedback

guide students to prepare to solicit feedback by generating questions about their project or the data they collected; to prepare to be skeptical of their proposal, of their hypotheses, and of their results, and invite others to be skeptical too. This is hard. Asking others for critique is a particularly vulnerable thing to do. We are transparent about these difficulties with students and create a supportive learning environment, coaching them through the emotional aspect of soliciting and receiving feedback. Despite the emotional hurdle, after the first feedback presentation – when everyone has invested in being vulnerable and questioning, most students find that it strengthens their science and their community of practice. In the words of one student, reflecting on the value of the first feedback presentation:

> With feedback from the teaching team and our peers, in discussions and informal feedback presentations, we were able to improve our skills. As I was trying to give useful feedback to my peers in lab, I also learned to think more critically about the process we use in lab to answer our own study question.

12.3 Curriculum Design, Implementation, and Evidence

Here we describe practical strategies for practitioners in a week-by-week implementation of a typical Biocore lab unit, highlighting opportunities for scientific discourse through the process of giving and receiving feedback. We have previously described how student-student and student-instructor feedback is incorporated through a 5–6-week CURE unit (Batzli et al., 2018). Table 12.1 in combination with the week-by-week description below are intended to serve as a practical guide summarizing teaching goals and instructor prompts eliciting scientific discourse through scaffolded feedback. These activities support the development of students' feedback skills within the context of the experimentation process in a consistent, timely, and developmentally appropriate way. Student quotes from oral interviews conducted as

Table 12.1 Examples of effective instructor approaches and discourse, scaffolded over a typical multi-week CURE unit (see also Batzli et al., 2018)[a]

Stage in research unit	Teacher cues and prompts for effective feedback and science discourse
Week 1 (*Question*, *Identify*) Intro to topic/model system; practice data collection	• Be curious; Pay attention. • Write down all of your brain-stormed questions (no 'dumb questions'). • Allow each member of the team to discuss their questions before coming to a consensus. • Do preliminary background search to generate key words and develop foundational understanding of underlying concepts.
Week 2 (*Plan*, *Communicate*) Group informal (non-graded) feedback presentations with Q&A	Before group informal feedback presentations: • Brainstorm list of pros/ cons to receiving feedback. • Prepare to explain concepts or reasoning that is specific to your team's question. • Show your expected results graph to another team and ask them to "guess your hypothesis". • Generate "Questions We Still Have" slide to encourage feedback.. • Be ready to take notes during the Q&A after your presentation. • Announce before presentations begin: "Our goal as a science community is to help each other with our science; to listen carefully, ask questions, offer suggestions, and share resources."
Week 3 (*Plan*, *Conduct*) Paper/poster peer review; pilot studies & data collection	During and after feedback presentation: • Students ask the first set of questions and provide feedback (instructors hang back before asking their own questions). • Gently steer students away from questions focused on methods/experimental design. • Model questions that focus on BioRationale (BR) underlying hypothesis such as: – What is the alignment between hypothesis & expected data? – Can you clarify how dependent variable X is directly influenced by your independent variable Y? – It appears that you are interested in measuring Z (Dependent variable) in your hypothesis, yet your expected results graph reports that you are measuring Y. Can you explain how Z and Y are connected? – (Directed to audience) How does this team's proposed research relate to each team's project? What references can you share?

(continued)

Table 12.1 (continued)

Stage in research unit	Teacher cues and prompts for effective feedback and science discourse
Week 4 (*Analyze*, *Conclude*) Group-instructor consults; complete data collection	As individuals prepare their research proposals: • Allow time for face-to-face meetings between peer review partners, requiring each student to bring their completed peer review form to the meeting, as a guide to peer review conference questions. • Focus peer review comments/suggestions on global issues (e.g. does the proposed research address the knowledge gap?) and not local concerns such as spelling or grammar. • Have students reflect on their incorporation of feedback by filling out a "Response to Peer Review" that is submitted along with their revised proposal paper, 2–3 days after peer review conferences.
Week 5 (*Conclude*, *Communicate*) Formal (graded) group presentation; one-on-one conferences between instructor and student	During data interpretation: Encourage teams to take notes during Q&A & use feedback to improve their final paper/poster. • How comfortable do you feel defending your conclusions? How do your data fill in the knowledge gap? • How would your suggested "next steps" further inform the knowledge gap? One –on –one conference feedback: – How will you incorporate feedback from your peer reviewer and instructors into your final paper/poster? – How do your data inform the conclusion? • How could you return to your BR and show how data inform the knowledge gap, assumptions, and logical next steps?
Week 6 (communicate) Final paper/poster submitted (capstone projects only)	As research teams wrap up their project: • How can you use what you learned from this research project to improve your next project? • What scientific questions would you like to explore based on your data or conclusions?

[a]The guidance and coaching we give students as they are learning how to give and receive feedback serves as a model of scientific discourse through feedback. The process of science corresponds to Fig. 12.1. Gray petal represents where feedback is given in the process of science each week of a research unit

part of a research study (Lee, 2020) are provided as insight into student development and the efficacy of feedback. Assignments, assessments, and instructional materials supporting each week are summarized in Table 12.2.

The plan outlined in Table 12.1 target the following teaching goals for fostering feedback-rich curriculum:

- Develop a classroom culture that focuses on feedback as a learning outcome.
- Establish and model norms to promote a collegial and productive cycle of soliciting, giving, receiving, and incorporating feedback.
- Use feedback as a form of formative assessment and a mode of entry into the scientific community.

Table 12.2 Aligned learning goals, assignments and instructional materials supporting activities in a typical multi-week research unit. Learning goals are adapted from Batzli et al. (2018)[a]

Stage in research unit (ACE-Bio Experimentation Competency)	Learning goals students should be able to:	Assignments announced at end of lab	Instructional materials
Week 1 (***<u>Question</u>***) Intro to topic/model system; practice data collection	Make and share observations with peers and instructors Formulate testable question(s) Ask for feedback regarding what and how to measure variables	Groups collaboratively: • Complete experimental Design worksheet • Begin search for relevant literature • Prepare PowerPoint slides for informal week 2 feedback presentation	Constructive & Destructive Group Behavior handout Experimental Design worksheet w/ embedded figure facts assignment and figure facts template Expectations for feedback presentation slides (POS companion)
Week 2 (***<u>Identify</u>***, ***<u>Plan</u>***) Group informal (non-graded) feedback presentations with Q&A	Find and sort relevant scientific literature associated with testable question (i.e. dependent and independent variables) Work with team to come to consensus on id of knowledge gap Co-create a logical biorationale and hypotheses with research team Anticipate expected and alternative results and implications Exchange ideas, give and receive feedback (orally) with peers & instructors	Research proposal paper Peer review of classmate's paper	Proposal paper rubric (POS companion) Biorationale rubric Peer review template Peer review rubric (POS companion)
Week 3 (***<u>Conduct</u>***) Paper/poster peer review; pilot studies & data collection	Give and receive written feedback through peer and instructor review of research proposal Sort and evaluate peer reviewer feedback, and incorporate suggestions based on relevancy and merit	Teams prepare informal data summaries including data graphs, summary statistics, and formal statistical analyses (if applicable)	Biocore statistics primer with RStudio tutorials (POS companion) Response to reviewers template File management plan template

(continued)

Table 12.2 (continued)

Stage in research unit (ACE-Bio Experimentation Competency)	Learning goals students should be able to:	Assignments announced at end of lab	Instructional materials
Week 4 ***(Analyze)*** Group-instructor consults; complete data collection	*Analyze* data and make logical conclusions utilizing statistical reasoning Evaluate assumptions associated with experimental design and biological system Give and receive written feedback	Prepare formal group presentation Group presentation practice run-through with undergraduate teaching assistant prior to week 5 lab	Instructor paper feedback and grade summary template
Week 5 **(*Conclude*, *Communicate*)** Formal (graded) group presentation; one-on-one conferences between instructor and student	Give effective oral research presentations Ask relevant questions and share information learned through research Write and communicate about scientific research	Final research paper or poster (for capstone projects)	Rubric for formal presentation (POS companion) Final paper and poster rubrics (POS companion) Group effort analysis (GEA) form
Week 6 **(*Communicate*)** Final paper/poster submitted (capstone projects only)	Write and communicate about scientific research		

[a]We use several curricular tools including clear guidelines and rubrics in our Process of Science (POS) Companion e-text available through PressBooks https://wisc.pb.unizin.org. The POS Companion integrates four Books: (1) Biocore Writing Manual, (2) Statistics Primer, (3) Group Learning and Collaboration guide, and (4) Tools & Techniques guide

- Scaffold the development of providing and receiving feedback through an iterative process, including numerous low-stakes practice opportunities.
- Help students to achieve intellectual equality by the end of the semester.

12.3.1 Week 1: Student Meet Research Team; Instructors Introduce Topic/Model System; Students Practice Data Collection

At the beginning of the first unit project, newly formed research teams are asked to first reflect on their constructive and destructive group behaviors individually, then respectfully listen to each teammate share their reflections and goals for improving group skills. Instructors introduce students to potential model systems and concepts for their first research project, encouraging them to practice feedback through small group discussions as they brainstorm and discuss potential novel research questions. This trading and swapping of research questions within groups is the first informal peer review experience. Below, we present an example from a set of oral interviews with students done by Lee (2020) of how one student's group navigated this first peer review experience.

> We kind of ran with a hypothesis until someone found a paper that said otherwise, and then, this was the last one that we ended up with, and no one had found anything that said otherwise. And this one made sense. I don't think that people had necessarily… put this string of ideas together before, because I combed Google Scholar for every article about intraspecific competition in these plants.

This student describes their research team's discussions about the hypotheses, which were mediated by information from peer-reviewed articles. This peer-reviewed information helps to transition students' discussions into scientific discourse. As group members bring more information into the discussion, the hypothesis matures until it becomes a new idea that has not been tested before. In this way, students are given the opportunity to use discourse that aligns well with the kinds of discourse that occur in professional scientific research.

Before week 1 lab adjourns, we introduce an Experimental Design Worksheet (see supplemental materials) that groups are expected to complete before week 2 lab (Parts A & B), and after week 2 lab (Part C). This worksheet contains questions and prompts designed to help teams formulate a testable question, conduct a literature search, and write a research proposal. Part B of the worksheet includes an embedded Figure Facts activity (Round & Campbell, 2017) which introduces students to reading and interpreting primary scientific literature.

12.3.2 Week 2: Group Informal Feedback Presentations with Q & A

Low stakes, non-graded informal group feedback presentations are the most valuable formative assessment piece of our curriculum and are modeled after typical meetings in STEM research labs. Each research team presents 8–10 PowerPoint slides to the class describing their research goals, hypotheses, methods, expected and alternative (unexpected) results in graphical form, implications, and questions they still have. Each member of the group is required to contribute to the presentation and all teammates are encouraged to answer audience questions after the presentation. Although these presentations and Q&A are non-graded, they are, perhaps, the most valuable learning activities that students can practice. As supported by Ngar-Fun and Carless (2006) non-graded opportunities for peer feedback allow students to practice discourse, listen carefully, think critically, and develop disciplinary language for feedback in a low stakes, high cognitive level, and potentially high-reward context.

During an oral interview in the first semester lab, one student describes how their group constructed an experimental plan by discussing with another group that was working with similar conditions (Lee, 2020). This discussion led this student to consider an experimental treatment of intermittent cold flashes, an experimental plan that they had not considered previously and, according to their literature search, was a knowledge gap to the scientific community.

> During our feedback presentation, we presented to another group that was also looking at temperature, but they were looking at warmer temperatures. We were discussing back and forth what would be better, cold flashes or straight (cold), ... Like all the time in the cold (...) We thought cold flashes. You put them in the cold for a day, they build up a lot of anthocyanin, and then you take them out and they can grow normally, but that anthocyanin is still there. That could give us our hypothesis. It could give us evidence to support our hypothesis.

For the student, this feedback presentation allowed for discourse that guided the development of their experiment. As the student and their team pivoted and improved their project through peer feedback, audience members reflected on and revised their own projects. At the same time, instructors were given a unique window into students' progress and scientific reasoning by listening to student questions and presenting teams' reasoning (accurate or flawed). Finally, instructors modeled respectful, collegial scientific norms through projecting a professional tone and challenged thinking through the substance of the questions they asked of students. When students are first introduced to feedback presentations, instructors hang back and allow the student audience to ask questions first. When instructors step in, they model questions at a higher level, often focusing on the "biorationale", (the knowledge gap, or reasoning underlying hypotheses), as well as clarifying connections or misalignment between expected data, hypotheses, and stated implications (see Table 12.1 for more examples of instructor cues and questions).

Before week 2 lab adjourns, instructors assign peer review partners by pairing students from different research teams. Students are directed to writing and peer review guidelines and rubrics in the *Biocore Writing Manual* (Book 1) of the Process of Science Companion e-text series (Batzli & Harris, 2020). The biological rationale figure and legend rubric (supplemental materials) describes expectations for this model-based reasoning assignment summarizing each team's scientific reasoning.

12.3.3 Week 3: Paper/Poster Peer Review; Pilot Studies & Data Collection

Based on the feedback received during week 2 presentations, each student writes an independent research proposal paper that is exchanged with a classmate in a different group for formal (graded) peer review (for introductory assignment see

Experimental Design Worksheet Part C). At least once per semester students also have the opportunity to send their proposal paper draft to an undergraduate teaching assistant (uTA) for peer review. Students use feedback from their peer reviewers to revise their proposal paper before it is submitted 2–3 days later for grading by an instructor. Students are also required to fill out a "Response to Reviewers" template as an opportunity to explain to their instructor how they incorporated feedback from their peer review partner, or conversely, why they chose to ignore feedback. Each team completes a File Management Plan intended to prompt important conversations about effective data management and security practices.

Following peer review conferences, student research teams *Conduct* pilot studies and/or collect data for their experiment, incorporating what they learn through these three layers of formative feedback and revision. Instructors often lead a short discussion during lab focusing on statistical analyses that may be appropriate for the kinds of data collected for that particular unit. These kinds of discourse help to establish a public set of standards for the generation of acceptable knowledge in this scientific community. Below, in an oral interview with a student during their first semester lab, they explain how instructors discuss the presentation of statistical data, and their perception of why presenting data in this way is important (Lee, 2020).

> We put error bars on them because it's one of the requirements. Once again, to be blunt. Whenever we're doing statistics, (instructor 1) and (instructor 2) both tell us--Make sure it's within a certain amount of deviation. Make sure you have error bars in order to determine if your data is statistically correct. Because it can seem like there's no correlation, but once you look at the whole picture and do some statistical analysis on Excel, the numbers can tell you otherwise.

The excerpt above highlights how this student values feedback and their current understanding of data presentation. They first list error bars as a requirement they are told to do by instructors and a measure of "correctness", but cite instructor feedback about not only the requirement of, but the importance of including error bars to assist the reader as they evaluate whether trends are statistically significant. While instructors' guidance is important at this early stage, it is equally important to listen carefully to students – in this case, to this student's valuation of statistics as a tool. In this short statement, the student reveals their incomplete understanding of how scientists use statistical analysis. Through discourse opportunities such as feedback presentations or consultations with instructors, we often hear students describe scientific hypotheses as 'correct or incorrect'. They refer to biological variation as 'error', and their ideas about 'good' data as 'proof' of a scientific conception. As students become more comfortable with feedback and the perspectives of others, they develop a more nuanced and less absolute view of their hypotheses as either right or wrong. They also develop a learning mindset when it comes to science as a way of knowing and become more tolerant of (and excited about) the uncertainty that science presents. This student ends the excerpt with a hopeful statement of growth: that statistical analysis can reveal relationships between variables that are not apparent at first glance.

Before week 3 lab adjourns, we encourage students to carefully examine their raw data before preparing overall data summary analyses, to share in week 4.

12.3.4 Week 4: Research Team-Instructor Consultations; Data Analysis and Interpretation Feedback Presentations

Once data is collected and analyzed, teams informally present a tentative interpretation of their data to their peers/ fellow researchers for another round of feedback and exchange of ideas. This is another low-stakes, formative assessment opportunity for students to practice feedback solicitation and incorporation. Not surprisingly, teams collect data that often does not support their hypothesis (alternative results) or data that teams cannot trust to address their hypothesis (e.g., due to lower than expected sample sizes, sampling errors, etc.). Novice CURE students often consider either of these outcomes as a failure. During consultations with individual teams, instructor feedback is key to helping students become comfortable with messy data, or their perceived "failure" as a normal part of the experimental process. As science mentors/ coaches, instructors encourage students through data analysis consultations to carefully inspect the variation in their data to look for clues of biological relevance or interesting patterns even if there is no statistically relevant difference in mean values. This is a key step in the feedback process – encouraging students to carefully reflect on what they learned from the data they gathered and use it to propose the next logical steps in experimentation (e.g. repeat experiment, adjust protocol, measure covariates, revise hypothesis).

By the end of week 4, students receive detailed feedback from their graduate teaching assistants (TAs) in the form of proposal paper grades and comments aligned with proposal paper rubric expectations. Teams are expected to incorporate this TA feedback into their final presentation. Before week 4 lab adjourns, instructors' direct students to guidelines and the rubric for their final group presentation assignment in the Biocore Writing Manual (Batzli & Harris, 2020). This oral presentation rubric describes expectations for content as well as presentation mechanics criteria such as time limits, equitable contributions from each group member, and effective visuals.

12.3.5 Week 5: Formal Group Presentation; One-on-One Conferences Between Instructor and Student

Finally, after four rounds of feedback and revision, teams prepare a formal presentation using expectations detailed in the Biocore Writing Manual (Batzli & Harris, 2020) final research presentation rubric. Prior to lab, teams often schedule an office hour with instructors to discuss their data analyses and conclusions based on the data. A day or two prior to presenting in week 5 lab, teams also meet with an undergraduate TA to do a practice presentation run-through. The uTAs provide formative feedback which student teams incorporate into the final graded version of their presentation. The Q&A, feedback, and discourse that follows formal group presentations differs from informal feedback presentations in week 2, in that, audience members focus on the strength of conclusions and share information learned through their own research.

In our third semester (capstone) lab course, we provide one last opportunity for students to incorporate feedback from the Q&A after their final formal presentation. Students incorporate feedback from their presentation into a final, revised paper, research poster, or re-proposal paper/poster.

Finally, as students wrap up their project, and submit their presentations for evaluation, we provide an opportunity for students to practice feedback associated with group work/ research team skills. Each individual is asked to complete a Group Effort Analysis form (GEA) (supplemental materials) where they refer to a rubric to rate themselves and their teammates on all aspects of their work together including: attendance and punctuality at meetings, preparedness in planning and in conduction of experiments, participation, ability to listen and cooperate, and – more recently added, students' ability to be inclusive and equitable in group work, and their cultural humility as they work and communicate with others. They then add a brief comment to their teammate and have the option to add confidential comments to the instructor. The instructor then compiles comments and shares them with each individual with feedback, encouragement, and support for how to improve.

12.3.6 Competencies and Feedback as Emphasized in Subsequent Semesters

Some aspects of experimentation competencies emphasized in the first semester lab are more salient than aspects emphasized in the second or third semester lab where students experience iterative cycles of the process of science. For instance, the first semester may emphasize <u>*Question*</u>, <u>*Identify*</u>, and <u>*Communicate*</u> competencies (Fig. 12.2a), which are then scaffolded, layered and expanded on in the second and third semesters in a developmentally appropriate way (Fig. 12.2b, c). In the first semester, through study of ecology, genetics and evolution, feedback is focused on

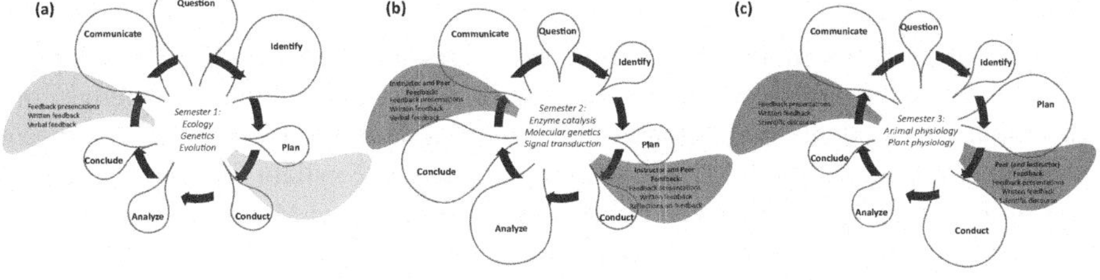

Fig. 12.2 Process of science cycle overlaid by experimentation competencies and feedback learning goals over three semesters (**a**, **b**, and **c** – left to right) in Biocore. The size of the 'petal' indicates the relative emphasis of the competency in each semester and strategic feedback 'checkpoints' are highlighted in gray. Student feedback develops over the course of three semesters as students gain more autonomy and progressively give and receive more feedback

students' scientific reasoning as they develop novel research questions and identify a logical biological rationale. Structured feedback (in the form of both graded and ungraded assessments) at the proposal checkpoint is critical because it provides students the opportunity to intentionally reflect on their process and make adjustments before committing to a concrete experimental plan. Proposal phase feedback is largely modeled by instructors during the first semester while students are given opportunities to practice giving and receiving feedback through feedback presentations and peer review of papers. In subsequent semesters, students are given more responsibility to take ownership in giving, receiving and negotiating feedback with their peers while the instructors take on more of a coaching, clarifying, and facilitation role.

In semester 2 (Fig. 12.2b), we place more emphasis on *Analyze*, *Conclude*, and *Communicate* competencies as we move through three units focusing on enzyme catalysis, molecular genetics, and signal transduction. For instance, we emphasize *Analyze* through introduction of formal hypothesis testing statistics (two-sample independent t-tests, 1- and 2-way ANOVA, chi-square test of independence) and RStudio. *Conclude* and *Communicate* are emphasized through learning goals, activities, and assessments that support students' choice of data analyses that are aligned with their experimental design. Feedback in this semester focuses most heavily on communicating the subtle yet crucial distinction between statistical conclusions and biological conclusions. Students begin to gain autonomy and confidence to give and receive feedback in this semester. While structured opportunities to give feedback are similar to semester 1, students progressively participate by giving more feedback during these structured opportunities and begin to actively seek unstructured feedback from peers throughout the process of science.

The third semester capstone lab emphasizes the *Plan*, *Conduct*, and *Communicate* competencies. Here we give students the most autonomy in the choice of experimental question, model system, and experimental design as teams move through two expanded independent projects of their choice in animal and plant physiology. Teams seek out and provide abundant feedback through numerous informal and formal presentations, peer review assignments, and revisions of paper and poster drafts. Here students display the most agency in their ability to give and receive feedback and ultimately engage in spontaneous scientific discourse throughout the process of science.

Regardless of how many iterations of the process of science and how much practice students have with experimentation, feedback is integrated into knowledge construction, into students' valuation of collaboration, and into their understanding of standards/ norms within the scientific process itself. The syllabi for all three semesters of the Biocore lab series can be found in Supplemental Materials, and can also be accessed under the "Courses" tab at https://biocore.wisc.edu.

12.4 Implications

Biocore students' reflections show how they recognize and value the role of feedback in their development as scientists. Over three semesters, students recognize their growth and progressive autonomy in their capacity to use, solicit, negotiate, and value feedback in their development as scientists. From student oral interviews during our first semester lab, one student reflected (Lee, 2020):

> And whoever I go talk to a (Biocore) professor or a teacher or something, I'll go up to them with a question, which another great thing is you're never afraid to ask anybody questions, we love questions. And they'll be like, "Okay, well here's how you might think about it." Or they'll suggest a pathway or a different way to find more knowledge about your topic. And kind of validate your knowledge or give you a new way to change the way you're thinking of it.

In reflections on written course evaluations from students completing the second semester lab:

> I always learn so much from informal feedback presentations. I think that they provide a really powerful opportunity to learn. The direct feedback and open discussion is really when I start to understand a project and I appreciate that it gives the presenters an opportunity to talk things out and practice their presentation skills and it gives those in the audience an opportunity to be inspired by their peers and also use their own skills in evaluation to better understand the project as well.

> I think informal feedback presentations are so beneficial, they have really taught me a lot about how to improve my work and ensure my logic is reasonable. They have taught me to explain science comprehensively. They have taught me to give and receive constructive feedback in a healthy beneficial way.

The kinds of scientific discourse mediated by the feedback mechanisms employed in Biocore satisfy the social norms for scientific knowledge construction established by Longino (2002). Through a feedback-rich curriculum, we are able to help students develop ideas about scientific knowledge construction that are aligned with scientific practice.

Providing high quality feedback to students is challenging, as is mentoring students to provide effective feedback to each other. It takes time, focus, and experience for both students and instructors. If you choose to invest, here are some affordances and challenges to consider.

12.4.1 Intellectual Confidence and Ownership

At first, students are excited but timid (and sometimes even defensive) about opening their ideas to feedback since they have either invested a great deal of time thinking about their project and sometimes cling to questionable reasoning or they have not quite done their homework. With some coaching and modeling, students gain confidence, they become more engaged, and experience a rise in intellectual equality as a group of practicing scientists. With iterative practice, they also build intellectual equality with their instructors, which further fuels their confidence, ownership in their science, and helps grow their identity as a scientist.

It is challenging for instructors to keep pace with students' projects and think deeply, critically, and constructively enough to provide high quality feedback. In addition, as teachers, it can be a challenge to learn to be comfortable not knowing and to learn from students. In addition, we have found it hard to balance how much support versus challenge we dish out to students as science mentors. When we challenge students with too much feedback and notice they are flagging, we need to ramp up the support with more suggested pathways to help them get 'unstuck'. When they are getting too much support, in terms of attention and guidance, we back off and let them work until their questions naturally emerge. We continually adjust our feedback (the level and pace), as students require less support and seek out more challenging projects and questions. This process can be difficult for the instructor's own intellectual confidence as well as students. As instructors, we must become comfortable with admitting 'I don't know'. Indeed, when instructors are transparent about their own lack of knowledge or understanding, they are modeling uncertainty and questioning. This is a valuable opportunity to encourage curiosity and model how to find answers. Instructors can provide feedback with humility and transparency in the form of questions and prompts that encourage students to go about addressing their own questions.

12.4.2 Equity and Collaboration

Exposing ideas to criticism is an act of trust and vulnerability. When students are in a research team, they are asked to collaborate and rely on one another for support and as partners in the research process. Early in the first iteration of an experimentation cycle, dominant students typically emerge who are vocal and empowered to articulate their ideas. Inserting an opportunity to discuss collaboration and assigning group roles such as appraiser, skeptic, questioner along with process checker, organizer, reporter, and recorder, gives each student agency to present their questions and practice feedback in an alias mode for the first time. On a stage where scientific reasoning, vulnerability, and humility has higher value than who speaks first, feedback and discourse can be a great equalizer. In addition, peer feedback can sometimes be a cathartic and bonding experience (e.g. "you don't understand this either?") and a place where barriers to learning can dissolve. With multiple venues to participate in feedback, pairs, small group, large group, orally or in written form, all students have a role (and a voice) in the experimentation process.

A challenge is to understand that all of us (instructors and students) are dealing with bias, and stereotypes of who a scientist is. If feedback is graded and given too much power, it can be very damaging to learning and to the community. Students who are doubly disadvantaged – students of color who are either first generation or from socioeconomic disadvantaged backgrounds may be particularly at risk if they are doubted or not given enough time to feel seen and heard. Instructors need to model how to encourage and value contributions from all team members while providing feedback. Quietly monitoring team dynamics and choosing the right time to intervene when destructive group behaviors are impeding research progress and/or harming individuals is key. Anecdotally, it takes 2–3 iterations of feedback before trust and investment in community takes place. Being mindful of best practices for inclusion, equity, and cultural humility particularly for students of color, first generation students, and those with cultural norms that eschew questioning authority is of utmost importance when using a feedback-rich curriculum. Ultimately, it is important to help students trust the process that moves their science forward. It requires that every student and instructor push through their imposter syndromes, consider classmates as curious colleagues that are valuable resources, and increase their confidence as members of this community.

12.4.3 Learning Mindset

In our experience, students want feedback and expect high quality and timely feedback to benefit their learning. That said, feedback within the context of experimentation is an exercise in uncertainty. When hypotheses are exposed to feedback and filtered through a lens of potential results- from predicted or expected results to a range of unexpected or alternative results- students begin to feel untethered. Many students have been conditioned to view science as confirmatory, where unexpected or alternative results mean error or failure. Using a feedback-enhanced experimentation approach, students soon discover that experiments yield alternative results most of the time. When students realize that unexpected or alternative results are the norm, we can better foster curiosity, a learning mindset, and greater tolerance for uncertainty.

A challenge is that teaching students to give and receive feedback is asking for a type of resiliency. Expect variation in how long it takes each student to value feedback and achieve experimentation skills, over the course of one semester, and over multiple semesters.

12.4.4 Economics of Feedback

With a multi-semester CURE curriculum and enrollment of ~100 students in each lab course, we teach in teams of two lead instructors, one lab manager, 4–5 graduate TAs and a small army of uTAs. Students benefit from a large teaching team by way of many layers of support and feedback from diverse perspectives and expertise. If you are a team of one with many students and limited resources, you need to make choices regarding where, when, and how feedback is dosed out, and is most beneficial for achieving your learning outcomes. If you are limited in either time or money, we highly suggest you invest in at least one of the two feedback checkpoints (i.e. proposal feedback in the form of informal feedback presentations or analysis feedback) with informal feedback presentations taking priority. As a practical note, we have found that a focus on feedback as a centerpiece in our lab curriculum has saved us in supplies and equipment expenses. Moving to three experimentation units per semester, rather than a weekly reset of lab equipment and prep for 10–13 confirmatory labs, actually saves money in capital, supplies and disposables costs. In a feedback enhanced curriculum, where discourse learning goals are prioritized along with a few important techniques rather than, for instance, the full spectrum of molecular genetics technical protocols, instructors may recognize savings in supplies/equipment expenses and time spent in lab prep. Instead, the time investment is spent on laser focused in-class feedback, student consultation and conferences. Once you set up a feedback enhanced curriculum, invite experienced students to serve as uTAs who can help model feedback, provide additional eyes and ears on student work, and another round of feedback.

As instructors, it is challenging to stay on top of the many disparate hypotheses, group projects, different logistical needs (e.g. right sized beaker, calculating concentrations, pipette malfunction) and intellectual progress of each group as they are developing experimentation skills. Yet this same intellectual challenge, keeping track of students' questions, often sparks new ideas and inspires us to ask new questions ourselves – which make it more stimulating and exciting to teach.

12.5 Summary

In summary, affordances and challenges for instructors and students relate to the following:

- Intellectual Confidence and Ownership
- Equity and Collaboration
- Learning Mindset
- Economics of feedback

On balance, the affordances of a heavy dose of feedback outweigh the challenges, with many students achieving a high level of intellectual maturity in their science, even within one semester. With multiple semesters, students become conditioned to seek feedback whenever approaching a new hypothesis. The discourse and exchange of ideas and questions that result from asking for feedback, creates a culture of curiosity and humility that are at the heart of science. Iterative feedback integrated into the process of science cycle and practice of experimentation competencies helps to establish disciplinary norms for students. With a focus of 'hard skills' or techniques only, students tend to go through the motions without understanding the scientific underpinnings of questions, biological rationale, and implications for their research. Helping students to solicit, receive, discuss, and value the process of feedback is a primary learning outcome in Biocore and fundamental for developing experimentation competencies. Scientific discourse through feedback ensures that experimentation competencies and practices are grounded in disciplinary language, context, and standards of the scientific community.

Acknowledgments We thank current and former students, undergraduate and graduate teaching assistants, and postdoctoral teaching fellows who contributed to our understanding of the value of feedback and scientific discourse in our learning community. We are grateful to Seth McGee who models and inspires rich feedback through his deep knowledge of model systems and a teaching philosophy of always asking 'why.' All authors contributed to the writing of this manuscript. Batzli: conceptualization, supervision; Harris: conceptualization, resources; Lee: data curation; Horn: visualization.

References

American Association for the Advancement of Science (AAAS). (2011). Vision and change in undergraduate biology education: A call to action. Washington, DC. Retrieved October 18, 2021 from https://live-visionandchange.pantheonsite.io/wp-content/uploads/2013/11/aaas-VISchange-web1113.pdf

Baranczyk, M. C., & Best, C. A. (2020). How does instructor feedback help and hurt students' academic performance? *Scholarship of Teaching and Learning in Psychology, 6*(1), 1–14.

Batzli, J. M., & Harris, M. A. (2020). *Process of science companion book 1: Science communication*. University of Wisconsin Pressbooks. https://wisc.pb.unizin.org/biocorewriting/

Batzli, J. M., Harris, M. A., & McGee, S. A. (2018). It takes time: Learning process of science through an integrative, multi-semester lab curriculum. Tested studies for laboratory teaching. *Proceedings of the Association for Biology Laboratory Education, 39*(21).

Corwin, L. A., Graham, M. J., & Dolan, E. L. (2014). Modeling Course-Based Research Experiences: An Agenda for Future Research and Evaluation. *CBE-Life Science Education, 14*(1). https://doi.org/10.1187/cbe.14-10-0167

Grinath, A., & Southerland, S. (2019). Applying the ambitious science teaching framework in undergraduate biology: Responsive talk moves that support explanatory rigor. *Science Teacher Education, 103*, 92–122. https://doi.org/10.1002/sce.21484

Hora, M. T., Benbow, R. J., & Oleson, A. K. (2016). *Beyond the skills gap* (p. 263). Harvard Education Press.

Lave, J., & Wenger, E. (1991). *Situated learning: Legitimate peripheral participation*. Cambridge University Press.

Lee, D. (2020). *Understanding students' biology epistemology through argument*. Doctoral Dissertation, Clemson University.

Leupen, S., Kephart, K., & Hodges, L. (2020). Factors influencing quality of team discussion: Discourse analysis in an undergraduate team-based learning biology course. *CBE-Life Science Education, 19*(1), 1–13.

Longino, H. E. (2002). *The fate of knowledge*. Princeton University Press.

Magnus, D. (2000). Down the primrose path: Competing epistemologies in early twentieth-century biology. In R. Creath & J. Maienschein (Eds.), *Biology and epistemology* (pp. 91–121). Cambridge University Press.

Ngar-Fun, L., & Carless, D. (2006). Peer feedback: The learning element of peer assessment. *Teaching in Higher Education., 11*(3), 279–290. https://doi.org/10.1080/13562510600680582

Pelaez, N., Anderson, T. R., Gardner, S. M., Yin, Y., Abraham, J. K., Bartlett, E, Gormally, C, Hill, J. P., Hoover, M., Hurney, C., & Long, T., Newman, D. L., Sirum, K., & Stevens, M. (2017). The basic competencies of bio-logical experimentation: Concept-skill statements. *PIBERG Instructional Innovation Materials. Paper 4*. Retrieved October 18, 2021 from https://docs.lib.purdue.edu/pibergiim/4/

Round, J. E., & Campbell, A. M. (2017). Figure facts: Encouraging undergraduates to take a data-centered approach to reading primary literature. *CBE-Life Science Education, 12*(1), 39–46. https://doi.org/10.1187/cbe.11-07-0057

Russ, R., Sherin, B., & Sherin, M. (2016). What constitutes teacher learning. In D. H. Gitomer & C. A. Bell (Eds.), *Handbook of research on teaching* (pp. 391–438). American Educational Research Association.

Walsh, E. M., & McGowan, V. C. (2017). 'Let your data tell a story:' Climate change experts and students navigating disciplinary argumentations in the classroom. *International Journal of Science Education, 39*(1), 20–43. https://doi.org/10.1080/09500693.2016.1264033

Chapter 13
Engaging Students with Experimentation in an Introductory Biology Laboratory Module

Annwesa Dasgupta, Swapnalee Sarmah, James A. Marrs, and Kathleen A. Marrs

13.1 Introduction to Teaching Science Through Experimentation

To prepare our undergraduate students, both biology majors and non-majors, to be authentic scientific thinkers, it is important to demonstrate that biological knowledge relies on experimentation, and is tentative in nature. One way to do this is to design biology courses and lab activities not as a collection of "facts" but as a collection of "findings" from experiments performed by past researchers, and helping student to see themselves as being able to contribute to new findings through their laboratory courses or undergraduate research experiences. Numerous policy reports have identified experimental design and processes of science as a core competency for biology undergraduates, including recommendations that "science should be taught as science is practiced" (AAAS, 2010; NRC, 2003).

The assessment triangle triangle rests on the foundation of specific learning outcomes that describe some type of cognition expected of the students and building out to the observations of their performance and interpretation of their performance in terms of the expected competence (Pellegrino et al., 2001). In a course, the foundation of cognitive learning outcomes informs the development of appropriate assessment and instructional strategies (Anderson, 2007; AAAS, 2010). A

A. Dasgupta
STEM Education Innovation and Research Institute (SEIRI), Office of the Vice Chancellor for Research, Indiana University Purdue University, Indianapolis (IUPUI), Indianapolis, IN, USA

S. Sarmah · J. A. Marrs · K. A. Marrs (✉)
Department of Biology, Indiana University Purdue University, Indianapolis (IUPUI), Indianapolis, IN, USA
e-mail: ssarmah@iupui.edu; jmarrs@iupui.edu; kmarrs@iupui.edu

N. J. Pelaez et al. (eds.), *Trends in Teaching Experimentation in the Life Sciences*, Contributions from Biology Education Research,
https://doi.org/10.1007/978-3-030-98592-9_13

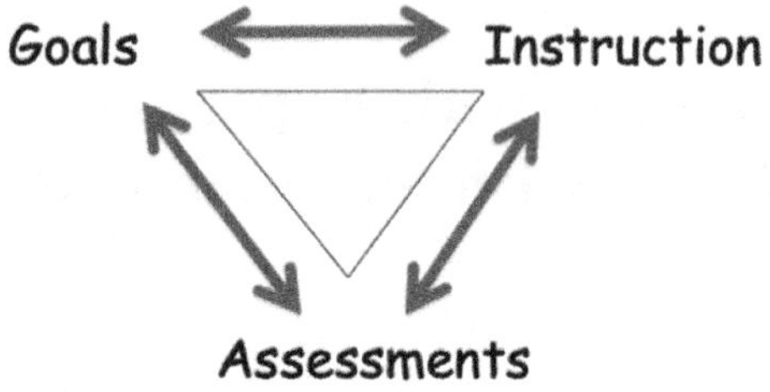

Fig. 13.1 The assessment triangle that shows that tight connections must exist between learning, goals and assessments (modified from AAAS, 2010). In this example, the function of an experimental design assessment is to help engage students with instructional experiences and innovations and to give students feedback to help them think about biological research practices in authentic ways

well-designed course will support students by revealing to them the tight connections between assessments, goals and instruction (Fig. 13.1). Such a course will emphasize the conceptual nature of science as well as the scientific practices that led to the content students learn. Assessment of the performance students are encouraged to demonstrate takes a center stage in this process because ongoing, frequent and multiple forms of feedback and assessment can not only inform progress on student learning outcomes but can allow an instructor the flexibility to modify educational experiences the students are engaged in as indicated by the interpretation of their performance on assessments. Thus, assessments drive changes in goals and instruction.

With these two ideas in mind: (1) biology content is best learned when instruction involves biological experimentation connected to that content (AAAS, 2010; Allen & Tanner, 2003; DiCarlo, 2006) and (2) instruction and learning goals of a biology course are driven by assessments (Pellegrino et al., 2001; Anderson, 2007), we now aim to guide readers about strategies to engage students as they think through the various components of experimentation as part of the research process. Several experimental design-based assessments published in the past characterize student knowledge in the terms of a few common elements they perform when students engage in experimental design: experimental hypothesis, experimental treatment and control variables, experimental results (including visualizations), inference, and interpretation of those results (Gormally, et al., 2012; Dasgupta et al., 2014; Sirum & Humburg, 2011; Dasgupta, et al., 2016; Killpack & Fulmer, 2018). Further details about assessment and grading of students are detailed in Part IV of this book. While these published assessments have similar features, each places a slightly different emphasis on the basic competencies of biological experimentation. However, in order to engage students effectively, we need to have tools that provide them formative feedback so they can reflect on their own progress.

Next, we provide an account of assessment tools to elicit and engage students in demonstrating a performance based on an instructors' experimentation-based expected learning outcomes.

13.2 Validated Assessments of Experimental Design

We describe four published assessments below simply as examples, but we recognize that there are other assessment instruments faculty might use in their classrooms. To elaborate, the Test of Scientific Literacy Skills, or TOSLS (Gormally, et al., 2012) examines scientific literacy skills for general education biology students using real world problems. It uses nine scientific literacy skills, gathered from a faculty survey, to explore how students evaluate the scientific credibility of a range of real-world contexts as well as demonstrate their knowhow of research investigations. The Rubric for Experimental Design, or RED (Dasgupta et al., 2014) presents five major categories of experimental design and describes multiple difficulties associated with each category exemplified with student responses. The RED also presents propositional knowledge for each experimental design category which indicate the correct ideas for each category. The Experimental Design Ability Test, or EDAT assessment (Sirum & Humburg, 2011) explores biology non majors' ability to accept a scientific claim for a commercial product using a 10-item rubric. The Neuron Assessment (Dasgupta et al., 2016) examines student abilities to design an experiment to explore the cause of a neuronal disorder in humans. The Tool for Interrelated Experimental Design, or TIED (Killpack & Fulmer, 2018) has a rubric designed to explore experimental design skills in an introductory laboratory course. The tool measures student changes in knowledge of individual experimental design aspects as well as looks at their interrelatedness.

13.3 Considerations for Selecting an Experimental Design Assessment

The examples mentioned above demonstrate a variety of ways an instructor can explore student engagement to observe their performance in biological experimentation. However, we recognize that there are additional context-based considerations that might be useful to evaluate when making choices for assessments, taking into consideration practicalities based on the characteristics of the students, instructors and departmental or course structure. Such additional considerations are detailed below, and were key to developing the course based experimental design module we implemented in an introductory biology laboratory course.

13.3.1 Understand Student Background and Prior Knowledge

To provide feedback from the experimental design assessments in a meaningful way, it is worthwhile to consider students' prior knowledge about the subject matter context for a given assessment. In other words, if a student exposes a difficulty because they were unaware of the model organism suggested in an assessment, the assessment will not end up being of use to reveal their performance of the competence of interest. Thus, it is key to minimize differences in students' prior knowledge for this assessment. One way to do this can be to provide examples of biological scenarios where all required concepts are defined for the students.

13.3.2 *Identify the Kind of Data You Need*

13.3.2.1 Assessments Can Be Open Ended or Multiple Choice

Open ended assessments can take longer and more effort from faculty to provide feedback to students, but they reveal rich information about student areas of difficulty that need explicit support. Open-ended experimental design-based assessments come with scaffolds and supporting visualizations in order to gradually probe student knowledge. Multiple choice assessment, on the other hand, can provide quick insights into common problems students experience with experiments and help an instructor modify their instruction or refine learning goals on a weekly basis. To make multiple-choice questions more informative, they can also be converted to "two-tier" wherein students pick a response as well as present a reason for their answer of choice.

13.3.2.2 Pre and Post Experimental Design Assessments

Pre and post experimental design assessments administered at the beginning and end of semester can inform instructors about knowledge gained in a course. Alternatively, research studies assessed with the EDAT and the RED have shown comparisons between two sections of a course, newly designed vs traditional, using the same pre-post assessments for both course formats (for a review, see Shortlidge & Brownell, 2016), which may be helpful in guiding decisions about engaging students in a particular experimental design-based instructional activity.

13.3.2.3 Delivery of Assessments

You might also consider whether the assessment delivered will be in person, online or hybrid (a blend of in person and online delivery). An online format allows students to complete assessments at their own time and also serves as a homework assignment. With hybrid, assessments typically direct students to topics they need to self-study prior to coming in class for the in-person component.

13.3.3 Account for Instructor Teaching Experience and Research Considerations

Whether designing or using previously designed experimentation assessments, a faculty member or course coordinator must take into account who will be doing the instruction and the characteristics of the lab instructors. CUREs (Course-based Undergraduate Research Experiences) that engage students in authentic experimentation-based practices have been shown to be very effective in promoting student thinking about practices of science (Auchincloss et al., 2014; Brownell, et al., 2012; Hicks et al., 2020). A typical course designed as a semester-long CURE may overlap with the professional scope and interests of a tenured faculty member, teaching professor, or senior postdoc who already has considerable experience in research and mentoring students in the lab. In a large research university, however, labs are often taught by graduate students pursuing a master's degree or PhD. Some experienced graduate students may have taught an introductory class a number of times through their training; some may be brand new graduate teaching assistants (TA) just out of their undergraduate experience. At this stage in their development many graduate students are not only learning the process of teaching but are themselves learning the research process. Graduate students may become future college instructors and research faculty, or may eventually work in industry, but in all circumstances, guiding students through the research process based on what they learn from students based on their performance on an assessment will be a benefit to their own professional development as well as contribute greatly to the teaching mission of their current or future department or research appointment where there will be an expectation to mentor students or employees new to research.

13.3.4 The Scope of Your Course and/or Department

The level of the course (freshman to senior level), majors' course or non-majors, the constraints of how much time in a semester can be spent on experimental design process, and the instructional goals will determine which of the individual elements of experimental design may be more beneficial to examine. Alternatively, since all aspects of experimental design are interrelated, students thinking about experiments might also be examined as a whole. To explain, experimental variables and experimental results are two different aspects of experimental design. However, students' manipulation of experimental variables ultimately affects experimental results. Results might suggest a correlation instead of a causal association based on how the variables were manipulated and how natural variation in sample size, the environment, or genetics were handled as a factor in the analysis of results (Hicks et al., 2020). Learning how to generate and frame testable questions, do a literature review, set up controls, work as a member of a lab team and discuss, write up and present results are again inter-related, and the emphasis will influence the nature of the feedback students need to get from an assessment.

These considerations (13.3.1–13.3.4) may intersect. For example, instructor experience or constraints such as large class size (consideration 13.3.3) might influence assessment format (consideration 13.3.2) as they might select a multiple-choice assessment or brief 'checklist' form of assessment owing to grading time constraints. Similarly, student prior knowledge (consideration 13.3.1) might influence assessment format (consideration 13.3.2); for example, using an open-ended assessment if students do not carry prior knowledge about the experimental scenario. Ideally, faculty are interested in not only how students develop their scientific thinking and practices over the course of a single semester, but how students in the major collectively progress over the course of their 4-year major and beyond. Expectations for first year students can set the foundation for the conceptual knowledge and initial research skills that can be further developed in advanced courses. Such increasing sophistication in their abilities to ***Plan*** and *Conduct* research can be assessed in the freshman year as a baseline, can be assessed multiple times over the course of a semester, or can be monitored at a department level as students skills advance from the freshman to the senior year (Marbach-Ad et al., 2007). To guide non-majors to be authentic scientific thinkers, it is just as important to give students an appreciation of current topics in research as understanding of how new knowledge is generated.

With these assessment considerations in mind, we share an originally designed zebrafish experimental design activity in a publically available Dasgupta_et.al_2022_ZebrafishExperimentalDesignFiles folder: https://tinyurl.com/ZebrafishExperimentalDesign

13.4 Description of a Zebrafish Experimental Design Lab Activity

The zebrafish (*Danio rerio*) is a model organism utilized by scientists and students to study basic questions about vertebrate development as well as questions about the behavior of fish when exposed to various environmental factors like ethanol. This research leads to a better understanding of early human embryonic development, brain development, and behavior. In this lab, students were asked to design a testable hypothesis and an experiment to examine an aspect of zebrafish development or behavior as a consequence of a treatment of their choice.

13.5 Practical Design Considerations for the Zebrafish Experimental Design Lab Activity

13.5.1 Our Students' Background and Prior Knowledge Drove the Lab Activity Design

In development of the zebrafish experimental design lab activity, we took into account student characteristics. Introductory Biology is a 100-level large-enrollment, "gateway" course required for Biology and other science majors and is taken by about 900 students per year, mainly freshmen. For many, it is among their first university course taken after high school, and a key course for the department as it forms an essential basis for future learning in subsequent courses. Each student brings prior knowledge – background knowledge (such as facts and definitions) as well as procedural knowledge (experiences and practice with how to integrate and apply their knowledge). As part of a pre-lab activity, students were asked, "In your own words (2–3 sentences), explain what a Model Organism is, and the advantages of working with zebrafish as a model organism." This allowed us to capture prior knowledge, followed by leveling of student prior background knowledge about specific model organisms, with the emphasis on zebrafish, as a short discussion at the start of the lab.

13.5.2 Our Lab Activity Learning Outcomes Guided our Strategies to Engage Learners

For the zebrafish experimental design lab activity, the goal was not to complete a research project from start to finish over the course of a semester, but to engage students in "experimental thinking" in one lab session, followed outside of lab by an independent write-up including visual representations. The goals of the lab were to (1) help students recognize the relevance of model organisms in research. (2) have students gain familiarity working with zebrafish as an experimental organism through handling and manipulating zebrafish, and documenting their work through imaging and visualizations, (3) have students gain an understanding of the role of environmental chemicals in embryonic development, using Fetal Alcohol Spectrum Disorder (FASD) as an initial example, and (4) practice "experimental thinking" through the design of an experiment. This last goal was complex, and involved researching a topic of interest, reading scientific papers, and developing a testable hypothesis, designing an experiment with attention to positive and negative controls, addressing independent and dependent variables, confounding factors, statistical tests, and a prediction of expected results based in reading of the literature. Based on those goals, and our design considerations, we used the Rubric for Experimental Design (assessment) to guide the development of the lab handout and instructional materials (instruction).

13.5.3 Instructor Experience and Constraints Driving Learning Activity and Scope of Course in Alignment with Department Goals

For the zebrafish experimental design lab activity, the initial choice to focus on using zebrafish and Fetal Alcohol Spectrum Disorder (FASD) was driven by the research expertise of two of the authors (S.S. and J.M.). We had previously initiated several CURE-based lab semester projects for honors students in introductory biology, taught by K.M and S.S, and had published results of ordinal research using zebrafish, co-authored with undergraduates in the honors labs (Muralidharan et al., 2017; Sarmah et al., 2016). Bringing the same semester-long original research experience to almost 600 students in 20 different lab sections, each taught by a graduate TA, was not practical or possible. An overarching intent of doing this laboratory module was to give all students a sense of the excitement and fascination of working with live vertebrates, show them the power of using model organisms in research, allow them to conceptualize and design their own experiment, and predict the results even if they did not get to follow the experiment through to the results stage. As the majority of the class are pre-health majors, we also wanted them to understand the serious medical consequences of alcohol and other chemicals that can cause birth defects during early embryo development. Hence, we designed the single lab session, and the lab write up that followed, in three stages.

13.6 Practical Design Considerations for the Zebrafish Experimental Design Lab Activity, and Alignment with the ACE-BIO Competencies (Pelaez et al., 2017; Chap. 1 in this Volume)

13.6.1 Prelab

As explained above, students were asked as part of a pre-lab activity: In your own words (2–3 sentences), explain what a Model Organism is, and the advantages of working with zebrafish as a model organism. After completing that assignment, students read two short articles from the *New York Times* about the effects of environmental chemicals and of alcohol use during pregnancy (Cernansky, 2018; Gunter, 2019). Finally, students read the abstract and introduction of a paper about children with FASD, and they looked carefully at two tables in the paper, one about drinking during pregnancy and one about the consequences of alcohol exposure on behavior and development when children with FASD are followed through in their pre-school, elementary, adolescent and early adult lives (Nash & Davies, 2017).

13.6.2 Lab Session with Instructor

Labs are taught by graduate TAs with a range of experience teaching the lab course before. Each TA taught two lab sections throughout the week. Some TAs had taught several times before, and some were new to the course or the university. About 25% of the TAs were international students, and typically none of the TAs had been undergraduates at IUPUI, meaning that they had taken introductory biology 4 or more years earlier, and somewhere other than where they were teaching. The zebrafish experimental design lab activity goals, activities and assignment was discussed among the TAs as a pre-view 2 weeks ahead of time at a regular TA lab meeting, led by K.M., and discussed again as the week of the zebrafish experimental design lab activity began. Graduate students who were not familiar themselves with working with zebrafish had time to practice the lab. Labs were 3 h in length. When teaching the lab with their students, TAs went through an overview of model organisms, FASD, and led a short discussion of the two tables in the manuscript mentioned above and gave an overview of the zebrafish experimental design lab activity. Students then spent about 45 min on each of these three activities.

13.6.2.1 Student Lab Activity 1: Visualization Skills and "a Feeling for the Organism"

Nobel Laureate Barbara McClintock encouraged biologists to "Develop a feeling for the organism" they studied, to understand at a deep level and see patterns that others may miss (Keller, 1984). As mentioned above, an overarching intent in developing this this zebrafish experimental design lab, was to create lab experiences that support students' self-confidence in the process of experimental design, but also cultivate their interest and curiosity in the research using zebrafish, a vertebrate with many physical similarities to human that can clearly be observed through the microscope such a beating heart, blood flow, large eyes, and behaviors like swimming. While students would not have a semester-long experience working with zebrafish, our design allowed for a single lab experience to help promote a "feeling for the organism."

Our experimental design assessment module was developed using the model organism zebrafish embryos as grown in two different conditions (with or without alcohol) and students were asked to first observe and determine the variability in the embryos which is an experimental design ability (Dasgupta et al., 2014; Killpack & Fulmer, 2018). Students were given a single lab session to observe live zebrafish at two different stages of development (one stage still developing in the egg and one stage a young fish, 3–5 days old), and with and without exposure to ethanol. They learned to transfer a small number of live zebrafish into different wells of a 24-well plate and to visualize the fish under a dissecting scope. They were asked to record the zebrafish movements (in the egg) or swimming behaviors (in the young fish), examine eye movements, record heartbeats per minute, and observe other characteristics of their choosing. They also imaged the fish with a microscope camera (still photos or video) and were asked to make two detailed drawings that they would submit with their assignment. By observing wild-type fish as well as fish treated with moderate concentrations of ethanol, students were able to see first-hand the effects of alcohol exposure on development. As students learned to transfer fish to a microtiter plate to observe under the microscope and return them to the instructor bench for other students to observe, they were able to practice some aspects of zebrafish care.

While examining the zebrafish and observing their behaviors, students also picked up an index card or two with scenarios to consider such as "Thinking about the zebrafish you are observing now, what do you think would happen if you added a defined amount of (caffeine, nicotine, ATP) to the medium?" There were about a dozen different scenarios to consider. Such prompts were to generate discussion between lab partners and to think about what compounds they might want to test, and how they would measure the effect (e.g., on heart rate, eye movements, hatching, or effect on embryonic development if exposed for a long period of time).

13.6.2.2 Student Lab Activity 2: Literature Search, Gap Analysis, and Creativity

Through the use of primary literature, students were guided to recognize the global burden of diseases and realize teratogenic exposure as one of the causes in developing disorders. Students next spent time in lab doing a literature search in PubMed on a compound they were interested in testing by first seeing what prior work had been done with this compound with zebrafish or other organisms (ACE-BIO Competency *Identify* gaps) (Pelaez et al., 2017; Chap. 1 in this volume). Students were able to find appropriate primary literature and evaluate the related information to identify research problems that were highly relevant to today's society. This component of the module was aimed at levelling prior student knowledge differences and as well helped students see connections of the module with real world scenarios. Following this, students generated research questions to inquire about the effects of commonly used substances (i.e., psychoactive drugs, over-the-counter medicines, vitamins or exposure to pesticides, chemicals present in make-up) by pregnant mothers on their developing babies that could be modeled be setting up a controlled experiment using zebrafish.

Due to resource limitations (providing enough zebrafish for every student, gaining access to experimental chemicals or drugs, requirements for protocols to be approved by the Institutional Animal Care and Use Committee), students knew they would not actually be doing the experiment they proposed, but rather mapping out how they would test their hypothesis.

13.6.2.3 Student Lab Activity 3 Experimental Design Thinking

Students were then asked to think about developing their own testable hypothesis (ACE-Bio Competency: Generate a *Question* and Hypotheses) (Pelaez et al., 2017; Chap. 1 in this volume) and mapping out an experiment using this prompt:

- Research: How would you design a simple experiment to determine the effects of a particular compound of interest on zebrafish development or behavior?
- Your goal is to design, on paper, an experiment to test some aspect of zebrafish development or behavior as a consequence of a treatment of your choice.
- Based on your research, identify a hypothesis that you could test.

A design constraint for students was to set up the experiment using a 24-well plate format. A 24-well plate provides an ideal template for experimental design, as there are numerous combinations and ways to set up rows or columns to allow room for negative controls, positive controls, and experimental treatments (Fig. 13.2). A dose-response configuration could be set up, or a scenario where fish at different stages of development were in each experimental row, each exposed to the same dose of compound. While in some ways using a 24-well plate may seem to limit options to design an experiment, in reality the flexible design possibilities actually free a student to think of creative ways to test a compound in a finite number of arrangements to keep their design focused and organized.

13.6.3 Post Lab: Experimental Design Proposal Write Up

For this final stage, which was completed after the lab period ended, students were asked to complete a 2–3 page write up (single-spaced) using guidelines we provided, which addressed the ACE-Bio Framework Competency *Communicate*. An

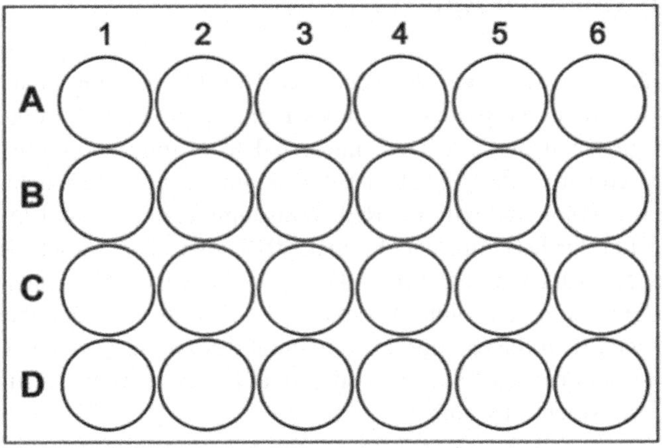

Fig. 13.2 Standard 24-well microtiter plate format used for zebrafish experimental design activity

additional constraint placed on the students was to not spend any longer than 2 h of focused time on their write up. We put this constraint in place as we were interested in uncovering students' initial thinking about the zebrafish experiments and thought processes as they designed the experiment and not necessarily going through the review, editing, and iterative process needed to develop a well-polished experimental design lab report.

To complete the lab activity, students mapped out their design containing their experimental setup and controls and replicates of the experiments. Students determined both positive and negative controls, different concentrations of the tested compounds (independent variables), different treatment periods (confounding factors) and predicted outcomes in terms of the phenotypes (dependent variables). Students developed a timeline and were asked to think about statistical tests that could help them analyze their results. In this way, students were able to address the ACE-Bio Competency *Plan* and test hypothesis, attempt the ACE-Bio Competency *Analyze* Process, and simulate ACE-Bio Competency *Conduct* investigation and ACE-BIO Competency *Conclude* without performing the experiment. Once submitted, student submissions were then examined and scored by Graduate TAs, using the Canvas Learning Management System "Speed Grader" tool. Scoring of each category in the rubric below was discussed generally at the in-person lab meeting, and the actual grading was done easily and reproducibly based on grading criteria uploaded into the Canvas Speed Grader in advance. The goal in grading was to use the rubric developmentally for the Graduate TAs to identify difficulties with which students might need help in each area, defined by the Rubric for Experimental Design (RED) criteria below (Dasgupta et al., 2014) and not to penalize students for having a 'wrong' approach or answer.

13.7 Results: Aligning the Zebrafish Experimental Design Lab with the ACE-Bio Competencies and Validated Rubrics

The zebrafish lab module was designed based on the learning goals and performances that could be interpreted with assessments appropriate for our introductory biology lab. Subsequently, we were interested to examine how our module and rubric compared with other published experimental design assessment goals. Thus, our module was compared with the RED (Dasgupta et al., 2014), EDAT (Sirum & Humburg, 2011), TIED (Killpack & Fulmer, 2018), as well as with the Advancement of Competence in Experimentation – Biology (ACE-Bio) Competencies (Pelaez et al., 2017; Chap. 1 in this volume) (Table 13.1). Findings revealed that our module was well aligned with the existing experimental design assessments as each of the goals and instructional scaffolds mapped with at least one of the abilities measured by the four assessment instruments.

Table 13.1 Comparison of zebrafish lab module with published experimental design assessments

Zebrafish experimental design criteria rubric	RED	EDAT	TIED	ACE-BIO
Background Research Select two papers that seem relevant and recent to cite in your assignment; Cite these in APA format at the end of your report. Explain in 2-3 sentences why you selected these two papers – what is your interest in this topic? (Up to 3 total points possible)		**EDAT.1** Recognition that an experiment can be done to test the claim		**ACE-BIO** *Identify* gaps
Hypothesis A one-sentence statement that is testable, such as "Alcohol exposure during early development causes developmental and functional defects in the zebrafish heart". include 1 sentence that mentions prior knowledge such as "This will build on previous work showing that alcohol exposure causes developmental defects in the zebrafish eye." (Up to 2 total points)	**Difficulty** (2) Manipulation of Variables: Testable hypothesis	**EDAT.1** Recognition that an experiment can be done to test the claim	**TIED.A.** Develop and state a hypothesis for your experiment.	**ACE-BIO** Generate *Question* & Hypothesis
Experimental Variables Independent variable Dependent variable – Explain in 2–3 sentences how your dependent variable will be measured: (direct or indirect)? What kinds of lab equipment would you need to measure these variables? (Up to 2 total pts possible for this category)	**Difficulty** (2) Manipulation of Variables: Treatment group (3) Measurement of Outcome	**EDAT.2** Identification of variable manipulated (independent) **EDAT.3** Identification of variable measured (dependent) **EDAT.4** Description of how dependent variable is measured	**TIED.E.** What data will you collect, and how will you collect it?	**ACE-BIO** *Plan*; test hypothesis

(continued)

Table 13.1 (continued)

Zebrafish experimental design criteria rubric	RED	EDAT	TIED	ACE-BIO
Controls Negative control / placebo group – Positive control group. Explain why you selected the component used as the positive control. (Up to 2 total points possible for this category)	**Difficulty** (2) Manipulation of Variables: Combinatorial reasoning Controlling outside variables Control group	**EDAT.5** Realization that there is one other variable that must be held constant EDAT.6. Understanding of the placebo effect	**TIED.C.** What are the control group(s)? TIED.D. What are the experimental group(s)?	**ACE-BIO** *Plan*; test hypothesis **ACE-BIO** *Conduct* investigation
Experimental Setup Appropriate sample size and selection – Doses or range of doses tested – Possible confounding factors – Replicates	**Difficulty** (4) Accounting for Variability Selection of random sample Randomized design Replication of treatments	**EDAT.7** Realization that there are many variables that must be held constant **EDAT.8** Understanding that the larger the sample size or # of subjects, the better the data **EDAT.9** Understanding that the experiment needs to be repeated		**ACE-BIO** *Plan*; test hypothesis **ACE-BIO** *Conduct* investigation
Simple statistical tests to analyze the data In 2-3 sentences, state the experimental findings you might expect to see, based on your background research? (Up to 3 total points possible for this category)	**Difficulties** (3) Measurement of Outcome Categorical and/or quantitative variables treatments	**EDAT.10** Awareness that one can only disprove the hypothesis, that there are possible sources of error, that there are limits to generalizing the conclusions	**TIED.E.** What data will you collect, and how will you collect it? "Data Collection" **TIED.F.** What observations would support your hypothesis? "Observations"	**ACE-BIO** *Analyze* Process

(continued)

Table 13.1 (continued)

Zebrafish experimental design criteria rubric	RED	EDAT	TIED	ACE-BIO
Clinical Significance Effect of the compound during human embryonic development or in early childhood / adolescence. In 2-3 sentences, make some recommendations to parents/ pediatricians/ public health practitioners / legislators alerting them the effect of these compounds on human health. (Up to 3 total points possible for this category)				**ACE-BIO** *Communicate*
Simple statistical tests to analyze the data In 2-3 sentences, state the experimental findings you might expect to see, based on your background research? (Up to 3 total points possible for this category)				**ACE-BIO** *Analyze* Process

13.8 Conclusions – Selection of Experimental Design Assessments, and Other Practical Considerations, Can Inform Module Design and Assessment Design

Selection of experimental design assessments, and other practical considerations, can inform design of instruction to engage students in practicing the targeted competence. Use of assessments as diagnostic tools can inform students about their experimental design thinking as well as direct their attention towards the goals of a lab activity. Findings from assessments can also inform instructors about certain aspects of experimental design to engage students with and that need attention (Dasgupta et al., 2014, 2016). In the first years of undergraduate courses, students might not have opportunities to engage in hands-on authentic research experiences such as CUREs (Auchincloss et al., 2014). However, a short lab module such as the one proposed here can present students with opportunities to engage in experimental investigations in the format of a course with written lab worksheets along with short and quick hands-on lab activities.

We illustrate here one example of how this experimental design activity worked in our particular department but emphasize that this approach can be generalized to fit any course or department. By selecting a compelling theme meaningful to a course or department (FASD, in our case, given existing research expertise in the department) we were able to generate meaningful connections to the research culture of the department. This model can drive the choice of the project in any department. By selecting an activity with a shorter time frame than a typical semester-long or multi-week CURE, we were still able to foster a wide range of experimentation skills. As a final note: as the spring 2020 semester progressed into a nationwide transition of college instruction moving quickly to an on-line format due to the coronavirus pandemic, we were able to convert the zebrafish experimental design lab in Summer 2020 into an all on-line format. Videos of zebrafish growth and development, movement, and care of the zebrafish were shown to student through the on-line course site. Students completed the same activity, even though they did not have hands-on experience with the fish. While not ideal in any sense, it still allowed students to experience a way to practice the experimental design process. Given a small class size in the on-line summer semester, we extended ACE-Bio Competency *Communicate* (Pelaez et al., 2017; Chap. 1 in this volume) to include a 3-min presentation to the class over Zoom, with each student sharing a single PowerPoint slide that included a drawing of a zebrafish and their 24-well plate design. In this way, students were able to communicate their ideas to others.

We emphasize the value of instruction in biology courses with the underlying foundation of scientific practices. Often, biology courses are instructed as a collection of facts instead of as findings from experimental investigations. Alternatively, if students are trained to think about biology using the lens of the "scientific method" or experiments – they can become authentic scientific thinkers. Authentic research experiences within a course have also been demonstrated to carry positive impacts in student confidence and persistence to pursue biology research and graduate studies in biology (Brownell et al., 2012; Jones et al., 2010).

In this chapter, we illustrated the use of experimental design-based assessments as the driving force that motivated instruction and learning goals of our lab module. Further, we proposed a guide for considering how to engage students according to selected published assessments matched with your laboratory activity goals by highlighting the similarities and differences between the ACE-Bio Competencies and other existing experimental design assessments (the RED, EDAT and TIED rubrics). Additionally, we also shared an originally designed module and accompanying rubric that practitioners can use as a template to design their own modules relying on the ACE-Bio Competencies to engage learners with experimentation and the use of research process skills, and to engage instructors to guide and revise the design or implementation of lab activities to address practical skills according to the ACE-Bio Competencies framework (according to Chap. 1 of this book).

As advice to others, in summary, the following are aspects of relevance:

- The assessment triangle (Pellegrino, 2001) is a useful guide to engage students in an activity and provide them feedback to support improvements in their experimentation skills.
- Based on the assessment triangle, it is pertinent to have
  - a framework (like the ACE-Bio Competencies in Chap. 1 of this book) to define the expected learning outcomes,
  - a plan to help students demonstrate a performance to determine if they meet the learning outcomes (Part II of this book), and
  - visual representations (like the 24-well plate in the module presented here) as a scaffold to engage students in their performance.

In summary, assessments can be used to elicit a performance so instructors can be informed how well students are engaged in and achieving what is expected from a lab activity.

Acknowledgments This work was supported by a SEIRI Seed grant from IUPUI to Dr. Kathleen Marrs and Dr. James Marrs (2016-2019).

References

Allen, D., & Tanner, K. (2003). Approaches to cell biology teaching: Learning content in context—Problem-based learning. *CBE-Life Sciences Education, 2*(2), 73–81.

American Association for the Advancement of Science (AAAS). (2010). Brewer C, Smith D. (Editors). *Vision and change in undergraduate biology education: A call to action*. AAAS.

Anderson, T. R. (2007). Bridging the educational research-teaching practice gap. *Biochemistry and Molecular Biology Education, 35*, 471–477. https://doi.org/10.1002/bmb.20135

Auchincloss, L. C., Laursen, S. L., Branchaw, J. L., Eagan, K., Graham, M., Hanauer, D. I., … Towns, M. (2014). Assessment of course-based undergraduate research experiences: A meeting report. *CBE-Life Sciences Education, 13*(1), 29–40.

Brownell, S. E., Kloser, M. J., Fukami, T., & Shavelson, R. (2012). Undergraduate biology lab courses: Comparing the impact of traditionally based "cookbook" and authentic research-based courses on student lab experiences. *Journal of College Science Teaching, 41*(4).

Cernansky, R. (2018, August 4). Environmental toxins – Risks during pregnancy. *The Washington Post*. https://www.washingtonpost.com/national/health-science/environmental-toxins-are-seen-as-posing-risks-during-pregnancy/2018/08/03/770fab98-7643-11e8-9780-b1dd6a09b549_story.html. Accessed 4 Jan 2021.

Dasgupta, A. P., Anderson, T. R., & Pelaez, N. (2014). Development and validation of a rubric for diagnosing students' experimental design knowledge and difficulties. *CBE-Life Sciences Education, 13*(2), 265–284.

Dasgupta, A. P., Anderson, T. R., & Pelaez, N. J. (2016). Development of the neuron assessment for measuring biology students' use of experimental design concepts and representations. *CBE-Life Sciences Education, 15*(2), ar10.

DiCarlo, S. E. (2006). Cell biology should be taught as science is practised. *Nature Reviews Molecular Cell Biology, 7*(4), 290–296.

Gormally, C., Brickman, P., & Lutz, M. (2012). Developing a test of scientific literacy skills (TOSLS): Measuring undergraduates' evaluation of scientific information and arguments. *CBE-Life Sciences Education, 11*(4), 364–377.

Gunter, J. (2019). Drinking while pregnant: An inconvenient truth. *The New York Times*, February 5.
Hicks, J., Dewey, J., Brandvain, Y., & Schuchardt, A. (2020). Development of the biological variation in experimental design and analysis (BioVEDA) assessment. *PLoS One, 15*(7), e0236098.
Jones, M. T., Barlow, A. E., & Villarejo, M. (2010). Importance of undergraduate research for minority persistence and achievement in biology. *The Journal of Higher Education, 81*(1), 82–115.
Keller, E. F. (1984). *A feeling for the organism, 10th anniversary edition: The life and work of Barbara McClintock*. Macmillan.
Killpack, T. L., & Fulmer, S. M. (2018). Development of a tool to assess interrelated experimental design in introductory biology. *Journal of Microbiology & Biology Education, 19*(3), 19.3.98. https://doi.org/10.1128/jmbe.v19i3.1627. PMID: 30377472.
Marbach-Ad, G., Briken, V., Frauwirth, K., Gao, L. Y., Hutcheson, S. W., Joseph, S. W., … Stein, D. C. (2007). A faculty team works to create content linkages among various courses to increase meaningful learning of targeted concepts of microbiology. *CBE-Life Sciences Education, 6*(2), 155–162.
Muralidharan, P., Connors, C. T., Mohammed, A. S., Sarmah, S., Marrs, K., Marrs, J. A., & Chism, G. W. (2017). Turmeric extract rescues ethanol-induced developmental defect in the zebrafish model for fetal alcohol spectrum disorder (FASD). *Journal of Food Science, 82*(9), 2221–2225.
Nash, A., & Davies, L. (2017). Fetal alcohol Spectrum disorders: What Pediatric providers need to know. *Journal of Pediatric Health Care, 31*(5), 594–606.
National Research Council (NRC). (2003). *BIO2010: Transforming undergraduate education for future research biologists*. National Academies Press.
Pelaez, N., Anderson, T. R., Gardner, S. M., Yin, Y., Abraham, J. K., Bartlett, E. L., Gormally, C., Hill, J. P., Hoover, M., Hurney, C. A., Long, T. M., Newman, D. L., Sirum, K., & Stevens, M. T. (2017). The basic competencies of biological experimentation: Concept-skill statements. *PIBERG Instructional Innovation Material Paper 4*. https://docs.lib.purdue.edu/pibergiim/4/. Accessed 8 Nov 2021
Pellegrino, J. W., Chudowsky, N., & Glaser, R. (2001). *Knowing what students know: The science and design of educational assessment*. National Academy Press.
Sarmah, S., Chism, G. W., III, Vaughan, M. A., Muralidharan, P., Marrs, J. A., & Marrs, K. A. (2016). Using zebrafish to implement a course-based undergraduate research experience to study teratogenesis in two biology laboratory courses. *Zebrafish, 13*(4), 293–304.
Shortlidge, E. E., & Brownell, S. E. (2016). How to assess your CURE: A practical guide for instructors of course-based undergraduate research experiences. *Journal of Microbiology & Biology Education, 17*(3), 399–408.
Sirum, K., & Humburg, J. (2011). The experimental design ability test (EDAT). *Bioscene: Journal of College Biology Teaching, 37*(1), 8–16.

Part IV
Assessment, Evaluation, and Grading What Students Learn About Biological Experimentation

Chapter 14
Comparison of Published Assessments of Biological Experimentation as Mapped to the ACE-Bio Competence Areas

Anna J. Zelaya, Lawrence S. Blumer, and Christopher W. Beck

14.1 Introduction

Experimentation is fundamental to our work as life scientists. It is the core source of new knowledge in the life sciences and experimentation incorporates skills found in any list of undergraduate biology learning outcomes (American Association for the Advancement of Science, 2011; Clemmons et al., 2020). During the past two decades, increased focus on evidence-based learning and teaching has put increased emphasis on learning science by doing science, which means experimentation (American Association for the Advancement of Science, 2011; Boyer, 1998; National Research Council (NRC), 2003; Project Kaleidoscope (PKAL), 2002). Consequently, learning experimentation and assessing the effectiveness of teaching experimentation is essential for undergraduate life sciences education to gauge what students actually learn. Yet, the effectiveness of curricula in teaching experimentation is rarely assessed in courses, such as laboratory courses (Beck et al., 2014). Furthermore, even when experimentation is assessed, published assessment tools are not often used (Beck et al., 2014). Using published assessments improves our understanding of student learning of experimentation, as these assessments generally have been validated. In addition, when multiple studies use the same assessment, comparisons of approaches to teaching experimentation can be compared explicitly.

A. J. Zelaya · C. W. Beck
Department of Biology, Emory University, Atlanta, GA, USA
e-mail: anna.zelaya@emory.edu; christopher.beck@emory.edu

L. S. Blumer (✉)
Department of Biology, Morehouse College, Atlanta, GA, USA
e-mail: lawrence.blumer@morehouse.edu

N. J. Pelaez et al. (eds.), *Trends in Teaching Experimentation in the Life Sciences*, Contributions from Biology Education Research,
https://doi.org/10.1007/978-3-030-98592-9_14

Once assessments are identified, they need to match learning outcomes. This provides evidence of what students know and can do as well as provides timely feedback to students during a course (Handelsman et al., 2007). However, identifying existing assessment tools that match the learning outcomes sought by an instructor might be the greatest barrier to genuine evidence-based teaching of biological experimentation. Current published assessments commonly used in biology courses represent a diverse array of tools ranging from descriptions of learning activities, self-reporting of student opinions, to multiple choice problem solving, and free response to prompts and assessment rubrics (Shortlidge & Brownell, 2016). This review is an attempt to align the commonly used assessments with defined competencies in biological experimentation.

The Basic Competencies of Biological Experimentation developed by the ACE-Bio Network (Pelaez et al., 2017; Chap. 1 in this volume) are a valuable starting point for biology educators to identify core competencies and assess the achievement of those outcomes in students. The network identified seven basic Competence Areas (*Identify*, *Question*, *Plan*, *Conduct*, *Analyze*, *Conclude*, and *Communicate*) that are components of experimentation. Each Competence Area contains two to ten Concepts and each concept contains one to nine Skill Statements. This framework of basic competencies in biological experimentation overlaps with some of the course-level learning outcomes of the BioSkills guide (Clemmons et al., 2020) that are based on the core competencies in the *Vision and Change* report (AAAS, 2011). However, the ACE-Bio framework is more detailed in elaborating Competence Areas, Concepts, and Skill Statements that describe biological experimentation. Here, we used this framework to categorize individual assessment items from assessments that address aspects of experimentation currently used in undergraduate biology courses. Mapping of assessments on this framework will allow instructors to better understand what is actually being assessed and education researchers to identify gaps in our arsenal of assessments related to experimentation.

14.2 Methods

We surveyed assessments of different aspects of experimentation currently used in undergraduate biology courses and categorized the assessment items using the framework of the Basic Competencies of Biological Experimentation (Pelaez et al., 2017; Chap. 1 in this volume). We limited our review to assessments that are freely available and documented in the biology education literature, starting with those suggested by Shortlidge and Brownell (2016) for the assessment of course-based undergraduate research experiences. We supplemented those references with additional assessments related to biological experimentation that have been published, including those collected by ACE-Bio participants in 2014. Our goal was not to include all possible assessments of biological experimentation, but to include a range of possible assessments that are used in biology courses. The complete list of assessments surveyed can be found in Table 14.1. In some cases, the references are

Table 14.1 Assessments reviewed in this study categorized by type, student class level, and instrument availability

Assessment	Number of items surveyed	Type	Student class level	Availability	References
Instructional Practices Survey	24	Description of class activities – Survey	Non-majors; Introductory majors; Upper-level majors	Main article or supplementary materials	Beck and Blumer (2016)
LCAS	17	Description of class activities – Survey	Introductory majors; Upper-level majors	Main article or supplementary materials	Corwin et al. (2015)
BEDCI	14	Measurement of learning – Multiple choice	Introductory majors; upper-level majors	Download via link provided in main article	Deane et al. (2014)
TOSLS	28	Measurement of learning – Multiple choice	Majors and non-majors	Main article or supplementary materials	Gormally et al. (2012)
Modified Classroom Test of Scientific Reasoning	24	Measurement of learning – Multiple choice	Non-majors; Introductory majors	Main article or supplementary materials	Benford and Lawson (2001)
Molecular Biology Data Analysis Test	20	Measurement of learning – Multiple choice	Introductory level; Upper-level majors	Main article or supplementary materials	Rybarczyk et al. (2014)
SRBCI	12	Measurement of learning – Multiple choice	Introductory majors; Upper-level majors	Download via link provided in main article	Deane et al. (2016)
BioSQuaRE	29	Measurement of learning – Multiple choice	Introductory majors	Online through author	Stanhope et al. (2017)
RED	18	Measurement of learning – Prompt with rubric	Introductory majors	Main article or supplementary materials	Dasgupta et al. (2014)
RED Bird Assessment	3	Measurement of learning – Prompt with rubric	Introductory majors	Main article or supplementary materials	Dasgupta et al. (2014)
RED Drug Assessment	6	Measurement of learning – Prompt with rubric	Introductory majors	Main article or supplementary materials	Dasgupta et al. (2014)

(continued)

Table 14.1 (continued)

Assessment	Number of items surveyed	Type	Student class level	Availability	References
RED Shrimp Assessment	4	Measurement of learning – Prompt with rubric	Introductory majors	Main article or supplementary materials	Dasgupta et al. (2014)
EDAT	10	Measurement of learning – Prompt with rubric	Non-majors	Main article or supplementary materials	Sirum and Humburg (2011)
E-EDAT	10	Measurement of learning – Prompt with rubric	Introductory majors; Upper-level majors	Main article or supplementary materials	Brownell et al. (2013)
TIED	20	Measurement of learning – Prompt with rubric	Introductory majors	Main article or supplementary materials	Killpack and Fulmer (2018)
CRBS	20	Measurement of learning – Research assignment with rubric	Non-majors; Introductory majors; Upper-level majors	Main article or supplementary materials	Kishbaugh et al. (2012)
The Rubric for Science Writing	15	Measurement of learning – Research assignment with rubric	Introductory majors; upper-level majors	Main article or supplementary materials	Timmerman et al. (2011)
Quantitative Literacy Rubric	9	Measurement of learning – Prompt with rubric	Introductory majors	Main article or supplementary materials	Speth et al. (2010)
BioTAP	15	Measurement of learning – Research assignment with rubric	Upper-level majors	Online through author	Reynolds et al. (2009)
Graph Rubric	12	Measurement of learning – Research assignment with rubric	Non-majors; Introductory majors; Upper-level majors	Main article or supplementary materials	Angra and Gardner (2018)
SPFA	18	Measurement of learning – Prompt with rubric	Not explicitly stated, but included major and non-majors	Main article or supplementary materials	Wilson and Rigakos (2016)

(continued)

Table 14.1 (continued)

Assessment	Number of items surveyed	Type	Student class level	Availability	References
EDR	4	Measurement of learning – Prompt with rubic	Introductory majors	Main article or supplementary materials	Dirks and Cunningham (2006)
Graphing Quiz	12	Measurement of learning – Short response and multiple choice	Introductory majors	Main article or supplementary materials	Dirks and Cunningham (2006)
TIPS	34	Measurement of learning – Multiple choice	Introductory majors	Main article or supplementary materials	Dirks and Cunningham (2006)
[a]URSSA	45	Self-report of learning – Survey	Not explicitly stated	Download via link provided in main article	Weston and Laursen (2015)
BioVEDA	12	Measurement of learning – Multiple choice	Introductory majors	Main article or supplementary materials	Hicks et al. (2020)
[a]CURE-Survey	46	Self-report of learning – Survey	Not explicitly stated	Online through author	Lopatto (2008)
Self-Efficacy, Science Identity, Science Community	15	Self-report of learning – Survey	Non-majors; Introductory majors; Upper-level majors	Main article or supplementary materials	Estrada et al. (2011)
Experimental Control Exercises	7	Short answer and multiple-choice quiz, no rubric	Upper-level majors	Main article or supplementary materials	Shi et al. (2011)
Pre-Post Test for Analytical Skills	12	Short answer and multiple-choice quiz, no rubric	Non-majors; Introductory majors; Upper-level majors	Main article or supplementary materials	Picone et al. (2007)

The number of individual items surveyed in each assessment instrument is shown. The number of items surveyed in an assessment varied from 3 to 52

[a]Denotes assessment that contained multiple components, some of which were deemed not applicable e.g., related to affect or specific programmatic or demographic information. These sections were excluded, therefore the number of items surveyed may be different from the total number of items in the assessment

for the assessment instruments themselves while in others, the assessments are supplementary and used for a study that examined student competence area in biological experimentation as an outcome measure.

Each assessment instrument was reviewed first to determine whether its items related to the Basic Competencies of Biological Experimentation. Instruments that do not assess biological experimentation (e.g., assessments of student affect with no items explicitly related to biological experimentation (Chemers et al., 2011; Glynn et al., 2011; Hanauer & Dolan, 2014; Hanauer & Hatfull, 2015; Semsar et al., 2011), student views of the nature of science (Halloun & Hestenes, 1998; Lederman et al., 2002)) were excluded since they do not measure students understanding, skills, or knowledge related to biological experimentation. For assessment instruments that were retained, we categorized each item in one (or more) of the seven Basic Competence Areas or "None of the above". Furthermore, we identified the Concepts and Skill Statements that are being assessed, when possible. To deal with the fact that assessment items might not map to specific Concepts and Skill Statements, we added an "Other" category within each of the seven Basic Competence Areas to represent additional Concepts and also within each of the subsidiary Concepts to represent additional Skill Statements.

To align our codings of assessments using the Basic Competencies of Biological Experimentation framework, all three authors coded items from three assessments (Corwin et al., 2015; Gormally et al., 2012; Sirum & Humburg, 2011) that included the range of types of assessments in our dataset (see Table 14.1). Based on discussion of preliminary coding, we agreed to code in a hierarchical fashion such that we first determined whether an assessment item fit one or more Basic Competence Areas, then whether it fit one or more Concepts within those Competence Areas, and finally whether it fit one or more Skill Statements within those Concepts. The remainder of the instruments were coded by two of the three authors, with each author coding approximately two-thirds of the instruments. When coders disagreed in their coding of a particular item at the level of the Basic Competencies, the coders discussed the item to determine a consensus coding. We included differences among coders at the level of Concepts and Skill Statements as they were reflective of the ambiguity in coding many of the items in the assessments at these levels.

14.3 Results and Discussion

14.3.1 Instruments for Assessing Competence Areas in Biological Experimentation

The majority of assessments included in our study aimed to measure learning via multiple choice assignments or short answer writing prompts (with or without a rubric), while three were survey type assessments that measured affect or self-reported learning gains with some items explicitly related to biological experimentation (Table 14.1). The LCAS (Corwin et al., 2015) and the instructional practices survey (Beck & Blumer, 2016) explore student perceptions on the types of activities they performed in class. Many of the assessments have been used with students in both introductory and upper-level courses for biology majors, suggesting that they can be used to assess aspects of experimentation in a wide range of students.

14.3.2 Mapping Assessments to Competence Areas

The assessments that we mapped varied considerably in the number of Competence Areas covered by the assessment, ranging from one to all seven Competence Areas (Table 14.2). The URSSA (Weston & Laursen, 2015), CURE-Survey (Lopatto, 2008), CRBS (Kishbaugh et al., 2012), and Rubric for Science Writing (Timmerman et al., 2011) assess all 7 competencies. The URSSA (Weston & Laursen, 2015) and the CURE-Survey (Lopatto, 2008) are student self-reports and are designed for programmatic assessment by considering a large number of areas. In contrast, CRBS (Kishbaugh et al., 2012) and Rubric for Science Writing (Timmerman et al., 2011) are rubric banks or rubrics that instructors can use for assessing a broad range of competencies in student products, such as paper, posters, and presentations. At the other end of the spectrum, the assessments that covered the least number of competencies were the Shrimp Assessment of the RED (Dasgupta et al., 2014), which only covered 1 Competence Area, followed by the Modified CTSR (Benford & Lawson, 2001), EDAT (Sirum & Humburg, 2011), SRBCI (Deane et al., 2016), E-EDAT (Brownell et al., 2013), Experimental Control (Shi et al., 2011), TIED (Killpack & Fulmer, 2018), and Graph Rubric (Angra & Gardner, 2018), all of which only covered 2 of the 7 Competence Areas (Table 14.2). In general, these assessments focus on *Plan* and *Conclude*, except for the Graphic Rubric (Angra & Gardner, 2018), which focuses on *Analyze* (Fig. 14.1). It is possible that some assessments, like the CURE-survey (Lopatto, 2008), covered a high percentage of the Competence Areas, because the items tended to be phrased in broad or generic terms (e.g., "Write a research proposal"), which subsequently was coded as having the potential to cover many skills within the framework. Others that had a lower total percent coverage of the competency framework (e.g., E-EDAT (Brownell et al., 2013)), had more narrowly phrased questions that encompassed a specific skill (e.g., "Develop a hypothesis about what causes changes in poppy growth rate") and subsequently was only categorized into one or fewer of the seven categories.

In most cases, when an assessment was scored as measuring a Competence Area, it considered multiple Concepts within each Competence Area (Table 14.2). Not surprisingly, however, we note a trade-off between the number of Competence Areas covered by an assessment and the proportion of items associated with a particular Competence Area. Assessments that covered a large number of Competence Areas tend to have fewer items associated with a particular Competence Area (Fig. 14.1). In contrast, assessments that focus on one or two Competence Areas had a high proportion of items concentrated in those Competence Areas.

From the perspective of individual Competence Areas, *Plan* and *Conclude* are covered by the most assessments (27 and 24 out of 30 instruments, respectively), indicating an emphasis on experimental design skills and drawing inferences from data in current assessments. *Identify* and *Conduct* are the least assessed of the Competence Areas (8 and 7 of 30 assessments, respectively) (Fig. 14.1). The nature of the Concepts and Skill Statements within *Identify* and *Conduct* might make them particularly difficult to assess. For example, many of the Skill Statements in *Conduct*

Table 14.2 Assessment coverage of Competence Areas, Concepts, and Skill Statements

Assessment	Competence Areas	Concepts	Skill Statements	Competence Areas with multiple Concepts	Concepts with multiple Skill Statements
Instructional Practice Survey	4 (57%)	6 (17%)	8 (6%)	2 (29%)	2 (7%)
LCAS	6 (86%)	14 (39%)	20 (14%)	5 (71%)	5 (17%)
BEDCI	4 (57%)	13 (36%)	27 (20%)	3 (43%)	7 (24%)
TOSLS	4 (57%)	13 (36%)	32 (23%)	4 (57%)	8 (28%)
Modified Classroom Test of Scientific Reasoning	2 (29%)	6 (17%)	10 (7%)	2 (29%)	4 (14%)
Molecular Biology Data Analysis Test	3 (43%)	9 (25%)	19 (14%)	2 (29%)	7 (24%)
SRBCI	2 (29%)	5 (14%)	15 (11%)	2 (29%)	4 (14%)
BioSQuaRE	5 (71%)	7 (19%)	12 (9%)	2 (29%)	2 (7%)
RED	3 (43%)	10 (28%)	23 (17%)	2 (29%)	7 (24%)
RED (Bird Assessment)	2 (29%)	6 (17%)	9 (6%)	2 (29%)	3 (10%)
RED (Drug Assessment)	2 (29%)	6 (17%)	14 (10%)	1 (14%)	4 (14%)
RED (Shrimp Assessment)	1 (14%)	5 (14%)	8 (6%)	1 (14%)	2 (7%)
EDAT	2 (29%)	10 (28%)	21 (15%)	1 (14%)	6 (21%)
E-EDAT	2 (29%)	8 (22%)	20 (14%)	1 (14%)	6 (21%)
TIED	2 (29%)	8 (22%)	15 (11%)	2 (29%)	5 (17%)
CRBS	7 (100%)	26 (72%)	45 (33%)	7 (100%)	12 (41%)
The Rubric for Science Writing	7 (100%)	19 (53%)	43 (31%)	7 (100%)	10 (34%)
Quantitative Literacy Rubric	3 (43%)	6 (17%)	17 (12%)	2 (29%)	5 (17%)
BioTAP	6 (86%)	15 (42%)	35 (25%)	6 (86%)	8 (28%)
Graph Rubric	2 (29%)	5 (14%)	9 (7%)	2 (29%)	2 (7%)
SPFA	5 (71%)	22 (61%)	30 (22%)	5 (71%)	6 (21%)
TIPS	3 (43%)	6 (17%)	11 (8%)	1 (14%)	4 (14%)
EDR	2 (29%)	8 (22%)	12 (9%)	2 (29%)	2 (7%)
Graphing Quiz	2 (29%)	3 (8%)	4 (3%)	0 (0%)	2 (7%)
URSSA	7 (100%)	16 (44%)	23 (17%)	5 (71%)	5 (17%)
BioVEDA	3 (43%)	10 (28%)	21 (15%)	2 (29%)	7 (24%)
CURE-Survey	7 (100%)	30 (83%)	56 (41%)	7 (100%)	12 (41%)
Self-Efficacy, Science Identity, Science Community	5 (71%)	6 (17%)	11 (8%)	2 (29%)	3 (10%)

(continued)

Table 14.2 (continued)

Assessment	Competence Areas	Concepts	Skill Statements	Competence Areas with multiple Concepts	Concepts with multiple Skill Statements
Experimental Control Exercises	2 (29%)	4 (11%)	8 (6%)	2 (29%)	2 (7%)
Pre-Post Test for Analytical Skills	3 (43%)	6 (17%)	12 (9%)	3 (43%)	4 (14%)

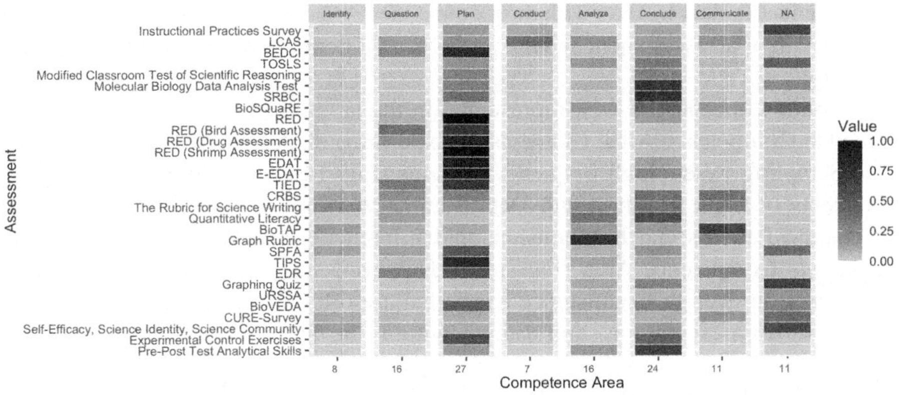

Fig. 14.1 Heatmap of the coverage of Competence Areas by each assessment. The values are the proportion of items in each assessment that address a given Competence Area. "NA" was assigned to items in an assessment if they could not be categorized in any of the Competence Area. The values at the bottom of each column are the total number of assessment instruments that addressed a given Competence Area

could only be observed by an instructor in a laboratory course or mentored research context.

Some assessments show a high proportion of items that do not fit into the ACE-Bio framework (Fig. 14.1). In some cases, items are related to student affect, student metacognition, faculty assessment practices, computational quantitative literacy, or are too general (Beck & Blumer, 2016; Estrada et al., 2011; Gormally et al., 2012; Lopatto, 2008; Stanhope et al., 2017; Wilson & Rigakos, 2016). Other assessments include items that are not currently considered in the ACE-Bio framework, but perhaps should be included (see below), such as collaboration skills and aspects of statistical literacy (Corwin et al., 2015; Gormally et al., 2012).

14.3.3 Mapping Assessments to Concepts

Similar to our mapping of assessments to Competence Areas, assessments were quite variable in the number of Concepts that are considered (Table 14.2). Some assessments focused on very few Concepts (4 or 5 out of 22 for the Experimental Control Exercise (Shi et al., 2011) and the Graph Rubric (Angra & Gardner, 2018), respectively). In contrast, the assessments that cover a broad range of Competence Areas also incorporate a high percentage of Concepts (e.g., CURE-Survey (Lopatto, 2008) and CRBS (Kishbaugh et al., 2012)). For most assessments, only a single Skill Statement was assessed for a particular Concept rather than multiple Skill Statements (Table 14.2). As with the Competence Areas, there is a trade-off between the breadth of an assessment and the proportion of items associated with a particular Concept.

Certain Concepts are well-represented in the assessments we surveyed. Within the Plan Competence Area, Concepts of Experimental Design, Variables, Controls, and Sampling have a high frequency of items (Fig. 14.2). The same is true for the Concepts of Data Curation and Data Summary within the *Analyze* Competence Area, and Patterns and Relationships and Inferences and Conclusions within the *Conclude* Competence Area (Fig. 14.2). However, some Concepts are conspicuously absent, even in Competence Areas that are often included. For example, the Concepts Representations and Ethics within the *Plan* Competence Area are

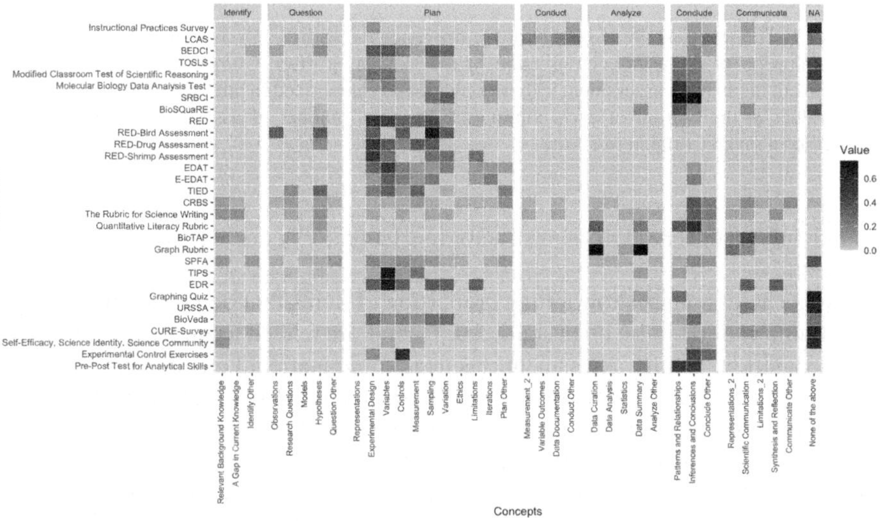

Fig. 14.2 Heatmap of the coverage of Concepts within each Competence Area by each assessment. The values are the proportion of items in each assessment instrument that addressed a given Concept. "NA" was assigned to items in an assessment if they could not be categorized in any of the Competence Areas. Within each Competence Area, "Other" tabulates items in an assessment that were categorized in a given Competence Area but did not address any of the specified Concepts in that Competence Area

infrequent even though *Plan* is commonly assessed (Figs. 14.1 and 14.2). Likewise, Models is uncommon within *Question* (Figs. 14.1 and 14.2).

Because items clearly fit within a particular Competence Area, but not necessarily within a Concept in that Competence Area, we created an "Other" category for each Competence Area. The frequency of items coded in these "Other" categories, especially within *Conclude* (Fig. 14.2), suggests the potential for expanding the ACE-Bio framework (see below).

14.3.3.1 Gaps in Existing Assessments of Biological Experimentation

None of the assessments we reviewed were developed with the ACE-Bio framework as a guide. Consequently, the match between assessment items and the Competence Areas, Concepts, and Skill Statements is not perfect and is subject to interpretation. We have therefore limited most of our reporting of gaps to the level of Competence Areas, to the most general level of categorization. Among the seven basic Competence Areas, two are not well addressed by the assessments we surveyed, *Identify* and *Conduct* (Fig. 14.1). Fewer than one-half of the assessments include items that were categorized in the *Identify* or *Conduct* Competence Areas, and among the assessments that include items in these Competence Areas, the proportion of assessment items in either Competence Area is small. Similarly, among those assessments that include six or all seven Competence Areas (Table 14.2), few items assess Skill Statements in *Identify* or *Conduct* (Fig. 14.1). One exception is the LCAS (Corwin et al., 2015) for *Conduct*, but this assessment is limited to descriptions of class activities rather than students' skills and knowledge. Another exception is the Rubric for Science Writing (Timmerman et al., 2011) for *Identify*. In both cases, the percentage of items addressing the Competence Area is 30–40%, but only for one of these two Competence Areas in each case. It is worth noting that the Concepts and Skill Statements in the *Identify* Competence Area (Pelaez et al., 2017; Table 1.3 in Chap. 1 in this volume) are relatively high-order (the ability to identify gaps and limitations in current knowledge) that require experiences uncommon among undergraduates and are infrequently expected learning outcomes in undergraduate courses (Cole & Beck's Chap. 3 in this volume). The *Conduct* Competence Area may be assessed more readily by using in-class methods than by using assessments reviewed here, such as a laboratory practical, mid-experiment discussions with students, direct observation of students while conducting an experiment, or checking extemporaneous documentation in laboratory notebooks (Moore & Lynn, 2020). Within the other Competence Areas, even those that are well covered by assessment items, there are noticeable gaps. Within the *Plan* Competence Area, the Concepts of Representations, Ethics, and Limitations are not well addressed by any assessments (Fig. 14.2), even though these Concepts are considered important in undergraduate teaching of experimentation (Clemmons et al., 2020; Cole & Beck's Chap. 3 in this volume; Diaz-Martinez et al., 2019). In *Analyze*, the Concept of Statistics (choosing and conducting the appropriate statistical test and others) also is not well addressed by assessments, as assessments of statistical literacy like SRBCI (Deane et al., 2016) focus on Conclude (Fig. 14.2).

14.3.4 Gaps in ACE-Bio Framework of Competence Areas

One of the most striking findings in our analysis is the frequency of assessment items that do not fit nicely in one of the ACE-Bio Competence Areas, and the number of assessment items that we categorized in a given Competence Area, but could not assign to a specific Concept or Skill Statement (Fig. 14.2). Some of this apparent mismatch is a result of assessments items that focus on quantitative literacy, but not experimentation. Similarly, many assessment items that we could not categorize in the framework address student affect (e.g., self-efficacy) in domains not directly related to biological experimentation. We did not code assessments that focused exclusively on the nature of science (e.g., Lederman et al., 2002), because they do not address experimentation. Although both quantitative literacy (Clemmons et al., 2020) and student affect (Trujillo & Tanner, 2014) are important student outcomes, they do not necessarily fit within the framework of biological experimentation. None-the-less, we found five aspects of experimentation that are not an explicit part of the ACE-Bio framework that appear in assessments and represent potential gaps in the existing framework. We do not present these as criticisms of the framework but note that the framework should be viewed as a document that requires interpretation and therefore thoughtful clarification and modification. The Concept of creativity is a pre-cursor to or facilitator of the *Question* and *Plan* Competence Areas at the very least and plays an underlying role in *Conclude* and *Communicate*. Creativity could be addressed as an aspect of experimentation (Beno & Tucker's Chap. 20 in this volume). Similarly, modern biological research often requires or is greatly enhanced by collaboration. In addition, collaboration is a core competency in the *Vision and Change* report (AAAS, 2011) and program-level learning outcome in the BioSkills guide (Clemmons et al., 2020). Collaboration is assessed in the LCAS in the context of course-based undergraduate research experiences (CURES) (Corwin et al., 2015). Yet, collaboration is not explicitly in the framework, but could be incorporated in a number of the Competence Areas, much like creativity. This gap and the possibility of incorporating collaboration in the existing Competence Areas is explored in more detail later in this volume (Chaps. 20 and 22).

The other potential gaps in the framework are more specific to individual Competence Areas. The articulation of hypotheses is well described in *Question* but making falsifiable predictions for each hypothesis is not. This important feature of experimentation appears in some assessments but is not part of the framework. Lastly, the Concept of Statistics is part of the *Analyze* Competence Area. However, interpretation of statistical tests is missing from the framework. Statistical interpretation is addressed in some assessments and could be more explicitly incorporated in the *Analyze* or *Conclude* Competence Areas.

14.4 Recommendations

14.4.1 Recommendations for Instructors

Choosing an assessment on experimentation that will be used in a course or program requires that an instructor first decide on the learning outcomes to be assessed. That is not a trivial issue since no one assessment will address every aspect of experimentation and the format of an assessment may limit its usefulness. The assessments that are available can be categorized in two groups – those that are narrowly focused and those that address the breadth of the experimentation Competence Areas. Narrowly focused assessments are best used as formative assessments or assessments for education research rather than for assigning grades in a course. Measuring learning with a prompt or narrowly focused assignment and a rubric (Angra & Gardner, 2018; Brownell et al., 2013; Speth et al., 2010) will permit instructors to assess specific aspects of experimentation, mainly in the *Plan* and *Conclude* Competence Areas. Objective tests of learning, such as multiple-choice tests (Bedford & Lawson, 2001; Deane et al., 2014, 2016; Dirks & Cunningham, 2006; Gormally et al., 2012; Picone et al., 2007; Rybarczyk et al., 2014; Shi et al., 2011; Stanhope et al., 2017), also may be used as measures of very specific learning outcomes related to experimentation. It is very tempting to use a rapidly scored test as means of assigning grades, but we recommend against that because tests are not authentic assessments of experimentation, scientific research is not assessed in this manner. Matching the assessment used to the learning outcome set for students is essential. If the learning outcome is student achievement in the ability to perform experimentation, then having them perform the activities that comprise the process of biological experimentation is the most authentic assessment (papers, posters, proposals, research seminars scored with a rubric (Kishbaugh et al., 2012; Reynolds et al., 2009; Timmerman et al., 2011). Measuring learning with a research assignment and a rubric will permit instructors address the broadest range of experimentation Competence Areas (Table 14.1) and also could be used as a means of assigning grades.

Instructors should ensure that any assessment that they use was designed for the level of their students. Those assessments that were developed for introductory students could be used with upper-level students (e.g., EDAT (Sirum & Humburg, 2011) with several caveats. First, instructors should be sure to administer the assessment at the beginning of the semester to determine whether there is a likelihood of a ceiling effect. Second, instructors should consider differences in the expectations of Competence Areas in experimentation for introductory and upper-level students (Cole & Beck's Chap. 3 in this volume). These differences in expectations also make assessments designed for upper-level biology majors unlikely to be useful for assessing experimentation in introductory courses. Finally, instructors need to remember that these assessments were validated with introductory students.

The timing of the use of specific assessments also matters, both within a course and within an undergraduate curriculum. Instructors might reasonably begin a course with very narrow learning outcomes and focus on specific skills and build to more comprehensive learning outcomes (and more authentic assessments such as papers, posters, proposals, research seminars) as the course developed during a semester. In this case, starting with less authentic assessments may be completely appropriate if they were used to create the scaffolding for more authentic assignments in that course. However, more advanced undergraduate courses should focus on the most authentic assessments (assessments that are closest to the activities performed by working scientists) and score them with rubrics to cover a broad range of experimentation competencies. A summary of these recommendations is given in bulleted form below the discussion.

14.4.2 Recommendations for Education Researchers

Our analysis of current assessments for biological experimentation leads to several recommendations for education researchers (summarized as a bulleted list below). The gaps in assessments that address the basic Competencies of Experimentation provide an opportunity to develop new assessment tools or modify existing tools. The Competence Areas of *Identify* and *Conduct* are essential aspects of the experimentation process, but we need the tools to assess them. Authors of other chapters in this volume provided examples of work to address this deficiency that we have identified, as described in the Preface to this book. Similarly, there are opportunities to develop assessment tools to address the Concepts of Representations, Ethics, and Limitations within the *Plan* Competence Area and the Concept of Statistics (choosing and conducting the appropriate statistical test and others) within the *Analyze* Competence Area. Using the ACE-Bio framework can be an important starting point for developing general or more discipline-specific assessments in these areas (Dasgupta et al., 2016). In addition, by using the framework as a basis for assessment, the aspects of biological experimentation that are being assessed will be clearer.

Aligning expectations of student competencies in experimentation for students at different levels with assessments designed for students at those levels is essential for rigorous studies of student learning on experimentation. While some assessments are applicable to students across multiple levels, others are specific to students at either the introductory or upper-level (Table 14.1). Therefore, education researchers can develop new assessments, or validate existing assessments for students at different levels, that align with the expectations for students at those levels (Cole & Beck's Chap. 3 in this volume). For example, the EDAT (Sirum & Humburg, 2011) was designed for non-majors introductory biology. However, faculty do not necessarily expect students to have much first-hand experience with the Competence Area *Plan* (Cole & Beck's Chap. 3 in this volume), which the EDAT covers extensively (Fig. 14.1). Even rubrics for student assignments could be refined to better articulate the expectations for students at different levels. The CRBS (Kishbaugh et al., 2012) is an example of where this has been done effectively.

Finally, how student learning of one competence area in biological experimentation relates to their learning of other competence areas is unclear. Linkages and correlations between learning of different experimentation competencies would be informative for both teaching and assessing experimentation. From the perspective of assessment, high correlations between learning of different competence areas would allow researchers and instructors to assess fewer competence areas while at the same time getting a complete picture of student understanding of experimentation.

In summary, consider the following recommendations for instructors and education researchers:

Recommendations for Instructors:

- Choose assessment instruments that best match the learning outcome expectations for a course.
- Use narrowly focused assignments as formative assessments but not for grading.
- Use broad based authentic assessments of learning, research assignment with a rubric for grading.
- Scaffold learning outcomes and assessments within course and within curriculum.

Recommendations for Education Researchers:

- Develop new assessments to fill current gaps in the *Identify* and *Conduct* Competence Areas.
- Develop new assessments to fill current gaps in the Concepts of Representations, Ethics, and Limitations within the *Plan* Competence Area.
- Develop new assessments to fill current gaps in the Concept of Statistics (choosing and conducting the appropriate statistical test and others) within the *Analyze* Competence Area.
- Develop new assessments, or validate existing assessments for students at different levels, so that expectations of students and assessments align.
- Explore linkages and correlations between learning of different experimentation competencies.

14.5 Conclusions

By mapping current assessments in biological experimentation on the ACE-Bio competence areas, we have provided a tool for instructors to select the best available assessments to examine student learning of experimentation in their classes and identified avenues of future research related to the development of new assessments on experimentation. Through appropriate application of current assessments and development of new assessments, we hope to advance our understanding of how students become competent at experimentation.

Finally, how student learning of one competence area in biological experimentation relates to their learning of other competence areas is unclear. Linkages and correlations between learning of different experimentation competencies would be informative for both teaching and assessing experimentation. From the perspective of assessment, high correlations between learning of different competence areas would allow researchers and instructors to assess fewer competence areas while at the same time getting a complete picture of student understanding of experimentation.

Acknowledgments This work was partially funded by a stipend from the National Science Foundation (NSF) grant #1346567 for the Advancing Competencies in Experimentation–Biology (ACE-Bio) Network. We thank Nancy Pelaez, Trevor Anderson and Stephanie Gardner for facilitating the work of the ACE-Bio Network. Preliminary feedback from Trevor Anderson, Annwesa Dasgupta, and others in the ACE-Bio Network helped improve this chapter. For this, we are grateful. This study and publication were funded by NSF grants DUE-1821184, HRD-1818458 and HRD-2010676 to Morehouse College and DUE-1821533 to Emory University. Any opinions, findings, and conclusions or recommendations expressed in this article are those of the authors and do not necessarily reflect the views of the NSF.

References

American Association for the Advancement of Science (AAAS). (2011). *Vision and change in undergraduate biology education: A call to action*. Retrieved from Washington, DC https://live-visionandchange.pantheonsite.io/wp-content/uploads/2013/11/aaas-VISchange-web1113.pdf

Angra, A., & Gardner, S. M. (2018). The graph rubric: Development *of a teaching, learning, and research tool. CBE Life Sciences Education, 17*(4), ar65. https://doi.org/10.1187/cbe.18-01-0007

Beck, C. W., & Blumer, L. S. (2016). Alternative realities: Faculty and student perceptions of instructional practices in laboratory courses. *CBE Life Sciences Education, 15*(4), ar52. https://doi.org/10.1187/cbe.16-03-0139

Beck, C. W., Butler, A., & Burke da Silva, K. (2014). Promoting inquiry-based teaching in laboratory courses: Are we meeting the grade? *CBE Life Sciences Education, 13*, 444–452.

Benford, & Lawson, A. E. (2001). *Relationships between effective inquiry use and the development of scientific reasoning skills in college biology labs*. Retrieved from https://files.eric.ed.gov/fulltext/ED456157.pdf. Accessed 3 Jan 2021.

Boyer, E. L. (1998). *The Boyer Commission on educating undergraduates in the research university, reinventing undergraduate education: A blueprint for America's research universities*. Stony Brook.

Brownell, S. E., Wenderoth, M. P., Theobald, R., Okoroafor, N., Koval, M., Freeman, S., … Crowe, A. J. (2013). How students think about experimental design: Novel conceptions revealed by in-class activities. *Bioscience, 64*(2), 125–137. https://doi.org/10.1093/biosci/bit016

Chemers, M. M., Zurbriggen, E. L., Syed, M., Goza, B. K., & Bearman, S. (2011). The role of efficacy and identity in science career commitment among underrepresented minority students. *Journal of Social Issues, 67*(3), 469–491. https://doi.org/10.1111/j.1540-4560.2011.01710.x

Clemmons, A., Timbrook, J., Herron, J., & Crowe, A. (2020). BioSkills guide: Development and national validation of a tool for interpreting the Vision and Change Core Competencies. *bioRxiv*, 2020. https://doi.org/10.1101/2020.01.11.902882

Corwin, L. A., Runyon, C., Robinson, A., & Dolan, E. L. (2015). The laboratory course assessment survey: A tool to measure three dimensions of research-course design. *CBE Life Sciences Education, 14*(4). https://doi.org/10.1187/cbe.15-03-0073

Dasgupta, A. P., Anderson, T. R., & Pelaez, N. J. (2014). Development and validation of a rubric for diagnosing students' experimental design knowledge and difficulties. *CBE Life Sciences Education, 13*(2), 265–284.

Dasgupta, A. P., Anderson, T. R., & Pelaez, N. J. (2016). Development of the Neuron Assessment for measuring biology students' use of experimental design concepts and representations. *CBE Life Sciences Education, 15*(2), ar10. https://doi.org/10.1187/cbe.15-03-0077

Deane, T., Nomme, K., Jeffery, E., Pollock, C., & Birol, G. (2014). Development of the Biological Experimental Design Concept Inventory (BEDCI). *CBE Life Sciences Education, 13*(3), 540–551. https://doi.org/10.1187/cbe.13-11-0218

Deane, T., Nomme, K., Jeffery, E., Pollock, C., & Birol, G. (2016). Development of the Statistical Reasoning in Biology Concept Inventory (SRBCI). *CBE Life Sciences Education, 15*(1), ar5-ar5. https://doi.org/10.1187/cbe.15-06-0131

Diaz-Martinez, L. A., Fisher, G. R., Esparza, D., Bhatt, J. M., D'Arcy, C. E., Apodaca, J., … Olimpo, J. T. (2019). Recommendations for effective integration of ethics and responsible conduct of research (E/RCR) education into course-based undergraduate research experiences: A meeting report. *CBE Life Sciences Education, 18*(2), mr2. https://doi.org/10.1187/cbe.18-10-0203

Dirks, C., & Cunningham, M. (2006). Enhancing diversity in science: Is teaching science process skills the answer? *CBE Life Sciences Education, 5*(3), 218–226. https://doi.org/10.1187/cbe.05-10-0121

Estrada, M., Woodcock, A., Hernandez, P. R., & Schultz, P. W. (2011). Toward a model of social influence that explains minority student integration into the scientific community. *Journal of Educational Psychology, 103*(1), 206–222. https://doi.org/10.1037/a0020743

Glynn, S. M., Brickman, P., Armstrong, N., & Taasoobshirazi, G. (2011). Science motivation questionnaire II: Validation with science majors and nonscience majors. *Journal of Research in Science Teaching, 48*(10), 1159–1176. https://doi.org/10.1002/tea.20442

Gormally, C., Brickman, P., & Lutz, A. (2012). Developing a Test of Scientific Literacy Skills (TOSLS): Measuring undergraduates' evaluation of scientific information and arguments. *CBE Life Sciences Education, 11*(4), 364–377. https://doi.org/10.1187/cbe.12-03-0026

Halloun, I., & Hestenes, D. (1998). Interpreting VASS dimensions and profiles for physics students. *Science Education, 7*, 553–577.

Hanauer, D. I., & Dolan, E. L. (2014). The project ownership survey: Measuring differences in scientific inquiry experiences. *CBE Life Sciences Education, 13*(1), 149–158. https://doi.org/10.1187/cbe.13-06-0123

Hanauer, D. I., & Hatfull, G. (2015). Measuring networking as an outcome variable in undergraduate research experiences. *CBE Life Sciences Education, 14*(4). https://doi.org/10.1187/cbe.15-03-0061

Handelsman, J., Miller, S., & Pfund, C. (2007). *Scientific teaching*. WH Freeman and Co.

Hicks, J., Dewey, J., Brandvain, Y., & Schuchardt, A. (2020). Development of the biological variation in experimental design and analysis (BioVEDA) assessment. *PLoS One, 15*(7), e0236098. https://doi.org/10.1371/journal.pone.0236098

Killpack, T. L., & Fulmer, S. M. (2018). Development of a tool to assess interrelated experimental design in introductory biology. *Journal of Microbiology & Biology Education, 19*(3). https://doi.org/10.1128/jmbe.v19i3.1627

Kishbaugh, T. L. S., Cessna, S., Jeanne Horst, S., Leaman, L., Flanagan, T., Graber Neufeld, D., & Siderhurst, M. (2012). Measuring beyond content: A rubric bank for assessing skills in authentic research assignments in the sciences. *Chemistry Education Research and Practice, 13*(3), 268–276. https://doi.org/10.1039/C2RP00023G

Lederman, N. G., Abd-El-Khalick, F., Bell, R. L., & Schwartz, R. S. (2002). Views of nature of science questionnaire: Toward valid and meaningful assessment of learners' conceptions of nature of science. *Journal of Research in Science Teaching, 39*(6), 497–521. https://doi.org/10.1002/tea.10034

Lopatto, D. (2008). Exploring the benefits of undergraduate research: The SURE survey. In R. Taraban & R. L. Blanton (Eds.), *Creating effective undergraduate research programs in science* (pp. 112–132). Teacher's College Press.

Moore, J. E., & Lynn, T. W. (2020). Lab notebooks: Efficient grading and effective learning. Article 45. In K. McMahon (Ed.), *Advances in biology laboratory education*. Volume 41. Publication of the 41st Conference of the Association for Biology Laboratory Education (ABLE). https://doi.org/10.37590/able.v41.art45

National Research Council (NRC). (2003). *BIO 2010. Transforming undergraduate education for future research biologists*. National Academy Press.

Pelaez, N., Anderson, T. R., Gardner, S. M., Yin, Y., Abraham, J. K., Bartlett, E., Gormally, C., Hill, J. P., Hoover, M., Hurney, C., Long, T., Newman, D. L., Sirum, K., & Stevens, M. (2017). The basic competencies of biological experimentation: Concept-skill statements. *PIBERG Instructional Innovation Materials. Paper 4*. Retrieved October 18, 2021 from https://docs.lib.purdue.edu/pibergiim/4/

Picone, C., Rhode, J., Hyatt, L., & Parshall, T. (2007). Assessing gains in undergraduate students' abilities to analyze graphical data. *Teaching Issues and Experiments in Ecology, 5*, Research #1. http://tiee.esa.org/vol/v5/research/picone/abstract.html

Project Kaleidoscope (PKAL). (2002). *Recommendations for action in support of undergraduate science, technology, engineering, and mathematics: Report on reports*. Washington, DC.

Reynolds, J., Smith, R., Moskovitz, C., & Sayle, A. (2009). BioTAP: A systematic approach to teaching scientific writing and evaluating undergraduate theses. *BioScience, 59*(10), 896–903, 898. Retrieved from https://doi.org/10.1525/bio.2009.59.10.11

Rybarczyk, B. J., Walton, K. L., & Grillo, W. H. (2014). The development and implementation of an instrument to assess students' data analysis skills in molecular biology. *Journal of Microbiology and Biology Education, 15*, 259–267.

Semsar, K., Knight, J. K., Birol, G., & Smith, M. K. (2011). The Colorado Learning Attitudes about Science Survey (CLASS) for use in biology. *CBE Life Sciences Education, 10*(3), 268–278. https://doi.org/10.1187/cbe.10-10-0133

Shi, J., Power, J. M., & Klymkowsky, M. W. (2011). Revealing student thinking about experimental design and the roles of control experiments. *International Journal for the Scholarship of Teaching and Learning, 5*, 1–16.

Shortlidge, E. E., & Brownell, S. E. (2016). How to assess your CURE: A practical guide for instructors of course-based undergraduate research experiences. *Journal of Microbiology & Biology Education, 17*(3). https://doi.org/10.1128/jmbe.v17i3.1103

Sirum, K., & Humburg, J. (2011). The experimental design ability test (EDAT). *Bioscene, 37*, 8–16.

Speth, E. B., Momsen, J. L., Moyerbrailean, G. A., Ebert-May, D., Long, T. M., Wyse, S., & Linton, D. (2010). 1, 2, 3, 4: Infusing quantitative literacy into introductory biology. *CBE Life Sciences Education, 9*(3), 323–332. https://doi.org/10.1187/cbe.10-03-0033

Stanhope, L., Ziegler, L., Haque, T., Le, L., Vinces, M., Davis, G. K., … Overvoorde, P. J. (2017). Development of a Biological Science Quantitative Reasoning Exam (BioSQuaRE). *CBE Life Sciences Education, 16*(4), ar66. https://doi.org/10.1187/cbe.16-10-0301

Timmerman, B. E. C., Strickland, D. C., Johnson, R. L., & Payne, J. R. (2011). Development of a 'universal' rubric for assessing undergraduates' scientific reasoning skills using scientific writing. *Assessment & Evaluation in Higher Education, 36*(5), 509–547. https://doi.org/10.1080/02602930903540991

Trujillo, G., & Tanner, K. D. (2014). Considering the role of affect in learning: Monitoring students' self-efficacy, sense of belonging, and science identity. *CBE Life Sciences Education, 13*(1), 6–15. https://doi.org/10.1187/cbe.13-12-0241

Weston, T. J., & Laursen, S. L. (2015). The Undergraduate Research Student Self-Assessment (URSSA): Validation for use in program evaluation. *CBE Life Sciences Education, 14*(3). https://doi.org/10.1187/cbe.14-11-0206

Wilson, K. J., & Rigakos, B. (2016). Scientific Process Flowchart Assessment (SPFA): A method for evaluating changes in understanding and visualization of the scientific process in a multidisciplinary student population. *CBE Life Sciences Education, 15*(4), ar63. https://doi.org/10.1187/cbe.15-10-0212

Chapter 15
Research Across the Curriculum Rubric (RAC-R): An Adaptable Rubric for the Evaluation of Journal Article Style Lab Reports

Karla B. Kinkade and Kristy J. Wilson

15.1 Introduction

In our biology department, we are committed to integrating research across the curriculum to progressively enhance student ability to think like a scientist. The laboratory component of multiple required and elective courses utilizes course-based undergraduate research experiences (CUREs) to teach biology students how real scientific experimentation is conducted. CUREs have been defined "as a course wherein students engage in activities resembling those done by scientists in a particular field to conduct novel investigations about relevant phenomena that are currently unknown" (Irby et al., 2018). CUREs must include five key elements: (1) Use of scientific practices such as asking questions, developing hypotheses, designing and performing experiments, and analyzing data; (2) Discovery of new knowledge or insights; (3) Involvement of relevant work that may contribute to a body of knowledge; (4) Involvement of collaboration; and (5) Involvement of iteration (Auchincloss et al., 2014).

CUREs have been shown to significantly increase undergraduate students' knowledge, experience, and confidence in authentic research activities (Irby et al., 2020), and satisfaction with their educational experience (Lopato, 2004). Further, CURE participation increases students' abilities to analyze and interpret data (Brownell et al., 2013, 2015), to use primary scientific literature to validate their arguments and improve scientific writing (Ward et al., 2014), and to "think like a scientist" (Brownell & Kloser, 2015). While these benefits can also be obtained by student research internships, an advantage afforded by CUREs is that many more students can be involved in the research experience (Staub et al., 2016). Additionally,

K. B. Kinkade (✉) · K. J. Wilson
Biology Department, College of Arts and Sciences, Marian University, Indianapolis, IN, USA
e-mail: kkinkade@marian.edu; kjwilson@marian.edu

N. J. Pelaez et al. (eds.), *Trends in Teaching Experimentation in the Life Sciences*, Contributions from Biology Education Research,
https://doi.org/10.1007/978-3-030-98592-9_15

CUREs can make scientific research more inclusive by overcoming barriers that may restrict research opportunities to select groups of people (Bangera & Brownell, 2014; Lopatto, 2004). Further, CUREs have had a positive impact on the number of students who intend to pursue graduate education or careers in science (Lopato, 2004; Harrison et al., 2011).

While there are multiple learning outcomes (Corwin et al., 2015) and methods of student assessment associated with CUREs (Dasgupta et al., 2014; Brownell et al., 2014; Corwin et al., 2015; Shortlidge & Brownell, 2016), practitioners of the CURE model often evaluate student learning through journal article style lab reports (Brownell et al., 2015; Resendes, 2015; Knutson et al., 2010). This writing is an authentic evaluation of skill as it represents an important currency for scientists and health professionals (AAAS, 2011; Kuhn, 1993). We utilize journal article style lab reports to assess student learning in both formative and summative scientific writing assignments. Faculty at our institution and others (Ruscetti et al., 2018) have found that student writing does not always improve as students' progress toward graduation and that we find ourselves teaching fundamentals of scientific writing to advanced students that should have been mastered earlier in their academic careers. Furthermore, biology students fail to grasp that even though the biological subdiscipline is different (e.g. ecology vs cell biology) the skills and content associated with the writing are similar.

To address these concerns, we have developed a Research Across the Curriculum Rubric (RAC-R) for assessing student scientific writing. The use of rubrics to assess scientific writing in undergraduate students is not novel. The "Biology Thesis Assessment Protocol (BioTAP)" (Reynolds et al., 2009) and the "Universal Rubric for Understanding Undergraduates' Scientific Reasoning and Writing Skills" (Timmerman et al., 2011) are outstanding examples; however, the utility of any standard rubric depends on its potential to be adapted to meet the local needs and resources of diverse colleges and universities. The rubric that we envisioned for our department would be broad enough to be useful across the curriculum, from entry-level classes, through required core classes and electives, to the senior biology capstone course, yet narrow enough to focus on specific skills. We also wanted to design our rubric to be flexible enough to incorporate the needs of multiple subdisciplines within our biology department. To address these needs, we created a modular rubric. Herein, we present our rubric as well as the process and timeline we used for its development, the intra-departmental communication required to ensure that the rubric would meet the needs of multiple instructors across several biology disciplines, and examples of how we have implemented the rubric in both upper and lower-level courses. Further, we offer guidelines on how the rubric can be adapted by practitioners to meet the need for specific teaching, learning, and assessment of research competencies in different classrooms.

15.2 Development of Research Across Curriculum Rubric (RAC-R)

15.2.1 Articulating Departmental and Student Needs

In our biology department at a small private university in the midwestern region of the United States, CUREs are utilized to teach students how real scientific experimentation is performed and communicated. In most required and some elective courses for biology majors, performance assessment includes submission of a formal piece of scientific writing. An informal survey of faculty was conducted to understand how student writing was currently evaluated in different biology courses. Faculty, in general, were dissatisfied with their writing assessment tools. Furthermore, students have noted that different professors utilize different criteria and vocabulary for assessing student writing. Hence, a team of faculty was selected to attend an ACE-Bio workshop to develop a rubric that could be used throughout the biology curriculum to assess student scientific writing.

Our process for the development of the Research Across the Curriculum Rubric (RAC-R) is described (Fig. 15.1). Briefly, the need for a new tool to assess student writing, and the recruitment of a team of faculty to lead the project occurred at a departmental faculty meeting. The team then met to brainstorm ideas and to articulate goals for developing a rubric that would be flexible enough to be used in multiple sub-disciplines within the biology curriculum and adaptable enough to be used for both introductory and advanced students.

Our instructional objectives for this flexible and adaptable rubric included the following:

1. To articulate to our students the skills required to become an "accomplished" scientific writer;
2. To develop consistent vocabulary among different biology disciplines so that students understand that skills taught and practiced in one biology course are transferrable to other biology courses
3. To map sufficient practice into the curriculum to afford student progression through different milestones
4. To scaffold instruction of scientific writing so students realize higher expectations over time
5. To provide evidence and consistent assessment of student achievement and performance in biology courses.

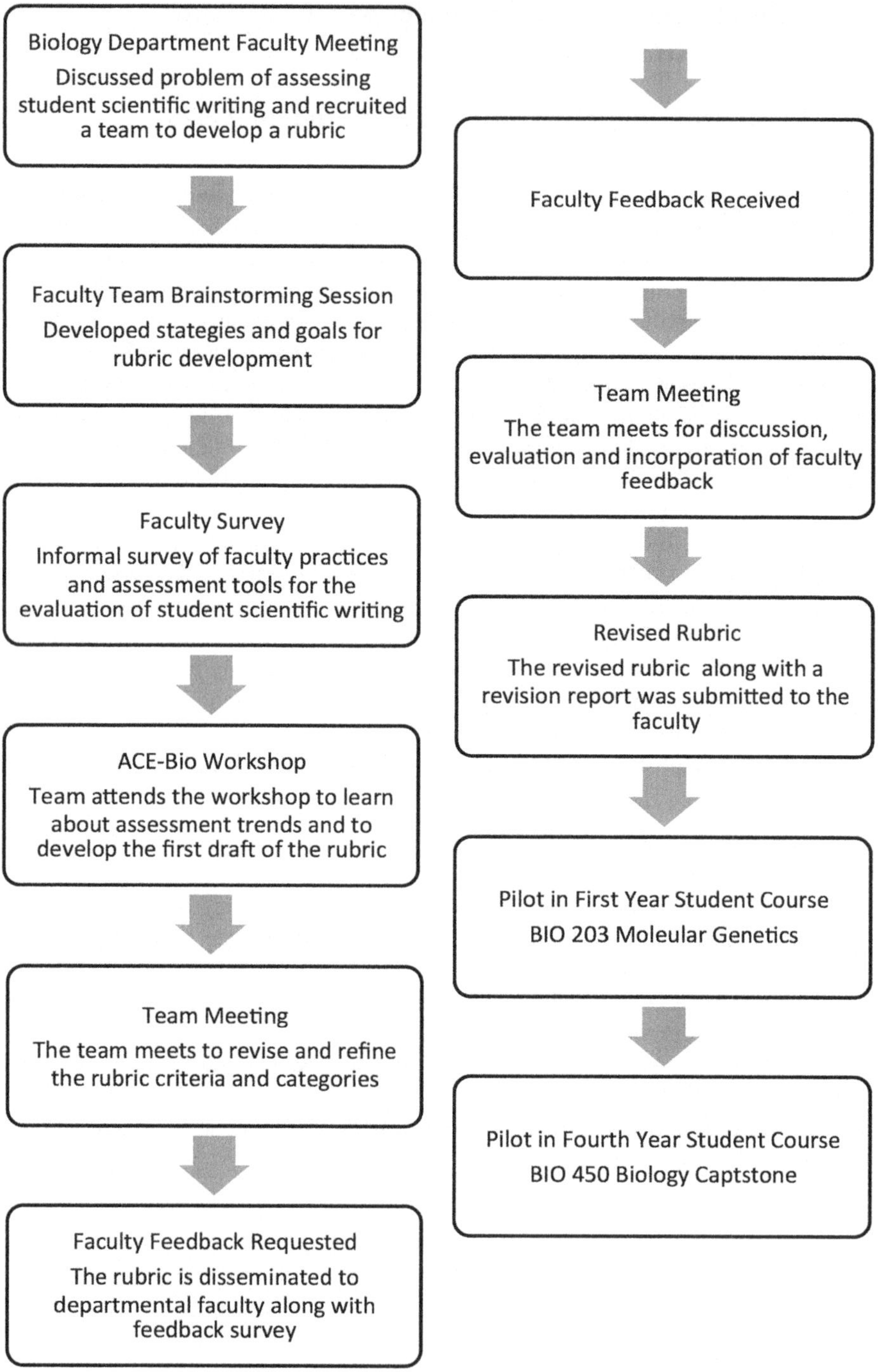

Fig. 15.1 Steps for the creation of a rubric to be used across the curriculum. The process for construction of the rubric required approximately 6 months. The rubric was piloted in a first-year course the following semester and in a fourth-year course the next academic year

15.2.2 Development of an Assessment at an ACE-Bio Workshop

A team of departmental faculty to develop the RAC-R at an ACE-Bio Network Assessment Workshop. The workshop began with an introduction and discussion of the elements of a quality assessment tool and continued with general suggestions for assessment tool development. To develop a tool for assessing student scientific writing, the faculty team brainstormed what should be expected of an accomplished, undergraduate, scientific writer and then employed a backward design approach to create the rubric while at the workshop. The faculty team formed pairs to develop the different sections of the rubric and utilized a variety of published rubrics and support materials like the Competencies of Biological Experimentation to serve as a foundation for developing rubric criteria and performance expectations (Dasgupta et al., 2014; Feldon et al., 2011; Mourots, 2018; Pelaez et al., 2017; Soneral & Wyse, 2015; Timmerman et al., 2011). Google Docs was used to enable simultaneous editing, and groups checked in periodically to discuss progress, address problems, and trade sections for review. This iterative process used by departmental faculty to design and critique the rubric was employed by the team to construct an initial draft during the workshop. After the workshop, the faculty team worked to clean the draft and complete missing criteria and performance objectives. We found that working face to face facilitated the completion of the rubric and built consensus among the team before the completed rubric was shared with the rest of the department.

The competencies identified by the ACE-Bio Network as The Basic Competencies of Biological Experimentation: Concept-Skill Statements (Pelaez et al., 2017; Chap. 1 in this volume) were used deliberately as an actual framework to guide the development, revision, and refinement of the rubric first created to assess scientific writing, RAC-R. Overall, RAC-R was intended to be used to assess many of the concepts and skills a competent biologist utilizes when doing experimentation in biology as identified by the ACE-Bio Network (Pelaez et al., 2017; Chap. 1 in this volume). The ratio of concepts within each segment of The Basic Competencies of Biological Experimentation that are assessed in the RAC-R is shown (Fig. 15.2). While RAC-R was created as a tool for assessment of scientific writing in undergraduate biology students, it utilizes 215/29 or 79% of concepts identified by the ACE-Bio Network as a basic foundation for biological experimentation (Pelaez et al., 2017; Chap. 1 in this volume). For example, within the Introduction section of RAC-R, accomplished student scientific writing is expected to encompass two concepts from the Identify competency, one concept from the Communicate competency, and four concepts from the Question competency (Table 15.1). Each of the incorporated ACE-Bio competences has been adapted to fit the criteria and rubric levels of performance. For example, the ACE-Bio concept "Identify A Gap in Current Knowledge", with the skills of "recognizing a gap in current scientific knowledge that can be addressed with experimentation" was utilized in the RAC-R rubric criteria, "Communicate How the Research Makes a Contribution to the Field." This rubric criterion was to

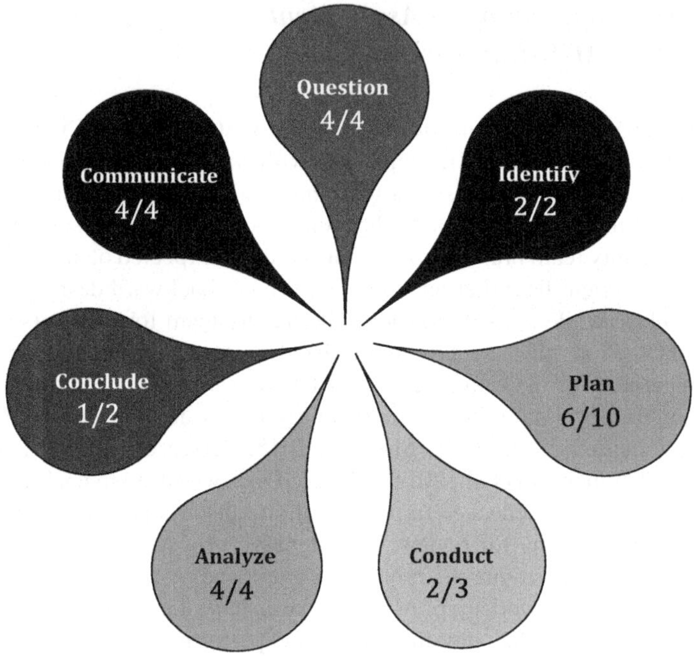

Fig. 15.2 A model of the seven areas a competent biologist utilizes when doing experimentation in biology identified by the ACE-Bio Network (Pelaez et al., 2017; Chap. 1 in this volume). Each competency is represented by a summary word on a uniquely colored segment of the model. The fraction within each segment identifies the ratio of concepts identified by ACE-Bio Network that are assessed by the final version of the RAC-R

Table 15.1 Alignment of the final version of RAC-R with the ACE-Bio Framework Concepts

RAC Rubric Section	ACE-BIO Framework Concepts (Skills)
General Criteria	*Communicate* (Scientific Communication), *Plan* (Iteration), *Conduct* (Data Documentation)
Introduction and Background	*Identify* (Relevant Background Knowledge, A Gap in Current Knowledge), *Communicate* (Synthesis and Reflection), *Question* (Observation, Research Questions, Models, Hypotheses)
Experimental Design and Methods	*Conduct* (Data Documentation), *Plan* (Experimental Design, Variables, Controls, Sampling, Variation)
Results	*Analyze* (Data analysis, Data Curation, Statistics, Data Summary), *Conduct* (Measurement, Data Documentation)
Discussion	*Communicate* (Synthesis and Reflection, Limitations, Representations), *Conclude* (Inferences and Conclusions)

The sections of RAC-R (in the left column) were designed to assess the concepts (in the right column) and associated skills (noted in parentheses) identified by Pelaez and colleagues (Pelaez et al., 2017; Chap. 1 in this volume)

be achieved at the accomplished level if it "thoroughly explores knowledge gaps in the field, logically leading reader to research question." A specific alignment of ACE-Bio competences to RAC-R criteria is available upon request.

15.2.3 *Feedback from Departmental Faculty and Revision*

Communication between the rubric development team and the departmental faculty who did not attend the ACE-Bio Network workshop was important for ensuring that the rubric met the needs of various department members. Thus, after completion of the rubric, the team disseminated it to the rest of the department faculty along with a questionnaire that solicited feedback on the RAC-R. Specifically, we asked the faculty to comment on the goals, versatility, evaluation criteria (rows), and performance expectations (columns) of the rubric (Table 15.2). Additionally, faculty were asked if any additional assessment criteria were needed, or if any of the assessment criteria needed further refining and/or focus. Finally, faculty were asked if they would use the rubric for evaluation of student reports of experimentation and if they had other comments, concerns, or suggestions for improvement of the rubric (Table 15.2).

The questionnaires were returned by 100% of the departmental faculty. Based on the feedback received, the rubric was revised. One of the largest changes to the rubric made in response to this feedback was the addition of a new "General" section. This section was added to assess the student's incorporation of instructor comments, the writing quality (including grammar, sentence structure, and active voice), and the collaborative efforts of students (as demonstrated by the paper, peer review, and instructor observations). One purpose of this section was to build a culture of revision so that students (novices) would work more similarly to scientific

Table 15.2 Biology Department Faculty Feedback Report

Please type your responses in this document and return to ______________by _________. Thanks!!!
Are the goals outlined for the rubric clear and complete? Why or why not?
Is the versatility of the rubric clear from the proposed uses? Why or why not?
Are there any rubric criteria (rows) or performance expectations (columns) that are vague or confusing? (Please list and provide comments about your confusion.)
Are there any missing criteria (rows)? If so, what are they and why do you think they are necessary for inclusion?
Do any of the criteria (rows) seem too broad? Do they need to be split? If so, what are they and why should the criteria (rows) be split?
Do you think that this rubric could be used to evaluate student work?
Do you have any other comments or suggestions for improvement?

The survey was distributed to the Biology Department faculty along with the RAC-R to solicit feedback concerning the flexibility, adaptability, and completeness of the document. Faculty members were asked to complete the evaluation form and return to the rubric development team within 2 weeks professionals (experts) by writing and extensively revising their journal articles in response to peer and instructor feedback. One faculty member supported this addition for the following reasons: "With writing, I spend a lot of time providing feedback and (1) it's exhausting when the grammar and spelling are terrible but they are technically hitting all the scientific points and (2) it's been nice to include points that incentivize students to make changes that show an effort to read and synthesize the critiques that I am providing." This section also allowed for us to document and assess student collaboration in performance and documentation of their research. Collaboration is a key element of the nature of science and having our rubric assess this communicates its importance to students.

Other changes made in response to faculty feedback included the addition of a criterion describing and justifying the choice of a model system/organism, and for describing the area under investigation. Based on faculty suggestions we also separated criteria for presentation of the results (graphing/figures) from criteria for statistical testing while adding mathematical treatment of data.

After careful consideration, the team chose not to include some of the faculties' suggestions for improvement. For example, a colleague suggested that we add a section on references to the rubric; however, we chose not to select a specific reference/citation style because different sub-disciplines in biology may require students to use different reference styles, and flexibility in reference/citation style might be especially important if students are preparing manuscripts for publication. In cases such as this, we reported our rationale to the faculty.

In some cases, faculty recommended removing a criterion of the rubric. For example, one colleague did not think that an assessment of explaining student assumptions or choice of statistical tests was necessary. Nevertheless, we chose to retain this criterion because it is consistent with expected trends in scientific publications and because it allows evaluation of a student's distinction between and understanding of statistical tests. We see this as adding to the versatility and flexibility of the rubric. If a faculty does not want to assess on a particular criterion, that criterion need not be included. Rather, faculty could select only the criteria (rows) that applied to a given course. In our curriculum, for instance, students take genetics before taking biostatistics; hence, the entire section on the statistical analysis of research would not be included in the rubric for the assessment of scientific writing in genetics.

Likewise, we chose to keep the criteria (row) called 'Summarizing and interpreting findings' in the results section. Some instructors said that they put that in their discussion section with the results only presented and not interpreted in the results. This format is more consistent with lab reports than the scientific literature. For these faculty then they might choose not to have a discussion section at all. For our rubric, the discussion section is the place to integrate what was found in the study presented with the literature looking at implications and significance, limitations, and future directions.

In general, the departmental faculty were pleased with the final rubric and envisioned themselves using it to assess students' scientific writing in their classes. One faculty member commented, "(I) think that (the rubric) will be versatile. I can envision using selected rows for assignments with the intent of skill development", while another added, "This is an AMAZING piece of documentation and I can see myself using this regularly."

15.3 Research Across Curriculum Rubric (RAC-R)

The RAC-R (Table 15.3) is a flexible rubric that can be utilized by instructors in classes in several ways (1) summative assessment of research projects, (2) formative growth of writing skill and communications, and (3) short practice assignments. The short practice assignments that work on a specific skill could use as little as a single rubric criterion (row). This allows students to practice a narrowly focused skill scaffolding deliberate practice to allow student progression through milestones to become 'accomplished' scientific writers.

For instructors to use the rubric, we recommend they first decide which rubric criteria (rows) are appropriate for their assignments to align with course/curriculum learning goals. Rubric criteria can be altered to focus on additional concepts and skills that a competent biologist utilizes when doing experimentation in biology as identified by the ACE-Bio Network (Pelaez et al., 2017; Chap. 1 in this volume). Then, instructors could identify what level of accomplishment (column) is expected. Finally, instructors can decide on the relative weight of individual criterion by assigning point or percentage values for each. Within our department, we want to articulate to students how specific courses/assignments build into the overall curriculum expectations.

15.4 Adapting Research Across Curriculum Rubric (RAC-R)

15.4.1 Pilot Utilization in Freshman Level Molecular Genetics Course

We piloted the RAC-R in both freshman and senior-level courses to exemplify the ways the rubric can be adapted to multiple levels of courses. For example, in our department, a required core class for biology majors is a 200 level Molecular Genetics lecture and lab which functions as a course-based undergraduate research experience (CURE). The students enrolled in this course are primarily freshmen and some sophomores who learn standard molecular biology procedures while cloning a student-selected gene of interest. Both formative and summative assessments in the CURE focus on writing a scientific journal-style article with particular attention to the development of a defined research question and hypothesis, an introduction based on a review of the relevant scientific literature, and an explanation of the methods utilized in the research. These learning goals, and thus RAC-R, align with

Table 15.3 The Research Across the Curriculum Rubric (RAC-R) for Assessing Undergraduate Scientific Writing

	Missing 1	Emerging 2	Developing 3	Proficient 4	Accomplished 5
General criteria **Show improvement of writing by incorporating instructor and peer comments for revisions**	The writer(s) has not obtained and/or incorporated any of the instructor's or peers' comments.	The writer(s) has incorporated some comments from the instructor or peers. The writer(s) revision does not significantly improve the content, focus, structure, and clarity, of an earlier draft. The mechanics (e.g. grammar, punctuation, spelling) appear to be the only focus of the revision.	The writer(s) has incorporated most comments from instructor or peer. The writer(s) has rewritten individual sentences (attempting to add clarity and variety). The writer provides some improved transitions that clarify the focus.	The writer(s) has addressed all of the instructor or peer comments. The writer (s) has deleted weak or irrelevant material. The writer improved transitions that clarify the focus.	The writer (s) has made revisions above and beyond what is suggested. The writer has substituted more relevant material for less relevant material. The writer improved transitions that clarify the focus.
Utilize appropriate grammar, sentence structure, and active voice to make paper clear and compelling	The reader has to make considerable effort to understand the underlying logic and flow of ideas. Grammatical and spelling errors make it difficult for the reader to interpret the text in places.	The reader has to make considerable effort to understand the underlying logic and flow of ideas. Sentences are mostly grammatical and/or only a few spelling errors are present but they do not hinder the reader.	The paper is clearly written for the most part. Words are well chosen with some minor exceptions. Sentences are mostly grammatical and/or only a few spelling errors are present but they do not hinder the reader.	The paper is clearly written. Words are chosen that precisely express the intended meaning and support reader comprehension. Sentences are grammatical and free from errors.	The paper is written in a compelling way that makes it interesting to read by using an active voice. Words are chosen that precisely express the intended meaning and support reader comprehension. Sentences are grammatical and free from errors with a variety of sentence structures.

Demonstrate effective collaboration during the research and writing phases of the project	The team did not collaborate or communicate well. A lack of respect and regard was frequently noted.	Writing indicates independent work with little effort to synthesize and increase flow. The team did not collaborate or communicate well. Some members would work independently, without regard to objectives or priorities.	Writing indicates some effort was made as a team to increase flow. The team worked well together most of the time, with only a few occurrences of communication breakdown or failure to collaborate when appropriate. Members were mostly respectful of each other.	The paper flows very well together. The team worked well together to achieve objectives. Each member contributed in a valuable way to the project. All data sources indicated a high level of mutual respect and collaboration.	The paper flows very well together and there is a sophisticated level of synthesis between sections. The team worked well together to achieve objectives. Each member contributed in a valuable way to the project. All data sources indicated a high level of mutual respect and collaboration.
Introduction section criteria					
Describe the research topic and give relevant background from the scientific literature	Too few primary literature references are used and those utilized are not relevant (i.e. largely misconstrued or irrelevant). The topic introduction provides no logical progression.	Too few primary literature references are used or those utilized are not relevant (i.e. largely misconstrued or irrelevant). The topic introduction is unclear.	The minimum number of primary references is included and most are relevant and interpreted accurately. The topic introduction is coherent and logical but may require the reader to make connections not presented.	Many primary literature references are included and they are all relevant and interpreted accurately. The topic is introduced logically, but minor details are irrelevant or unclear.	Primary literature references are extensive and relevant and interpreted accurately. The topic is introduced logically using relevant research specific to the project.

(continued)

Table 15.3 (continued)

	Missing 1	Emerging 2	Developing 3	Proficient 4	Accomplished 5
Define motivation and significance of the research project	The significance of the study is unclear.	The significance of the study can be determined but is not obvious.	The significance of the study is clear but is not necessarily well developed. Significance is supported by primary literature but could be improved or strengthened.	The significance of the study is clear and fairly well developed. Significance is supported by primary literature.	The significance of the study is clear and well developed. Significance is well supported by primary literature.
Communicate how research contributes to the field	Knowledge gaps are not well identified or unclear.	Knowledge gaps are superficially identified but not narrowly defined.	Knowledge gaps are identified and logically introduced, but the reader may not be able to deduce a specific research question from the information provided.	Knowledge gaps are identified and logically introduced, leading the reader to the research question.	Thoroughly explores knowledge gaps in the field, logically leading the reader to the research question.
Define a research question that communicates a gap in the current knowledge	The question is presented but is too broad to be investigated.	The question is answerable but not narrowly focused enough to be practical.	The question is answerable, focused, and practical.	The question is answerable, focused, practical, relevant, and important.	The question is answerable, focused, practical, relevant, important, and interesting/novel.

Describe and justify the model system/organism or study area under investigation	Does not address the nature of the model system/organism or the study area under investigation.	Incompletely addresses the nature of the model system/organism or the study area under investigation.	Addresses and superficially justifies the selection of the model system/ organism or the study area under investigation.	Addresses and justifies the selection of the model system/ organism or the study area under investigation using references not specific to hypotheses and precedents of previous studies.	Addresses and justifies the selection of the model system/organism or the study area under investigation using references specific to hypotheses and precedents of previous studies.
Propose narrowly defined and testable hypotheses that include predictions	Hypotheses are not narrowly defined enough to be testable. Predictions are absent.	Hypotheses are narrowly defined but do not address the research question. Predictions are either unreasonable or illogical.	Hypotheses include both independent and dependent (outcome) variables. Predictions are reasonable, but overgeneralized.	Hypotheses are based on theory and include both independent and dependent (outcome) variables but may not be written in the proper format. Predictions are reasonable and logical.	Hypotheses are based on theory, and are included for all dependent variables, and are written in the proper format. Predictions are reasonable, logical, and insightful.
Experimental design section criteria					
Propose experiments that address the research question	The methods are not appropriate to address the research question.	The methods are appropriate but may be incomplete or insufficient to fully address the research question.	The methods are appropriate and complete but insufficiently address the research question.	The methods are appropriate measures to examine hypotheses and address the research question.	The methods are optimized to be as direct as possible to test hypotheses and address the research question.

(continued)

Table 15.3 (continued)

	Missing 1	Emerging 2	Developing 3	Proficient 4	Accomplished 5
Identify and describe relevant variables	Identifies variables but does not distinguish between dependent and independent. Attempted to describe standardized variables but methods to control were inappropriate.	Identifies both dependent and independent variables but did not identify range (e.g. dose, treatment conditions). Describes some but not all standardized variables and methods to control were appropriate.	Identifies both dependent and independent variables and the range (e.g. dose, treatment conditions). Describes most but not all standardized variables and the methods to control are appropriate.	Identifies both dependent and independent variables and the range (e.g. dose, treatment conditions). Includes literature citations for the independent variable range. Describes most of the standardized variables and methods to control are appropriate.	Identifies both dependent and independent variables and the range (e.g. dose, treatment conditions). Includes literature citations and justification of the independent variable range. Describes standardized variables and methods to control are appropriate and justified.
Describe appropriate comparison Group(s) (i.e. Positive control group, Negative control group, Concurrent comparison group, historic comparison group)	An attempt was made to include a comparison group, but the group chosen did not provide an appropriate baseline.	The comparison group provides an appropriate baseline measurement/ indication of the behavior/ conditions being analyzed, however, the independent variable was not isolated.	The comparison group provides an appropriate baseline measurement/ indication of the behavior/ conditions being analyzed, and the independent variable was isolated. Terminology relating to controls may be absent or not appropriate.	The comparison group provides an appropriate baseline measurement/ indication of the behavior/ conditions being analyzed, and the independent variable was isolated. Terminology relating to controls is used appropriately (positive, negative, concurrent, historic, etc.).	The comparison group is thoroughly explained/ justified and provides an appropriate baseline measurement/ indication of the behavior/ conditions being analyzed, and the independent variable was isolated. Terminology relating to controls is used appropriately (positive, negative, concurrent, historic, etc.).

Include a discussion of sampling method with appropriate levels of replicates, and replication	Fails to describe sampling procedures, replicates, or replication.	Explains how to sample subjects randomly for control and treatment groups to reduce the effect of unanticipated variables.	Explains how to sample subjects randomly for control and treatment groups to reduce the effect of unanticipated variables. Differentiates between measurement variability and system variability (natural variation or heterogeneous populations).	Explains how to sample subjects randomly for control and treatment groups to reduce the effect of unanticipated variables. Designs the sampling strategy to expose and account for measurement variability and system variability (natural variation or heterogeneous populations) Includes a discussion of replicates and repeatability.	Explains how to sample subjects randomly for control and treatment groups to reduce the effect of unanticipated variables. Designs the sampling strategy to expose and account for measurement variability and system variability (natural variation or heterogeneous populations) Determines the amount of replication or repeatability needed to quantify variation.

(continued)

Table 15.3 (continued)

	Missing 1	Emerging 2	Developing 3	Proficient 4	Accomplished 5
Explain and justify methods	The explanations of methods employed have numerous and/or substantial flaws in scientific interpretation. The methods are not based on cited peer-reviewed protocols and procedures.	The explanations of methods employed have 1–2 minor flaws in scientific interpretation or are underdeveloped. The methods are minimally referenced, poorly cited, or inappropriate. Methods lack justification to address the hypothesis.	The explanations of methods are scientifically accurate. The methods and protocols are referenced. Justification of the methods employed are included but lack precision and insight.	The explanations of methods are scientifically accurate and concise. The methods and protocols are referenced and appropriate for both investigation and model system/ organism or study area. Justification of the methods employed are complete.	The explanations of methods are scientifically accurate and concise. The methods and protocols are extensively referenced and appropriate for both investigation and model system/organism or study area. Justification of the methods employed are complete and insightful.

Include detailed methods to facilitate external replication	The methods are missing the majority of essential elements and/or inadequately described and therefore the experiment cannot be replicated.	The methods are missing several details, the audience can get the idea of the experiments performed, but too little information is provided to allow the audience to evaluate the materials and methods.	The methods are well described but not concise and missing a few minor details (vendor, size of the experimental container, etc.) and/or provided in step-by-step format with subheadings. Can be easily replicated by other investigators.	The methods are described completely and concisely. Subheadings are used to separate components of experimentation, The sources of materials are identified and referenced, The methods are cited and included in enough detail to be evaluated by the audience and replicated by other investigators while minimizing extraneous details.	The methods are described thoroughly and concisely, effective and descriptive. Subheadings are used to separate components of experimentation. The sources of materials are identified and referenced. The methods are cited and included in enough detail to be evaluated by the audience and replicated by other investigators while minimizing extraneous details.

(continued)

Table 15.3 (continued)

	Missing 1	Emerging 2	Developing 3	Proficient 4	Accomplished 5
Results section criteria					
Identify and organize relevant results	Raw data presented. The presentation of results is unorganized and difficult to follow.	Data are analyzed mildly, but without relevant findings presented. The major project hypothesis is stated however individual results are not guided or explained by smaller hypotheses.	Data are analyzed and some relevant findings are presented. The major project hypothesis is stated and some individual results are guided or explained by smaller hypotheses. Each results section contains a descriptive header.	Data are thoroughly analyzed and some relevant findings are presented. The major project hypothesis is stated and most individual results are guided or explained by smaller hypotheses. Each results section contains a descriptive header with a declarative statement of the outcome.	Data are thoroughly analyzed and all relevant findings are presented. The major project hypothesis is stated and all individual results are guided or explained by smaller hypotheses. Each results section contains a descriptive header with a declarative statement of the outcome and a rationale.

Construct appropriate ways to organize and/or present results	Figures/graphs required to convey results or summarize (mathematical treatment, e.g. normalization) data are incomplete, not appropriate. Contain numerous inaccurate, incomplete, or unclear title, legend, labels, and/or units.	Figures/graphs required to convey results or summarize (mathematical treatment, e.g. normalization) data are complete but may not be the most appropriate. Some title, legend, labels or units for graphs/figures missing or inaccurate.	Figures/graphs required to convey results or summarize (mathematical treatment, e.g. normalization) data are complete and appropriate. A few (2–3) title, legend, labels or units for graphs/figures missing or inaccurate. Figure titles and legends are accurate and logical.	Figures/graphs required to convey results or summarize (mathematical treatment, e.g. normalization) data are complete and appropriate. Almost all components accurately, appropriately, and clearly labeled. Figure titles and legends are accurate and logical.	Figures/graphs required to convey Results or summarize (mathematical treatment, e.g. normalization) data are complete and appropriate. Figures, graphs, representations (e.g. bar graphs, histograms, scatterplots, etc.) were carefully considered and were used appropriately and consistently with the type of data/ information collected. All components accurately, appropriately, and clearly labeled. Figure titles and legends are accurate, concise, and logical.

(continued)

Table 15.3 (continued)

	Missing 1	Emerging 2	Developing 3	Proficient 4	Accomplished 5
Choose and conduct statistical tests that are appropriate for the type/nature of data	Inferential and/or descriptive statistics are incomplete for the experimental design and data collection.	Inferential and/or descriptive statistics are inappropriate for the experimental design (e.g., repeated measures, number of comparison groups, etc.) and data collected.	Inferential and/or descriptive statistics complete and appropriate for the experimental design (e.g., repeated measures, number of comparison groups, etc.) and data collected. Assumptions of statistical tests have not been addressed.	Inferential and descriptive statistics complete and appropriate for the experimental design (e.g., repeated measures, number of comparison groups, etc.) and data collected. Assumptions of statistical tests have been addressed. Statistical analyses effectively summarize trends and major findings.	Inferential and descriptive statistics and any data processing complete and appropriate for the experimental design (e.g., repeated measures, number of comparison groups, etc.), data collected. Assumptions of statistical tests have been addressed. Statistical analyses effectively summarize trends and major findings.
Summarize and interpret findings	The summary is unclear and illogical. Patterns and interpretations in data are not clearly identified.	The summary of the data is unclear or incomplete. Patterns and interpretations of data unclear or incomplete.	A reasonable summary of data (may be lacking in clarity, conciseness, or flow) is presented. Patterns in data or interpretations are not clearly identified.	Data summarized in a clear, logical manner. Patterns are identified and interpreted.	Data summarized in a clear, concise, logical manner. Patterns identified and findings were interpreted concisely and insightfully.

Discussion section criteria					
Discuss the implications and significance of the research	Significance and implications are not discussed.	Mentions of significance and implications are vague or inappropriate.	Evaluates, analyzes, and explains the significance and implications of the research.	Evaluates, analyzes, and explains the significance and implications of the research. Compares results and findings with previously published scientific work.	Evaluates, analyzes, and explains the significance and implications of the research. Compares results and findings with previously published scientific work. Evaluates the scope of influence for implications (research or practical).
Discuss the limitations of the study	Limitations are not discussed.	Limitations are discussed trivially (e.g. 'human error' is the major limitation invoked).	Discusses the limitations and uncertainty of methods, data analysis (sources of error, inaccurate measurement, sample bias), and/or statistical significance vs. biological relevance. Describes an alternative explanation of the results.	Discusses the limitations and uncertainty of methods, data analysis (sources of error, inaccurate measurement, sample bias), and/or statistical significance vs. biological relevance. Describes and analyzes alternative explanations for the results.	Discusses the limitations and uncertainty of methods, data analysis (sources of error, inaccurate measurement, sample bias), and/or statistical significance vs. biological relevance. Describes and analyzes alternative explanations for the results. Considering alternatives clearly describes a persuasive explanation

(continued)

Table 15.3 (continued)

	Missing 1	Emerging 2	Developing 3	Proficient 4	Accomplished 5
Propose and justify future directions	Does not describe future directions.	Proposes follow up experiments that do not necessarily flow from predicted or actual results. Future directions are vague or implausible.	Proposes follow up experiments based on predicted or actual results. Future directions are useful but indicate an incomplete knowledge of the field. A few citations are included but not necessarily the most relevant.	Proposes follow up experiments based on predicted or actual results. Articulates unanswered question(s). Justifies follow up questions and experiments with relevant citations.	Proposes follow up experiments based on predicted or actual results. Articulates significant and important unanswered questions and how those questions can be answered. Justifies follow up questions and experiments with multiple relevant citations.
Propose a model (an abstraction used to explain and predict patterns or properties) to show conclusions, future directions, and/or observations from the literatur	No model is included.	An explanatory or predictive model was attempted but does not accurately represent research findings or future directions.	Develops an explanatory or predictive model to summarize research findings and/or future directions.	Develops an explanatory or predictive model to summarize research findings and/or future directions. Partial representation of the literature that aligns model with experimental context, question, and/or hypothesis.	Develops an explanatory or predictive model to summarize research findings and/or future directions. Includes literature that aligns model with experimental context, question, and/or hypothesis. Notes areas of conflict between literature and findings.

The final version of the rubric that incorporated edits and recommendations that were made from the biology department faculty members the following ACE-Bio Framework Concepts and Skills: *Identify* (Relevant Background Knowledge, A Gap in Current Knowledge), *Communicate* (Synthesis and Reflection), *Question* (Observation, Research Questions, Models, Hypotheses) (Table 15.1). In this course, less emphasis is placed on the analysis of results. Hence, RAC-R was modified for this entry-level course (Table 15.4).

Students in this course were expected to reach column three (a "developing" level of performance) for criteria in the "Introduction", "Methods", and "Results" sections of the rubric while reaching column four (a "proficient" level of performance) in the "General" section. The performance levels that are higher than the levels expected of a class can be shaded from the students' view. For this particular class, the unused levels of the rubric were replaced with detailed lists of the elements the instructors required for inclusion in this assignment as scaffolding for first-year students who likely had little to no experience with formal scientific writing. Further, in this modified rubric (Table 15.4) we chose to assess only three of the six criteria that were in the RAC-R Introduction section. Additional versatility can be built into the rubric by determining the relative worth of each section in the rubric to the overall score. For example, while we expected a level four proficiency on the "General" section criteria, criteria from this section only accounted for 20% of the grade. Likewise, since we did not focus on analysis and discussion or results, points for this section only accounted for 20% of the grade. The Introduction and Methods sections each accounted for 40% of the total grade in this assignment.

15.4.2 Pilot Utilization in Senior Capstone Course

An assignment in a 400 level course required students to find and interpret scientific literature, problem solve, collaborate, and communicate a narrowly defined contemporary problem related to the field of biology and to propose possible solutions to that problem. For this writing assignment, we utilized the General section of RAC-R, and modified portions of the Introduction (not shown) and Discussion (Table 15.5) sections by slightly changing the wording of some criteria to better align with the expectations for the assignment. The criteria chosen for this assignment aligned with the following ACE-Bio Framework Concepts and Skills: *Communicate* (Synthesis and Reflection, Limitations, Representations), *Conclude* (Inferences and Conclusions) (Table 15.1). Since this was a senior-level course, students were expected to reach column five (an "accomplished" level of performance) for each criterion. For this assignment, we adjusted the weight of the sections such that the General and Discussion criteria each accounted for 40% of the grade while the Introduction only accounted for 20% of the grade. The use of the RAC-R in these two classes exemplified the versatility and flexibility of this assessment tool.

Table 15.4 Modified RAC-R for a 1st-year course

Introduction section criteria	Missing	Emerging	Developing	What elements are necessary for full credit?
Describe the research topic and give relevant background from the scientific literature	7 pts Too few primary literature references are used and those utilized are not relevant (i.e. largely misconstrued or irrelevant). The topic introduction provides no logical progression.	15 pts Too few primary literature references are used or those utilized are not relevant (i.e. largely misconstrued or irrelevant). The topic introduction is unclear.	25 pts The minimum number of primary references is included and most are relevant and interpreted accurately. The topic introduction is coherent and logical but may require the reader to make connections not presented.	Explain published material on your protein including: Yeastmine- published papers Yeastmine- phenotypes observed At least 4 primary literature references Domain map figure(s) Contains amino acid numbers for each domain and the end of the protein, and domain name Contains a figure legend that specifies the source of sequence and program that was used to calculate domain map Discusses domain(s) function and the relevance of this information (what it tells you)
Define motivation and significance of the research project	4 pts The significance of the study is unclear.	6 pts The significance of the study can be determined but is not obvious.	10 pts The significance of the study is clear but is not necessarily well developed. Significance is supported by primary literature but could be improved or strengthened.	Explain published material on the human homolog using papers from Pubmed references including information on human disease relevance and prevalence Describes the overall objectives of the work. Alignment figure(s) with the human homolog Specify which sequence was gene and the human homolog Specify the gene name for the human homolog Explain what alignment tells us including %coverage, %identity, and %similarity Contains figure legend that specifies the source of sequence and program that was used to obtain alignment?
Describe and justify the model system/organism or study area under investigation	0 pts Does not address the nature of the model system/organism or the study area under investigation.	3 pts Incompletely addresses the nature of the model system/organism or the study area under investigation.	5 pts Addresses and superficially justifies the selection of the model system/organism or the study area under investigation.	Explain yeast as a model system and why it is a good model for studying your gene?

The RAC-R was modified for use in a first-year biology lab. The Introduction section was limited to three levels of performance, and a detailed list of items that should be included in this section was included to scaffold expectations for first-year students. The point values of each level of performance for each criterion were set to reflect instructor expectations and to the relative weight of each section to the overall assignment value

15.4.3 Proposed Adaptations of RAC-R

We envision that RAC-R can be adapted to meet the needs of a particular class or curriculum. As demonstrated above, instructors can elect to use only certain sections of the rubric or select specific criteria (rows) within each section. Further, instructors or departments can select what level of accomplishment is expected for beginning, intermediate, or advanced students, while maintaining consistency across various sub-disciplines within biology. Also, scaffolding support for students can be built by providing additional assignment-specific information in particular sections as described in the modified rubric used in a first-year course (Table 15.4). The flexibility of the rubric can be facilitated with changes to keywords within criteria to better align with the intent of an assignment while maintaining the original goal of the criterion as described in the modified rubric used in the fourth-year course (Table 15.5). One faculty member from our department commented that she would like to use the rubric for assessing student grant proposals; however, she noted that she would have to add certain elements to the rubric, "like the inclusion of specific aims and probably expanding out the alignment of the experiments with the research question". We believe that an addition such as this is an exciting possibility. Another faculty member offered, "Having students turn in a self-assessment using this rubric along with their paper might be a good way to get them to think about their writing."

While students should aspire to become accomplished at all criteria, certain assignments could focus on developing particular skills important in scientific writing such as data analysis, summarizing published studies, or developing hypotheses. These writing skills could be mapped to specific courses within a curriculum. Formative assignments might focus on just one criterion at a time, with multiple rounds of instructor feedback and student revision to promote mastery of a skill. Likewise, multiple assignments covering the same criteria could facilitate the deliberate practice of particular skills within a course or across a curriculum.

15.5 Discussion

According to the AAAS document *Vision and Change in Undergraduate Biology Education*, effective communication is an essential skill of scientists, and formal methods of written communication should be a standard part of undergraduate biology education (AAAS, 2011). In our biology department, we have developed a method of research across the curriculum to teach biology students how real scientific experimentation is conducted. In many courses, an assessment incorporates submission of a formal piece of scientific writing; however, a survey of departmental faculty found that multiple different assessment tools were being used and that satisfaction with assessment tools was lacking. Thus, we recognized the need for a consistent rubric that could be used across the curriculum of our biology department.

Table 15.5 Modified RAC-R for a 4th-year course

Discussion section criteria	Missing	Emerging	Developing	Proficient	Accomplished
Discuss the significance and implications (affordances) of the solution using appropriate citations	0 pts Affordances (significance and implications) are not discussed.	6 pts Mentions of affordances (significance and implications) are vague or inappropriate.	9 pts Evaluates, analyzes, and explains the affordances (significance and implications) of the solution.	12 pts Evaluates, analyzes, and explains the affordances (significance and implications) of the solution. Compares and justifies solutions with appropriate citations from both the scientific literature as well as other relevant sources.	15 pts Evaluates, analyzes, and explains the affordances (significance and implications) of the solution. Compares and justifies solutions with appropriate citations from both the scientific literature as well as other relevant sources. Evaluates the scope of impact of the solution.
Discuss the barriers and limitations of the solution	0 pts Barriers and limitations are not discussed.	6 pts Barriers and limitations are discussed trivially (e.g. hard to pass laws).	9 pts Discusses a few barriers and limitations Does not describe how barriers could be addressed.	12 pts Discusses a comprehensive set of barriers and limitations Describes and analyzes barriers and limitations by starting to assess how they could be addressed and their impact.	15 pts Discusses a comprehensive set of barriers and limitations Describes and analyzes barriers and limitations by assessing how they could be addressed and their impact. A persuasive explanation is included about the solution despite any barriers and limitations

Propose a Model (an abstraction used to explain and predict patterns or properties) to show conclusions, future directions, and/or observations from the literature	0 pts No model is included.	4 pts An explanatory or predictive model was attempted but does not accurately represent research findings or future directions.	6 pts Develops an explanatory or predictive model to summarize research findings and/or future directions.	8 pts Develops an explanatory or predictive model to summarize research findings and/or future directions. Partial representation of the literature that aligns model with problem context, solution, and/or affordances/ barriers.	10 pts Develops an explanatory or predictive model to summarize research findings and/or future direction. Includes literature that aligns model with problem context, solution, and/or affordances/ barriers. Notes areas of conflict between literature and findings.

The RAC-R was modified for use on an essay assigned to students in a fourth-year capstone class. For this assignment, three of the four criteria in the Discussion section were selected from RAC-R, and students were expected to reach and "Accomplished" level of performance for each. The wording of the criteria was modified slightly to better align with the particular assignment

Herein, we present the modular rubric that we developed (Table 15.3), as well as the process and timeline we used for its development (Fig. 15.1), the intra-departmental communication required to ensure that the rubric would meet the needs of multiple instructors across several biology disciplines (Table 15.2), and the alignment of the rubric criteria with the ACE-Bio Network framework concepts and skills required by a competent biologist (Table 15.1). We include examples of how we have adapted the rubric for use in both upper (Table 15.5) and lower-level courses (Table 15.4) to meet the need for specific teaching, learning, and assessment of research competencies in different classrooms. We believe that practitioners could use RAC-R "as is" or could modify criteria (rows) or levels of accomplishment (columns), or add or remove entire sections or criteria as needed to meet instructional and assessment needs. Alternatively, we encourage faculty departments to construct their own rubrics by following the steps we used for rubric development (Fig. 15.1) or the following checklist.

Recommendations for Instructors:

- Identify a team of faculty members to work on rubric development
- Survey all faculty members to identify the practices and assessment tools currently used for the evaluation of student scientific writing
- Develop a draft of the rubric
- Disseminate the rubric to the department faculty along with feedback survey
- Discuss, evaluate and incorporate the faculty feedback into the rubric
- Disseminate the revised rubric to the department faculty
- Pilot the rubric is selected courses
- Evaluate the reliability of the rubric

We caution that inter-rater reliability of the RAC-R as an assessment tool has not been rigorously tested for cumulative assessment or individual criterion scores. We recommend that practitioners test the reliability of this rubric using exemplar student papers before its adoption when multiple assessors are employed (Cockett & Jackson, 2018). However, as noted in a previously reported meta-analysis, inter-rater reliability varies greatly, even among professional peer reviewers (Timmerman et al., 2011). The reliability of another rubric to assess student writing was shown to increase with the number of criteria assessed (Timmerman et al., 2011). Although picking and choosing which criteria will be assessed for a given assignment adds to the flexibility of the rubric, it may lessen the opportunity for students to practice and improve upon certain skills. We stress that the alignment of anticipated learning outcomes, assignment details, rubric criteria, and verified learning outcomes must be established by practitioners.

We believe that consistent use of RAC-R across the curriculum maintains consistent vocabulary among different biology disciplines so that students understand that skills taught and practiced in one biology course are transferrable to other biology courses. Additionally, consistent use of the rubric enables the department to scaffold instruction of scientific writing so students realize higher expectations over time, and to map sufficient practice into the curriculum to afford student progression through different milestones. RAC-R potentially benefits students by articulating the skills required to become an "accomplished" scientific writer and benefits the biology department by providing evidence and consistent assessment of student achievement and performance in biology courses. We envision that RAC-R will be useful for the identification of gaps and improperly sequenced expectations in the curriculum. Accordingly, it may be a useful tool for both internal and external evaluation of the biology program. Next, we plan to assess whether student's scientific writing skills improve with consistent use of the RAC-R.

In summary, RAC-R was developed to meet the needs of multiple instructors across several sub-disciplines within our biology department. This rubric was developed using the ACE-Bio Network Basic Competencies of Biological Experimentation (Pelaez et al., 2017; Chap. 1 in this volume) as a framework and utilizes a majority (79%) of the basic concepts and skills identified by Pelaez and colleagues within its criteria of assessment. We believe that it is flexible and adaptable to individual needs, yet universal enough to provide consistent guidelines for improvement of undergraduate students' scientific writing skills. We present the process, timeline, and communication tools we utilized in the rubric development, and provide examples of how the rubric was modified for use in both upper- and lower-level courses. We predict that it will be a useful tool for student learning, instructor evaluation of student work, and departmental self-assessment.

Acknowledgments We gratefully acknowledge the expert contributions of the other faculty members of the rubric development team, Dr. Kimberly Vogt and Dr. Joyce Horton. We thank members of the Biology department, Dr. Colleen Doci, Dr. David Benson, Professor G. Steven Doughtery, and Dr. Azeem Ahmad, and the Chemical and Physical Sciences department, Professor Tanja Greene, who provided feedback and suggestions for improvement and adaptation of RAC-R. Finally, we acknowledge, Dr. Trisha Staab, Dr. Colleen Doci, and Dr. Joyce Horton who piloted the RAC-R rubric in their courses.

References

American Association for the Advancement of Science (AAAS). (2011). *Vision and change in undergraduate biology education: A call to action.* Washington, DC. Retrieved October 18, 2021 from https://live-visionandchange.pantheonsite.io/wp-content/uploads/2013/11/aaas-VISchange-web1113.pdf

Auchincloss, L. C., Laursen, S. L., Branchaw, J. L., Eagan, K., Graham, M., Hanauer, D. I., Lawrie, G., McLinn, C. M., Pelaez, N., Rowland, S., Towns, M., Trautmann, N. M., Varma-Nelson, P., Weston, T. J., & Dolan, E. L. (2014). Assessment of course-based undergraduate research experiences: A meeting report. *CBE Life Sciences Education, 13*(1), 29–40. https://doi.org/10.1187/cbe.14-01-0004

Bangera, G., & Brownell, S. E. (2014). Course-based undergraduate research experiences can make scientific research more inclusive. *CBE Life Sciences Education, 13*(4), 602–606. https://doi.org/10.1187/cbe.14-06-0099

Brownell, S. E., & Kloser, M. J. (2015). Toward a conceptual framework for measuring the effectiveness of course-based undergraduate research experiences in undergraduate biology. *Studies in Higher Education, 40*(3), 525–544. https://doi.org/10.1080/03075079.2015.1004234

Brownell, S. E., Kloser, M. J., Fukami, T., & Shavelson, R. J. (2013). Context matters: Volunteer bias, small sample size, and the value of comparison groups in the assessment of research-based undergraduate introductory biology lab courses. *Journal of Microbiology & Biology Education, 14*(2), 176–182. https://doi.org/10.1128/jmbe.v14i2.609

Brownell, S. E., Wenderoth, M. P., Theobald, R., Okoroafor, N., Koval, M., Freeman, S., Walcher-Chevillet, C. L., & Crowe, A. J. (2014). How students think about experimental design: Novel conceptions revealed by in-class activities. *Bioscience, 64*(2), 125–127. https://doi.org/10.1093/biosci/bit016

Brownell, S. E., Hekmat-Scafe, D. S., Singla, V., Chandler Seawell, P., Conklin Imam, J. F., Eddy, S. L., Stearns, T., & Cyert, M. S. (2015). A high-enrollment course-based undergraduate research experience improves student conceptions of scientific thinking and ability to interpret data. *CBE Life Sciences Education, 14*(2), ar21. https://doi.org/10.1187/cbe.14-05-0092

Cockett, A., & Jackson, C. (2018). The use of assessment rubrics to enhance feedback in higher education: An integrative literature review. *Nurse Education Today, 69*, 8–13. https://doi.org/10.1016/j.nedt.2018.06.022

Corwin, L. A., Graham, M. J., & Dolan, E. L. (2015). Modeling course-based undergraduate research experiences: An agenda for future research and evaluation. *CBE Life Sciences Education, 14*(1), es1. https://doi.org/10.1187/cbe.14-10-0167

Dasgupta, A. P., Anderson, T. R., & Pelaez, N. (2014). Development and validation of a rubric for diagnosing students' experimental design knowledge and difficulties. *CBE Life Sciences Education, 13*(2), 265–284. https://doi.org/10.1187/cbe.13-09-0192

Feldon, D. F., Peugh, J., Timmerman, B. E., Maher, M. A., Hurst, M., Strickland, D., Gilmore, J. A., & Stiegelmeyer, C. (2011). Graduate students' teaching experiences improve their methodological research skills. *Science (New York, N.Y.), 333*(6045), 1037–1039. https://doi.org/10.1126/science.1204109

Harrison, M., Dunbar, D., Ratmansky, L., Boyd, K., & Lopatto, D. (2011). Classroom-based science research at the introductory level: Changes in career choices and attitude. *CBE Life Sciences Education, 10*(3), 279–286. https://doi.org/10.1187/cbe.10-12-0151

Irby, S. M., Pelaez, N. J., & Anderson, T. R. (2018). How to identify the research abilities instructors anticipate students will develop in a biochemistry Course-Based Undergraduate Research Experience (CURE). *CBE Life Sciences Education, 17*(2), es4. https://doi.org/10.1187/cbe.17-12-0250

Irby, S. M., Pelaez, N. J., & Anderson, T. R. (2020). Student perceptions of their gains in course-based undergraduate research abilities identified as the anticipated learning outcomes for a biochemistry CURE. *Journal of Chemical Education, 97*(1), 56–65. https://doi.org/10.1021/acs.jchemed.9b00440

Knutson, K., Smith, J., Wallert, M. A., & Provost, J. J. (2010). Bringing the excitement and motivation of research to students; Using inquiry and research-based learning in a year-long biochemistry laboratory: Part I-guided inquiry-purification and characterization of a fusion protein: Histidine tag, malate dehydrogenase, and green fluorescent protein. *Biochemistry and Molecular Biology Education, 38*(5), 317–323. https://doi.org/10.1002/bmb.20400

Kuhn, D. (1993). Science as argument: Implications for teaching and learning scientific thinking. *Science Education, 77*(3), 319–337. https://doi.org/10.1002/sce.3730770306

Lopatto, D. (2004). Survey of Undergraduate Research Experiences (SURE): First findings. *CBE Life Sciences Education, 3*(4), 270–277. https://doi.org/10.1187/cbe.04-07-0045

Mourots, N. J. (2018). *Assessment tools: Design of experiment rubric*. Retrieved September 10, 2018, from http://www.sjsu.edu/people/nikos.mourtos/assessment/

Pelaez, N., Anderson, T. R., Gardner, S. M., Yin, Y., Abraham, J. K., Bartlett, E., Gormally, C., Hill, J. P., Hoover, M., Hurney, C., Long, T., Newman, D. L., Sirum, K., & Stevens, M. (2017). The basic competencies of biological experimentation: Concept-skill statements. *PIBERG Instructional Innovation Materials. Paper, 4*, 2017. http://docs.lib.purdue.edu/pibergiim/4

Resendes, K. K. (2015). Using HeLa cell stress response to introduce first year students to the scientific method, laboratory techniques, primary literature, and scientific writing. *Biochemistry and Molecular Biology Education, 43*(2), 110–120. https://doi.org/10.1002/bmb.20852

Reynolds, J., Smith, R., Moskovitz, C., & Sayle, A. (2009). BioTAP: A systematic approach to teaching scientific writing and evaluating undergraduate theses. *Bioscience, 59*(10), 896–903. https://doi.org/10.1025/bio.2009.59.10.11

Ruscetti, T., Krueger, K., & Sabatier, C. (2018). Improving quantitative writing one sentence at a time. *PLoS One, 13*(9), e0203109. https://doi.org/10.1371/journal.pone.0203109

Shortlidge, E. E., & Brownell, S. E. (2016). How to assess your CURE: A practical guide for instructors of course-based undergraduate research experiences. *Journal of Microbiology & Biology Education, 17*(3), 399–408. https://doi.org/10.1128/jmbe.v17i3.1103

Soneral, P. A. G., & Wyse, S. (2015). Your tax dollars at work: A mock grant writing experience centered on scientific process skills. *CourseSource*. https://doi.org/10.24918/cs.2015.16

Staub, N., Blumer, L., Beck, C., Delesalle, V., Griffin, G., Merritt, R., Hennington, B., Grillo, W., Hollowell, G., White, S., & Mader, C. (2016). Course-based science research promotes learning in diverse students at diverse institutions. *CUR Quarterly, 38*(2), 36–46. https://doi.org/10.18833/curq/37/2/11

Timmermann, B. E. C., Strickland, D. C., Johnson, R. L., & Payne, J. R. (2011). Development of a 'universal' rubric for assessing undergraduates' scientific reasoning skills using scientific writing. *Assessment & Evaluation in Higher Education, 36*(5), 509–547. https://doi.org/10.1080/02602930903540991

Ward, J. R., Clarke, H. D., & Horton, J. L. (2014). Effects of a research-infused botanical curriculum on undergraduates' content knowledge, STEM competencies, and attitudes toward plant sciences. *CBE Life Sciences Education, 13*(3), 387–396. https://doi.org/10.1187/cbe.13-12-0231

Chapter 16
Assessing Undergraduate Research, a High Impact Practice: Using Aligned Outcomes to Detail Student Achievement to Multiple Stakeholders

Jill Rulfs and Jessica Caron

16.1 Introduction

In 2008 George Kuh identified ten learning practices demonstrated to have high impact on a variety of aspects related to individual student learning. Benefits ascribed to these practices include supporting students' deep and cumulative learning gains and enhancing engagement (Eagan et al., 2013; Kuh, 2008; Laursen et al., 2010; Russell et al., 2007). These practices have also been shown to benefit institutions and programs by increasing retention, strengthening persistence in the discipline, especially in STEM fields, and providing positive benefits to traditionally underserved populations (Eagan et al., 2013; Nagda et al., 1998; Rodenbusch et al., 2016). As a result, many institutions are adding high impact practices to their undergraduate programs either at the general curricular or specific disciplinary level.

One such high impact practice is providing undergraduate students with research experience, including individual mentoring. However, especially in STEM fields, research experience for undergraduates comes at a high cost, both in real financial expenditures for materials and supplies, and in faculty, post-doctoral and graduate student instructional and supervisory time. Scaling of these opportunities must have initial and sustained administrative support, and, in order to justify this continuing expense, benefits to the institution beyond student retention must be identified.

J. Rulfs (✉)
Biology & Biotechnology Department, Worcester Polytechnic Institute (WPI), Worcester, MA, USA
e-mail: jrulfs@wpi.edu

J. Caron
Northeastern University, Boston, MA, USA
e-mail: j.caron@northeastern.edu

N. J. Pelaez et al. (eds.), *Trends in Teaching Experimentation in the Life Sciences*, Contributions from Biology Education Research,
https://doi.org/10.1007/978-3-030-98592-9_16

In recent years, higher education has been under pressure to demonstrate that it has prepared students with relevant skills and knowledge for the real world. A variety of stakeholders, including legislatures, parents, and employers, is demanding evidence of achievement of learning outcomes that demonstrate transferable skills that translate to employability in the workplace (Chiang et al., 2020).

Toward those ends, it is not sufficient to simply make high impact practice opportunities, such as undergraduate research, available. Both student learning and the quality of the practice must be assessed (Finley, 2019; Kuh, 2008). While it can be challenging to collect and evaluate student learning artifacts, processes and procedures need to be put into place to do so in order to systematically assess student progress (Finley, 2011). To do this effectively, the intended outcomes of the practice must be identified at the outset to guide design of the practice, and subsequently to assure progress and support improvements identified by their assessment.

As experiences will vary across institutions and even across disciplines within institutions, the data must be disaggregated to allow fine-grained analysis. At the curriculum level, high impact learning experiences are often individually assessed for grading purposes, using pre-determined outcomes as the benchmark. But these practices may also be identified as potential data sources for higher organizational levels, such as departmental and institutional learning outcomes assessment matrices. Done well, these assessments can not only guide curricular improvement, they can also help make the case that they are worth the institution's investment and demonstrate to the public the institution's commitment to both relevant and transformative education. Using them for these broader purposes is the focus of this chapter.

Beyond assessing student learning, aligning departmental and institutional learning outcomes assessment matrices with one another, as well as with other expected outcomes, such as those articulated by national, regional and specialty accrediting boards and bodies, has important value for institutional and programmatic assessment. It can provide an integrated view not only of the impact of the individual practices, but also of the utility of the benchmark outcome statements being used for evaluation. Is there alignment among the outcomes for the different constituencies? Are there significant differences and, if so, why? Do these need to be addressed? Should they be the basis for curricular realignment or changes in institutional priorities?

16.2 The Process

16.2.1 Identifying Stakeholders

Before any of this can be done, as a first step in the assessment process, the outcomes of interest to each constituency must be defined. This may take the form of clearly articulated departmental goals, institutional or programmatic learning objectives, or accreditation board standards for example. Development and clear articulation of these outcomes are essential to any evaluation process. At the disciplinary level, they may be based on published competencies often in the form of concept or skills inventories. In biology, subdisciplines may have their own set of essential concepts and practices that demonstrate mastery (Smith & Marbach-Ad, 2010). For biology experimentation, the ACE-Bio competencies provide a useful framework specific to biologic experimentation (Pelaez et al., 2017; Chap. 1 in this volume). Finding and aligning these outcomes with those at the program, department, school, institutional or even professional organizational level can provide clarity as to the intent and value of including these high impact practices at each level of the organization.

At our institution, in order to qualify for graduation, every student must complete a senior level research experience. This project meets all the requirements for effective student-faculty research, a practice recognized as a high impact (Kuh, 2008). Students participate in the entire inquiry process. In biology, this entails each student, with faculty support, guidance and mentoring, identifying and fleshing out research questions or hypotheses, reading the relevant literature, designing experiments, collecting, analyzing and interpreting data, and presenting their findings both in an oral and a written format. Given that a similarly structured project experience is required by all departments and programs, the institution as a whole has defined learning outcomes specific to this project. In addition, each individual department or program has articulated and published learning outcomes to which this project maps. As the institution has many STEM programs, we are also mindful of the standards published by relevant accreditation organizations. What we report here is a process by which these outcomes, from the level of the accreditation agency down to the individual department, can be used to assess the value of undergraduate research in supporting these outcomes.

16.2.2 Aligning the Outcomes

Recently, with some institutional support, we developed a process which allowed us to successfully align institutional and departmental objectives for a single department (Biology & Biotechnology) and to use the aligned objectives to assess student progress on those outcomes as evidenced by the senior level research experience. Additionally, we also separately evaluated the alignment of the ABET (Accrediting Board for Engineering and Technology – the national accrediting agency for our institution) student outcomes criteria with our institutional student outcomes. Finally, using the departmental data, the summary evaluation of progress on all of the outcomes collectively was compared to individually assigned student grades. The process by which this was done is elaborated here.

To assure the objectivity of the process, an outside evaluator (JC) was hired to do the assessment of progress on learning outcomes using our identified high impact practice, the senior level research project. The outside assessor had access to the institutional and department learning outcomes as well as to the final report for each of the senior projects completed in the department in the previous 3 years. These reports were the artifacts used as the basis for this evaluation.

As a first step in the process, the outcomes were aligned. There are eight departmental program outcomes and seven institutional outcomes. Because the artifact being used for the departmental evaluation was the final senior project report, some outcomes were omitted as they could not be evaluated from a written document. Additionally, institutional outcomes were aligned with the ABET criteria for student outcomes. Alignment with accreditation board outcomes, while not always necessary at the department level, can be important at the institutional level for reasons mentioned earlier, including as evidence both directly during accreditation reviews and for other stakeholder groups.

16.2.3 Defining the Evidence

Next, working together, the assessor and the associate department head (JR), who is involved in both departmental and institutional assessment efforts, reviewed each of the outcomes and agreed upon the kinds of evidence that could be identified for each. These were compiled into a scoring rubric. Validity of the rubric for this project was established through a process of interrater reliability and baseline establishment. Using the rubric, the assessor, the associate department chair and one departmental faculty member independently scored an initial set of project reports which were not included in the set to be evaluated, and then met to discuss their individual scores and comments. This process was repeated twice to ensure alignment at which point there was nearly complete agreement on the final scores and the assessor continued using the agreed upon criteria. All of the reports used for this process were stripped of identifying information to avoid reviewer bias.

16.2.4 Selecting the Artifacts

For our purposes it was determined that one third of the senior project reports (hereafter referred to as "reports") for each year would be evaluated. In smaller departments, it might be feasible to evaluate every report. In larger departments, a power analysis could be used to determine an appropriately representative number of reports that should be included in the process (Lenth, 2001).

Completed reports authored by students in Biology & Biotechnology in the past 3 years were compiled from the institutional repository. Again, identifying information, such as student and advisor names, was removed from the reports before scoring began. Each report was assigned a number and, using a random number generator, a representative set of project reports was selected for assessment. The assessor then read and scored each of the reports using the departmental program outcomes and the agreed upon rubric. These data were recorded by individual outcome for each report.

16.2.5 Scoring and Reporting

A final overall score was determined for each report giving all outcomes equal weight. A four-point scoring scale was used: (1) Not acceptable (2) Marginally acceptable/does not meet expectations at the senior level (3) Acceptable/Meets expectations at the senior level (4) Exemplary/Exceeds expectations at the senior level. Later, because the institution's published grading scheme for projects is a three-point scale, for some purposes these data were collapsed to a three-point scale, combining marginally acceptable and acceptable into a single category. The final grade (A, B or C) that was given to each report by the faculty advisor at the time the project was completed was also recorded.

16.3 The Results

16.3.1 Aligning Outcomes

As a first step in the process, institutional and department learning outcomes related to the research experience were aligned. Of the seven institutional and eight departmental outcomes, five were identified as describing equivalent outcomes that could be identified using the same criteria. One departmental outcome (Students can explain and give examples of the five unifying themes of biology) was written in specific language that was incongruent with a research experience and one (Students can function in a collaborative environment) would not have been evident in the written report. One institutional outcome (lifelong learning) seemed too broadly defined to be specifically identifiable in the project report. (see Table 16.1).

During the alignment process, although it was not directly relevant to the assessment being done, the institutional outcomes were also aligned with the ABET criteria for student outcomes, often referred to colloquially as "ABET a-k". Because accreditation boards are increasingly asking for evidence of progress on learning outcomes, this exercise had potential benefit both to the institution and to programs accredited by ABET for reporting purposes. That alignment can be seen in Table 16.2.

Table 16.1 Alignment of institutional outcomes for the senior research experience and programmatic learning outcomes for the department (Worcester Polytechnic Institute, 2018)

Institutional outcomes Students who complete a senior research experience will:	**Department outcomes** Graduates in Biology & Biotechnology:
Apply fundamental and disciplinary concepts and methods in ways appropriate to their principal areas of study.	*Will know and understand the five unifying themes and can provide and explain examples of each from each of the three divisions of biology.*
Demonstrate skill and knowledge of current information and technological tools and techniques specific to the professional field of study.	Can demonstrate mastery of a range of quantitative and procedural skills applicable to research and practice in biology & biotechnology.
Use effectively oral, written and visual communication.	Demonstrate oral and written communication skills relevant to the discipline.
Identify, analyze, and solve problems creatively through sustained critical investigation.	Are able to generate hypotheses, design approaches to test them, and interpret data to reach valid conclusions.
Integrate information from multiple sources.	Can find, read and critically evaluate the scientific literature.
Demonstrate an awareness and application of appropriate personal, societal, and professional ethical standards.	Can describe the broader scientific or societal context of their work or that of others. Understand and can adhere to accepted standards of intellectual honesty in formulating, con-ducting and presenting their work.
Practice the skills, diligence, and commitment to excellence needed to engage in lifelong learning.	
	Can function effectively in a collaborative scientific environment

Italics statements identify areas where alignment was absent or where evidence of progress could not be identified through an assessment of the written project reports

16.3.2 Defining the Evidence

The process of identifying evidence that would demonstrate progress on each outcome resulted in framing questions that defined the evidence relative to that outcome. For example, for the outcome, which refers to mastering procedural skills, the question asks whether experiments were carried out using appropriate controls (see Tables 16.3 and 16.4). In determining what the assessor should find in each area of the report, the decision was made to use the institutional outcome related to fundamental disciplinary concepts to contextualize research, even though it did not have a direct cognate in the departmental outcomes (see Table 16.3). The original scale for scoring was (1) Not acceptable (2) Marginally acceptable/does not meet expectations at the senior level (3) Acceptable/Meets expectations at the senior level (4) Exemplary/Exceeds expectations at the senior level. As explained previously, later, because the institution's published grading scheme for projects is a three-point scale, for some purposes these data were collapsed to a three-point scale, combining marginally acceptable and acceptable into a single category.

Table 16.2 Alignment of institutional senior research project outcomes with ABET criteria, a-k (ABET, 2017)

Institutional outcomes Students who complete a senior research experience will:	**ABET criteria** By the time of graduation students will have:
Apply fundamental and disciplinary concepts and methods in ways appropriate to their principal areas of study.	The ability to apply mathematics, science and engineering principles.
Demonstrate skill and knowledge of current information and technological tools and techniques specific to the professional field of study.	The ability to use the techniques, skills and modern engineering tools necessary for engineering practice.
Use effectively oral, written and visual communication.	The ability to communicate effectively.
Identify, analyze, and solve problems creatively through sustained critical investigation.	The ability to identify, formulate and solve engineering problems The ability to design a system, component, or process to meet desired needs. The ability to design and conduct experiments, analyze and interpret data.
Integrate information from multiple sources.	
Demonstrate an awareness and application of appropriate personal, societal, and professional ethical standards.	The broad education necessary to understand the impact of engineering solutions in a global and societal context. Knowledge of contemporary issues. An understanding of professional and ethical responsibility.
Practice the skills, diligence, and commitment to excellence needed to engage in lifelong learning.	Recognition of the need for and an ability to engage in life-long learning.
	The ability to function on multidisciplinary teams.

Italics identify areas where alignment was absent and where evidence of progress could not be identified through an assessment of the written project reports

Table 16.3 Evidence for aligned outcomes 1 and 2

University Outcomes	Apply fundamental and disciplinary concepts in ways appropriate to their principal areas of study and contextualizes research.	Demonstrate skill and knowledge of current information and technological tools and techniques specific to the professional field of study.
Evidence	Focus on conceptual application – Introduction should demonstrate this. Do you know what you are talking about? and Why do I care?	Focus on application – this should be methods. Did they choose the correct method? Did you do an experiment? Controls?
Department Outcomes	Will know and understand the five unifying themes and can provide and explain examples of each from each of the three divisions of biology.	Can demonstrate mastery of a range of quantitative and procedural skills applicable to research and practice in biology & biotechnology.

Although departmental outcome 1 was deemed to be inappropriate for this assessment, institutional outcome 1 articulated an important outcome for contextualizing the work and so was included in the final assessment rubric

Table 16.4 Notes on assessment

	University outcome	Departmental outcome	Notes on assessment
1	Apply fundamental and disciplinary concepts in ways appropriate to their principal areas of study and contextualizes research.	Will know and understand the five unifying themes and can provide and explain examples of each from each of the three divisions of biology.	Focus on conceptual application.
2	Demonstrate skill and knowledge of current information and technological tools and techniques specific to the professional field of study.	Can demonstrate mastery of a range of quantitative and procedural skills applicable to research and practice in biology & biotechnology.	Focus on application – This should be methods. Did they choose the correct method? Did they do an experiment? Controls?
3	Use effectively oral, written and visual communication.	Demonstrate oral and written communication skills relevant to the discipline.	Almost acceptable for publication with an organizational schema and writing convention that is adhered to.
4	Identify, analyze, and solve problems creatively through sustained critical investigation.	Are able to generate hypotheses, design approaches to test them, and interpret data to reach valid conclusions.	Hypothesis or an inferred hypothesis is evident through text with work (including data analysis) being done towards addressing this hypothesis.
5	Integrate information from multiple sources.	Can find, read and critically evaluate the scientific literature.	Number, quality and type of resources evaluated holistically.
6	Demonstrate an awareness and application of appropriate personal, societal, and professional ethical standards.	Can describe the broader scientific or societal context of their work or that of others Understand and can adhere to accepted standards of intellectual honesty in formulating, conducting and presenting their work.	Ethical standards in writing and data collection are evident
7	Practice the skills, diligence, and commitment to excellence needed to engage in lifelong learning.	Can function effectively in a collaborative scientific environment.	

Questions that should be addressed, information that should be present, and practices that should be identifiable for each of the outcomes were determined as the basis for scoring each project relative to each outcome being used

16.3.3 Scoring and Reporting

A total of 33 reports were reviewed and scored using the six outcomes and evidence shown in Table 16.4. Each paper being assessed was read at least two times, once as an overview and a second time for the purposes of scoring. Because the institution has as one of its undergraduate learning outcomes "graduates will function

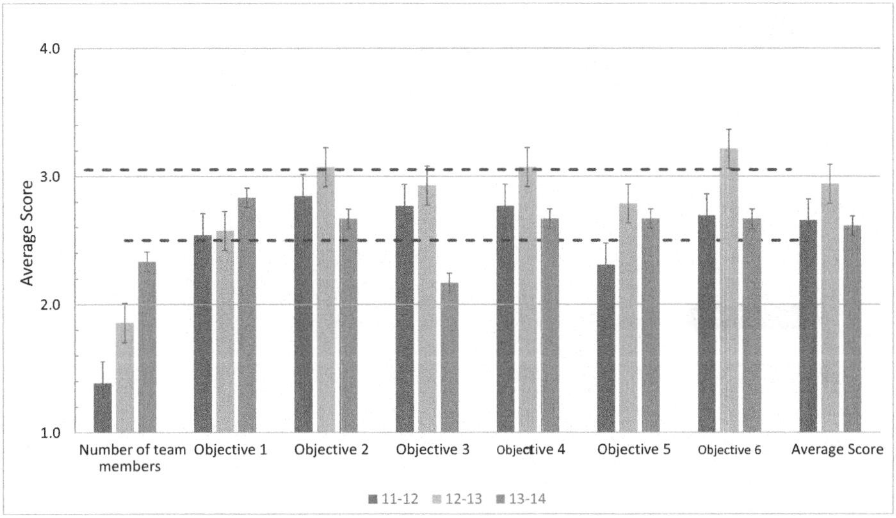

Fig. 16.1 Achievement of learning outcomes. Average score for each of the departmental learning outcomes 1–6 (see Tables 16.1 and 16.2) and a summary average score (calculated as (Objective 1 Score + Objective 2 Score + … Objective 6 Score)/6) is shown for each academic year. Dashed line indicates the scale midpoint. Error bars are standard error of the mean (SEM)

effectively both individually and on teams", the number of students who worked on each research project was also recorded. The overall results of scoring on each objective and the final overall score for each project are shown in Fig. 16.1. For this analysis, the three-point scale which corresponds to the university grading scheme for projects was used. Using the university grading system, an A project is one where the product and process meets all expectations and exceeds them in several areas; a B project is one in which all of the project expectations have been met but generally not exceeded; a C project is one where some but not all of the expectations have been met. Projects that failed to meet the requirements for a C project were deemed unacceptable for graduation and so would not be included among archived project reports. To mirror this grading scheme, the scoring scale was collapsed to three categories which generally correspond to exceptional (a score of 3), acceptable (a score of 2) and not acceptable (a score of 1). This scale is one that the Undergraduate Outcomes Assessment Committee at the university has developed.

The projects evaluated were also identified by the academic year in which they were completed. In general, the achievement of all objectives as evidenced in the final project reports falls within the acceptable range, with no real statistically significant differences across the 3 years that were part of this evaluation. Over time, the average number of students on a team has risen with fewer students completing projects individually.

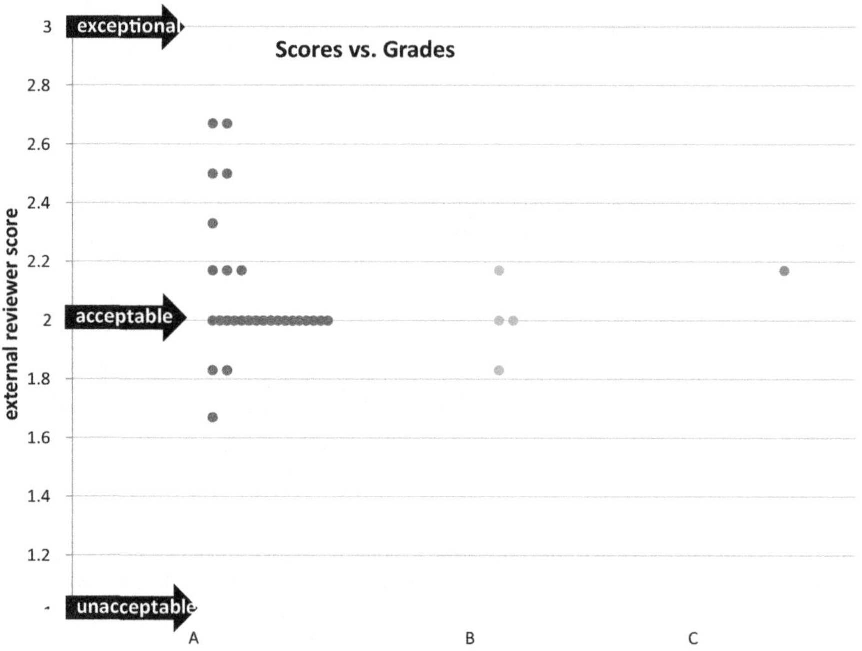

Fig. 16.2 Scores vs. grades. Final grades given by the faculty advisor for each project are shown relative to the final average score for the project determined by the external evaluator

16.4 Extending the Project

As a final exercise, for departmental purposes only, final grades recorded for each project were compared with the average score given to that project. The results are shown in Fig. 16.2. Final grades are given by the faculty member who advises the project. Of all thirty-three projects evaluated, only five received grades other than A. In the scoring scheme articulated by the university this translates into 85% of the projects being characterized as work which has exceeded several of the expectations set by the advisor.

The process described here used the final report of a yearlong undergraduate research project as the source of information for assessment. This has obvious limitations as the report, which is the final product of the practice, may not fully represent the learning process, which occurred across the year. Faculty members or other supervisory personnel have insight into the progress a student has made that would not be reflected in the final report. This may account for the discrepancy between the assessor's scoring and grades assigned by the faculty. Nonetheless, these data were helpful in convincing department faculty that we needed to do some grade norming relative to research project grading to be consistent among faculty members and with the university's standards.

16.4.1 Using ACE-Bio Competencies

In the event the integration of undergraduate research is new to a department or institution, certainly as a starting point the ACE-Bio Competencies would work well as both the learning outcomes and the evidence for each as the concepts and skills for each are elaborated. We have mapped them to our outcomes (Table 16.5) and find that the alignment is quite good, with differences at the level of detail. This detail and elaboration of skills and concepts for each (Table 16.1) would actually facilitate their use in this assessment procedure and further support the use of the ACE Bio Competencies as the outcomes for undergraduate research experiences.

No matter the specific outcomes used as the basis for evaluation, the steps for developing an assessment process remain fundamentally the same and are shown in the flow diagram in Fig. 16.3.

Table 16.5 Alignment of ACE-Bio Competencies (Pelaez et al., 2017; Chap. 1 in this volume) with WPI Biology & Biotechnology Department Outcomes

ACE-Bio Competencies What a competent biologist doing experimentation has the ability to do:	**WPI Department Outcomes** What a graduate in Biology & Biotechnology the ability to do:
<u>*Conduct*</u> an investigation to achieve research goals	Can demonstrate mastery of a range of quantitative and procedural skills applicable to research and practice in biology & biotechnology.
<u>*Communicate*</u> research work in professionally appropriate modes, including visual, written and oral formats.	Demonstrate oral and written communication skills relevant to the discipline.
Generate a research <u>*Question*</u> and formulate hypotheses; <u>*Analyze*</u> and process data; <u>*Conclude*</u> about data with inferences that are limited to the scope inherent in the experimental design; <u>*Plan*</u> feasible and ethical experiments to answer research questions or test hypotheses.	Are able to generate hypotheses, design approaches to test them, and interpret data to reach valid conclusions.
<u>*Identify*</u> gaps or limitations in current research knowledge through review, filtering and synthesis of relevant literature.	Can find, read and critically evaluate the scientific literature.
<u>*Plan*</u> feasible and ethical experiments to answer research questions or test hypotheses.	Can describe the broader scientific or societal context of their work or that of others. Understand and can adhere to accepted standards of intellectual honesty in formulating, conducting and presenting their work.
	Can function effectively in a collaborative scientific environment.

Italics identify areas where alignment was absent and where evidence of progress could not be identified using the research report

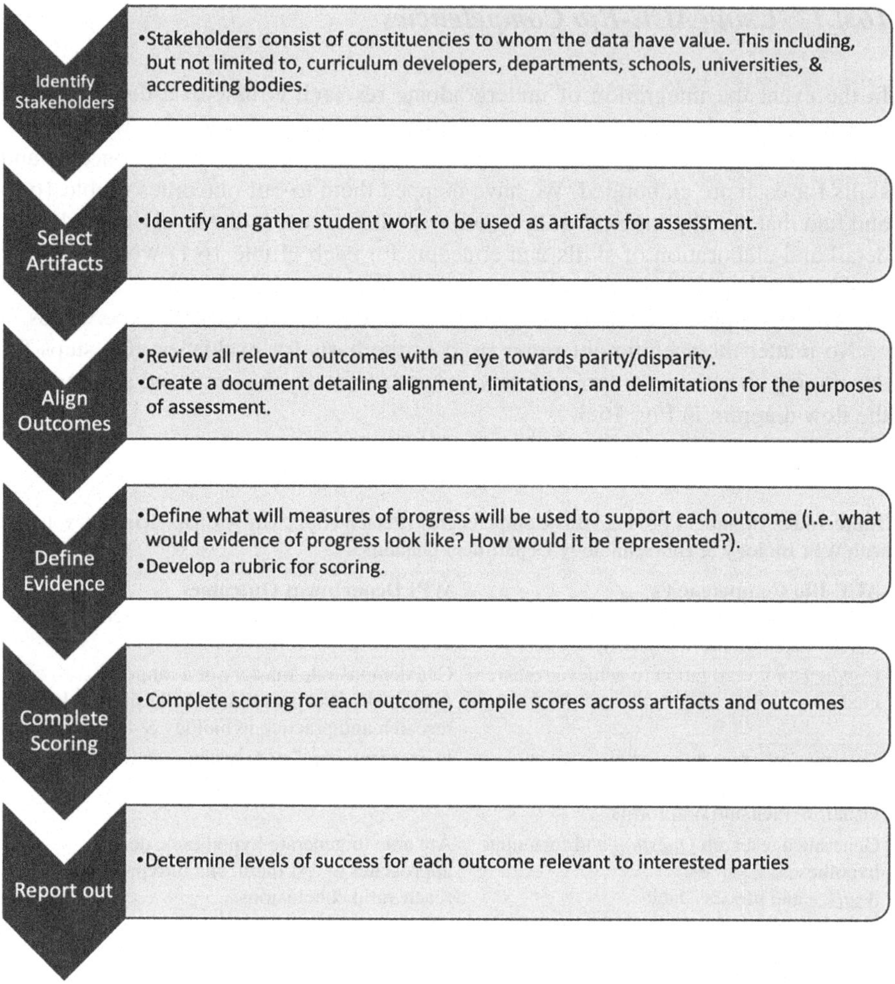

Fig. 16.3 The process for developing assessment plans to evaluate undergraduate re-search experiences. Shown is a summary of the process described in the preceding text

16.5 Discussion

As educators, having embraced the need for and value of assessment of student learning, our intent here is to provide practical guidance on assessing undergraduate research for non-student constituencies, through alignment of programmatic, institutional and accreditation learning outcomes.

High impact practices have been evaluated by a number of studies, generally with respect to their impact on student engagement, attitude, and retention. But as part of a curriculum, they also must be assessed for their value as measures of achievement relative to learning outcomes. Clearly in order to do this in a meaningful way, those outcomes must be agreed upon and clearly articulated. This can be challenging, especially in establishing broad, university level outcomes which satisfy all the constituencies each of whom is concerned that their outcomes are appropriately and proportionally represented. However, as the value of higher education is increasingly scrutinized, the ability to both articulate those outcomes and demonstrate progress or achievement by students is critical.

Beyond the public relations value to an institution, as high impact practices are increasingly required by undergraduate programs, their true value lies in identifying the strengths and weaknesses associated with the practices and determining where improvements can be made. These purposes can appear to be at odds with one another, and at the program or department level, care must be taken to assure the validity of the process to identify areas for improvement. After all, improvement is the true purpose of assessment, not to make current practice appear to be ideal. Thus, establishing a clear process to be followed and assigning a set of measures to serve as evidence for each outcome is essential. A large part of our effort in outcomes alignment and rubric development was determining what constituted evidence and what could be reasonably measured given the artifacts being used for assessment.

Here we recognize two clear short comings of this process. The first, as mentioned previously, is that we were using a written document, which was the culminating product of an experience. The value of these high impact practices may well be as much the process which would not necessarily be captured in a written report. For example, in aligning our outcomes for this assessment, we recognized that while the research experience certainly should have included opportunities for students to "function effectively in a collaborative scientific environment" (departmental outcome 7), progress on this outcome could not necessarily be demonstrated in a written document and so we eliminated it from our assessment. The second issue with our process is that it was a *post hoc* analysis. Perhaps this is really an extension of the first issue, that is what we are seeing is a final picture of a process that took place over time. When and how achievement of learning outcomes happens cannot be assigned to a specific experience, especially at the senior level. Care must also be taken to recognize that when the report does provide evidence of achievement, it is not evidence that the high impact practice caused or was responsible for that level of achievement.

At a more practical level, the process itself can provide insights for improvement of the learning outcomes. If it is difficult to identify what might be used as evidence, perhaps the outcome needs to be more clearly articulated. In our case, one of the learning outcomes (will know and understand the five unifying themes and can provide and explain examples of each from each of the three divisions of biology) was defined narrowly as a performance objective, making it not as broadly useful as it might have been. By contrast, the institutional outcome, apply fundamental and disciplinary concepts and methods in ways appropriate to their principal areas of study, was more useful in evaluating the artifact we were using. So, the alignment exercise itself has value in looking at the big picture. Beyond that, alignment, once accomplished and codified, should bring cohesion to the assessment process and thus minimize the amount of work in future iterations.

At the department level, a more fine-grained scoring scale had more value in clearly identifying areas where more attention is needed. By collapsing the acceptable and marginally acceptable categories, some of the areas that might need attention would be lost. However, from an institutional point of view, this level of detail was not as meaningful and a broader look at the overall picture had more value. Determining the purpose of the process before starting will allow the data to be more precisely useful.

Scores for individual outcomes can help identify areas where interventions might lead to improvement. While overall the data suggest that we are "hitting the mark", the scores for "can find, read and critically evaluate the scientific literature" are consistently the lowest. Perhaps we should consider requiring a session with a research and instruction librarian as part of each research experience. Instituting such a requirement and subsequent evaluation would help determine the value of the intervention. Again, having an established process should increase the value of the assessment and minimize the amount of work in future iterations.

Although the comparison of the scores given by the assessor to the project grades assigned by faculty advisors was not specifically part of the project, having the data provided by an objective external evaluator shed real light on what is likely grade inflation among the faculty evaluations of the senior research experience. While the reasons for and the impact of grade inflation are beyond the scope of this work, it is certainly a topic worth considering and having some data to support the discussion makes it difficult to dismiss the need for discussion among the faculty.

All of the processes here have potential value beyond the direct assessment of student learning. While our specific focus was on biology, and more specifically biology research, at the institutional level, it can be used as a model that, with different disciplinarily defined outcomes, would be useful across departments. Done systematically, it is valuable evidence for accreditation renewal. From a broader educational viewpoint, it provides a framework to support the inclusion of high impact practices such as undergraduate research in the educational paradigm of the future.

References

ABET. (2017). *Criteria for Accrediting Engineering Programs, 2018–2019*. https://www.abet.org/accreditation/accreditation-criteria/criteria-for-accrediting-engineering-programs-2018-2019/

Chiang, T.-M., Altman, J., Haeger, H., & Gunnels, C. (2020). *Beyond the CUR Whitepaper, The Role of Assessment for Advocacy, CUR Virtual Biennial Conference*, June 28–July 1, 2020.

Eagan, M. K., Jr., Hurtado, S., Chang, M. J., Garcia, G. A., Herrera, F. A., & Garibay, J. C. (2013). Making a difference in science education: The impact of undergraduate research programs. *American Educational Research Journal, 50*(4), 683–713.

Finley, A. (2011). Assessment of high-impact practices: Using findings to drive change in the compass project. *Peer Review, 13*(2).

Finley, A. (2019). *A comprehensive approach to assessment of high-impact practices* (Occasional Paper No. 41). University of Illinois and Indiana University, National Institute for Learning Outcomes (NILOA).

Kuh, G. D. (2008). *High-impact educational practices: What they are, who has access to them, and why they matter*. Association of American Colleges and Universities.

Laursen, S., Hunter, A.-B., Seymour, E., Thiry, E. H., & Melton, G. (2010). *Undergraduate research in the sciences: Engaging students in real science*. Jossey-Bass.

Lenth, R. (2001). Some practical guidelines for effective sample size determination. *Amer. Statist., 55*(3), 187–193.

Nagda, B. A., Gregerman, S. R., Jonides, J., von Hippel, W., & Lerner, J. S. (1998). Undergraduate student-faculty research partnerships affect student retention. *The Review of Higher Education, 22*, 55–72.

Pelaez, N., Anderson, T. R., Gardner, S. M., Yin, Y., Abraham, J. K., Bartlett, E., Gormally, C., Hill, J. P., Hoover, M., Hurney, C., Long, T., Newman, D. L., Sirum, K., & Stevens, M. (2017). The basic competencies of biological experimentation: Concept-skill statements. *PIBERG Instructional Innovation Materials. Paper 4*. Retrieved October 18, 2021 from https://docs.lib.purdue.edu/pibergiim/4/

Rodenbusch, S. E., Hernandez, P. R., Simmons, S. L., & Dolan, E. L. (2016). Early engagement in course-based research increases graduation rates and completion of science, engineering, and mathematics degrees. *CBE Life Sciences Education, 15*(2), 20.

Russell, S. H., Hancock, M. P., & McCullough, J. (2007). Benefits of undergraduate research experiences. *Science, 316*, 548–549.

Smith, A. C., & Marbach-Ad, G. (2010). Learning outcomes with linked assessments – An essential part of our regular teaching practice. *Journal of Microbiology & Biology Education, 11*(2), 123–129.

Worcester Polytechnic Institute. (2018). *Undergraduate catalog (2018–2019)*. https://www.wpi.edu/sites/default/files/2018/12/05/WPI_UGCat18-19WEB-Final3.pdf

Chapter 17
Assessment of Evidentiary Reasoning in Undergraduate Biology: A Lit Review and Application of the Conceptual Analysis of Disciplinary Evidence (CADE) Framework

Shiyao Liu, Chao Cai, Chaonan Liu, Ala Samarapungavan, Stephanie M. Gardner, Kari L. Clase, and Nancy J. Pelaez

17.1 Introduction

The knowledge of concepts and development of the competence to use that knowledge are foundational for students learning biology as a disciplinary practice. In accordance with the ACE-Bio Competencies framework (Pelaez et al., 2017; Chap. 1 in this volume), the AAAS (2011) *Vision and Change in Undergraduate Biology Education* report emphasizes that all post-secondary biology students need to develop core competencies applied to biology research practice. To understand how the design of scientific processes reveals what is known about living systems, competent students must demonstrate observational strategies, hypothesis testing, experimental design, evaluation of experimental evidence, and problem-solving strategies (AAAS, 2011). However, in this description of teaching and learning biology as an evidence-based discipline, the notion of evidence remains obscure. The

S. Liu · C. Liu · S. M. Gardner · N. J. Pelaez (✉)
Department of Biological Sciences, Purdue University, West Lafayette, IN, USA
e-mail: liu2313@purdue.edu; sgardne@purdue.edu; npelaez@purdue.edu

C. Cai
Libraries and School of Information Studies, Purdue University, West Lafayette, IN, USA
e-mail: caic@purdue.edu

A. Samarapungavan
Department of Educational Studies, Purdue University, West Lafayette, IN, USA
e-mail: alasam@purdue.edu

K. L. Clase
Department of Agricultural and Biological Engineering, Biotechnology Innovation and Regulatory Science (BIRS) Center, Purdue University, West Lafayette, IN, USA
e-mail: klclase@purdue.edu

N. J. Pelaez et al. (eds.), *Trends in Teaching Experimentation in the Life Sciences*, Contributions from Biology Education Research,
https://doi.org/10.1007/978-3-030-98592-9_17

monitoring of students' developing competence for reasoning with and about evidence in the context of biology disciplinary knowledge of relevance to an investigation is another challenge in teaching biology students to understand and do research. With the aim to facilitate appropriate choice of assessment tools and to identify gaps for the development of new assessments that reveal evidentiary reasoning difficulties in post-secondary biology laboratory classrooms, here we present a comprehensive literature review with existing assessments categorized using the Conceptual Analysis of Disciplinary Evidence (CADE) framework, which links biological knowledge with epistemic considerations, in addition to the Basic Competencies of Biological Experimentation (ACE-Bio) framework (Pelaez et al., 2017). The ACE-Bio Competencies and CADE frameworks partially overlap. Findings with the CADE show that some assessments fail to link disciplinary knowledge with epistemic reasoning processes while assessing students' evidentiary reasoning. To address the gaps revealed by the literature review and to extend our study of evidentiary reasoning beyond experimentation, two assessments were designed to identify difficulties that students have in reasoning about evidence in the context of research that involves evolutionary tree-thinking.

The performance expectations of the ACE-Bio Competencies, the CADE framework, or the *Vision and Change* (AAAS, 2011) report serve as a foundation for the development of assessments. However, such reform documents do not give enough detail to construct or use an assessment. As background for constructing new assessments, we first review literature on aspects to consider: (1) how to prompt or elicit a performance that reveals evidence of students' abilities; (2) what format is ideal for eliciting students' thoughts in a way that is feasible for the intended use of that information; (3) the need to situate the assessment tool or task in a relevant disciplinary context; and (4) what difficulties or competent performances are expected to be observed in the students (National Research Council, 2014).

17.1.1 Assessment Triangle

As a framework, we considered the assessment triangle as a process of reasoning from evidence about what students know and can do with their knowledge. The assessment triangle is defined as a "theory or set of beliefs about how students represent knowledge and develop competence in a subject domain" (National Research Council, 2001, p. 44). The triangle has three main points: *cognition* is the foundation which refers to a set of knowledge and abilities to use that knowledge that are important for a competent student; *observation* is the process of using an assessment instrument or a specific assignment to elicit a performance that reveals a student's cognition abilities; *interpretation* is the process of comparing that performance to a standard that would be expected for a competent student, which also involves identifying students' difficulties that can then be addressed. To monitor a student's developing competence for reasoning from evidence, our study employed the assessment triangle according to the National Research Council (2001) and Pellegrino (2012).

Cognition serves as the foundation of the triangle, meaning that as a starting point, there is a need for a clearly defined set of knowledge and skills that are important for a competent student. A clear cognitive model like the ACE-Bio Competencies framework (Pelaez et al., 2017; Chap. 1 in this volume) provides understanding of how a student typically demonstrates domain expertise.

Observation stands on the second corner of the assessment triangle. Observation involves the process of eliciting a performance such as a writing assignment, research poster presentation, or a test item response designed to reveal a student's ability in the context of specific tasks. In particular, the assessment must have a precisely defined target for cognitive competence. For example, the assessment examples provided later in this chapter aim to gather information about how students use or apply the notion of evidence.

Interpretation is the last corner of the assessment triangle. Interpretation has been defined as the "methods and tools used to reason from fallible observations" (National Research Council, 2001, p. 48). Also, consider the audience who will engage in the interpretation. One important audience is the instructor who might modify instruction to help students address their difficulties according to what the instructor has observed in student responses to various assessment tasks. Instructors and administrators might use sample student responses to assessment tasks to track progress on anticipated learning outcomes, to identify the quality and range of student performances, and to determine if what was anticipated can be verified as a learning outcome resulting from a particular course or learning experience. This sort of summative assessment refers to the use of assessment data to evaluate students' knowledge upon completion of a learning sequence (Phelps, 2011).

Perhaps the most important interpretation is done by the student, who gets feedback and, as a result, may increase their own effort or they may abandon their goals or settle for lower personal expectations. Assessment used for individualized feedback to help students address their difficulties, also referred to as formative assessment, can also target motivating and helping students to develop their own improvement strategies. Student motivation involves changing their beliefs about themselves so that they can appropriately respond in ways that will advance their competence (Black & Wiliam, 1998). In this study we are interested in both formative and summative assessment. By formative assessment, we refer to the use of methods to encourage students to express what they are thinking so that they can adapt to the teaching flow and adopt strategies to achieve the anticipated learning outcomes (Black et al., 2003).

17.1.2 The CADE Framework

In this study our focus was on assessment tools, to identify gaps in existing assessments and to develop and implement new assessments that reveal evidentiary reasoning difficulties in post-secondary biology laboratory classrooms. But what are scientific evidence and evidentiary reasoning and why are these topics so important for students? First, various consensus reports have shifted instructional emphasis toward these notions in biology. According to the AAAS (2011) *Vision and Change* report mentioned earlier, undergraduate students should learn biology by applying the process of science, which involves getting data and evaluating it as experimental evidence. The ACE-Bio Competencies framework is not explicit about how data is used as evidence, but this can be inferred, for example, when the *Plan* competency item C. mentions Variables, which points out that a competent scientist will identify relevant, measurable variables for testing the hypothesis (Pelaez et al., 2017; Chap. 1 in this volume). According to Sandoval et al. (2004) and the *Next Generation Science Standards* (NGSS Lead States, 2013), scientific evidence is defined as data for addressing a question or supporting a claim. Thus, here we refer to evidentiary reasoning as reasoning with and about evidence throughout the entire research process. More specifically, this means applying evidence generated from a set of theoretical and methodological frameworks to assess the consistency or fit between potential theories and the reality (Giere, 2010). It is important for students to get a better understanding of evidence to help them make better decisions in their future when faced with issues like vaccines and climate change. Furthermore, an interest in biological evidence may even encourage some students to choose a biology career.

Instructors generally recognize that students struggle with understanding, using, and evaluating the evidence underpinning scientific knowledge, but the nature of those problems is not entirely clear. When Sandoval and Millwood (2005) examined the quality of secondary school students' use of evidence in written scientific explanations of natural selection, they found that students often failed to cite sufficient evidence. Despite a significant body of literature in science education focusing on issues like students' use of evidence, epistemic understandings about the nature of science, and development of scientific knowledge, educators consistently find that both K-12 and undergraduate students struggle with understanding the evidence to support their advanced science knowledge, as well as applying and evaluating this evidence by using scientific practices (Abd-El-Khalick et al., 2004; Duncan et al., 2018; Duschl, 2008; Furtak et al., 2010; Manz et al., 2020; McNeill & Berland, 2017; Tytler & Peterson, 2005). Since these problems may also relate to the complex nature of evidence (Samarapungavan, 2018), it is useful to consider what students should be doing when they are testing hypotheses and generating evidence to draw conclusions in terms of the processes that professional scientists engage in when they discover new knowledge. Scientific research practices involve decisions about what is worth investigating in addition to considering the limitations, uncertainties, and strength of any conclusions. Thus, by evidentiary reasoning, we include the evaluation of theories or models based on evidence, referred to as evidence-based reasoning in recent studies built upon Toulmin's (1958) *The Uses of Argument*, according to Erduran et al. (2015) and Furtak et al. (2010) who have focused on students' reasoning about science phenomena and their use of evidence for backing their claims. In our approach to evidentiary reasoning, we also include scientists' theoretical and disciplinary knowledge applied to designing, executing, and analyzing investigations based on norms and procedures that are shared by members of their discipline, and considering the nature, scope, quality and sufficiency of the data, approaches, and theories according to what is already known of relevance to the research evidence.

The CADE framework aims to promote evidentiary reasoning by unpacking the notions of evidence described above into component parts (Samarapungavan, 2018). It is a wholistic framework that explicitly examines both the disciplinary knowledge as well as epistemological considerations of relevance to students' use of evidence at all stages of the research process by deconstructing evidence into four research practice component relationships: (1) *Theory->Evidence (T->E)* relationships are of relevance to formulating testable models; (2) *Evidence<=>Data (E<=>D)* relationships relate to the design, execution, and analysis of investigation findings; (3) *Evidence->Theory (E->T)* relationships refer to evaluation of evidence to draw and justify conclusions; (4) *Social Dimension* relationships refer to communicating with and about evidence to the public. By linking disciplinary and epistemic knowledge, the CADE draws attention to the knowledge and practices of the discipline as well as the scientific skepticism for justifying the nature, scope, and quality of the data, approaches, theories, and claims that underpin the evidence. With the CADE as a comprehensive and practical framework, it may be feasible to address students' difficulties in evidentiary reasoning among students who have struggled with understanding, using, and evaluating the evidence underpinning scientific knowledge.

In the first part of our study, assessments were examined using the CADE framework as a lens to monitor both domain-general and discipline-specific aspects of evidentiary reasoning that is a target of assessment. When the CADE framework components were mapped to current established assessments and rubrics, it served as the cognitive model to provide an explicit target for how students and experts represent the notion of evidence when they conduct evidentiary reasoning. In the second part of our study, the CADE framework was further used to identify students' difficulties with evidentiary reasoning in the context of biological science research practices.

17.1.3 Research Goals

The overarching goal of this study was to gain an understanding of established assessments that are being used to evaluate students evidentiary reasoning, to identify gaps, and then to address these gaps using the CADE to inform the design of new assessment items for revealing students' difficulties in evidentiary reasoning in the context of evolutionary tree-thinking as an example.

(1) What assessments are being used to reveal evidentiary reasoning difficulties among students in post-secondary biology laboratory classrooms where students conduct practical research? (2) What assessment gaps remain for new development of useful assessments? and (3) How did two CADE informed assessments of evolutionary tree-thinking used as undergraduate biology lab class test items reveal students' difficulties with evidentiary reasoning and address the gaps from the literature review?

17.2 Published Assessments Target Reasoning About Evidence

Our first study was a comprehensive literature review to identify a range of assessments used to monitor students' progress in understanding and using evidence as they learn to conduct biological research. Mapping of existing assessments to the CADE (Samarapungavan, 2018) and ACE-Bio Competencies (Pelaez et al., 2017) frameworks made it possible to identify gaps that remain in the assessments that are being used to reveal students' difficulties in evidentiary reasoning.

17.2.1 Literature Review

To find out what assessments are being used to reveal evidentiary reasoning difficulties among students in post-secondary biology laboratory classrooms where students conduct practical research and what assessment gaps remain for the new development of useful assessments, we first conducted a comprehensive literature review. We searched for assessments of students' evidence reasoning that have been used or adapted in the context of experimental/practical work in undergraduate biology laboratory classrooms. According to the National Research Council (2005), *America's Lab Report*, practical work includes experiences where learners interact with data about the natural world gathered by the learners themselves and with data about the natural world provided to them. Again, evidentiary reasoning in this chapter refers to the use of shared disciplinary norms to generate and evaluate evidence to reach scientific consensus (Giere, 2010; Manz et al., 2020; Samarapungavan, 2018). With the literature review, we were interested in both formative and summative assessments. We include formative assessments such as a coding rubric to understand students' classroom discussions. Summative assessments include pre- and post-test assessment items, proposals, and surveys measuring students' difficulties with evidentiary reasoning.

17.2.1.1 Search Procedure

We included peer-reviewed journal articles, proceedings, books, and dissertations to gain a comprehensive understanding of the assessments that are being used to reveal students' difficulties in evidentiary reasoning. We first identified 19 articles we thought must be included based on our experience with assessment of student learning about biology research. We expanded and refined the searching key words by reading through the 19 articles. Comprehensive literature searches were conducted in six databases by our second author, a librarian and a biological sciences specialist. These databases include education research related databases: ERIC, Education Sources, Education Full Text, and APA PsycINFO, which were searched in the EBSCO interface; a general database: Web of Science Core Collection, which was searched in the Web of Science interface; a dissertation database: ProQuest Dissertations and Theses in the ProQuest interface. The search was performed in EBSCO using the search string: (assess* OR evaluat* OR measur* OR test* OR effective* OR rubric*) AND (reasoning OR "critical thinking" OR "scientific writing*" OR "scientific literac*" OR "research concept*" OR "biolog* concept*" OR "experimental design*" OR "hypothesis testing" OR "test* hypothesis" OR "variability" OR "variation") AND (lab* OR experiment* OR "practical work*" OR "investigation*" OR "research experience*" OR "scientific practi*") AND (bio*) AND (undergrad* OR post-secondary). The same search string was adapted to fit the syntax for searches in the Web of Science and ProQuest. Additional articles were also obtained using hand searching in Google Scholar. The search was performed on September 20, 2021, and was limited to articles published after January 1, 2001. We selected 2001 as the beginning date range in order to include a decade before the report on *Vision and Change in Undergraduate Biology Education* (AAAS, 2011), which emphasizes the essential role of thinking with and about evidence in undergraduate biology education. Our method was designed to capture a comprehensive picture of assessments focusing on evidentiary reasoning.

Based on the search criteria described above, the number of results retrieved in the initial online searches was 719 articles. Among these 719 articles, only 10 out of 19 articles we thought must be included were found with the search strategy. This indicates the difficulty in conducting an educational literature review on this topic, as in disciplinary biology education, people tend to use different terms to describe one concept. Therefore, although we may not have found all publications of relevance to our study, the number of articles in our sample was sufficient plus we decided to include the 9 articles that we had already identified to make the literature review more comprehensive.

17.2.1.2 Screening the Search List

The second author uploaded the 719 articles in the search lists plus the 9 additional articles we had identified using Rayyan (https://www.rayyan.ai/), a collaborative platform. The first and last authors carried out a preliminary review of the articles in the list together. Criteria used for including an article in our review were focused on the purpose of this literature review. To be more specific, as we defined inclusion criteria for the screening, we decided that the following categories were excluded: (1) Articles that do not contain an assessment, such as articles about curriculum designed for improving students evidentiary reasoning as a learning outcome without measuring the effectiveness of the curriculum design or with measuring the effectiveness by using student self-reports, which excluded 331 articles (e.g., Fry & Burr, 2011); (2) Articles that do not focus on measuring students' evidentiary reasoning, including articles that are trying to assess students' understanding of the nature of science (NOS) (e.g., Bautista et al., 2014), content knowledge understanding/retention (e.g., Gauthier et al., 2019), moral reasoning (e.g., Stransky et al., 2021), and self-efficacy (e.g., Beck & Blumer, 2021), which excluded 93 articles; (3) Articles not targeting the undergraduate level, which excluded 14 articles; (4) Articles not in the context of education, such as experiments about phycology, which excluded 44 articles; (5) Articles not in the context of biology or that would not ever be taught in a biological sciences department or in biology classrooms, such as studies about clinical reasoning for diseases diagnosis, analytical chemistry, and evidence reasoning in a domain-general context (e.g., Bhavana, 2009), which excluded 155 articles; (6) Articles that do not target students, such as studies of GTAs or instructor groups (e.g., Gardner & Jones, 2011), which excluded 3 articles; (7) Articles not in English, which excluded 3 articles; (8) Articles that could result in the same assessment being included twice because the authors used established assessments or adapted established assessments without much change (e.g., Auerbach & Schussler, 2017), which excluded 20 articles; (9) Articles with an assessment without a rubric or scoring structure with the assessment (e.g., Bugarcic et al., 2012), which excluded 26 articles; (10) Scientific literacy reading skills without evidentiary reasoning (e.g., Krontiris-Litowitz, 2013), which excluded 1 article; (11) and 93 duplicate articles. Since an article may be excluded by multiple exclusion criteria, a total of 46 articles were included.

17.2.2 Coding

The 46 included articles were coded both into the four relationships of the CADE and the seven scientific practice competencies of ACE-Bio theoretical frameworks (Pelaez et al., 2017; Chap. 1 in this volume). First, the seven ACE-Bio competencies were mapped into the four relationships of the CADE framework (See columns 1 and 2 in Tables 17.1, 17.2, 17.3, and 17.4). Then, the coding scheme was further divided into both disciplinary knowledge and epistemic considerations. The subcategories of each relationship were divided into different scientific practice competencies, as shown in column two of Tables 17.1, 17.2, 17.3, and 17.4. Using the <u>*Plan*</u> competence as an example, we divided and mapped it from the ACE-Bio Competencies framework into two different relationships within the CADE framework, which are the *Theory->Evidence* and the *Evidence<=>Data* relationships.

Table 17.1 Theory to evidence relationships (Codes T -> E)

CADE Evidentiary Practices	ACE-Bio Competencies/CADE questions	Rubric/Worksheets for Presentations/ Proposals/Reports	Assessment that can be used on an Exam
T -> E **Knowledge** Identify, Question, Plan	*Identify*: the ability to identify gaps or limitations in current research knowledge through the review, filtering and synthesis of relevant literature. **T->E** What are the key domain phenomena? What are the important unsolved problems?	Blair, 2014[a, b] Fisher, 2018 (IWCR) Full, 2015 King, 2018 Klein, 2014 (Written Communication Rubric)[b] Kowalski, 2016 Lansverk, 2020[b] Martinez-Vaz, 2020[b] Resendes, 2015[b] Seixas Mello, 2021 (Scheme representing the epistemic levels) Simmons, 2014 Sorte, 2020 Spence, 2020 (Proposal, Poster) St. Onge, 2007 Timmerman, 2011 Ward, 2014 Weaver, 2016 (adapted from AAC&U VALUE rubrics) Younkin & Romano, 2018[b]	Deane, 2014 (BEDCI)[b] Killpack, 2018[b]
T -> E **Epistemology** Identify, Question, Plan	*Identify*: the ability to identify gaps or limitations in current research knowledge through the review, filtering and synthesis of relevant literature. **T->E** Are alternative models or theories considered?	Fisher, 2018 (IWCR) Full, 2015 King, 2018 Kowalski, 2016 Seixas Mello, 2021 (Scheme representing the epistemic levels) Simmons, 2014 Sorte, 2020 Spence, 2020 (Poster) St. Onge, 2007 Timmerman, 2011 Ward, 2014 Weaver, 2016 (adapted from AAC&U VALUE rubrics)	

(continued)

Table 17.1 (continued)

CADE Evidentiary Practices	ACE-Bio Competencies/CADE questions	Rubric/Worksheets for Presentations/ Proposals/Reports	Assessment that can be used on an Exam
T -> E **Knowledge** Identify, Question, Plan	*Question*: the ability to generate research questions and formulate hypotheses. **T->E** What are the possible mechanisms, causal relationships, and processes?	Blair, 2014[b] Boomer, 2021 Full, 2015 King, 2018 Kowalski, 2016 Lansverk, 2020 Martinez-Vaz, 2020 Ott & Carson, 2014[b] Resendes, 2015[b] Simmons, 2014 Sorte, 2020 Spence, 2020 (Proposal, Poster) St. Onge, 2007 Timmerman, 2011 Ward, 2014 Younkin, 2018[b]	Anderson, 2011 (IPSA) Dasgupta, 2016 (neuron)[b] Dasgupta, 2014 (RED Shrimp, Drug, Bird) Deane, 2014 (BEDCI) Kiillpack, 2018 (TIED) Kowalski, 2016 (BIOCHEM, NEURO SRQ) Robertson, 2008[b]
T -> E **Epistemology** Identify, Question, Plan	*Question*: the ability to generate research questions and formulate hypotheses. **T->E** Have relationships between variables been clearly specified?	Boomer, 2021 Full, 2015 King, 2018 Kowalski, 2016 Lansverk, 2020 Martinez-Vaz, 2020 Simmons et al., 2014 Sorte, 2020 Spence, 2020 (Poster) St. Onge, 2007 Timmerman et al., 2011 Ward, 2014	Dasgupta, 2014 (RED Shrimp, Drug, Bird) Deane, 2014 (BEDCI) Kiillpack, 2018 (TIED) Kowalski, 2016 (BIOCHEM, NEURO SRQ)

(continued)

Table 17.1 (continued)

CADE Evidentiary Practices	ACE-Bio Competencies/CADE questions	Rubric/Worksheets for Presentations/ Proposals/Reports	Assessment that can be used on an Exam
T -> E **Knowledge** Identify, Question, Plan	_Plan_: the ability to plan feasible and ethical experiments to answer research questions or test hypotheses. **T->E** What variables are relevant? Why did you decide to look at those variables?	Blair, 2014 Boomer, 2021[b] Full, 2015 Kowalski, 2016 Lansverk, 2020 Martinez-Vaz, 2020[b] Resendes, 2015[b] Reyynders, 2020 Simmons, 2014 Sorte, 2020 Spence, 2020 (Proposal, Poster) St. Onge, 2007 Timmerman, 2011 Ward, 2014 Weaver, 2016 (adapted from AAC&U VALUE rubrics) Younkin & Romano, 2018	Brownell, 2014[b] Dasgupta, 2016 (neuron) Dasgupta, 2014 (RED Shrimp, Drug, Bird) Deane, 2014 (BEDCI) Deane, 2016 (SRBCI) Irby, 2018 Kiillpack, 2018 (TIED) Kowalski, 2016 (CHEMBIO, BIOCHEM, NEURO SRQ) Robertson, 2008[b] Rybarczyk, 2014 (MBDAT) Shi, 2011 (Experimental Control Exercises)
T -> E **Epistemology** Identify, Question, Plan	_Plan_: the ability to plan feasible and ethical experiments to answer research questions or test hypotheses. **T->E** Is relevant evidence used to render the question, hypotheses, plausible? Is an articulated model complete, specific, and internally consistent?	Boomer, 2021 Full, 2015 Kowalski, 2016 Lansverk, 2020 Reynders, 2020 Simmons, 2014 Sorte, 2020 Spence, 2020 (Poster) St. Onge, 2007 Timmerman, 2011 Ward, 2014 Weaver, 2016 (adapted from AAC&U VALUE rubrics) Younkin, 2018	Dasgupta, 2016 (neuron) Dasgupta, 2014 (RED Shrimp, Drug, Bird) Deane, 2014 (BEDCI) Deane, 2016 (SRBCI) Irby, 2018 Kiillpack, 2018 (TIED) Kowalski, 2016 (CHEMBIO, BIOCHEM, NEURO SRQ) Rybarczyk, 2014 Shi, 2011 (Experimental Control Exercises) Sirum, 2011 (EDAT)[b]

[a]To save space only the first author was used in the table

[b]This assessment or rubric fails to link disciplinary knowledge with epistemic considerations in examining students' evidentiary reasoning ability

Table 17.2 Evidence ⇔ Data Relationships (Codes E ⇔ D)

CADE Evidentiary Practices	ACE-Bio Competencies/ CADE questions	Rubric/Worksheets for Presentations/ Proposals/Reports	Assessment that can be used on an Exam
E ⇔ D **Knowledge** Analyze, Conduct, Plan	<u>*Analyze*</u>: the ability to apply analytical reasoning to data processing. **E ⇔D** What data models are used to organize/analyze data (e.g., graphs, statistical models)? What are known sources of error and how will they be accounted for?	Angra, 2017 (graph rubric)[b] Boomer, 2021[a, b] Brunnauer, 2016 (Summary Report)[b] Fisher, 2018 (IWCR) Full, 2015 Klein, 2014 (Written Communication Rubric)[b] Kowalski, 2016 Lansverk, 2020 Martinez-Vaz, 2020 Ott, 2014 Reynders, 2020 Seixas Mello, 2021 (Scheme representing the epistemic levels) Simmons, 2014 Sorte, 2020 Spence, 2020 (Proposal, Poster) St. Onge, 2007 Timmerman, 2011 Volz, 2009 Ward, 2014 Weaver, 2016	Anderson, 2011 (IPSA)[b] Dasgupta, 2016 (neuron)[b] Dasgupta, 2014 (RED Shrimp, Drug, Bird) Deane, 2014 (BEDCI) Deane, 2016 (SRBCI) Fisher, 2018 (QLR) Gormally, 2012 (TOSLS) Hester, 2014 (adapt IMCA)[b] Irby, 2018[b] Rybarczyk, 2014 (MBDAT)[b]

(continued)

Table 17.2 (continued)

CADE Evidentiary Practices	ACE-Bio Competencies/ CADE questions	Rubric/Worksheets for Presentations/ Proposals/Reports	Assessment that can be used on an Exam
E ⇔ D **Epistemology** Analyze, Conduct, Plan	*Analyze*: the ability to apply analytical reasoning to data processing. **E ⇔D** Are the models used appropriate? Have potential sources of error and confounding factors been evaluated?	Angra, 2017 (graph rubric)[b] Fisher, 2018 (IWCR) Full, 2015 Kowalski, 2016 Lansverk, 2020 Martinez-Vaz, 2020 Ott, 2014 Reynders, 2020 Seixas Mello, 2021 (Scheme representing the epistemic levels) Simmons, 2014 Sorte, 2020 Spence, 2020 (Proposal) St. Onge, 2007 Timmerman, 2011 Ward, 2014 Weaver, 2016 Younkin, 2018	Brunnauer, 2016 [b] Dasgupta, 2014 (RED Shrimp, Drug, Bird) Deane, 2014 (BEDCI) Deane, 2016 (SRBCI) Fisher, 2018 (QLR) Gormally, 2012 (TOSLS) Hicks, 2020 (BioVEDA)[b]
E ⇔ D **Knowledge** Analyze, Conduct, Plan	*Conduct*: the ability to conduct an investigation to achieve research goals. **E ⇔ D** Have relevant investigations been conducted? Are diverse relevant data types collected? Has an investigation been replicating with enough trials?	Brunnauer, 2016 (Summary Report)[b] Full, 2015 St. Onge, 2007 Ott, 2014[b] Simmons, 2014 Sorte, 2020[b] Spence, 2020[b] Volz, 2009[b] Ward, 2014[b] Younkin, 2018[b]	Brownell, 2014 Irby, 2018[b] Robertson, 2008[b]
E ⇔ D **Epistemology** Analyze, Conduct, Plan	*Conduct*: the ability to conduct an investigation to achieve research goals. **E ⇔ D** Are different types of data collected from diverse measures to provide support? Are sufficient trials conducted to identify data variability?	Full, 2015 Simmons, 2014 St. Onge, 2007	Brownell, 2014 Dasgupta, 2014 (RED)

(continued)

Table 17.2 (continued)

CADE Evidentiary Practices	ACE-Bio Competencies/ CADE questions	Rubric/Worksheets for Presentations/ Proposals/Reports	Assessment that can be used on an Exam
E ⇔ D **Knowledge** Analyze, Conduct, Plan	<u>*Plan*</u>: the ability to plan feasible and ethical experiments to answer research questions or test hypotheses. **E ⇔ D** Deciding what to observe or measure: ·How are variables defined? ·Continuous or categorical, ·Independent, dependent, controlled etc. ·Intervals, range sampled What instruments, techniques, apparatus will be used to collect/record data and why are these appropriate? What sampling procedures are used?	Angra, 2017 (graph rubric)[b] Blair, 2014[b] Fisher, 2018 (IWCR) Full, 2015[b] Klein, 2014 (Written Communication Rubric)[b] Kowalski, 2016 Lansverk, 2020 Martinez-Vaz, 2020 Ott, 2014 Simmons, 2014 Sorte, 2020 Spence, 2020[b] St. Onge, 2007 Timmerman, 2011 Volz, 2009[b] Weaver, 2016	Brownell, 2014 Dasgupta, 2014 (RED Shrimp, Drug, Bird) Deane, 2014 (BEDCI) Deane, 2016 (SRBCI) Gormally, 2012 (TOSLS) Hicks, 2020 (BioVEDA) Kiillpack, 2018 (TIED) Kowalski, 2016 (CHEMBIO, BIOCHEM, NEURO SRQ) Rybarczyk, 2014 (MBDAT)[b] Shi, 2011
E ⇔ D **Epistemology** Analyze, Conduct, Plan	<u>*Plan*</u>: the ability to plan feasible and ethical experiments to answer research questions or test hypotheses. **E ⇔ D** Are variables clearly defined? Are variables defined in a way that is consistent with what is known (similarity versus differences)? Is technical precision, power, sensitivity, reliability, of data collection procedures adequate?	Angra, 2017 (graph rubric)[b] Fisher, 2018 (IWCR) Full, 2015 Kowalski, 2016 Lansverk, 2020 Martinez-Vaz, 2020 Simmons, 2014 Sorte, 2020 St. Onge, 2007 Timmerman, 2011 Weaver et al., 2016 (adapted from AAC&U VALUE rubrics)	Brownell, 2014 Dasgupta, 2014 (RED)[b] Dasgupta, 2014 (RED Shrimp, Drug, Bird) Deane, 2014 (BEDCI) Deane, 2016 (SRBCI) Gormally, 2012 (TOSLS) Hicks, 2020 (BioVEDA) Kiillpack, 2018 (TIED) Kowalski, 2016 (CHEMBIO, BIOCHEM, NEURO SRQ) Shi, 2011 Sirum, 2011 (EDAT)[b]

[a]To save space only the first author was used in the table

[b]This assessment or rubric fails to link disciplinary knowledge with epistemic considerations in examining students' evidentiary reasoning ability

Table 17.3 Evidence -> Theory Relationships (Codes E -> T)

CADE Evidentiary Practices	ACE-Bio Competencies/CADE questions	Rubric/Worksheets for Presentations/ Proposals/Reports	Assessment that can be used on an Exam
Evidence -> Theory **Knowledge** Conclude	<u>*Conclude*</u>: the ability to draw conclusions about data that are limited to the scope inherent in the experimental design. **E->T** What data reporting standards apply? (e.g., attrition, error rates, outliers)? What has been learned from the evidence? Are findings/conclusions explained in terms of what is already known in biology? What other conclusions are compatible with the evidence?	Angra, 2017 (graph rubric)[b] Blair, 2014[b] Fisher, 2018 (IWCR) Full, 2015[a] Klein, 2014 (Written Communication Rubric) King, 2018 Kowalski, 2016 Lansverk, 2020 Martinez-Vaz, 2020 Ott, 2014[b] Reynders, 2020 Seixas Mello, 2021 (Scheme representing the epistemic levels) Simmons, 2014 Sorte, 2020 Spence, 2020 St. Onge, 2007 Timmerman, 2011 Volz, 2009 Ward, 2014 Weaver, 2016 Younkin, 2018[b]	Anderson, 2011 (IPSA) Brownell, 2014 Coleman, 2015 (correlation) Dasgupta, 2016 (neuron) Dasgupta, 2014 (RED Shrimp, Drug, Bird) Deane, 2016 (SRBCI) Gormally, 2012 (TOSLS) King, 2018 Rybarczyk, 2014 (MBDAT) Schen, 2007 Shi, 2011[b] Terry, 2007 (CEAT)

(continued)

Table 17.3 (continued)

CADE Evidentiary Practices	ACE-Bio Competencies/CADE questions	Rubric/Worksheets for Presentations/ Proposals/Reports	Assessment that can be used on an Exam
Evidence -> Theory **Epistemology** Conclude	<u>*Conclude*</u>: the ability to draw conclusions about data that are limited to the scope inherent in the experimental design. E->T Are data reports fair/complete? Are the conclusions internally consistent? Are the conclusions aligned with what is known? How does the evidence distinguish between multiple interpretations or hypotheses? Have alternative conclusions been explored and rebutted? Are limitations and uncertainties explicitly acknowledged/ addressed?	Fisher, 2018 (IWCR) Full, 2015 Klein, 2014 (Written Communication Rubric) King, 2018 Kowalski, 2016 Lansverk, 2020 Martinez-Vaz, 2020 Reyynders, 2020 Seixas Mello, 2021 (Scheme representing the epistemic levels) Simmons, 2014 Sorte, 2020 Spence, 2020 (Proposal, Poster) St. Onge, 2007 Timmerman, 2011 Volz, 2009 Ward et al., 2014 Weaver, 2016	Brownell, 2014 Coleman, 2015 (correlation) Dasgupta, 2016 (neuron) Dasgupta, 2014 (RED) Dasgupta, 2014 (RED Shrimp, Drug, Bird) Deane, 2016 (SRBCI) Gormally, 2012 (TOSLS) King, 2018 Rybarczyk, 2014 Schen, 2007 Sirum, 2011 (EDAT)[b] Terry, 2007(CEAT)

[a]To save space only the first author was used in the table

[b]This assessment or rubric fails to link disciplinary knowledge with epistemic considerations in examining students' evidentiary reasoning ability

The <u>*Plan*</u> competence within the *Theory->Evidence* relationship considers the variables from the theory perspective, which includes the biological disciplinary knowledge of choosing relevant variables based on some established biological theories and epistemic justification of the variables and the model used to organize them. In contrast, the <u>*Plan*</u> competence within the *Evidence<=>Data* relationship considers the variables from the data perspective, which includes the biological disciplinary knowledge of how to define and measure the variables, what sampling procedures are used, and epistemic justification of the definition and techniques chosen.

17.2.2.1 Data Analysis Method

By reading the full text of the included articles in detail, we identified the assessments and rubrics linked to the assessments that aim to reveal students' evidentiary reasoning competence and difficulties. The mapped CADE and ACE-Bio theoretical frameworks were used as the coding scheme for data analysis. Each subcategory related to the notion of evidence that is measured by the assessment was coded as a unit. The first author inductively coded all the assessments for the first pass, and

Table 17.4 Social dimensions (Codes E -> T)

CADE Evidentiary Practices	ACE-Bio Competencies/CADE questions	Rubric/Worksheets for Presentations/ Proposals/Reports	Assessment that can be used on an Exam
Social Dimensions **Knowledge** Communicate	<u>*Communicate*</u>: the ability to communicate research work in professionally appropriate modes, including visual, written, and oral formats. **Social dimensions** Are the credentials/expertise described? Has work been peer-reviewed? Is research infrastructure adequate?	Angra, 2017 (graph rubric)[b] Blair, 2014[a, b] Boomer, 2021[b] Fisher, 2018 (IWCR) Full, 2015 Klein, 2014 (Written Communication Rubric)[b] King, 2018 Lansverk, 2020 Martinez-Vaz, 2020[b] Simmons, 2014 Sorte, 2020 Spence, 2020 (Poster)[b] St. Onge, 2007 Timmerman, 2011[b] Ward, 2014 Weaver, 2016	Brrunauer, 2016 (Graphical analyses) Fisher, 2018 (QLR)
Social Dimensions **Epistemology** Communicate	<u>*Communicate*</u>: the ability to communicate research work in professionally appropriate modes, including visual, written, and oral formats. **Social dimensions** Are researchers credible? Has there been expert critique? Was it feasible to do research well?	Angra, 2017 (graph rubric)[b] Fisher, 2018 (IWCR) Full, 2015 King, 2018 Lansverk, 2020 Simmons, 2014 Sorte, 2020 St. Onge, 2007 Ward et al., 2014 Weaver, 2016	Brrunauer, 2016 (Graphical analyses) Fisher, 2018 (QLR)

[a]To save space only the first author was used in the table
[b]This assessment or rubric fails to link disciplinary knowledge with epistemic considerations in examining students' evidentiary reasoning ability

"peer debriefing to enhance the accuracy of the account" strategy suggested by Creswell and Poth (2018) was used to enhance the accuracy of findings. The last author, an experienced researcher on assessment, consistently challenged the coding and played the role of peer debriefer. All disagreements raised during the second pass were discussed until reaching consensus. As both the CADE and the ACE-Bio frameworks unpack the complex notion of evidence and the meaning of scientific practice, instead of comparing interrater reliability, we chose to use peer consensus coding to discover complexities in the data (Richards & Hemphill, 2018).

17.2.2.2 Data Analysis Examples

As an example, in the neuron assessment where biological disciplinary knowledge related to mitochondria movement in neurons is provided as a scenario, Dasgupta et al. (2016) assess students' reasoning about visualization of experiments by letting students predict their expected key findings in diagrams and explain what improvement they could make in the data to become more certain of their diagrams. In comparing the components of this assessment as well as actual expert and student responses from the publication to the table of CADE categories and criteria at https://tinyurl.com/CADE2022, we established, for example, that the neuron assessment measures students' evidentiary reasoning regarding the *Evidence<=>Data* relationship with emphasis on the *Analyze* competence, which includes knowledge about understanding the data models that were used to organize the data and epistemic considerations about the model's appropriateness and limitations. Disciplinary knowledge of experimentation research design is called for even if this assessment provides the relevant cell biology disciplinary knowledge in the form of a narrative scenario in the assessment with three diagrams to illustrate the mechanisms for moving mitochondria that can be modified in cells exposed to various drugs. Another example is from the Scheme Representing the Epistemic Levels framework of Seixas Mello et al. (2021) who measure students' arguments based on the quality of the justifications for conclusion validity in the context of the complement system in seven epistemic levels. In comparing components of the CADE categories and criteria at https://tinyurl.com/CADE2022 to the seventh epistemic level "statements incompatible with scientific knowledge" of Seixas Mello et al. (2021), we established that the authors measured students' conclusion competency under *Evidence->Theory* relationships regarding their use of established knowledge and theories linked to their justification of external consistency, which is an epistemic consideration.

17.2.3 Findings from a Review of Published Assessments

For our first research goal about how established assessments are being used to evaluate students' evidentiary reasoning, we discuss here the findings in terms of how the included assessments/rubrics evaluate students' difficulties in evidentiary reasoning. We then identify gaps that remain to be addressed according to the current established assessments/rubrics.

17.2.3.1 What Assessments Are Being Used to Reveal Evidentiary Reasoning Difficulties Among Students?

Tables 17.1, 17.2, 17.3, and 17.4 show which competencies of scientific practice and which relationships of the notion of evidence are being measured by the established assessments/rubrics that have been used in tracking the progress of post-secondary biology students in laboratory classrooms where students conduct practical research on a variety topics.

17.2.3.2 What Assessment Gaps Remain for Development of New and Useful Assessments?

As the coding results shown, first there are both assessments and rubrics that fail to link the disciplinary knowledge with epistemic considerations while assessing students' evidentiary reasoning ability (indicated by the superscript[b] in the tables). Some of these assessments/rubrics pay close attention to the important role of knowledge in evidentiary reasoning but fail to provide students with opportunities to justify the validity of their claims and less attention is directed to examining students' epistemic considerations. For example, Killpack & Fulmer (2018) assess students' experimental design skills, where students have to conduct evidentiary reasoning by designing an experiment to explore the factors that cause the diversity of feeding behavior in guppies. By using questions like "what are the control group(s)?" and "what data will you collect, and "how will you collect it?", the assessment evaluates students' biological disciplinary knowledge related to the experimental design, while ignoring the importance of assessing students' epistemic considerations by having students justify their decisions. Others measure students' evidentiary reasoning in a general context without linking epistemic considerations with specific biological disciplinary knowledge (see, for example, the EDAT by Sirum & Humburg, 2011).

Secondly, few assessments examine students' competence to <u>*Conduct*</u> an experiment within the *Evidence<=>Data* relationship. While only two assessments measure students evidentiary reasoning regarding reasoning about variation with replication (Brownell et al., 2014; Dasgupta et al., 2014), there is no assessment to evaluate students' evidentiary reasoning about the necessity of using diverse evidence in drawing conclusions or the use of convergent evidence for conclusions.

Finally, only two of the assessments measure students' competence to <u>*Identify*</u> a research problem to address in the *Theory->Evidence* relationship, where students need to reason through their decisions about the evidence to be examined with disciplinary knowledge of relevance to the investigation, and whether alternative models or theories are considered.

17.3 Assessment Gaps Addressed with CADE-Informed Test Questions

Two assessments informed by the CADE framework were developed to specifically target assessment gaps identified in the literature review but in the context of a lab activity on evolutionary tree-thinking, thus expanding our focus from experimentation in biology to include another research approach. The assessments were implemented as part of a biology lab classroom test guided by the assessment triangle to address several assessment gaps in order to track undergraduate students' evidentiary reasoning progress in biology: the linking of disciplinary knowledge with epistemic reasoning, use of disciplinary knowledge to inform a hypothesis or research goal, considering alternative models to test, and evaluating claims in terms of convergent evidence that could support or raise questions about the strength of an inference.

17.3.1 Design of the Assessments

The assessments presented below were designed to reveal post-secondary biology students' difficulties in evidentiary reasoning. To do so, assessment design was informed by both the CADE (Samarapungavan, 2018) and the assessment triangle frameworks (National Research Council, 2001; Pellegrino, 2012). Each assessment prompted a response that would link disciplinary knowledge with epistemic considerations and to target evidentiary reasoning according to *Theory->Evidence*, *Evidence<=>Data*, and *Evidence->Theory* science research practice relationships by using three open-ended questions.

Based on the assessment triangle framework (National Research Council, 2001; Pellegrino, 2012), a cognitive model with a rich psychological perspective provides detailed information to inform the assessment design. Thus, we linked each epistemic consideration that has been identified in the CADE framework we would like to assess with the correlated specific biology disciplinary knowledge in the context of evolutionary tree-thinking identified by the NGSS (NGSS Lead States, 2013) and an authoritative undergraduate evolution website (Thanukos et al., 2010) to establish the cognitive foundation of our assessments. To be more specific, in the *Theory->Evidence* relationship, students must consider if relationships between variables been clearly specified. To do this, students need to link their evolutionary tree-thinking disciplinary knowledge such as using convergent evidence from diverse sources to infer the relatedness of taxa, which includes the similarity and differences of unique DNA nucleotide sequences, anatomical evidence, variable features of fossils such as comparing the shape or number of bones, physical, chemical, and geological evidence to establish the age of fossils, etc. to reason about this knowledge in concert with epistemic considerations for justifications. For interpretation, this cognitive foundation also served as the rubric, "the methods and tools used to reason from fallible observations" (National Research Council, 2001, p. 48). In order to observe students' evidentiary reasoning competence and difficulties, the assessments provide students with rich conflict through open-ended scenarios where two scientists have different claims regarding the closest living relatives of whale/ echidna according to their different evidence. These open-ended scenarios aim to invite students to reason with and about evidence without worrying about the correct answers, since there is no one correct answer. In order to interpret students' evidentiary reasoning, we inductively coded each student's answer into the rubric we established (roughly based on the CADE table at https://tinyurl.com/CADE2022 or contact the first author for the rubrics). If the specific disciplinary knowledge correlated to the epistemic consideration is hard to define by referring to standard reports such as *Vison and Change* (AAAS, 2011), the cognitive foundation can also be established using expert answers.

17.3.2 Participants

The assessments were implemented in an introductory biology lab course at a large midwestern university with high research activity. Expert answers were from a graduate student in the Ecology and Evolutionary Biology program and a professor who teaches an upper division Evolutionary Biology course for teachers. All assessment responses were collected according to a protocol that was reviewed and approved by the Institutional Review Board (IRB#17020187760251). The graduate student had served as a graduate teaching assistant for the target course during three semesters without any CADE or ACE-Bio Competencies training. Student responses were collected from pre- and post-tests at the beginning and end of the target lab course.

17.3.3 Addressing Assessment Gaps to Reveal Students' Difficulties with Evidentiary Reasoning About Evolutionary Trees

To address the gaps, two assessments (Boxes 17.1 and 17.2) were designed to reveal students' difficulties in evidentiary reasoning by using scenarios where two scientists from different biological disciplines using different sets of convergent evidence draw different claims. With the aim to evaluate students' evidentiary reasoning in a comprehensive matter, there are no right or wrong answers to the questions. The questions are open-ended, inviting the students reasoning through the scenarios from *Theory->Evidence*, *Evidence<=>Data*, and *Evidence->Theory* relationships. All questions aimed to provide students with inquiries to link their biological disciplinary knowledge with epistemic considerations, by asking them to use reasoning and their biological disciplinary knowledge and to justify the answers in order to reveal their epistemic considerations. Below we provide the assessments, example expert answers, and then we discuss selected examples of students' answers that meet expectations and others that do not meet expectations to show how observation and interpretation of the variation in performance works with these assessments.

17.3.3.1 Assessment Items Informed by CADE

A question about whale evolution was used as a pre-test at the start of an undergraduate biology lab course and a question about echidna evolution was used as a post-test item on the final exam. In addition to a scenario, each assessment had three probing questions: "Why did the two scientists make different decisions about what types of evidence to gather" to assess students evidentiary reasoning under the *Theory ->Evidence* relationship; "Which scientist provides the strongest evidence for their claims" under the *Evidence->Theory* relationship; and "What additional kinds of evidence to consider and why" under the *Evidence<=>Data* relationship in terms of the CADE practices of reasoning with and about the evidence.

Box 17.1: Whale Assessment

It has been long established that whales are mammals, but scientists are not yet certain of their exact ancestry and which current species are their closest living relatives. Two scientists told our local news reporter their ideas about whale evolution:

Scientist Sandra Wells says:	Scientist Rosendo Pascual says:
The manatee is the closest living relative of the whale because manatees have flippers and tail structures more like whales and can spend long periods of time under the water like whales. We also found dozens of DNA sequences shared by whales and manatees.	The hippopotamus is the closest living relative of the whale. We found a fossil of the hippo's ancestor with a complete ancient skeletal remain like the backbone of a whale. It also had limb and teeth structures found in the modern hippopotamus. We even found one DNA sequence that whales and hippopotami share.

1. Why did these two scientists make different decisions about what types of evidence to gather and how did their assumptions influence the quality and the accuracy of their claims?
2. Which scientist (Dr. Wells or Dr. Smith) do you believe provides the strongest evidence for their claims about the closest living relative of the whale? Explain the reasons for your answer.
3. What additional kinds of evidence should the two scientists consider? Why you think these additional kinds of evidence might be useful to test their ideas?

Box 17.2: Echidna Assessment

It has been long established that there are many different types of mammals, but scientists are not yet certain of their exact ancestry and which groups are more closely related. Two scientists told our local news reporter their ideas about echidna evolution:

Conservationist Mandy Watson says:	Scientist Rosendo Pascual says:
The bandicoot is the closest living relative of the echidna. Both have long slender snouts that function as both mouth and nose and both feed primarily on earthworms. Both are found in Australia near a water supply. Throughout Australia we found four kinds of fossil bandicoots and also fossils of the echidna, both with short, strong limbs and claws for powerful digging. A collaborator found many DNA sequences shared by modern echidnas and bandicoots.	The duck-billed platypus is the closest living relative of the echidna. In both animals the upper appendage bones are held roughly parallel to the ground when the animal walks, more like most modern reptiles. The platypus has a cloaca through which eggs are laid and both liquid and solid waste is eliminated. The echidna also has one body cavity for the external openings of the urinary, digestive, and reproductive organs. We even found one DNA sequence that modern platypus and echidna share.

Animal Names	Eastern Barred Bandicoot	Long-beaked Echidna	Duck-billed Platypus
Average mass	640–766 g	11 kg	1.52 kg
Average basal metabolic rate	1.902 W	6.493 W	1.931 W

1. Why did these two scientists make different decisions about what types of evidence to gather and how did their assumptions influence the quality and the accuracy of their claims?
2. Which scientist (Dr. Watson or Dr. Pascual) do you believe provides the strongest evidence for their claims about the closest living relative of the echidna? Explain the reasons for your answer.
3. What additional kinds of evidence should the two investigators examine? Explain why they should consider that evidence and why you think this additional evidence is reasonable to consider.

17.3.3.2 Expert Answers for Whale and Echidna Questions

Expert answers to both the whale and echidna questions provide examples of how the three questions in each assessment were able to invite reasoning through the different subcategories within the research practice categories of the CADE, linking disciplinary knowledge with epistemic considerations. For example, regarding T -> E, the model articulation component of evidentiary reasoning is evident when decisions about the evidence to be examined are informed by disciplinary knowledge of relevance to the investigation. **Bold font** in the answers indicate that **alternative models or theories are considered**, which is an epistemological consideration for T -> E. Regarding E <=> D, expert responses to both assessments consider comparison of DNA sequences as a research method but they also bring up several ideas about different types of data to collect from diverse measures to provide additional support, which is an epistemological consideration for E <=> D. Regarding E -> T, epistemological considerations about using *convergent evidence to draw conclusions* are indicated with an *italics font,* and biological disciplinary knowledge of relevance to decisions about the evidence or conclusions is underlined.

WHALE ASSESSMENT EXPERT ANSWER

Whale Q1: **Dr. Wells is assuming that the morphological, behavioral, and some genetic similarities between manatees and whales are an indication of phylogenetic similarity.** These assumptions do not necessarily account for which DNA sequences are shared. Mammals generally have many homologous sequences within their genome, even taxa that are not closely related (e.g. there is genetic similarity between dogs and wallabies). Morphological and behavioral similarity could be a sign of convergent evolution rather than phylogenetic similarity (e.g. sugar gliders and flying squirrels share many morphological and behavioral traits even though they are not closely related). These assumptions could skew the interpretation of evidence by Dr. Wells whose conclusions may not be accurate.

Whale Q2: More data is needed to understand which claim is better supported by evidence. While shared physical and DNA traits could be an indicator of phylogenetic similarity, this evidence alone is not enough. Different animal lineages may share physical and behavioral traits while not being closely related. Similarly, common ancestors in the fossil record may give evidence for relatedness, but all mammals have a common ancestor if one looks back far enough.

Whale Q3: *Knowing which DNA sequences are shared between whales and manatees, the timing of the common ancestors (between the hippo and the whale) when whales and hippos diverged, and when whales and manatees diverged (evidence supporting/refuting convergent evolution) are some of the information that would help to fully understand which claim (if either) is more accurate. If the shared DNA sequences are unique to aquatic mammals (i.e. dealing with flipper formation), then this could be strong evidence for relatedness. However, if the DNA sequences are common among all mammals (i.e. general vertebrae formation), then this evidence may not be very robust. The time periods when the common ancestor existed and when whales and hippos diverged would be useful to know because if the ancestor is significantly more ancient than when these animals diverged, this common ancestor may not be a strong indicator of relatedness. Knowledge of when whales and manatees diverged could indicate whether similarities or differences suggest coevolution or speciation. If whales and manatees diverged very far back in time, then the common morphological and behavioral traits are likely due to coevolution. However, if they diverged recently, then the commonalities between whales and manatees may be strong evidence for phylogenetic similarity.* [Expert answer from graduate student in Ecology and Evolutionary Biology.]

ECHIDNA ASSESSMENT EXPERT ANSWER

An expert response to the echidna assessment shows disciplinary knowledge of relevance to the investigation including the depiction of two alternative cladogram models in this figure that was applied to decisions about evidence to be examined (Fig. 17.1).

The following expert response to the echidna assessment considers **alternative models or theories**, indicated as **bold font**, the use of *convergent evidence* to draw conclusions is indicated in *italics*, and underlined text indicates reasoning about evidence that is informed by biological disciplinary knowledge.

Echidna Q1. Mandy Watson is a Conservationist, so she studies where organisms live and what they eat in order to conserve living organisms along with their environment. She observed that both bandicoot and the echidna have long slender snouts as well as claws for powerful digging, and that both feed on earthworms and live near water. Historically, these types of observations were used to classify animals into groups when they were identified and named. As a conservationist, Mandy may not know much about evidence from a collaborator who found many DNA sequences shared by modern echidnas and bandicoots. Was she **wondering if the platypus shares those same sequences? Perhaps she has not considered that different animals generally have some homologous sequences within their DNA. More evidence is needed to determine if the structures she described represent homology (inherited from a shared ancestor) or homoplasy, which refers to structural similarity such as from convergent evolution rather than from a recent shared ancestor. Many examples of convergent evolution are found in animal morphology or the fossil record, where we find evidence that environments shape organisms. When some individuals from distantly related taxa are more likely to survive and reproduce, they become more similar because both are fit to eat similar food in similar environments.** Instead, an evolutionary biologist would use evidence to establish the chronology of evolution and common ancestry, not just with biogeographical evidence of fossils, but instead by using the fossils and other data to identify derived traits that distinguish organisms on one branch of their family tree from those on the other branches. An example in this case is that most mammals walk on four legs holding their body upright, unlike the duck-billed platypus and the echidna that walk like most modern reptiles, which is with their upper appendage bones being held roughly parallel to the ground, according to the Scientist, Rosendo Pascual. Rosendo may have been considering the chronology of their ancestry in noticing that the bandicoots have a trait like most modern mammals, unlike the duck-billed platypus and the echidna.

Echidna Q2. To decide who provides the strongest evidence, **consider a diagram of the different models being suggested (Fig. 17.1)**. Mandy's use of anatomical (snout and claws), food source, and biogeographical evidence leads her to believe that the bandicoot and echidna are sister taxa. In contrast, Rosenda places a more recent ancestor as one that

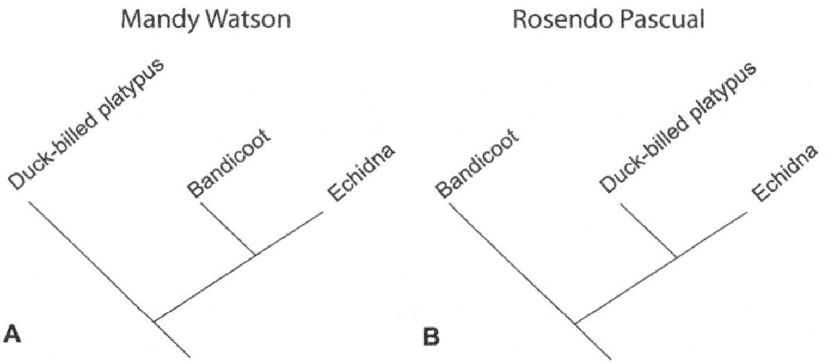

Fig. 17.1 According to an expert's answer, these cladograms depict alternative models for the chronology of ancestors shared among these three types of animals. The branch with bandicoot and echidna as sister taxa in panel A illustrates Mandy Watson's idea that bandicoot and echidna are more closely related whereas the branch with platypus and echidna connected by a more recent ancestor in Panel B illustrates Rosenda Pascual's claim. In both diagrams, all three share an ancient ancestor indicated by the branching point at the bottom of the tree, but the sister taxa share a branch through a connection to a more recent ancestor that the third group or outgroup does not share

is shared by the platypus and echidna. I agree with Rosendo because I know that the duck-billed platypus and the echidna are both egg-laying mammals. If I accept her description of *their upper appendage bone structures and walking behavior and with both having a cloaca for the urinary, digestive, and reproductive organ external openings, this is strong evidence for the sister taxa branch model in panel B. In fact, despite it's larger size, according to the data table provided, the higher metabolic rate of the bandicoot compared with the platypus and echidna provides additional evidence that it may be more closely related to other mammals like kangaroos or carnivorous mammals that are also quite warm-blooded. However,* ***this evidence does not rule out the possibility that echidna and platypus are paraphyletic, with other mammals being a monophyletic subgroup in which case Mandy's model could be right if data from fossil bones and DNA sequence homology both suggest that the platypus is an outgroup for a clade that includes all other mammals and the echidna according to the sister taxa branch model in panel A.***

Echidna Q3. As additional evidence, both could collaborate to more carefully examine the biogeographical and chronological history of the four kinds of fossil bandicoots and the fossil echidna. I would expect to find the upper appendage bones for the short, strong limbs and claws in the fossil echidna to be held roughly parallel to the ground, but do any of the fossil bandicoots have that anatomical feature? I would like to know whether features and the chronology of the fossil data suggests echidna and bandicoots share a more recent ancestor as in panel A and if additional data - not just one DNA sequence that modern platypus and echidna share, but instead a thorough comparison of homologous DNA sequences among all three animals - suggest that platypus and echidna share more recently derived traits. Neither Mandy nor Rosenda have stated whether DNA sequences shared by the two they group together are found to be more different in the third group. In summary, to rule out either model A or B, there is a need for evidence of a more recent ancestor linking the two that are closest living relatives, leaving the third as an outgroup. *Evidence of homologous traits shared among the two but that is missing from the third and from other living mammals, and ruling out the possibility that the trait could result from convergent evolution, are two uses of evidence that could converge in strengthening claims about which two could be closer living relatives.* [Expert answer from a professor who teaches an Evolutionary Biology course for science teachers.]

17.3.4 Findings from Typical Examples of Students' Answers to the Whale and Echidna Questions

For our second research goal about how well the CADE-informed assessments reveal students' difficulties in evidentiary reasoning, we provide some typical students answers as examples. The following student answer examples were selected to illustrate how the gaps that were found in our review of published assessments have been addressed by the whale and echidna questions used as pre- and post-test in an introductory level undergraduate biological lab course. The examples range from good answers that cover many aspects of the notion of evidence in the reasoning process of research practices as well as answers that fail to consider some aspect of the notion of evidence during reasoning. Again, the biological disciplinary knowledge within the reasoning is underlined.

17.3.4.1 The Assessments Probed Evidentiary Reasoning with Disciplinary Knowledge Linked to Epistemic Considerations

Student answer examples illustrate how disciplinary knowledge was linked with epistemic considerations in evidentiary reasoning.

> Whale Q3: These additional forms of evidence can reveal more detailed information about how closely the species are related. Comparing mitochondrial DNA is important because it is present in most cells in an organism, it evolves quickly but at a known rate, and it is passed down through the maternal line.

Above, the student applied specific disciplinary knowledge about mitochondrial DNA when using general epistemic considerations to justify the value of including mitochondrial DNA as additional evidence in determining how closely the species are related.

> Echidna Q3: The scientists should always examine DNA evidence further to establish an even stronger connection. Similarities in the genome of two species can greatly bolster any potential relationship and can point to the closeness of the two species. They should also delve deeper into the molecular use of proteins, enzymes, metabolic pathways, etc. to show that the two species utilize such compounds similarly. Also, examining the species habitats and niches can point to commonalities. If the two species live in similar environments and occupy the same niche and carry out the same functions in their ecosystems, that may be because they are closely related. Having the same homologous and vestigial structures can also be indicators of common ancestry.

Above, the student linked specific disciplinary knowledge about similarities in genome, molecular use of proteins, enzymes, metabolic pathways, species habitats and niches with the epistemic considerations to justify the value of those data as additional evidence in determining the closest living relatives.

17.3.4.2 Some Responses Described Disciplinary Knowledge But Failed to Link to Epistemic Reasoning About the Relevance or Quality of Evidence

Next we provide several examples of student answers that failed to link disciplinary knowledge with general epistemic considerations. Using guiding questions in the CADE table at https://tinyurl.com/CADE2022 for scoring, they failed to get full credit in all three Science Research Practices for reflective evaluation or critique of the evidence (epistemic considerations) in ways that link to the relevant disciplinary knowledge of biology.

> Whale Q2: I believe Dr. Smith provides the strongest evidence for her claims about the closest living relative of the whale because she compares skeletal structure.

This student correctly identified skeletal structure as relevant biological knowledge for determining evolutionary relationships. However, they failed to note specific properties of skeletal structures to examine or how the skeletal structure data might be used as evidence. They could have detailed the quality and scope of structural data to use as evidence for a better reasoning process (epistemic considerations).

Without considering the relevance or quality of the evidence, the next examples of student responses imply that having MORE data is better without justification, which was common among undergrad student responses that failed to link to epistemic reasoning to justify the relevance, scope, or quality of evidence.

> Whale Q1: Scientist Sandra Wells based her conclusion on external morphological observations between manatees and whales. Ashley Smith based her conclusions on internal skeletal observations, comparing bone structures of the whale and hippo. Both Scientists found DNA segments that link both the hippo and the manatee to the whale. They both have evidence that can suggest logical assumptions about whale evolution and have genetic evidence to support their claims. The biggest difference among their findings, however, is Wells discovered DOZENS of DNA sequences and Smith found A DNA sequence.
>
> Whale Q2: I believe Dr. Wells' findings are more persuasive due to the dozens of DNA sequences found. If she just based her evidence on morphology, I wouldn't have found her evidence strong. But, since she presented <u>DNA evidence to suggest linkage</u>, her evidence is stronger.
>
> Whale Q3: Some other types of evidence that could be useful is <u>dietary evidence</u>. Do they have similar diets? Other possible evidence is <u>behavioral aspects, life histories, mating strategies, geographical distributions, derived traits, and ancestral traits</u>. Use the evidence gathered to start developing theories of how that animal evolved. After more information is gathered, then start placing the animal in a phylogeny and determining plausible common ancestors. This evidence can start to lead to answers and develop more theories that explains whale evolution better.

The above response thoroughly describes some useful evidence by repeating the information provided in the assessment in their response and then also listing a few additional types of biological evidence. Although there is mention of the need to "start developing theories of how that animal evolved," it does not explain how evidence might be used to establish the chronology in a phylogeny.

The next example illustrates this same problem of not considering the nature, scope, and quality of data for use as evidence, and the response again illustrates the "MORE data is better" superficial reasoning about evidence that we often find among undergrads.

> Echidna Q3: Other sorts of evidence that the scientists could consider are <u>diets, lifespan, behavior (whether they live in groups, how long/if the offspring stay with their mother, how aggressive they are towards members of their own species or predators, level of activity etc.), body mass, migration patterns or whether or not they live in a singular area, feeding habits</u>, and many more. Additional pieces of evidence are always very useful ... More data that can be collected can help refine your conclusion and it can help support the claims that you have already made.

Above, the student lists quite a bit of biological disciplinary knowledge, such as diets, lifespan, behavior as relevant to determining closest relatives. However, the student failed to link the disciplinary knowledge with general epistemic considerations in justifying reasons for including the data as evidence.

17.3.4.3 Student Answer Examples Discuss Convergent Evidence That Could Support or Raise Questions About the Strength of an Inference

Many student answers illustrate how convergent evidence (italicized words) are used in drawing conclusions, which is an epistemic consideration worth mentioning in justifying ideas about using more data in evidentiary reasoning.

> Whale Q1: Scientist Sandra Wells decided to gather more observational evidence (noting the physical appearance and characteristics of the modern whales and manatees), whereas Scientist Ashley Smith decided to gather more concrete evidence from past skeletal remains. This highlights the difference between their two approaches: Scientist Sandra Wells wanted to look more at current day observations, but Scientist Ashley Smith wanted to look at the past (hence the archaeological evidence). Scientist Sandra Wells' approach of just focusing on present day observations means that the quality and accuracy of her claims are not strong because she only has one part of the picture. You cannot just look at observations and make the assumption that it implies to a different concept, which in this case was ancestry. Many animals look closely related, but that is not enough to make such a broad claim (she needs more concrete evidence). Although she refers to similar DNA sequences, they could just be extremely common sequences found between multiple organisms. Scientist Ashley Smith's approach leads to the higher quality and accuracy of her claims because she looked at similarities in fossils, which is more concrete evidence. In order to make such a broad claim, however, *various types of approaches should be used*.
>
> Whale Q2: I believe that Dr. Smith provides the strongest evidence for her claim that the hippopotamus is the closest living relative of the whale. Her evidence, mainly the fossil of the hippo's ancestor, is more concrete than Dr. Wells' observational evidence on characteristics. Although both methods are important in science, Dr. Smith's approach makes the most sense given that they are looking at ancestry. Additionally, both scientists mention similarities in DNA sequences, which is also good evidence.
>
> Whale Q3: Additional kinds of evidence that *the two scientists should consider is similarities in sequences of RNA, comparing organ systems, or looking more in depth at previous fossils and comparing their similarities (or the presence of homologous parts).* I think that *additional kinds of evidence are useful to test their ideas because it offers a different perspective that will provide additional insight on each of their claims.* The more evidence that supports a claim, the more valid it becomes. On the other hand, *if other types of evidence don't support a claim, it's a good indication that the claim needs to be reevaluated.*

As with the expert responses, disciplinary knowledge has been underlined and the italicized words highlight ideas about convergent evidence. Both the previous example and the next connect epistemological considerations about the scope and relevance of the evidence with their disciplinary knowledge to highlight the chronology of biological evolution.

> Echidna Q1: The conservationist and scientist relied on their different scientific educations and exposure to distinctive scientific literature within their specialties to identify and evaluate evidence. The conservationist is trained to analyze the ecological interactions of the two species under the umbrella of environmental sciences. On the other hand, the scientist is trained to recognize their similar anatomical and physiological features. Their assumptions are expected to introduce some bias into the process of evidence selection as they could, unintentionally or intentionally, gravitate towards familiar explanations in regards to their different backgrounds. This is likely to have an effect on the final claim and reduces its accuracy and quality.

Echidna Q2: In my opinion, the scientist provides the strongest evidence for their claims because of their precise comparisons between the echidna and platypus. Anatomical similarities are discussed and encompass the entire organism including musculoskeletal system, digestive system and reproductive system. These are precise observations rooted in biology that strongly suggest that both species evolved from a recent, shared common ancestor. Furthermore, the investigators themselves identified a shared DNA sequence. This is in contrast to the conservationist who relied on a "collaborator" to provide DNA testing results. This shows that, in the case of the scientist, the process of data collection was controlled and accountable. The conservationist's claims are too broad to establish a close evolutionary relationship between the bandicoot and echidna. The statement that "both are found in Australia near a water supply" does not indicate a relationship because a vast majority of animals are expected to live near water supplies. The conservationist also notes that both bandicoot and echidna fossils were found in Australia. However, considering the geographical size of the country, this is insufficient because a variety of fossils are also found in Australia that do not necessarily imply that they all came from the same lineage. Similarly, the claim that both species feed on earthworms is also too broad because diet similarities is a tenuous relationship especially compared to the anatomical similarities discussed by the scientist.

Echidna Q3: Both investigators should consider the evolutionary history of echidna and trace its lineage back to a shared common ancestor with either the platypus or the bandicoot. This additional evidence is reasonable because they are attempting to determine the closest living relative of the echidna therefore by drawing its phylogenetic tree, *the investigators should be able to identify similar morphology and molecular characteristics*. *Fossil records will also be useful in this case as they will aid in pinpointing the moment of divergence* of the bandicoot and platypus from the echidna. The closest living relative will therefore be the species that diverged most recently.

Here is another example from a student who was able to link their disciplinary knowledge about fossil structure and the type of creature with epistemic considerations about the use and quality of the evidence to test a model and draw conclusion.

Whale Q2: Dr. Smith, on the other hand, mainly looked at ancestral skeletal structures and these can easily evolve over time. Why yes, Smith found that the hippopotamus and whale may have been related quite a long time ago, but the two species have evolved greatly. The hippopotamus is both a terrestrial and aquatic creature, while a whale is only an aquatic creature. Also, Dr. Smith could only find one common DNA sequence between the hippo and the whale, which also shows that Wells' findings are more accurate.

In this case above, the student argued that considering the huge difference between the hippopotamus and whale nowadays, without knowing the age of the fossils, and using homologous structure of fossils and only one common DNA sequence alone would be insufficient evidence for testing the model.

Echidna Q2: Both scientists decided to take DNA into account. This is reasonable because shared sequences maybe link to ancestry. Both also decided to note the anatomy of each creature, citing similar features that could be a sign of homology. Watson looked at geographical location and fossil history which may point to a shared point of evolution considering conditions for the time period.

Above, the student used disciplinary knowledge related to DNA, anatomy, homology, geographical location and fossil history linked with their justification like similar features may indicate homology in evaluating the conclusion about closest relatives. The way the student organized different kinds of evidence illustrates how convergent evidence was used to draw conclusions. Specifically, they pointed out that similarity in fossil structure indicates homology and that geographical location and fossil history can further help the scientist to identify the time period for speciation.

Of course, there were many examples of student answers that failed to use convergent evidence in drawing conclusions.

> Whale Q2: I believe that she is correct due to the fact she found multiple DNA sequences shared between the two of them.

In this example, the student relied on one single kind of evidence, the similarity in DNA sequences, in drawing conclusions. In fact, quite a few students claimed that the best and most sufficient evidence is based on multiple DNA sequences. By doing so, the student ignored other kinds of evidence, such as homology in fossil structures and chronology of the fossil evidence in reasoning about the models being tested.

> Echidna Q2: Neither scientist provides the strongest answer since both only used one criteria and method to support their assumptions.

Above, the student claimed that both scientists use only one criterion to support their assumptions. They failed to identify that each scientist used multiple kinds of evidence in drawing conclusions.

17.3.4.4 Some Responses Failed to Use Appropriate Disciplinary Knowledge to Inform a Hypothesis or Research Goal

There were many good answers such as the following which illustrates how students reasoned from theory to inform their hypothesis.

> Whale Q1: These two scientists made different decisions about what types of evidence to gather because they both felt that there were different characteristics that determined whether or not a certain animal evolved from another certain animal. One scientist, Dr. Wells, thought that you could determine if two animals were related based on the structures of the outside parts of their bodies (flippers, tails, can breathe in water, etc.) and their DNA sequences. The other scientist, Dr. Smith, thought you could determine if two animals were related based on the structure of the inner parts of their bodies (bones/teeth) and their DNA sequences."

Above, the student successfully identified reasons for different scientists to gather different evidence informed by a foundation of different knowledge and theories. One scientist based their hypothesis upon similarities of outside structure and DNA, while the other considered similarities of inner structure and DNA.

> Echidna Q1: The conservationist and scientist relied on their different scientific educations and exposure to distinctive scientific literature within their specialties to identify and evaluate evidence. The conservationist is trained to analyze the ecological interactions of the two species under the umbrella of environmental sciences. On the other hand, the scientist is trained to recognize their similar anatomical and physiological features.

Above, the student identified that conservationists inform their decisions based on their training to analyze ecological interactions in the context of environmental sciences, whereas the scientist was trained to recognize similar anatomical and physiological features.

When limitations to the use of modern DNA evidence for establishing a phylogeny were discussed in class, some students did not realize that we were discussing how evidence might be used in combination with other evidence of shared derived traits that might suggest a recent ancestor shared by two sister taxa but that is not shared with an outgroup clade. For example, the next response shows that potential use of DNA sequence data was not understood by a student who complained that "both scientists use DNA sequencing to prove their point (bad science)." With both questions, many responses showed difficulty understanding relevant disciplinary knowledge or, as in this case, they revealed wrong ideas about how to apply disciplinary knowledge when deciding how data might be used as evidence.

> Echidna Q1: Those two scientists made different decisions because they most likely work in different fields, so depending on their specialty this would have lead them to their particular claims. Both claims could be accurate; however, the finding of shared DNA sequences is irrelevant when trying to connect the species.

> Echidna Q2: Disregarding that both scientists use DNA sequencing to prove their point (bad science), I believe Dr. Pascual provides the strongest evidence because he relates the anatomy and physiology of both species. Ultimately, no scientist is more right than the other one because they both use valid arguments and data.

> Echidna Q3: The scientist can look into where all three species originated. This information can provide crucial geographical information as to if the species is native to where it's living - whether it naturally has always eaten, reproduced, and looked that way - or if they are an invasive species or migrated over there and have had to adapt these characteristics.

We also found many examples where students failed to recognize that different hypotheses were informed by different disciplinary approaches to research.

> Whale Q1: They made different decisions because they both looked at different types of data and interpreted it in different ways.

> Echidna Q1: These two scientists made different assumptions based on the specific data they each had.

In these examples, the students failed to consider any theory or disciplinary knowledge that informed each hypothesis.

17.3.4.5 The Assessments Probed Evidentiary Reasoning About Whether Alternative Model Had Been Considered

Many responses failed to consider alternative models and how data might be used as evidence to rule out one of the models, as illustrated with the following example response.

> Echidna Q1: These two scientists made different decisions because each scientist has a different way of thinking based on the information given to them. Their assumptions influenced the quality and the accuracy because both tested their hypothesis on what they believed to be true and both made observations that they can back up with evidence.

> Echidna Q2: Dr. Pascual provides the strongest evidence because he found evidence based on the animals' reproduction and bone structures plus the fact that they found evidence of sharing a DNA sequence.

> Echidna Q3: They should consider their genotypes because this will show a tree how each generations' alleles could have altered to provide evidence of evolution.

However, several examples of student responses correctly illustrate skepticism toward data and they raised alternative models to consider as they attempted to rule out some alternatives with evidentiary reasoning.

> Whale Q2: Although she refers to similar DNA sequences, they could just be extremely common sequences found between multiple organisms.

> Echidna Q2: Humans share many DNA sequences with bananas, and we aren't related to fruits, so I don't think that is sufficient evidence.

Other examples of student responses failed to identify the alternative models.

> Whale Q2: I believe Dr. Wells provides the strongest argument, because her evidence is based on the physical characteristics of how the manatee and whales look now. Their similar structures and DNA sequences can infer that they can be related through evolution.

In the above example, the student failed to identify that the similar structures shared between manatee and whales may be caused by selection pressures (homoplasy) instead of inherited from a common ancestor (homology).

> Echidna: Data such as structure, diet, habitat, and fossil record are all important. With the DNA sharing more sequences, I feel they are closer in relation as well.

In this example, the student simply listed data that was provided in the scenario of the question without reasoning about the evidence.

17.4 Summary and Discussion

The CADE provided the cognition foundation for evidentiary reasoning as a target for assessing how well students understand and would be able to do more authentic biological research. Our first study aimed to gain an understanding of established assessments that are being used to evaluate students evidentiary reasoning. We found that there is a need for assessments to track progress as students learn to reason through their decisions about evidence. Areas that still need to be more carefully examined include how well the students link epistemic considerations with their disciplinary knowledge to inform decisions about evidence, their use of disciplinary knowledge of relevance to the investigation to inform a hypothesis or research goal, how diverse evidence is used to establish sufficiency of evidence such as by using convergent evidence to strengthen conclusions, and whether alternative models or theories are considered.

CADE-informed assessments helped to address these gaps by revealing competent reasoning about evidence and students' difficulties with evidentiary reasoning in a meaningful way in the context of evolutionary tree-thinking. Guiding questions in the CADE table at https://tinyurl.com/CADE2022 highlighted what to observe. The findings revealed difficulties that could be addressed by instructors and students.

According to the assessment triangle (National Research Council, 2001; Pellegrino, 2012), the foundation for assessment design is to identify a cognition model that shows how leaners typically represent information and build domain expertise. Here we adapted the CADE (Samarapungavan, 2018) to clarify how each research practice in biology involves disciplinary knowledge integrated with epistemic considerations as a cognitive foundation for understanding how students could demonstrate evidentiary reasoning throughout the process of biological research practice. A realistic research scenario as an assessment with three open-ended questions provided a useful test item that targeted three research practice relationships to provide students with opportunities to demonstrate comprehensive evidentiary reasoning with evolutionary tree-thinking as a biological context. In our implementation of the assessments, we were able to observe and compare biology experts' and students' evidentiary reasoning in terms of the targeted cognitive competencies. Examples of both expert and student answers show that assessments informed by the CADE framework were able to reveal aspects of the reasoning process of relevance to three types of science research practices: Theory to Evidence Relationships (T -> E) involve formulating testable models, hypotheses or explanations; Evidence <=> Data Relationships (E <=> D) relate to designing, executing, and analyzing data from investigations; and Evidence to Theory Relationships (E->T) relate to inferences and the sufficiency of conclusions. The E->T component reflects evidence-based reasoning in science (Erduran et al., 2015; Furtak et al., 2010; Toulmin, 1958; Tytler & Peterson, 2005) but the CADE more comprehensively highlights T->E and E <=> D relationships as well. Furthermore, the CADE framework provides a practical framework for interpreting student answers to reveal their difficulties with evidentiary reasoning. Example responses from undergraduate students show that some students have difficulty in using convergent evidence to draw conclusions. Instead of constructing a model to organize diverse evidence to draw conclusions, some students claim that more evidence is needed but without justifying why the evidence would be useful, some rely on one kind of evidence, or they emphasize the value of DNA sequence data while ignoring other evidence that would be useful for determining phylogenetic relationships among different animals.

The CADE-informed assessment examples should, however, be modified or new assessments developed for any other research context or subdiscipline in biology, such as experimental design, microbiology, or immunology. By altering the story about investigators with different disciplinary knowledge and by presenting a conflict that encourages the students to link their relevant knowledge of biology with epistemic considerations, the CADE framework could be tested as a cognitive foundation model in other biology learning situations.

17.5 Conclusions

Use of the CADE framework as a lens made it possible to delve deeply into evidentiary reasoning in a comprehensive way that includes identifying a research problem and planning and conducting research in addition to evidence-based reasoning about scientific evidence for backing claims. CADE-informed assessments revealed students' evidentiary reasoning difficulties by emphasizing the fundamental role of epistemic cognition, and by further linking epistemic cognition with disciplinary knowledge. The CADE was compatible with ACE-Bio Competencies to provide meaningful insight into the meaning of evidentiary practices throughout the research process and it helped inform how to evaluate students' evidentiary reasoning.

Several gaps were identified by the literature review: (1) Some assessments and rubrics failed to link disciplinary knowledge with epistemic considerations while assessing students' evidentiary reasoning; (2) Few assessments target student competence within the Evidence <=> Data relationship of relevance to *Conducting* a research study and no published assessment evaluated students evidentiary reasoning about the necessity of using diverse evidence in drawing conclusion, or how to use convergent evidence to draw conclusions; (3) Only two assessments in our study measure students competence to *Identify* a research problem with Theory->Evidence relationships where students must reason through the need to consider alternative models or theories and for making decisions about the evidence to be examined using disciplinary knowledge of relevance to the investigation. Our CADE-informed assessments addressed these gaps by revealing students' evidentiary reasoning competence and difficulties by emphasizing the fundamental role of epistemic cognition, and by further linking epistemic cognition with disciplinary knowledge.

The CADE framework proved to be a useful guide for revealing students' difficulties with evidentiary reasoning. In terms of the assessment triangle (National Research Council, 2001; Pellegrino, 2012), it provides a cognition target detailing several types of reasoning about evidence that the assessment task should elicit. It also targets components to notice in expert responses to the assessment and it facilitates creating new assessments where the key components can be observed in student responses. When those components were missing from student responses, it helped with interpretation to target those areas of difficulty to address in the future (i.e. the linking of biological knowledge with epistemic considerations while considering both domain-general and discipline-specific aspects of evidence).

As an implication, the findings of this research will benefit the teaching and assessment of learning about evolutionary trees by providing educators and students with a feasible way to deconstruct and unpack the notion of evidence for the various tree diagrams. The findings also provide insight into assessment instrument choices and the design of new assessment tools for use on tests to reveal students' difficulties with evidentiary reasoning in biology as a discipline. The CADE table at https://tinyurl.com/CADE2022 is provided in a digital format to be easily modified for use in defining the target cognition or in developing scoring rubrics so that others can use CADE for their own purposes. Future studies should continue to modify the framework as it is applied to other contexts to help develop additional understanding for how evidentiary reasoning can be assessed in science students.

In summary, the CADE framework unpacks the notions of evidence into component parts (Samarapungavan, 2018) that might be applied to assess learning about evidentiary reasoning in other research contexts together with the assessment triangle (National Research Council, 2001; Pellegrino, 2012). Based on our work and synthesis of the literature, we recommend considering the following ideas and approaches when developing assessments to monitor students' competence for reasoning with and about evidence in the context of science investigations:

- According to the CADE framework, comprehensive knowledge of the quality and use of scientific evidence involves several inter-related research practices: theoretical knowledge informs what evidence is relevant, disciplinary practical knowledge guides collection of data for use as evidence, and interpretation of evidence to refute, confirm, or advance knowledge is informed by existing disciplinary knowledge.
- The discipline-specific components of scientific research practices of relevance to an investigation are linked to the domain-general epistemology of research.
- The assessment triangle is a useful guide with the CADE for implementing discipline-specific assessments to track students' evidentiary reasoning: first, use the CADE to detail a cognition target by unpacking the various types of reasoning about evidence that an assessment task should elicit, then implement the task to observe the range of student behaviors the task reveals, and finally interpret students' competence or their difficulties that should be addressed according to a CADE table from https://tinyurl.com/CADE2022 that can be modified for a particular research task.
- Prompt students to link disciplinary knowledge with epistemic reasoning as they consider how a hypothesis or research goal was informed, what alternative models to consider, or to evaluate sufficiency of claims in terms of relevant and convergent evidence.

Acknowledgments We thank our research collaborators for joining us in Exploring Biological Evidence and for their contributions to the progress of our study. We especially thank our laboratory class teaching staff and students for willingly participating in this study. This work was supported by the National Science Foundation under Grant #1661124. Any opinions, findings, and conclusions or recommendations expressed in this material are those of the authors and do not necessarily reflect the views of the National Science Foundation.

References

Abd-El-Khalick, F., Boujaoude, S., Duschl, R., Lederman, N. G., Mamlok-Naaman, R., Hofstein, A., & Tuan, H. L. (2004). Inquiry in science education: International perspectives. *Science Education, 88*(3), 397–419.

American Association for the Advancement of Science (AAAS). (2011). *Vision and change in undergraduate biology education: A call to action* (C. Brewer & D. Smith (Eds.)). https://visionandchange.org/finalreport/. Accessed 18 Oct 2021.

Anderson, W. L., Sensibaugh, C. A., Osgood, M. P., & Mitchell, S. M. (2011). What really matters: Assessing individual problem-solving performance in the context of biological sciences. *International Journal for the Scholarship of Teaching and Learning, 5*(1), n1.

Angra, A., & Gardner, S. M. (2017). Reflecting on graphs: Attributes of graph choice and construction practices in biology. *CBE—Life Sciences Education, 16*(3), ar53.

Auerbach, A. J., & Schussler, E. E. (2017). Curriculum alignment with vision and change improves student scientific literacy. *CBE—Life Sciences Education, 16*(2), ar29.

Bautista, N. U., Schussler, E. E., & Rybczynski, S. M. (2014). Instructional experiences of graduate assistants implementing explicit and reflective introductory biology laboratories. *International Journal of Science Education, 36*(7), 1184–1209.

Beck, C. W., & Blumer, L. S. (2021). The relationship between perceptions of instructional practices and student self-efficacy in guided-inquiry laboratory courses. *CBE—Life Sciences Education, 20*(1), ar8.

Bhavana, A. R. (2009). Honing language skills using blogs. *Journal on School Educational Technology, 4*(3), 30–33.

Black, P., Harrison, C., Lee, C., Marshall, B., & Wiliam, D. (2003). *Assessment for learning: Putting it into practice*. Open University Press.

Black, P., & Wiliam, D. (1998). Assessment and classroom learning. *Assessment in Education: Principles, Policy and Practice, 5*(1), 7–74.

Blair, A. C., Peters, B. J., & Bendixen, C. W. (2014). Promoting student inquiry using Zea mays (Corn) cultivars for hypothesis-driven experimentation in a majors introductory biology course. *The American Biology Teacher, 76*(5), 333–336.

Boomer, S. M., Kumar, S., & Dutton, B. E. (2021). A CURE for meat: Comparing bacterial contaminants on different ground beef sources emphasizes process of science and quantitative reasoning. *Journal of Microbiology & Biology Education, 22*(1), ev22i1-2237.

Brownell, S. E., Wenderoth, M. P., Theobald, R., Okoroafor, N., Koval, M., Freeman, S., Walcher-Chevillet, C. L., & Crowe, A. J. (2014). How students think about experimental design: Novel conceptions revealed by in-class activities. *BioScience, 64*(2), 125–137.

Brunauer, L. S. (2016). Purification and electrophoretic characterization of lactate dehydrogenase from mammalian blood: A different twist on a classic experiment. *Journal of Chemical Education, 93*(6), 1108–1114.

Bugarcic, A., Zimbardi, K., Macaranas, J., & Thorn, P. (2012). An inquiry-based practical for a large, foundation-level undergraduate laboratory that enhances student understanding of basic cellular concepts and scientific experimental design. *Biochemistry and Molecular Biology Education, 40*(3), 174–180.

Creswell, J. W., & Poth, C. N. (2018). *Research design: Qualitative, quantitative, and mixed methods approaches* (5th ed.). SAGE.

Coleman, A. B., Lam, D. P., & Soowal, L. N. (2015). Correlation, necessity, and sufficiency: Common errors in the scientific reasoning of undergraduate students for interpreting experiments. *Biochemistry and Molecular Biology Education, 43*(5), 305–315.

Dasgupta, A. P., Anderson, T. R., & Pelaez, N. (2014). Development and validation of a rubric for diagnosing students' experimental design knowledge and difficulties. *CBE—Life Sciences Education, 13*(2), 265–284.

Dasgupta, A. P., Anderson, T. R., & Pelaez, N. J. (2016). Development of the neuron assessment for measuring biology students' use of experimental design concepts and representations. *CBE—Life Sciences Education, 15*(2), ar10.

Deane, T., Nomme, K., Jeffery, E., Pollock, C., & Birol, G. (2014). Development of the biological experimental design concept inventory (BEDCI). *CBE—Life Sciences Education, 13*(3), 540–551.

Deane, T., Nomme, K., Jeffery, E., Pollock, C., & Birol, G. (2016). Development of the statistical reasoning in biology concept inventory (SRBCI). *CBE—Life Sciences Education, 15*(1), ar5.

Duncan, R. G., Chinn, C. A., & Barzilai, S. (2018). Grasp of evidence: Problematizing and expanding the next generation science standards' conceptualization of evidence. *Journal of Research in Science Teaching, 55*(7), 907–937.

Duschl, R. (2008). Science education in three-part harmony: Balancing conceptual, epistemic, and social learning goals. *Review of Research in Education, 32*(1), 268–291.

Erduran, S., Ozdem, Y., & Park, J. Y. (2015). Research trends on argumentation in science education: A journal content analysis from 1998–2014. *International Journal of STEM Education, 2*, 5. https://doi.org/10.1186/s40594-015-0020-1

Fisher, G. R., Olimpo, J. T., McCabe, T. M., & Pevey, R. S. (2018). The Tigriopus CURE – A course-based undergraduate research experience with concomitant supplemental instruction. *Journal of Microbiology & Biology Education, 19*(1), 19–11.

Fry, J. R., & Burr, S. A. (2011). A double-blind atropine trial for active learning of autonomic function. *Advances in Physiology Education, 35*(4), 438–444.

Full, R. J., Dudley, R., Koehl, M. A. R., Libby, T., & Schwab, C. (2015). Interdisciplinary laboratory course facilitating knowledge integration, mutualistic teaming, and original discovery. *Integrative and Comparative Biology, 55*(5), 912–925.

Furtak, E. M., Hardy, I., Beinbrech, C., Shavelson, R. J., & Shemwell, J. T. (2010). A framework for analyzing evidence-based reasoning in science classroom discourse. *Educational Assessment, 15*(3–4), 175–196. https://doi.org/10.1080/10627197.2010.530553

Gardner, G., & Jones, G. (2011). Perceptions and practices: Biology graduate teaching assistants' framing of a controversial socioscientific issue. *International Journal of Science Education, 33*(8), 1031–1054.

Gauthier, A., Jantzen, S., McGill, G., & Jenkinson, J. (2019). Molecular concepts adaptive assessment (MCAA) characterizes undergraduate misconceptions about molecular emergence. *CBE—Life Sciences Education, 18*(1), ar4.

Giere, R. N. (2010). *Scientific perspectivism*. University of Chicago Press.

Gormally, C., Brickman, P., & Lutz, M. (2012). Developing a test of scientific literacy skills (TOSLS): Measuring undergraduates' evaluation of scientific information and arguments. *CBE—Life Sciences Education, 11*(4), 364–377.

Hester, S., Buxner, S., Elfring, L., & Nagy, L. (2014). Integrating quantitative thinking into an introductory biology course improves students' mathematical reasoning in biological contexts. *CBE—Life Sciences Education, 13*(1), 54–64.

Hicks, J., Dewey, J., Brandvain, Y., & Schuchardt, A. (2020). Development of the biological variation in experimental design and analysis (BioVEDA) assessment. *PLoS One, 15*(7), e0236098.

Irby, S. M., Pelaez, N. J., & Anderson, T. R. (2018). Anticipated learning outcomes for a biochemistry course-based undergraduate research experience aimed at predicting protein function from structure: Implications for assessment design. *Biochemistry and Molecular Biology Education, 46*(5), 478–492.

Killpack, T. L., & Fulmer, S. M. (2018). Development of a tool to assess interrelated experimental design in introductory biology. *Journal of Microbiology & Biology Education, 19*(3), 19–13.

King, S. L. (2018). *Teaching assistant pedagogical development: Evolution, biological change over time, and concept based instruction*. Doctoral dissertation, The Pennsylvania State University (Order No. 28778767). Available from Dissertations & Theses @ CIC Institutions; ProQuest Dissertations & Theses Global. (2572521567). https://www.proquest.com/dissertations-theses/teaching-assistant-pedagogical-development/docview/2572521567/se-2?accountid=13360

Klein, G. C., & Carney, J. M. (2014). Comprehensive approach to the development of communication and critical thinking: Bookend courses for third-and fourth-year chemistry majors. *Journal of Chemical Education, 91*(10), 1649–1654.

Kowalski, J. R., Hoops, G. C., & Johnson, R. J. (2016). Implementation of a collaborative series of classroom-based undergraduate research experiences spanning chemical biology, biochemistry, and neurobiology. *CBE—Life Sciences Education, 15*(4), ar55.

Krontiris-Litowitz, J. (2013). Using primary literature to teach science literacy to introductory biology students. *Journal of Microbiology & Biology Education, 14*(1), 66–77.

Lansverk, A. L., Lichti, D. A., Blinka, K. W., & Callis-Duehl, K. L. (2020). Comparing the outcomes of "Pre-CURE" compared to inquiry-based introductory biology labs. *Bioscene: Journal of College Biology Teaching, 46*(1), 10–22.

Martinez-Vaz, B. M., & Mickelson, M. M. (2020). In silico phage hunting: Bioinformatics exercises to identify and explore bacteriophage genomes. *Frontiers in Microbiology, 11*, 2294.

Manz, E., Lehrer, R., & Schauble, L. (2020). Rethinking the classroom science investigation. *Journal of Research in Science Teaching, 57*(7), 1148–1174.

McNeill, K. L., & Berland, L. (2017). What is (or should be) scientific evidence use in K-12 classrooms? *Journal of Research in Science Teaching, 54*(5), 672–689.

National Research Council. (2001). *Knowing what students know: The science and design of educational assessment*. The National Academies Press. https://doi.org/10.17226/10019

National Research Council US. (2005). In S. R. Singer, M. L. Hilton, & H. A. Schweingruber (Eds.), *America's lab report: Investigations in high school science*. National Academies Press.

National Research Council. (2014). In J. W. Pellegrino, M. R. Wilson, J. A. Koenig, & A. S. Beatty (Eds.), *Developing assessments for the next generation science standards*. National Academies Press.

NGSS Lead States. (2013). *Next generation science standards: For states, by states*. The National Academies Press.

Ott, L. E., & Carson, S. (2014). Immunological tools: Engaging students in the use and analysis of flow cytometry and enzyme-linked immunosorbent assay (ELISA). *Biochemistry and Molecular Biology Education, 42*(5), 382–397.

Pelaez, N., Anderson, T. R., Gardner, S. M., Yin, Y., Abraham, J. K., Bartlett, E. L., Gormally, C., Hill, J. P., Hoover, M., Hurney, C. A., Long, T. M., Newman, D. L., Sirum, K., & Stevens, M. T. (2017). *The basic competencies of biological experimentation: Concept-skill statements* (PIBERG instructional innovation material paper 4). https://docs.lib.purdue.edu/pibergiim/4/. Accessed 8 Nov 2021.

Pellegrino, J. W. (2012). Assessment of science learning: Living in interesting times. *Journal of Research in Science Teaching, 49*, 831–841. https://doi.org/10.1002/tea.21032

Phelps, R. P. (2011). *Standards for educational & psychological testing*. American Psychological Association.

Resendes, K. K. (2015). Using HeLa cell stress response to introduce first year students to the scientific method, laboratory techniques, primary literature, and scientific writing. *Biochemistry and Molecular Biology Education, 43*(2), 110–120.

Reynders, G., Lantz, J., Ruder, S. M., Stanford, C. L., & Cole, R. S. (2020). Rubrics to assess critical thinking and information processing in undergraduate STEM courses. *International Journal of STEM Education, 7*(1), 1–15.

Richards, K. A. R., & Hemphill, M. A. (2018). A practical guide to collaborative qualitative data analysis. *Journal of Teaching in Physical Education, 37*(2), 225–231.

Robertson, A. L., & Phillips, A. R. (2008). Integrating PCR theory and bioinformatics into a research-oriented primer design exercise. *CBE—Life Sciences Education, 7*(1), 89–95.

Rybarczyk, B. J., Walton, K. L., & Grillo, W. H. (2014). The development and implementation of an instrument to assess students' data analysis skills in molecular biology. *Journal of Microbiology & Biology Education, 15*(2), 259–267.

Samarapungavan, A. (2018). Chapter 4: Construing scientific evidence: The role of disciplinary knowledge in reasoning with and about evidence in scientific practice. In F. Fischer,

K. Englemann, J. Osborne, & C. A. Chinn (Eds.), *Scientific reasoning and argumentation: The roles of domain-specific and domain-general knowledge*. Routledge Taylor & Francis Group. https://doi.org/10.4324/9780203731826

Sandoval, W. A., Brian, J., & Reiser, B. J. (2004). Explanation-driven inquiry: Integrating conceptual and epistemic scaffolds for scientific inquiry. *Science Education, 88*(3), 345–372.

Sandoval, W. A., & Millwood, K. A. (2005). The quality of students' use of evidence in written scientific explanations. *Cognition and Instruction, 23*(1), 23–55.

Schen, M. S. (2007). *Scientific reasoning skills development in the introductory biology courses for undergraduates*. Doctoral dissertation, The Ohio State University.

Seixas Mello, P., Cotta Natale, C., Marzin-Janvier, P., Vieira, L. Q., & Manzoni-de-Almeida, D. (2021). Inquiry-based learning in immunology: Analysis of scientific argument construction by undergraduate students in biological science and health care classes. *Journal of Biological Education*, 1–15. https://doi.org/10.1080/00219266.2021.1877778

Shi, J., Power, J., & Klymkowsky, M. (2011). Revealing student thinking about experimental design and the roles of control experiments. *International Journal for the Scholarship of Teaching and Learning, 5*(2). https://digitalcommons.georgiasouthern.edu/ij-sotl/vol5/iss2/8/

Simmons, A. D., Larios-Sanz, M., Amin, S., & Rosell, R. C. (2014). Using mini-reports to teach scientific writing to biology students. *The American Biology Teacher, 76*(8), 551–555.

Sirum, K., & Humburg, J. (2011). The experimental design ability test (EDAT). *Bioscene: Journal of College Biology Teaching, 37*(1), 8–16.

Sorte, C. J., Aguilar-Roca, N. M., Henry, A. K., & Pratt, J. D. (2020). A hierarchical mentoring program increases confidence and effectiveness in data analysis and interpretation for undergraduate biology students. *CBE—Life Sciences Education, 19*(3), ar23.

Spence, P. L., Phillips, R. S., McAllister, A. R., White, S. L., & Hollowell, G. P. (2020). Student-scientist curriculum: Integrating inquiry-based research experiences and professional development activities into an introductory biology laboratory course. *Journal of Microbiology & Biology Education, 21*(3), 30.

St. Onge, S. (2007). *Long term urban wildlife research potential through course-based undergraduate research experiences*. Masters dissertation, University of Colorado at Denver (Order No. 10982264). Available from ProQuest Dissertations & Theses Global. (2150116451). https://www.proquest.com/dissertations-theses/long-term-urban-wildlife-research-potential/docview/2150116451/se-2?accountid=13360

Stransky, J., Bassett, L., Bodnar, C., Anastasio, D., Burkey, D., & Cooper, M. (2021). A retrospective analysis on the impacts of an immersive digital environment on chemical engineering students' moral reasoning. *Education for Chemical Engineers, 35*, 22–28.

Terry, D. R. (2007). *Using the case study teaching method to promote college students' critical thinking skills*. Doctoral dissertation, State University of New York at Buffalo (Order No. 3277766). Available from ProQuest Dissertations & Theses Global. (304778960). https://www.proquest.com/dissertations-theses/using-case-study-teaching-method-promote-college/docview/304778960/se-2?accountid=13360

Thanukos, A., Scotchmoor, J. G., Caldwell, R., & Lindberg, D. R. (2010). The real process of science: Understanding science at https://evolution.berkeley.edu/teach-evolution/conceptual-framework/ and http://www.understandingscience.org and Science 101: Building the foundations for real understanding. *Science, 330*(6012), 1764–1765. https://doi.org/10.1126/science.118699

Timmerman, B. E. C., Strickland, D. C., Johnson, R. L., & Payne, J. R. (2011). Development of a 'universal' rubric for assessing undergraduates' scientific reasoning skills using scientific writing. *Assessment & Evaluation in Higher Education, 36*(5), 509–547.

Toulmin, S. E. (1958). *The uses of argument*. Cambridge University Press.

Tytler, R., & Peterson, S. (2005). A longitudinal study of children's developing knowledge and reasoning in science. *Research in Science Education, 35*, 63–98.

Volz, T., & Saterbak, A. (2009). Students' strengths and weaknesses in evaluating technical arguments as revealed through implementing Calibrated Peer Review™ in a bioengineering laboratory. *Across the Disciplines, 6*(1).

Ward, J. R., Clarke, H. D., & Horton, J. L. (2014). Effects of a research-infused botanical curriculum on undergraduates' content knowledge, STEM competencies, and attitudes toward plant sciences. *CBE—Life Sciences Education, 13*(3), 387–396.

Weaver, K. F., Morales, V., Nelson, M., Weaver, P. F., Toledo, A., & Godde, K. (2016). The benefits of peer review and a multi-semester capstone writing series on inquiry and analysis skills in an undergraduate thesis. *CBE—Life Sciences Education, 15*(4), ar51.

Younkin, K., & Romano, C. (2018). Student-centered microbioassay laboratory activity utilizing bioluminescent bacteria. *Journal of Microbiology & Biology Education, 19*(2), 19–12.

Part V
Complementary Frameworks for Guiding Students' Experimentation Practice

Chapter 18
Hybrid Labs: How Students Use Computer Models to Motivate and Make Meaning from Experiments

Julia Gouvea, Aditi Wagh, Robert Hayes, and Matt Simon

18.1 Introduction

The focus of this volume is to promote the teaching and learning of experimentation in the life sciences. Experimentation has been the primary means of engaging undergraduate students in scientific practice. College lab courses, with their focus on designing, or at least conducting, experiments, serve as spaces where students can practice doing science and thinking scientifically (Hofstein & Lunetta, 2004). However, in many traditional laboratory courses, experiments are conducted as stand-alone activities, beginning with students designing or setting up an experiment, and ending when results are analyzed and conclusions are reported. This narrow version of experimentation isolates experiments from other scientific practices that give them purpose and meaning (Hodson, 1996; Lehrer et al., 2001; Manz et al., 2020; Wong & Hodson, 2009).

In science, the need to conduct an experiment often arises when gaps, limitations or inconsistencies are surfaced by prior theoretical or empirical work (Manz, 2015; Pickering, 1992). Theoretical arguments in the literature, model-based hypotheses, or unexplained patterns in complex data sets inspire questions that experiments can potentially answer. Once results are obtained, meaning-making continues well beyond the bounds of the original experiment (Gooding, 1990). Results are put into conversation with other data and theories, often raising new questions and

J. Gouvea (✉) · R. Hayes · M. Simon
Department of Education, Tufts University, Medford, MA, USA
e-mail: julia.gouvea@tufts.edu; Robert.Daniel.Hayes@gmail.com; msimon@ccsnh.edu

A. Wagh
Department of Comparative Media Studies & Writing, Massachusetts Institute of Technology, Cambridge, MA, USA
e-mail: awagh@education.mit.edu

N. J. Pelaez et al. (eds.), *Trends in Teaching Experimentation in the Life Sciences*, Contributions from Biology Education Research,
https://doi.org/10.1007/978-3-030-98592-9_18

suggesting next steps rather than settling on a general conclusion (Manz et al., 2020). At both beginning and end, experiments make contact with a longer trajectory of work (Ford, 2015; Gooding, 1990). However, in many classroom instantiations, experiments are presented in isolation from other practices (Hodson, 1996; Lehrer et al., 2001; Manz et al., 2020).

This isolation, we argue, is a problem not just because it misrepresents the complex and interconnected nature of scientific practice, but because it constrains students' engagement in experimental practices. Without connection to what has come before or what comes next, it can be difficult to identify a meaningful research question or to craft an appropriately scoped conclusion. Scientists' abilities to do these things well is partly explained by their participation in rich contexts that inspire and challenge their thinking (Ford, 2015; Rouse, 2007).

In our work, we have explored the potential for interactions between computational models and experiments to expand how students experience experimentation. Specifically, we focus on how this interaction can support students in identifying and articulating questions that motivate experimentation and in interpreting data to generate meaning. The choice to couple experiments and computational models was inspired by research describing this coupling as productive in scientific practice (MacLeod & Nersessian, 2013). We call our design "hybrid labs" to emphasize the importance of interaction between, not just co-occurrence of, the two modalities.

At a high-level our purpose in this chapter is to expand the boundaries of what is considered "experimentation" in university settings. In this respect our aims align with the ACE-Bio Competencies framework, which presents an expansive view of experimentation by defining it in terms of the myriad component understandings, skills, and representational competencies observed in the work of experts (Pelaez et al., 2017). In designing curricula, the ACE-Bio framework can be used (e.g. as described in Chaps. 3 and 8) to support "backward design" to promote the development of certain sets of competencies in students.

In this chapter, we take a different approach to expanding experimental practice. Rather than start from descriptions of expert knowledge or actions, our design process starts from attending to the socio-material contexts that evoke and shape expert practice (Engle & Conant, 2002; Ford, 2008; Manz, 2015). We focus on designing learning environments from which students' experimental practices can emerge and be refined. Underlying this choice is an assumption that students, especially at the college-level, have many productive resources for practicing science, but that these resources are not typically tapped into in traditional laboratory contexts (Lehrer et al., 2001; Hofstein & Lunetta, 2004). Our design practice also conceptualizes outcomes differently. Rather than expecting that all (or most) students should develop the same set of experimental competencies, we expect that what develops will be different for different students, depending on the specific idiosyncratic ideas and decisions they make. We see this variation as important for developing adaptive expertise with scientific practices – the ability to flexibly apply knowledge and skills to specific situations (Hatano & Inagaki, 1986; Ford, 2008, 2015).

In the next sections, we first describe how students' difficulties with experimentation can be explained in terms of a problem with the narrowness of traditional learning contexts. We then explain how the relationship between experimentation and computational modeling is one way of expanding experimental practice in science. Next we describe how this idea of integration inspired the design of a "hybrid lab" unit and present two examples of students' practices in this unit to illustrate how interactions between experiment and computer model helped them engage in foundational aspects of experimental practice (Pelaez et al., 2017): Identifying Questions to motivate experimental design and making meaningful Conclusions from experimental results.

18.1.1 Challenges Arising from Experimentation in Isolation

We begin with a short overview of why identifying research questions and making meaning can be difficult for students, particularly when experiments are experienced as isolated activities.

18.1.1.1 Motivating Experimental Design

Identifying an experimental question can be challenging even for experienced researchers. It can therefore be tempting to expect that articulating a research question is beyond the reach of introductory-level students. Indeed, researchers investigating the experimental design strategies of students and experts noticed differences in their approaches (Jordan et al., 2011). Experts (advanced graduate students) leveraged their theoretical understandings of the discipline to ask principled questions to motivate their experiments. In contrast, novice undergraduate students tended to rely on the available materials to guide experimental design, using them to identify dependent variables in a paired comparison (e.g. low or high acid, low or high salt, low or high light), a strategy researchers termed "designing by twos."

In our own context, we have observed students posing questions that seem designed to test ideas they already know to be true (e.g. The purpose of our experiment was to test if the bacteria with the damaged DNA repair mutated more). We have also seen students design experiments to demonstrate canonical principles (e.g. This experiment will show that mutations are adaptive). Neither strategy, "designing by twos" nor designing to demonstrate concepts, produces experiments that are motivated by the need to figure something out. Consequently, the results of these experiments are unlikely to push students to think in creative or critical ways about the discipline.

One interpretation of these observations is that students lack the expertise to be able to ask interesting scientific questions: In the absence of relevant expertise, they use available materials or canonical concepts to generate experimental questions. We locate the problem not in the students, but in the isolated nature of experimental design activities. In scientific practice, researchers do not simply "come up" with questions in isolation. Questions are identified and constructed out of other experiences – empirical or theoretical – that have brought gaps or contradictions in understanding to their attention (Phillips et al., 2017) or have made it necessary to test the validity of proposed ideas. In this sense, experts have an advantage because they are steeped in richer theoretical and empirical contexts where it is more likely for them to notice the problems and questions that would benefit from an experimental approach. Without this context, it makes sense that students' experimental designs are motivated by what is readily available.

18.1.1.2 Making Meaning of Experimental Results

Another challenging aspect of experimental practice is data interpretation. Many undergraduate students have come to expect that the goal of lab experiments is to *demonstrate* disciplinary concepts (Hayes & Gouvea, 2020; Hu et al., 2017; Smith et al., 2020). This expectation can undercut authentic data analysis and interpretation, sometimes leading students to engage in "questionable research practices" such as ignoring or "fudging" contradictory data or dismissing deviations from expected results as "experimental error" (Smith et al., 2020; Stein et al., 2018). Moreover, in lab courses, students are often instructed to come to a "conclusion" based on the results of a single experiment. This practice reinforces a simplistic role for experimental analysis as verifying (or rejecting) hypotheses and sets up students to think about data analysis using this binary.

In science, single experiments comprise a small part of broader research conversations, and scientists have the advantage of holistically engaging in those conversations (Bazerman, 1988; Gooding et al., 1989). While experimental analysis does sometimes play the role of supporting or refuting theoretical ideas, it does so rarely. More often, experiments are not conclusive, but instead generate partial understandings or generate new questions that inspire further inquiry (Koponen & Mäntylä, 2006). For these reasons, scientists tend to approach data analysis without the expectation of simple answers and with a readiness to follow where the data may lead next.

We argue that the difficulties students face in finding motivation for and making meaning from experiments stem largely from activity structures that promote a narrow version of experimentation as isolated from other scientific practices. In the next section, we examine the specific advantages of coupling experiments with computational modeling as a way to address these challenges.

18.1.2 Coupling Computational Modeling and Experimentation in Scientific Practice

In modern biology, experimentation and computation are two powerful and complementary approaches to conducting research (National Research Council, 2009). Progress in one approach can inspire the need to conduct work in the other. Most straightforwardly, modeling generates hypotheses that can be tested by experiments, and experiments yield information that can lead to model validation or revision.

Scholars have identified additional points of intersection between experiments and models. For example, modeling can play an exploratory role, allowing researchers to consider possible system behaviors over a range of parameters (Odenbaugh, 2005). This kind of analysis can give meaning to experimentally derived parameter estimates by indicating whether small changes to those parameters will lead to incremental or more drastic changes in system behavior. Experimental work can also play an exploratory role, generating unexpected patterns that call for novel modeling frameworks (Gooding et al., 1989; Koponen & Mäntylä, 2006). Thus, rather than a simple relationship of construct and test, the relationship between models and experiments can be generative in multiple and potentially unexpected ways.

Many scientific research groups tend to specialize in either experimental or computational approaches, meaning that much of the generative interaction takes place via collaborations or extended conversations in the literature (MacLeod & Nersessian, 2014). More recently, some subfields of biology, such as systems biology, have developed more integrative approaches (Macleod & Nersessian, 2013). Researchers who have studied these scientists describe how the coupling can be a productive strategy for finding a path of inquiry in the face of complex systems. As one scientist relayed to researchers, "I like the idea that I'm building my model, things are popping up in my head, oh wow this would be a good experiment. I plan out the experiment myself and then go into the lab and I do it." She also noted how analyzing data in concert with modeling allowed her to "discover and extract the relevant information" from complex experimental output (MacLeod & Nersessian, 2013). Inspired by such accounts, we designed a laboratory curriculum with the aim of providing opportunities for students to leverage these potentially productive intersections.

18.2 Design of Hybrid Labs

Building on the research on scientists, as well as work in science education (e.g. Blikstein et al., 2016), we developed a curriculum around the idea of engaging students in the interactions between experiments and modeling.

Specifically, we conjectured that:

1. Exploring a computational model along with an experimental system would help students notice interesting or confusing patterns that would motivate an authentic need to test their ideas experimentally.
2. Discrepancies between the computer model and the experimental system would provoke considerations of experimental data beyond confirmation – including reflection on assumptions, situating results in parameter space and the identification of open questions and new directions.

In the remainder of this section we describe the design of one "hybrid" unit organized around the phenomenon of variation in bacterial mutation rates. We use this unit to illustrate how we translated our design principles into practice.

18.2.1 Project Context

The work described in this chapter took place in the context of a multi-year design-based implementation research project (Fishman et al., 2011) in an introductory-level college biology laboratory course. The lab and associated lecture focus on organismal and population biology. Labs meet weekly for 3-h sessions that are led by a graduate and undergraduate teaching assistant pair. Lab units span multiple (typically 3) sessions.

As part of the larger research study, we recorded data in these lab sessions in each year from 2015 to 2020. Data collected included observational notes, video recordings of small group work and whole class discussions, and students' lab reports. We also interviewed focal students and teaching assistants (TAs).

18.2.2 An Example Hybrid Lab: Mutation Rate Unit

The example unit we present here is the first of the semester, a 3-week (9-h) set of labs on mutation rate variation.

18.2.2.1 The Phenomenon

The phenomenon that motivates this lab is variation in mutation rate: mutation rates within and across taxa vary by orders of magnitude. This led us to pose, as a launching question for the unit: *Is it better to mutate a lot or a little, and under what conditions?* This broad question provides a conceptual focus for the unit without suggesting what the answer should be. We purposefully prompt students to consider "conditions" to open space to investigate tradeoffs playing out over time and in different environments. To investigate this question, students use two methods, experimentation and computational modeling.

18.2.2.2 Experimental System

We use two strains of the bacterium *Escherichia coli* as the focal organisms for experimentation: One is a standard laboratory strain, and the other strain has a mutation in a component of the DNA mismatch repair system (MutH) that elevates its mutation rate by about 100-fold. Neither strain is initially resistant to antibiotics and neither can digest lactose. Prior to lab, each strain is transformed with a plasmid containing a color marker (fluorescence proteins or chromoproteins) so that they can be visually distinguished from one another if grown together. In addition to expressing color, the plasmids include a gene for ampicillin resistance and so transformed strains must be incubated and plated on media that contains ampicillin to prevent ejection of the plasmid.

Before designing their own experiments, all students conduct a simple introductory experiment in which they incubate each strain overnight in enriched media (LB) and plate each strain independently on agar and antibiotic (rifampicin) plates. Typically, the results of this experiment reveal that the strain with a higher mutation rate grows more colonies on the antibiotic plate, suggesting that mutation is advantageous in this context. This introductory experiment serves two purposes. First, it introduces students to the lab techniques (setting up bacterial cultures, serial dilutions, plating bacteria) that they will use in the next set of experiments that they design. Second, the results of this experiment present a simple interpretation (mutation is advantageous because it allows for adaptation to antibiotics) that can be challenged and extended in later work.

To design and conduct their own experiments, each group is given access to a set of materials: *E. coli* bacterial cells of the low and high mutator strains (both transformed and untransformed), culture tubes, and culture media, including the possibility of adding an antibiotic (rifampicin) or novel carbon source (lactose). Students choose how long to incubate cultures and then plate a subsample of the cultures on plates of their choosing, including LB agar plates, antibiotic plates (LB agar + rifampicin) to select for antibiotic resistance, and MacConkey agar plates that indicate lactose digesting colonies with red color.

This set of experimental materials was chosen to make it possible for students to observe variable outcomes. The different growth media vary in the extent that they are expected to favor high or low mutating strains. The introductory experiment reveals a higher rate of antibiotic resistance and hence survival in the higher mutating strain. Similarly, the higher mutator is more likely to develop mutations allowing it to digest lactose and thus tends to produce a higher proportion of lactose metabolizing colonies (indicated in red) on MacConkey plates (Fig. 18.1). However, whether or not the ability to digest lactose is an advantage depends on the environment, and it is not clear that a faster rate mutation to lactose metabolism outweighs the costs of increased deleterious mutations when other nutrients are available. Similarly, it is not obvious which strain should have an advantage when incubated in an LB environment. In our experience it is possible for either strain to show higher numbers.

These ambiguities are productive theoretically in that they evoke considerations of the mechanisms that allow high or low mutators to thrive under different conditions. In addition, the underlying randomness of mutation provokes ideas about probability, as students consider whether chance events (such as a higher mutator gaining an unknown advantage) could impact experimental outcomes. Finally, the context-dependence of results can create a need to consider generalizability. If students see a particular result such as a larger proportion of low mutators, would they

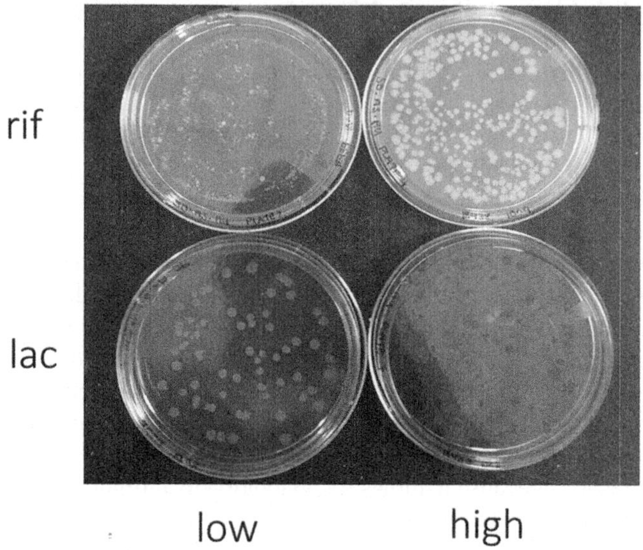

Fig. 18.1 **Typical plates from the introductory experiment.** Low and high mutating strains of E. coli plated on rifampicin (an antibiotic) and lactose media. The high mutating strain has more antibiotic resistant colonies (upper right plate) and more colonies that have mutated to digest lactose (red colored colonies in lower right)

expect to see that result if they replicated the experiment or changed the initial conditions slightly?

18.2.2.3 Computational Model

Our goal in designing a computer model was to provide students with information that would align with, contradict or expand what they could learn from experiments. To meet this goal, we designed an agent-based NetLogo model that included two virtual strains of bacteria and allowed students to compare their relative success under a range of conditions (Fig. 18.2).

As in the experiment, students could manipulate growth conditions in the model, choosing to grow bacteria in a standard nutrient environment or to add antibiotics. Students could also manipulate initial population sizes of higher and lower strains.

Unlike in the experimental setup, students could directly manipulate system parameters to explore how the populations behaved. For example, students could alter the relative mutation rate of each strain. They could also alter the incidence of specific kinds of mutations – metabolic benefit, lethal, antibiotic resistance, and

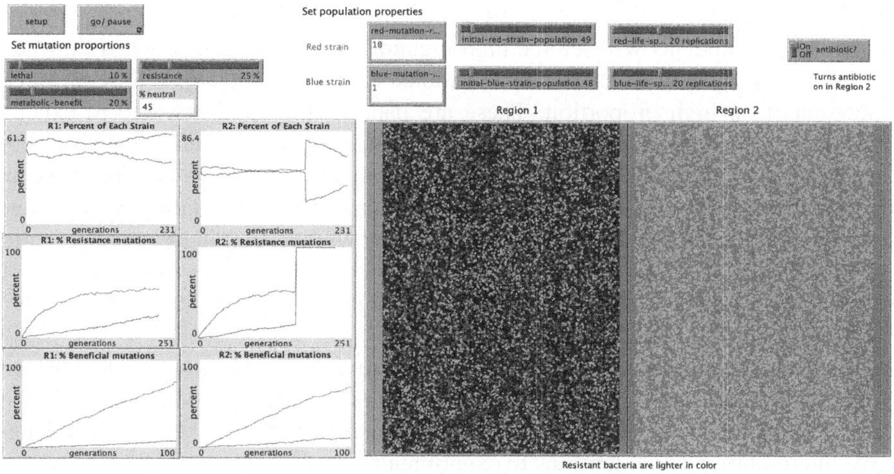

Fig. 18.2 The computer model interface. Two regions depict populations in two independent runs of the simulation. Individual bacteria are represented as blue or red dots. Lighter colors (light blue and pink) are cells that are resistant to antibiotic. Antibiotic can be switched on or off in Region 2 only. Population properties (relative mutation rate, initial population size, and life span of each strain) can be set before a run. The proportion of different types of mutations can be adjusted. Plots (top to bottom) show relative proportion of each strain in each region, proportion of each strain that has resistance mutation and proportion of each strain that has metabolic benefit mutation

neutral mutations.[1] The model also allowed students to simulate population dynamics over longer periods of time than was possible in the experiment. Both these features were important to supporting student investigations because they allowed an exploration of the system beyond what the experiment could reveal.

On the model interface, students had access to representations that showed the relative percentages of the low and high mutator strain, as well as the proportion of each strain that had either beneficial or antibiotic resistance mutations. These representations made it possible to immediately visualize how changes to the model parameters impacted the populations.

Finally, because the model is agent-based, it behaves stochastically. Students can run a simulation with the same set of parameter values and get different results. To facilitate their noticing of this stochasticity, each “run” of the simulation allows students to run two independent trials. This aspect of the interface allows students to see two replicates running side-by-side or to directly compare two environments by adding antibiotic to one run and comparing it to a run without antibiotic present.

[1] “Metabolic benefit” mutations are always beneficial in that they increase the amount of energy a bacterium acquires at each timestep. Lethal mutations immediately kill bacteria. Antibiotic resistance mutations confer complete resistance when antibiotic is present (otherwise antibiotic is lethal). Neutral mutations have no effect. All mutation types must sum to 100%.

The model output typically yields a mix of expected and unexpected patterns. For example, students will see the high mutator gain an advantage when they add antibiotics. They will tend to see the low mutator have an advantage if they set lethal mutations to a high proportion. These are patterns that students tend to expect. Simulations will also produce patterns that many students find surprising. For example, in an antibiotic environment a higher mutating strain will tend to have an initial advantage. However, if even a small number of lower mutating bacteria survive the initial addition of antibiotics, they will slowly increase in frequency eventually displacing the higher mutating strain. This pattern is due to the possibility of back-mutating in the model, which causes the higher mutator to lose resistance at a faster rate than the lower mutating strain.

As with the experimental system, the possibility of encountering unexpected trends is an important feature of the design of the computer model both because it can motivate questions that can be addressed experimentally and because it can suggest interpretations or extensions to experimental results.

18.2.2.4 Activity Structure

The structure of activities in the lab is designed to help students notice the intersections between experimentation and modeling. We design multiple opportunities for students to work with each method in close succession (Table 18.1).

Over the three-week unit, students work in the same group of 3–4 students. The unit begins in the first week with a presentation of the phenomenon: mutation rates in *E. coli* populations vary over orders of magnitude. In small group and whole class discussions, we ask students to consider why mutation rate might vary and under what conditions they would expect to see higher or lower rates of mutations. They then set up and discuss their predictions for the simple introductory experiment that measures relative growth and survival of high and low mutating strains on LB agar and antibiotic plates.

In the second week, students analyze and discuss the results of the introductory experiment. The computer model is then introduced as an activity that is meant to expand on this initial experiment. Students are encouraged to explore and make note of interesting or surprising findings that could potentially inspire next experiments. As groups begin designing their experiments, the computer model remains a resource to which they can return to test ideas.

Table 18.1 Outline of activities in mutation rates hybrid lab unit

Week 1	Week 2	Week 3
Introduce phenomenon of variation in mutation rates Discuss relative advantages of high v. low mutation rate Set up introductory antibiotic experiment	*Analyze* results of antibiotic experiment Initial exploration of computer model *Identify Questions* and *Plan* and *Conduct* experiments	*Analyze* results of designed experiments Revisit computer model (simultaneously) *Conclude* and then *Communicate* findings with group poster presentations

Finally, in the third week, groups analyze their experimental findings and are encouraged to revisit the computer model to attempt to replicate or extend their experimental investigations. The unit ends with a poster session in which groups share their research question and findings (from experimental results and model explorations) with other students in the class.

18.2.2.5 Instruction and Assessment

Each lab section is led by a pair of teaching assistants, one graduate and one undergraduate student. TAs receive training that emphasizes listening to and engaging with students' ideas rather than correcting them (e.g. Robertson et al., 2016).

Rather than a report of "results" from experiments, we position written lab reports in terms of making an argument, using evidence from both experiments and simulations to support claims about the relative fitness of high and low mutators. In students' written work, TAs are instructed to focus their feedback around the strength of students' arguments rather than formatting and style (Hill et al., 2018).

18.3 Students' Scientific Practice in Hybrid Labs

In this section, we illustrate how the coupling of experiments and computer simulations played out for students as they identified questions for experiments and interpreted their results. We present the work of two focal groups of students to illustrate, in different ways, how intersections between model and experiment enriched students' scientific practice. The first example summarizes a group of students whose experimental design was inspired by their noticing a surprising pattern that arose when they ran the simulation for many generations. Then, after obtaining unexpected experimental results, they once again used the simulation to help make sense of what could have happened. The second example features a group who identified an assumption in the model that they were unsure would hold in the real system. When their experimental results were inconclusive, they used the simulation to expand on their initial investigation, conducting an exploration of system behavior over a range of parameters.

18.3.1 Example 1: Attending to Time in Simulation and Experiment

18.3.1.1 An Experimental Design Motivated by Questions About Time

A pair of students we call M and T designed an experiment to explore how the relative proportions of low and high mutating bacterial strains would change over time. They planned to culture the strains together in two conditions – in an LB medium and in an antibiotic-containing LB medium. Each culture would be subsampled at two time points: after 24 h (T1) and again after 72 h (T2). At each sampling they would use LB agar plates to estimate the relative proportion of each strain and LB + antibiotic plates to estimate the relative proportion of antibiotic resistant bacteria in each strain (Fig. 18.3).

In their written work, M and T described their experimental design as inspired by their investigations in the computer model. They had run a simulation in which they added an antibiotic to a mixed population of the two strains. Initially, they saw an increase in high mutators, which made sense to them since they expected high mutators to be more likely to acquire resistance mutations. As the simulation continued to run however, they noticed that the lower mutating strain began to increase in frequency, eventually overtaking the high mutator. This result surprised them and led to their design of an experiment that featured sampling at multiple time points. In his lab report, M described how the simulation influenced their experimental design:

> In the simulations, we saw that as time went on, the higher mutating strain, which initially was out-competing the lower mutating strain, started to die much faster than the low-mutating strain. These observations made our lab group reconsider the advantages of having high rates of mutations when time is taken into account. Our new question was as follows: As time goes on, does the low mutating strain eventually outcompete the high mutating strain, and thus is better selected for the environment with antibiotic?

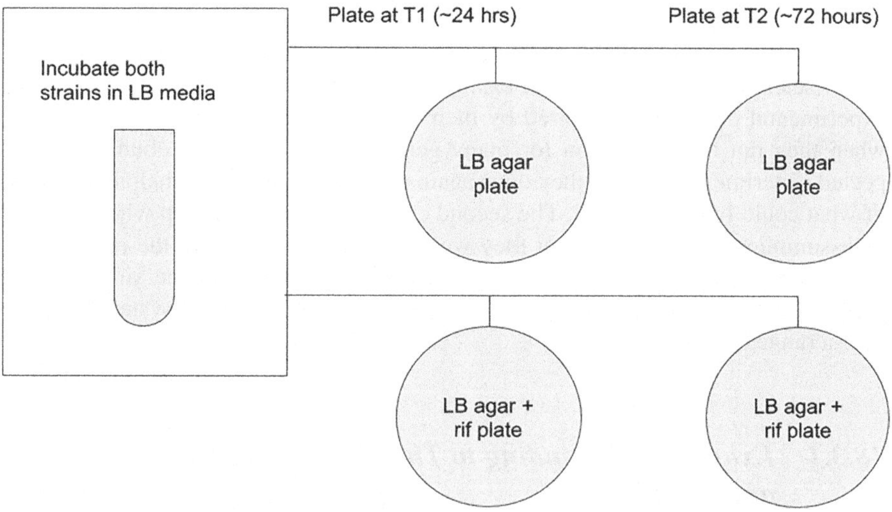

Fig. 18.3 A schematic of the experiment designed by M and T. They decided to incubate both strains together in an LB medium at 37C (each strain was transformed with a plasmid to appear as a different color on the plates). Then they planned to plate a subsample of the culture after about 24 h of incubation on both LB agar and LB agar + rifampicin plates. They planned to do the same plating again after about 72 h of incubation.

Similarly, T, in his report, framed the experiment around his wondering, "how long does it take for the low mutating bacteria to outgrow high mutating bacteria?" and "Was the simulation correct?" For M and T, the computer model drew their attention to the possibility of advantage changing over time, highlight uncertainty in the dynamics that became central to their experimental design.

18.3.1.2 Comparing Model and Experiment to Make Meaning and Ask New Questions

During data analysis M and T looked at their experimental plates while simultaneously continuing to run simulation trials over many generations. As they did this M and T made active comparisons between experimental and computational output. For example, T identified a pattern in the experiment that he found "weird." In the LB media, where they expected lower mutators to have an immediate advantage, they instead saw that the higher mutator was more prevalent at both time points, though slightly lower by T2, 72 h.

M, who was watching the low mutator win out in the model, proposed that they would have seen the low mutator increase in numbers if they had run their experiment for longer. M argued that the decline in the high mutator from T1 to T2 supported this possibility. T was not convinced and countered by noting that there was "no reason" for the high mutator to have an advantage in a neutral environment. The conversation continued with M and T considering other possible explanations for their experimental results including the possibility that the high mutating strain acquired some other sort of beneficial mutation.

In their final lab reports, both students talked about the relationships between experiment and model. T discussed the discrepancy between the model output and experiment as an open question and proposed that a, "feasible explanation for the T1 and T2 plates without the rifampicin is that a positive mutation occurred within the high mutating (strain) that boosted its growth." M attended to another unexpected experimental result – the failure of the low mutator to grow at all in the antibiotic medium. Pointing out that they set the initial population size to 1000 in the simulation, M argued that low numbers in the experiment may have made it possible for the low mutator to die out. M proposed that they repeat the experiment with a larger initial population size.

For both students, discrepancies between experiment and model prevented them from settling on a certain conclusion while also inspiring new thinking about possible mechanisms that could explain these discrepancies and the shape future work could take.

18.3.2 Example 2: Questioning the Nature of "Benefit"

18.3.2.1 A Question Arises from a Comparison of "Benefit" in Model and Experiment

Four students, A, D, N and W[2], began by talking about an experiment that would involve growing high and low mutating strains in lactose. At first, their interest appeared driven by the novelty of the material; they had already completed the whole class experiment that involved plating the strains on antibiotic-containing plates, and lactose was a new available material. In this sense, their early thinking resembled those of the undergraduates in Jordan et al.'s (2011) study. However, as they continued to consider how this new material fit into their work, uncertainties arose.

As they started to think through a possible experiment, they first considered mutation to lactose digestion as a benefit to the bacteria. N explained, "If the bacteria can digest lactose, in addition to the other nutrients, they will be able to grow more because they have more energy available to them." But later, when explaining their prediction to the TA, D noted that "there is nothing negative" in the environment, so perhaps both strains would do equally well. If the growth medium contains other nutrients, added N, then it is not clear that being able to digest lactose in addition is an actual advantage.

Over the next 20 min, N and D went back and forth considering the two alternatives. On the one hand, the mutator might gain an advantage by using "the lactose no one else is using." On the other hand, perhaps "there is enough glucose to go around" and the mutation doesn't confer an advantage. Frustrated by their lack of a clear prediction, D asked the group if they "want to use the thing? The computer program?"

The group turned to the computer and set up a trial in which the probability of beneficial metabolic mutations was high. Immediately, they saw the higher mutating strain increasing in proportion. Clustered around the screen, they had the following exchange:

D: Ok, now red's going up (red is the high mutating strain)
A: Ok
N: But the metabolic benefit like…
N: But is being able to digest lactose a metabolic benefit?
D: That is the question, I think it is
N: That- that is our question! Oh my god!

While initially unsure if their experiment was worth pursuing, seeing the output from the simulation helped the group articulate what they wanted to test. The simulation depicted a situation in which mutations had been programmed in as beneficial. Their question was, is lactose this sort of "beneficial" mutation? Would they see the same pattern of benefit that they see represented on the screen, or would

[2] A more in-depth analysis of this group's work can be found in Gouvea and Wagh (2018).

mutation to lactose digestion not matter? In this example, the simulation helped represent one possible outcome, allowing the group to question whether or not that outcome would happen in their experiment.

18.3.2.2 Using the Computational Model to Expand on and Rethink the Experiment

Due most likely to over-dilution of the bacterial cultures, the group saw too few colonies on their experimental plates to detect a pattern. While N and D attended to the plates, the other two members of the group, A and W, turned back to the computer model. Their first move was to again set the proportion of beneficial mutations higher than the lethal mutations. They quickly saw an advantage for the high mutating strain. However, after a while, W noticed that in one of the replicates the low mutator population was slowly accumulating beneficial mutations. As it did so, A noted that it began "making a comeback." Intrigued, W called for the TA, explaining that they did not know how to explain this pattern. The TA agreed that the pattern was interesting and encouraged them to keep thinking about it.

W and A then ran a series of 5 more trials in which they varied the relative proportion of beneficial and lethal mutations. In the end, they noticed that the high mutator only won out in one trial in which they had set the lethal rate very low relative to the rate of beneficial mutations. As A summarized to W: "if it's a high mutator, it needs a really low chance of a lethal mutation in order for it to really thrive."

In the end, the group used the model analysis to make a prediction about what they would have seen in the experiment. They now argued that it would be less likely that the high mutator would have an advantage, because, as N explained in their presentation, "it would have to be a benefit *and* overcome any bad mutations." Despite their poor-quality data, all four students made compelling arguments in their lab reports that combined their expectations with data from the simulation. For example, W argued:

> Although we thought that a higher mutator would be better adept at using lactose to survive, as the simulations demonstrated, the chance of it developing a lethal mutation was high enough where being a high mutator was actually harming the colonies' chances at surviving. We had overestimated the potential lactose had to act as a metabolic benefit in the presence of strains that could mutate to use it.

Moreover, students began to compare the relative validity of experimental and computational approaches. W wrote:

> It is important to note though that because the simulation presents more ideal conditions, it therefore does not represent a completely realistic scenario. A simulation cannot take into account all the factors involved that go into an experiment or how these processes occur naturally outside of a lab. It does not take into account that resources and environmental conditions can change consistently or periodically. Overall, both are valid in different ways.

For the students in this group, the simulation helped open up their thinking about their experiment, helping them make well-reasoned arguments about what to expect if they were able to repeat it, as well as to consider the question of how parameters in the experimental system might compare to the space they explored in the model.

18.4 Conclusions and Implications for Instructors

In this chapter we make the case for the potential of putting experiments into conversation with computer models. We argue that juxtaposing experimental work with a carefully designed computer model can help enrich two main aspects of experimental practice as articulated in the ACE-Bio Competencies for Experimentation in the Life Sciences (Pelaez et al., 2017; Chap. 1 in this volume):

1. Motivating the need to *Conduct* an experiment, which includes: *Identifying* gaps or problems, generating a research *Question*, and *Plan*ning relevant experiments. The examples we presented show how the act of comparing model and experiment helped students notice differences in assumptions or output that motivated additional study. This builds on prior work that has suggested discrepancies between experiments and models can motivate sense-making (Blikstein et al., 2016).
2. Making meaning from results, which includes: *Analyzing* data and generating *Conclusions* that integrate multiple forms of evidence, include appropriate expressions of uncertainty, and suggest possible future directions. We claim that having data from multiple methodologies helped disrupt simple conclusions, preserving real uncertainty and stimulating authentic next steps motivated by unresolved questions.

Crucially, our approach does not rely upon directly instructing students on how to identify a good research question or how to engage in data analysis or providing explicit scaffolding. Instead, our contribution is to show how these practices can emerge for students when they are in contexts designed to motivate and shape these practices.

For instructors interested in pursuing a "hybrid" approach we offer the following design considerations:

- Anchoring the lab in a phenomenon or study system that is complex enough that it is possible for students to have multiple different or conflicting theoretical ideas about what will happen.
- Using an experimental system that produces data that supports multiple interpretations to discourage confirmatory modes of data analysis (Manz, 2015).
- Designing a computer model to allow students to see and explore alignments and discrepancies with the experiment (see also Blikstein et al., 2016).
- Designing activity structures that enable students to experience and understand experiment and model as interrelated tools for inquiry.

- Framing and orienting the lab reports as an opportunity to make a theoretical argument to explain results to a research question instead of reporting on results from the experiments.
- Overall, communicating that lab is a space for students to explore patterns and consider questions that they have rather than come to settled conclusions.

More broadly, this chapter is an argument for putting experiments into richer contexts so that students can access some of the same richness of context that surrounds scientists. In doing so, students can find reasons to ask questions and experience the need to make meaning from the data they collect.

Acknowledgments This work was funded by a grant from the Davis Educational Foundation as well as through seed funding from the Tufts University Innovates program.

References

Bazerman, C. (1988). *Shaping written knowledge*. The University of Wisconsin Press. https://doi.org/10.1126/science.248.4957.877

Blikstein, P., Fuhrmann, T., & Salehi, S. (2016). Using the bifocal modeling framework to resolve "discrepant events" between physical experiments and virtual models in biology. *Journal of Science Education and Technology, 25*(4), 513–526. https://doi.org/10.1007/s10956-016-9623-7

Engle, R. A., & Conant, F. R. (2002). Guiding principles for fostering productive disciplinary engagement: Explaining an emergent argument in a community of learners classroom. *Cognition and Instruction, 20*(4), 399–483. https://doi.org/10.2307/3233901

Fishman, B. J., Penuel, W. R., Allen, A., & Sabelli, N. (2011). Design-based implementation research: An emerging model for transforming the relationship of research and practice. *National Society for the Study of Education, 112*(2), 136–156. https://doi.org/10.1177/016146811311501415

Ford, M. J. (2008). Disciplinary authority and accountability in scientific practice and learning. *Science Education, 92*(3), 404–423. https://doi.org/10.1002/sce.20263

Ford, M. J. (2015). Educational implications of choosing "practice" to describe science in the Next Generation Science Standards. *Science Education, 99*(6), 1041–1048. https://doi.org/10.1002/sce.21188

Gooding, D. (1990). *Experiment and the making of meaning: Human agency in scientific observation and experiment*. Kluwer Academic Publishers.

Gooding, D., Pinch, T., & Schaffer, S. (Eds.). (1989). *The uses of experiment: Studies in the natural sciences*. Cambridge University Press.

Gouvea, J. S., & Wagh, A. (2018). Exploring the unknown: Supporting students' navigation of scientific uncertainty with coupled methodologies. In J. Kay & R. Luckin (Eds.), *Rethinking learning in the digital age: Making the learning sciences count, 13th International Conference of the Learning Sciences (ICLS) 2018* (Vol. 1). International Society of the Learning Sciences.

Hatano, G., & Inagaki, K. (1986). Two courses of expertise. In H. W. Stevenson, H. Azuma, & K. Hakuta (Eds.), *Child development and education in Japan* (pp. 262–272). Freeman/Times Books/Henry Holt & Co. http://psycnet.apa.org/psycinfo/1986-97669-017

Hayes, R., & Gouvea, J. (2020). Destabilizing a target idea framing to expand epistemic agency. In M. Gresalfi & I. S. Horn (Eds.), *The interdisciplinarity of the learning sciences, 14th International Conference of the Learning Sciences (ICLS) 2020* (Vol. 3, pp. 1677–1680). International Society of the Learning Sciences.

Hill, C. F. C., Gouvea, J. S., & Hammer, D. (2018). Teaching assistant attention and responsiveness to student reasoning in written work. *CBE Life Sciences Education, 17*(2), ar25. https://doi.org/10.1187/cbe.17-04-0070

Hodson, D. (1996). Laboratory work as scientific method: three decades of confusion and distortion. *Journal of Curriculum Studies, 28*(2), 115–135. https://doi.org/10.1080/0022027980280201

Hofstein, A., & Lunetta, V. N. (2004). The laboratory in science education: Foundations for the twenty-first century. *Science Education, 88*(1), 28–54. https://doi.org/10.1002/sce.10106

Hu, D., Zwickl, B. M., Wilcox, B. R., & Lewandowski, H. J. (2017). Qualitative investigation of students' views about experimental physics. *Physical Review Physics Education Research, 13*(2), 1–12. https://doi.org/10.1103/PhysRevPhysEducRes.13.020134

Jordan, R. C., Ruibal-Villasenor, M., Hmelo-Silver, C. E., & Etkina, E. (2011). Laboratory materials: Affordances or constraints? *Journal of Research in Science Teaching, 48*(9), 1010–1025. https://doi.org/10.1002/tea.20418

Koponen, I. T., & Mäntylä, T. (2006). Generative role of experiments in physics and in teaching physics: A suggestion for epistemological reconstruction. *Science & Education, 15*(1), 31–54. https://doi.org/10.1007/s11191-005-3199-6

Lehrer, R., Schauble, L., & Petrosino, A. (2001). Reconsidering the role of experiment in science education. In K. Crowley, C. Schunn, & T. Okada (Eds.), *Designing for science: Implications from everyday, classroom, and professional settings* (pp. 251–278). Lawrence Erlbaum Associates.

Macleod, M., & Nersessian, N. J. (2013). Coupling simulation and experiment: The bimodal strategy in integrative systems biology. *Studies in History and Philosophy of Biological and Biomedical Sciences, 44*(4), 572–584. https://doi.org/10.1016/j.shpsc.2013.07.001

Macleod, M., & Nersessian, N. J. (2014). Strategies for coordinating experimentation and modeling in integrative systems biology. *Journal of Experimental Zoology Part B: Molecular and Developmental Evolution, 322*(4), 230–239. https://doi.org/10.1002/jez.b.22568

Manz, E. (2015). Resistance and the development of scientific practice: Designing the mangle into science instruction. *Cognition and Instruction, 33*(2), 89–124. https://doi.org/10.1080/07370008.2014.1000490

Manz, E., Lehrer, R., & Schauble, L. (2020). Rethinking the classroom science investigation. *Journal of Research in Science Teaching, 57*(7), 1–27. https://doi.org/10.1002/tea.21625

National Research Council. (2009). *A new biology for the 21st century*. National Academies Press. https://doi.org/10.17226/12764

Odenbaugh, J. (2005). Idealized, inaccurate but successful: A pragmatic approach to evaluating models in theoretical ecology. *Biology and Philosophy, 20*(2–3), 231–255. https://doi.org/10.1007/s10539-004-0478-6

Pelaez, N. J., Anderson, T. R., Gardner, S. M., Yin, Y., Abraham, J. K., Bartlett, E., Gormally, C., Hill, J. P., Hoover, M., Hurney, C., Long, T., Newman, D. L., Sirum, K., & Stevens, M. (2017). The basic competencies of biological experimentation: Concept-skill statements. *PIBERG Instructional Innovation Material Paper 4*. Retrieved from http://docs.lib.purdue.edu/pibergiim/4. Accessed 18 Oct 2021.

Phillips, A. M., Watkins, J. E., & Hammer, D. (2017). Problematizing as a scientific endeavor. *Physical Review Physics Education Research, 13*(2), 020107. https://doi.org/10.1103/PhysRevPhysEducRes.13.020107

Pickering, A. (1992). *Science as practice and culture*. University of Chicago Press.

Robertson, A. D., Scherr, R. E., & Hammer, D. (Eds.). (2016). *Responsive teaching in science and mathematics*. Routledge.

Rouse, J. (2007). Practice theory. In S. P. Turner & M. W. Risjord (Eds.), *Handbook of the philosophy of science. Philosophy and anthropology of science* (pp. 639–681). Elsevier.

Smith, E. M., Stein, M. M., & Holmes, N. G. (2020). How expectations of confirmation influence students' experimentation decisions in introductory labs. *Physical Review Physics Education Research, 16*(1), 10113. https://doi.org/10.1103/PhysRevPhysEducRes.16.010113

Stein, M. M., Smith, E. M., & Holmes, N. G. (2018). Confirming what we know: Understanding questionable research practices in intro physics labs. *Physics Education Research Conference Proceedings, 2018*, 1–4. https://doi.org/10.1119/perc.2018.pr.stein

Wong, S. L., & Hodson, D. (2009). From the horse's mouth: What scientists say about scientific investigation and scientific knowledge. *Science Education, 93*(1), 109–130. https://doi.org/10.1002/sce.20290

Chapter 19
Electronic Laboratory Notebook Use Supports Good Experimental Practice and Facilitates Data Sharing, Archiving and Analysis

Michael A. Buckholt and Jill Rulfs

19.1 Introduction and Background

Following the guidance embodied in the *Vision and Change* report (AAAS, 2011), the discipline has been re-envisioning and redesigning biology curricula including laboratory teaching. At Worcester Polytechnic Institute (WPI), as we have transitioned our undergraduate laboratory curriculum away from more traditional "cookbook" formats to authentic research-based experiences including CUREs and SUREs, we have also moved away from using traditional laboratory reports to using communication styles our graduates will see as they transition into the workplace. While we clearly value laboratory work for its "hands-on" training and practice, we also recognize that we have a responsibility to prepare students for other aspects of professional practice.

19.2 Electronic Laboratory Notebooks

Electronic laboratory notebooks have become widely used in commercial laboratories and specifically in the biotechnology research and development community. They are valued for advantages they provide over paper records including data security, back-up, archiving, workflow and data management. They are accepted as evidence in patent filings and other legal proceedings (Cardenas, 2014; Walsh & Cho,

M. A. Buckholt (✉) · J. Rulfs
Biology and Biotechnology Department, Worcester Polytechnic Institute (WIP), Worcester, MA, USA
e-mail: mbuckhol@wpi.edu; jrulfs@wpi.edu

N. J. Pelaez et al. (eds.), *Trends in Teaching Experimentation in the Life Sciences*, Contributions from Biology Education Research,
https://doi.org/10.1007/978-3-030-98592-9_19

2012). Thus, familiarity with electronic notebooks will likely be an advantage for our students going on to careers in the bio-technology or pharmaceutical industries.

Our students going on to practice in medical fields will find clinical practice has also moved from paper record keeping to electronic medical records. These require clear communication among professionals many of whom will access them remotely. Thus, the use of templates and prompts can begin to ensure the uniformity of information archived in these records. The ability to use images and annotated photos can also allow more accurate communication than descriptive text (Dood et al., 2018). These are all aspects of record keeping consistent with the electronic laboratory notebook, which allow the affordances of structured record keeping as well as non-text inputs.

As academic laboratories recognize the benefits of data sharing and archiving that electronic laboratory notebooks allow, they are becoming more widely used in academic practice, both in teaching and research (Okon & Nocera, 2017; Riley et al., 2017). In addition to the advantages in the research laboratory, both academic and industrial, there are other advantages which are more specific to the teaching laboratory. Four years ago, we adopted LabArchives (LabArchives, LLC) as our electronic notebook product. While our comments and observations are based on our experience with that product, most of the features we address are available in several other commercial products. Some examples of other ELNs are Labfolder, RSpace, Benchling, SciNote, ElabJournal, and Google Docs. Please note however that while LabArchives has developed a Classroom edition, some others are designed for the research lab setting and may not be easily adapted to a teaching lab setting.

19.3 Advantages in the Teaching Laboratory

In the biology teaching lab there are clear practical advantages over paper notebooks. They avoid any possible transfer of organisms out of the lab since there is no physical removal of documents from the laboratory. Often paper notebooks are collected a few times during a class, and generally returned at the next class meeting, perhaps a week later. Since electronic notebooks are electronically and remotely accessible, requiring only internet access, they allow instructors and teaching assistants to read and comment on students' entries on their own schedules, without limiting student access for continued data entry or analysis while grading is occurring. Because students can see the comments as soon as they are posted, electronic commenting also facilitates more prompt feedback, making it more useful to students (Lee, 2018). Students in an Ohio State University biomedical engineering lab, which was transitioned to using electronic notebooks felt it allowed them to incorporate that feedback in future submissions in a more timely way (Okon & Nocera, 2017), thus improving future submissions. Additionally, the use of the ELN resulted in higher scores for notebook content on a grading rubric when compared to paper notebooks.

Faculty find the ability to copy and paste comments related to common errors more convenient than the reiterative writing required by paper notebooks (Dood et al., 2018). If grading and commenting is done by teaching or learning assistants, the course instructor can also see the original student file and the comments to monitor the feedback being given to students and help assure consistency across the course (Johnston et al., 2014). Another very pragmatic advantage is that electronic entries eliminate the challenges of reading student handwriting in paper notebooks. Likewise, students can more easily read the comments and queries provided by instructors.

Another benefit of ELNs designed for teaching is that they can function as a learning management system, including options such as grading functions which are helpful if the course is not already managed using a management system. More importantly, because the students are, in general, sharing the instructor's notebook, class wide changes to assignments or protocols can be easily pushed out to all students.

Electronic notebooks also facilitate record keeping for group projects because they allow all group members working together on a notebook to access the same information. While all ELN formats are not the same, most notebooks can be shared with team or group members and can often also be shared with other teams on a "read only" basis, supporting collaborative learning but maintaining academic integrity (Riley et al., 2017). Generally, submissions can be date stamped and locked and, because each student can have an individual account, entries by individual students can be tracked. This allows easy audit trails, again supporting academic integrity, which can be helpful particularly when students are working in teams (Riley et al., 2017).

On an individual basis, faculty can see student progress through the course, tracking the number of logins, the time spent and the number of assignments completed (Dood et al., 2018). This allows faculty to reach out to individual students, providing timely intervention with students who may be struggling with personal or academic issues that are impeding their progress. If necessary, instructors can actually monitor the student's work in real time, which may be helpful in diagnosing potential problems (Dood et al., 2018; Riley et al., 2017).

19.4 Practicing Professional Practices

The use of electronic notebooks introduces students to practices in data curation and digital record keeping, which are now considered part of digital literacy in the sciences (Van Dyke & Smith-Carpenter, 2017). Biomedical Engineering students in the Ohio State University reported that using electronic notebooks enhanced their documentation skills, consistent with this expectation (Okon & Nocera, 2017). Some commercial products allow direct input of digital data from instrumentation in formats such as EXCEL or CSV files (Riley et al., 2017). Some widely used laboratory systems such as Vernier system can link directly to ELNs such as LabArchives.

The use of templates can reinforce consistency in formatting. Submissions can be linked to other portions of the notebook supporting reproducibility of methods or procedures (Guerrero et al., 2019; Lee, 2018). Real time entry during lab sessions supports honesty and transparency, preventing post hoc "corrections" student s may feel compelled to make to paper notebooks in response to perceived expectations. This supports students' critical understanding that the data are what they are and do not always meet expectations or support hypotheses. The honest recording of "errors" made to protocols can provide valuable learning opportunities including highlighting the occasional occurrence of scientific serendipity.

19.5 Templates and Frameworks

In the LabArchives format the course instructor has the master notebook and can mirror that for student notebooks. Most formats allow the instructor to adopt available or create custom frameworks with folders and templates that can be transferred to student notebooks. These can be practically and pedagogically helpful (Lee, 2018). In helping to assure that students are prepared for lab, pre-lab activities can include templates for students to generate experimental plans and objectives. Questions related to materials provided in reading, video and other material presented prior to the lab can be used to reinforce relevant concepts in advance of the laboratory session. Once students are in lab, templates for data entry can help both with organizational skills and with data sharing and archiving for instructor use (see Sect. 19.6). Post lab analysis, reflection and future planning can be done off line, allowing students to add images which can be annotated, references and links to other relevant sources including other sections of the notebook, protocols, lab manuals, or instructional videos all of which can also be provided via the electronic notebook (Johnston et al., 2014). Frameworks and templates included here can provide guidance on data curation and organization for students, and also facilitate grading by eliminating extraneous material.

19.6 Facilitation of Data Sharing and Archiving

Both students and instructors have the ability to archive notebooks after the course ends. For instructors, many years of notebooks can be archived without taking up physical space. Electronic archiving also eliminates data's being lost over time as can happen when paper notebooks are lost or discarded (Lee, 2018). Especially when submission formats or templates are specified, data from students or groups even across semesters or academic years can be collected and shared out as a single data set. This can provide students access to large enough data sets to allow valid statistical analysis, thus providing the instructor with an opportunity to focus on reproducibility, highlight the power of statistics and demonstrate the limitations of single experiments in reaching conclusions. In our recent move to remote laboratory teaching, when used in conjunction with laboratory demonstrations and computer simulations, these data collections have allowed us to provide students with "real" data to support understanding and analysis even when they cannot generate their own data. As data for ongoing curricular assessment of laboratory teaching and learning, individual student notebooks comprise authentic artifacts that can be used to track changes in student skills as the curriculum evolves.

Students can also archive their own notebooks after the course has ended. As they move through the laboratory curriculum, the notebooks can serve as archives of procedures or protocols they used in previous labs. The ability to access archived lab notebooks can also be helpful to students doing senior undergraduate research projects. We have heard from students who are using their archived notebooks to access and review procedures that they learned and used in courses and which are now both useful for their research project and easily accessible through their ELN. In addition to protocols or procedures, the ELN may also support review of concepts or model systems presented in earlier lab sequences. As students approach graduation, the notebooks could be used to demonstrate their experience to future employers, or help in the process of completing graduate school applications where they might support admission to a particular lab or program.

19.7 Concerns or Barriers

Concerns about cost and accessibility have been met with a number of homegrown analogues to the commercial products. Early efforts included the use of tablets and HTML versions to document design parameters in an engineering lab (Meyer et al., 2008). A number of software programs have been used to design in-house electronic notebooks, including OneNote, GoogleDocs, EverNote, an ePortfolio system (PebblePad) and the and learning and classroom management software Sakai and Blackboard Learn (Aoyagi, 2018; Guerrero et al., 2019; Hesser & Schwartz, 2013; Johnston et al., 2014; Lee, 2018; Meyer et al., 2008; Van Dyke & Smith-Carpenter, 2017). Riley et al., who also adopted the LabArchives product, report that the cost to students for subscriptions was the same as that to purchase paper laboratory specific notebooks (Dood et al., 2018; Riley et al., 2017). The LabArchives system allows individual student purchase or institutional purchase for a number of student notebooks.

We find that most students enjoy the ability to take photos of their results without having to draw reasonable facsimiles of things like bacterial plates or electrophoresis gels. We have already noted the advantages of typed over hand written material. However, some students report that they prefer to physically write out their notebook entries, and while some products include digital drawing tools, there are occasions where free hand drawings may be preferred. These preferences can be accommodated by allowing scanned (pdf) or photographed images to be inserted into the notebook, including those of hand written or hand drawn pages as long as they are clearly readable in the image. Instructors can decide whether or not to allow this option perhaps dependent on the reason for the accommodation request.

Although our current students are digital natives, comfortable with many electronic modes of communication, their ability to translate these skills to scientific communication is limited as they enter college. Teaching skills such as curation and organization of data can be facilitated with the use of electronic laboratory notebooks, which are becoming the norm in many professional scientific practices. What follows are illustrated examples of frameworks and templates, student entries, and extensions that have been allowed as a result of the data sharing and archiving aspects of an electronic laboratory notebook, here a product of LabArchives.

19.8 What Follows

This section will provide examples from labs we have taught using the LabArchives electronic laboratory notebook. The specific course from which most of these examples are taken is a WPI specific course called Searching for Solutions in Soil, but is our version of the crowd sourced CURE (course based research experience) called Tiny Earth (https://tinyearth.wisc.edu/). Students in this lab are searching for antibiotic producing microbes in soil samples they have gathered. The examples include both instructor guidance and prompts as well as example student entries to highlight the options and opportunities provided by use of the ELN.

We hesitate to identify anything as best practices as we are not aware of any studies done by us or others that compare specific ELN models to determine what works the best from a student or faculty viewpoint. But to provide some context for what follows, the practices we have found most useful include providing within the context of the ELN:

- templates for structure and organization
- examples of student work
- grading rubrics and checklists to guide students and facilitate grading
- guidance for the experimental design process
- on-going electronic feedback
- easy access to protocols and procedures
- notebook sharing for student teams

We also suggest archiving notebooks for subsequent use in data accumulation.

19.9 Organization, Planning, Data Curation and Entry

In order to help our students learn to keep and organize data they are given a notebook template that has some structure such as pre-labs, pages with prompts to remind them to write out what they do in lab each day etc. Figure 19.1 shows the

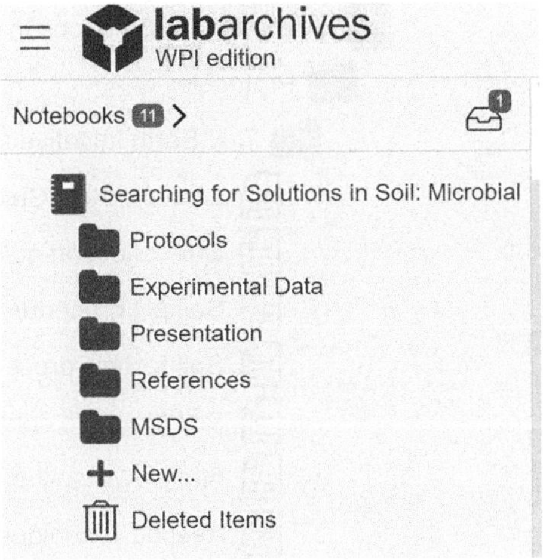

Fig. 19.1 Organizational scheme for the laboratory course Searching for Solutions in Soil using LabArchives. (Used with permission)

overall organization of the notebook. In this case it is only slightly different than the LabArchives preset organization.

Several levels of organization are possible. Figure 19.2 shows the ease with which procedures can be made available to students using the ELN. The easy availability of electronic procedures has both positive and negative aspects. While protocols and procedures are easily available as students need them in lab, this can also foster bad habits such as cutting and pasting procedures into the notebook rather than recording exactly what was done or failing to cite sources for published protocols. In both instances, instructors can take advantage of the opportunity to help students learn more about why recording what they do, including any deviations from a published protocol, is important or the need for proper attribution. The ability both to recognize what was done and to directly comment on it are advantages afforded by ELN use, which allows continual review and comment without the need for notebook collection.

Figures 19.3 and 19.4 below show the structure for notebook organization and data collection that we have imposed on the students. We do this for two reasons. One is to help students learn to organize their data. Students new to notebook keeping are often unsure of where to begin, what to include, and how to enter it so that it can be used later. The second reason is that it benefits the instructional staff by facilitating grading. With a weekly organizational scheme imposed, instructional staff do not have to hunt all over the notebook to find submissions for commenting and assessment. In class, there is a discussion explaining that this format is imposed to help them learn to organize their notebook, but that we recognize that it may not necessarily be the optimal organization for their own personal or research lab notebook. Examples of other organizational schemes such as keeping everything from a single experiment together (which is more easily done in an ELN than a paper notebook) or the more traditional format often used in paper notebooks where it is a continuous journal of what is done in lab, are mentioned, but we adhere to the scheme defined in the notebook for this course.

Fig. 19.2 List of relevant protocols available to students in the notebook. Each document opens to a specific protocol

Use of ELNs not only helps students better organize and present their data, it also helps them develop their experimental design skills. Here we provide an example of pre-lab designed to guide student their thinking about options for their design, adapted from the Original Small World Initiative and materials from Hernandez et al. (2018). Having the assignment in the notebook where students can also have access to the procedures aids in the design process by guiding their thinking about options for their design. Importantly because we don't have to collect the notebooks

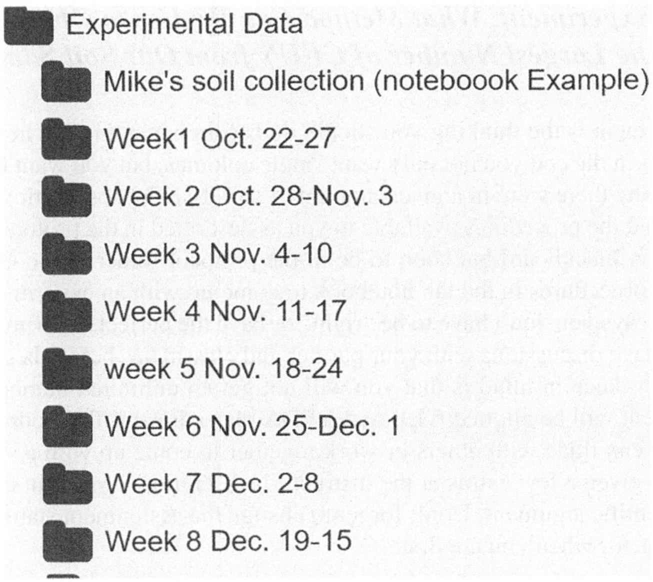

Fig. 19.3 Organizational scheme for data collection by week. The first folder contains an example provided by the lab faculty instructor

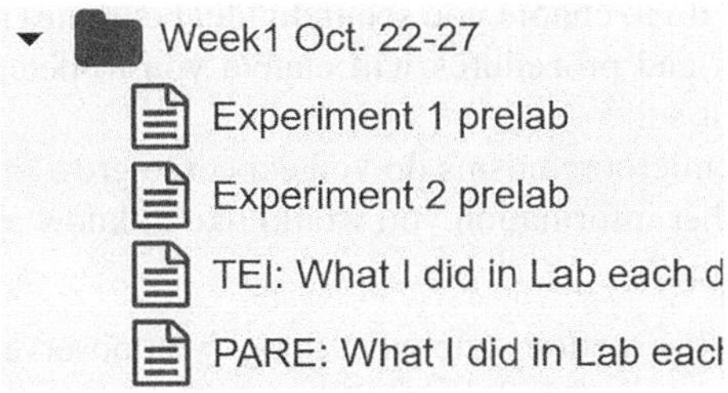

Fig. 19.4 Structured organization for weekly recording of pre-lab hypothesis and design, and in-lab documentation of experimental procedure and data gathering. TEI and PARE are acronyms for two parts of the course, Tiny Earth Initiative and Prevalence of Antibiotic Resistance in the Environment

we can give students feedback on their proposed experimental design and return it to them before they come to lab. This gives the instructional staff opportunity to intervene with comments, questions or suggestions before the experiment takes place. It also makes it easier for students to access comments and feedback while they are running the experiment since everything is in the same place. If students have done a good job in the planning aspect, they can even transfer the procedure from the pre-lab to where they are recording what they did in the lab and simply note any deviations that occur along the way.

19.9.1 Experiment: What Method Can We Use to Obtain the Largest Number of CFU's from Our Soil Samples?

This assignment is the thinking you should do before beginning the next lab. Keep in mind that in the end you not only want single colonies, but you want to be able to say how many there were in a given amount of soil. Use the observations you made as a class and the procedures available to you as described in the protocol document (currently on blackboard but soon to be in the protocol folder in the notebook) as well as the procedures in the lab notebook to come up with an experimental procedure. As always you don't have to be "right" or have the perfect experiment you will be able to have discussions with your partner and classmates before beginning. The only thing to keep in mind is that you will not get an unlimited number of plates. Every student will be allotted 6 LB and 6 PDA plates for this first round of experiments. You can trade with others or work together to come up with a strategy. You can also be given a few extras at the instructor's discretion if you can show a compelling scientific argument. Don't forget to change the assignment status to submitted to instructor when you are done

- *Biological Questions*: (Answering these questions will help you think about designing your experiment. You may find that some parts of your experiment will be a repetition of what you answered here but that is OK)
  - What will you do to enable you see individual colonies on a plate?
  - What methods and procedures will enable you to determine the number of bacteria in your soil?
  - What types of microorganisms do you expect to grow on your plates?
  - Is there any other information you would like to know about the microorganisms you isolated?
- *Observations*: In this section state what things you observed in previous experiments that might be useful or helpful in deciding what method to use. Think about what you saw in your own results and those presented by your classmates in "show and tell" sessions.
- *Objective*: State what you are attempting to accomplish with this particular experiment, not the overall goal of the entire project.
- *Procedure for obtaining single colonies on plate*: This should be a description of how you plan to get varying numbers of bacteria onto your plate. You can include exact detail in the procedure.
- *Procedural scheme*: Here you should draw a schematic of the procedure you will use. This can be a sketch or drawing or flow chart of how you plan to accomplish your objective. You can use the notebook drawing program, generate this using another drawing program or perhaps drawing it by hand and providing a picture by adding this a separate entry below this assignment.
- *Hypothesis*: In the formal scientific method the next three topics are separate events (see scientific method document in Course Materials) but in practice they

often happen as one process. This section can contain all three if it helps you to think about it that way, but you must at least state what you believe will happen/ what you think you are testing. In this case, for example, you might think about something related to plating densities and possibly different media

- *Rationale*: This is an explanation of how you generated your hypothesis and why you think your prediction is true. It is meant to reveal your thought processes and may be helpful later if you can't remember why you performed a specific experiment. Your rationale may relate back to your observations, to something you found in your literature research or what you discovered by answering the biological questions.
- *Experimental design*: This is a global explanation of your experiment, i.e. what you generally plan to do. This section should include an example of how you will collect and record your data (e.g. table) and what you think will be the best way to visualize/analyze it (e.g. type of graph).
  - *Controls*: These can include many things. Often in science we talk about positive and negative controls meaning things we know will definitely work or not work, but in this case you want to think of it as control groups and experimental groups. One way to think of these are what outcomes you will compare to determine whether your independent variables had an effect.
  - *Independent variables/treatments*: the things your will change in order to test your hypothesis
  - *Dependent variables*: the thing you will measure to test the effect of your treatment. In this case you need to decide which is more important to you.
  - *Confounding/controlled variables*: other things that could vary but that ideally you will hold constant between groups.
- *Protocol*: This is a detailed description of how you will perform the experiment, e.g., numbers of tubes, volumes, equipment, supplies etc. You can change it later but there should be enough detail to actually run the experiment. You can write this section by stating the details of your particular experiment and/or referring to a known protocol. If you do that be sure to in include a full citation that includes title of experiment and/or manual it comes from, who wrote it, and date written.

Table 19.1 shows a grading rubric provided to the students in the ELN in the experimental data folder and in the LMS, as a Word document so that they can see how their notebook entries will be assessed. It is detailed to try to normalize grading across multiple teaching assistants and laboratory sections, as well as to give students clear guidance about the level of detail and the quality of work expected.

In this course, notebook entries, termed "what I did in lab", are assessed twice a week so that students can get continuous feedback on their progress in data entry, curation and analysis. The ELN facilitates this continuous feedback because we don't have to collect notebooks to do it, so students can still be using and working on their notebook while assessment and feedback is occurring. Below, a student checklist for lab notebook entries (used with permission) lists requirements for each

Table 19.1 Grading rubric and point values for components of the notebook pre-labs

Biological questions (20 points)	Responses were well-thought out, detailed, and demonstrated a clear understanding of course and relevant outside materials. Correct citations were provided for all sources. (20)	Responses were well-thought out. Could use some further description and/or clarification. Correct citations were provided for all sources. (15)	A good effort was made to answer the questions; however, the response lacked some understanding and/or omitted a key point. Correct citations were provided for all sources. (10)	An attempt was made to answer the questions; however, the response lacked depth and true understanding (e.g. simple or one-worded responses). Correct citations were provided for all sources. (5)	No attempt was made to directly answer the questions. The responses demonstrated a lack of effort and/or were vague. Source citations were absent or incorrect. (0)
Observations/ background (10 points)			Experimental thought processes and/or observations were well presented and could be clearly understood without additional information. (10)	An attempt at explaining experimental thought processes and showing observations was apparent, but unclear, inconsistent or and needs further explanation. (5)	Part or all of the experimental thought processes and/or observations are lacking, unclear, and/or fragmented. (0)
Objective (10 points)			Goal of particular experiment was stated specifically and concisely so that what was being done was clear. (10)	A general statement alluding to what might be done in lab was provided but required more thought Flaws in experimental planning and preparation were evident. (5)	A stated objective was lacking, unclear, too broad, and/or fragmented, demonstrating a lack of effort and depth in thinking and preparation. (0)
Hypothesis (10 points)			A valid, specific, and well-supported hypothesis including what was being tested and predicted outcome was provided, setting the groundwork for a well-designed experiment. (10)	A general opinion regarding the experimental outcome was given but lacked rationale and experimental or observational support. (5)	A stated hypothesis was either lacking or poorly defined, resulting in poorly designed experiment. (0)

Rationale (10 points)			The thought rationale for the hypothesis was clear, valid, and well-supported by past observations and/or outside sources. It was clear why the experiment was set up the way it is. (10)	The rationale for the hypothesis relies on some invalid assumptions and/or was unclear. The relationship of the experimental design and the hypothesis requires more thought. (5)	The rationale for the hypothesis was lacking, unclear, fragmented and/or had no experimental or observational support. The relationship of the experimental design and hypothesis is unclear. (0)
Experimental design (10 points)			A global/general explanation of the experiment was well thought out, concise, and clear. It is apparent that the experiment is well-designed. (10)	A global/general explanation of the experiment is somewhat unclear and/or general understanding issues caused problems in experimental design. (5)	A global/general explanation of the experiment was either lacking or unclear. General understanding issues resulted in a flawed experimental design. (0)
Controls (5 points)				The definitions and use of positive and negative controls were clear and demonstrated in the experimental thought process. (5)	The definitions and use of positive and negative control lacked understanding as demonstrated in the experimental design (0)
Treatments (5 points)				The definitions and purposes of independent variables were well understood and demonstrated in the experimental design. (5)	The definitions and use of independent variables lacked understanding, and no effort or thought was put into their role in the experimental design. (0)

(continued)

Table 19.1 (continued)

Biological questions (20 points)	Responses were well--thought out, detailed, and demonstrated a clear understanding of course and relevant outside materials. Correct citations were provided for all sources. (20)	Responses were well-thought out. Could use some further description and/or clarification. Correct citations were provided for all sources. (15)	A good effort was made to answer the questions; however, the response lacked some understanding and/or omitted a key point. Correct citations were provided for all sources. (10)	An attempt was made to answer the questions; however, the response lacked depth and true understanding (e.g. simple or one-worded responses). Correct citations were provided for all sources. (5)	No attempt was made to directly answer the questions. The responses demonstrated a lack of effort and/or were vague. Source citations were absent or incorrect. (0)
Protocol (20 points)	The developed protocol was well-thought out and demonstrated a deep understanding of course and relevant outside materials. The protocol indicated an exemplary level of preparation. Correct citation was provided for all sources. (20)	The developed protocol was well-thought out; however, could use some further description and/or clarification. Correct citation was provided for all sources. (15)	A good effort was made to prepare a protocol; however, the protocol lacked some clarity and/or omitted a key point. The protocol was inconsistent with other aspects of the prelab write-up. Correct citation was provided for all sources. (10)	A general attempt was made to describe a protocol; however, it lacked detail, and displayed some significant misconceptions. The protocol was inconsistent with other aspects of the prelab write. Correct citation was provided for all sources. (5)	No attempt was made to develop a descriptive and detailed protocol. The responses demonstrated a lack of effort and/or were vague. Source citations were absent or incorrect. (0)

Used with permission

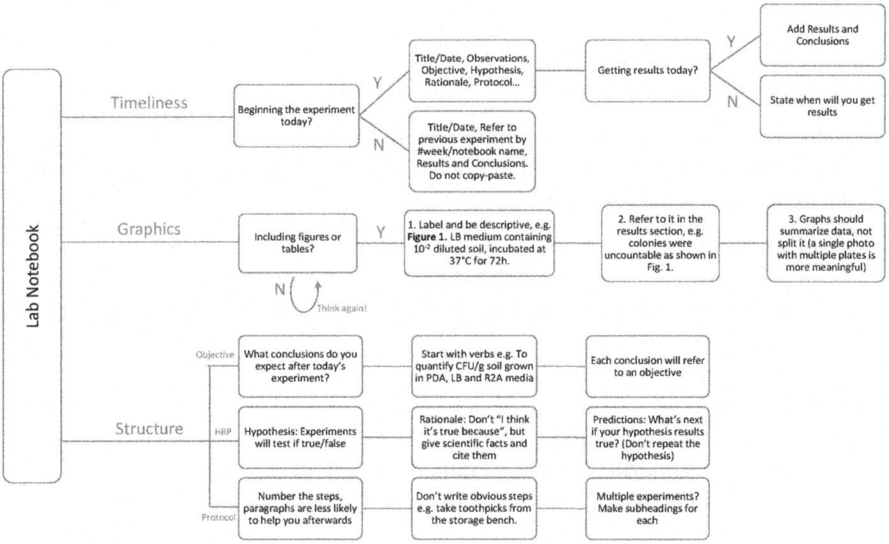

Fig. 19.5 Flow chart for lab notebook entries. (Used with permission)

part of the lab and Fig. 19.5 shows a flow chart provided to students to help them assess their entry for each experiment. These are each included in the experimental data section of the notebook, so they are readily accessible to the students. The checklist also serves as a grading template each week. We have purposely left it very general so that it can be flexible. Since most of our labs are designed around an authentic research paradigm, students are proceeding at different paces and conducting different experiments on any given day. The expectations for the lab entries in a given week are discussed and agreed on by the instructional staff at a weekly meeting. This gives us flexibility and a chance to be sure we have consistency among graders.

19.9.2 Example Checklist for Lab Notebook Entries (Used with Permission)

- Title
  - Includes Date
  - Unique to the experiment, not generic
- Purpose/Objective
  - Specific to the experiment, not about the entire project
  - Gives a clear and concise understanding of what will be discussed in the notebook

- Hypothesis
  - Clearly testable statement
  - Specific to the experiment, not about the entire project
- Rationale: Clearly explains thinking behind the hypothesis
- Prediction: Gives good detail about what they anticipate to see in their results
- Methods
  - In list format
  - In past tense
  - Talks about what was actually done, not just a copy of the experimental protocols
- Results-verbal
  - No interpretation present
  - Thoroughly describes results obtained through experiments
  - If figures are included, they are referenced in the text
- Results-visual
  - Figures and tables are labeled in the correct places
  - Figure legends are descriptive, and an understanding of the figure can be obtained from just looking at the legend
  - Figures and tables include all relevant units
  - Appropriate figures are used to represent the type of data shown
- Interpretations
  - Discusses how results can be taken in context of the experiment
  - Creative and insightful, not just a repetition of the results section
- Conclusions
  - Sums up what happened in the experiment
  - Talks about future directions, next steps

In addition to the checklist and the flow chart, students are given a chance in lab to discuss what should go into the lab notebook. They have also been provided with a set of guidelines about what constitutes an acceptable entry, an example of an entry by the instructor, and an anonymous example notebook such as from a student previously in the course or a similar one. Another advantage of the ability to archive student notebooks is allowing the instructor to provide such examples. Not only does the example notebook show what can and should be done, it also has the feedback in it from the instructors, so students also get a sense of what instructors look for and comment on over and above what they get from the rubric. These are included in an examples lab notebook folder in the experimental data folder as well as the LMS so that it is always accessible to the students. They can easily see the examples while still in the notebook environment.

19.10 Examples of Student Entries

A few examples of student entries in Figs. 19.6, 19.7 and Table 19.2 show how the ELN can enhance and extend the lab notebook experience. Exactly what can be done will vary by the ELN platform but the ability to upload content digitally allows for creation of a more complete notebook entry than is often possible in a physical paper notebook. Below are several examples of various kinds of annotated photos along with a photo of an on-the-fly diagram created in lab and a data table created in the notebook environment. Other possible entries not shown here are uploading and working with Microsoft or Google files. Again, the extent to which you can work with these files within the ELN environment may vary with different products,

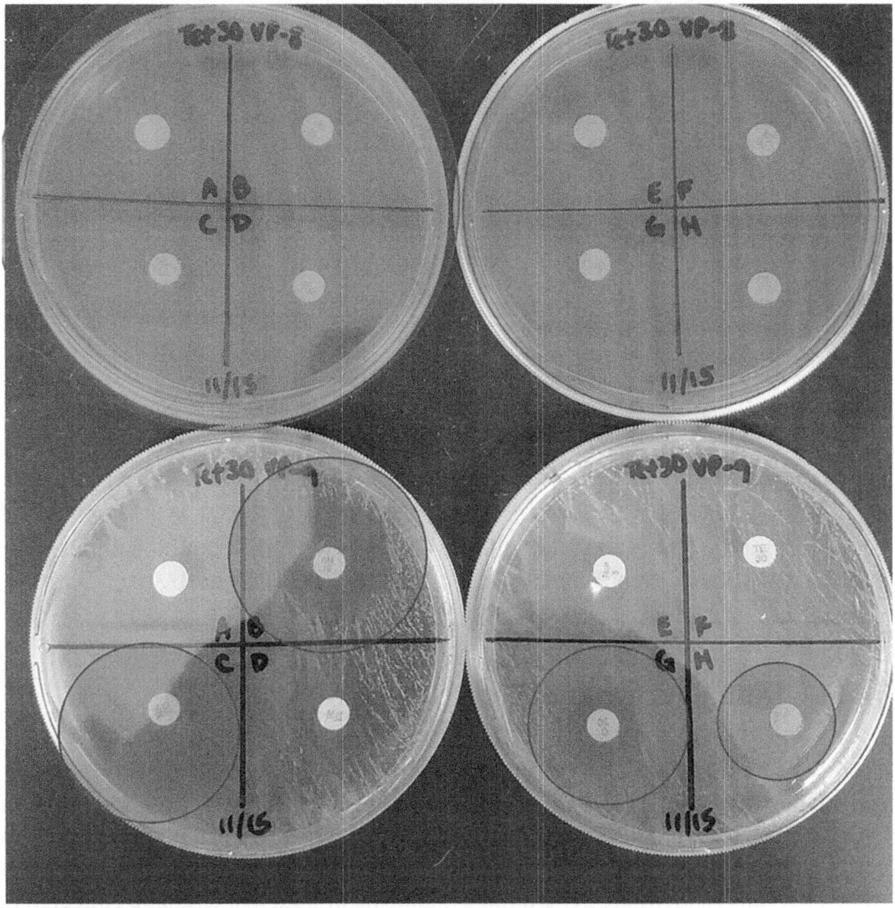

Fig. 19.6 Photos from student notebooks (Used with permission): (**a**) Filter disc assays showing putative antibiotic response, (**b**) Agarose gel with PCR products, (**c**) Annotated photo of soil collection site and (**d**) Bacterial colonies demonstrating zones of inhibition indicated using the drawing capability

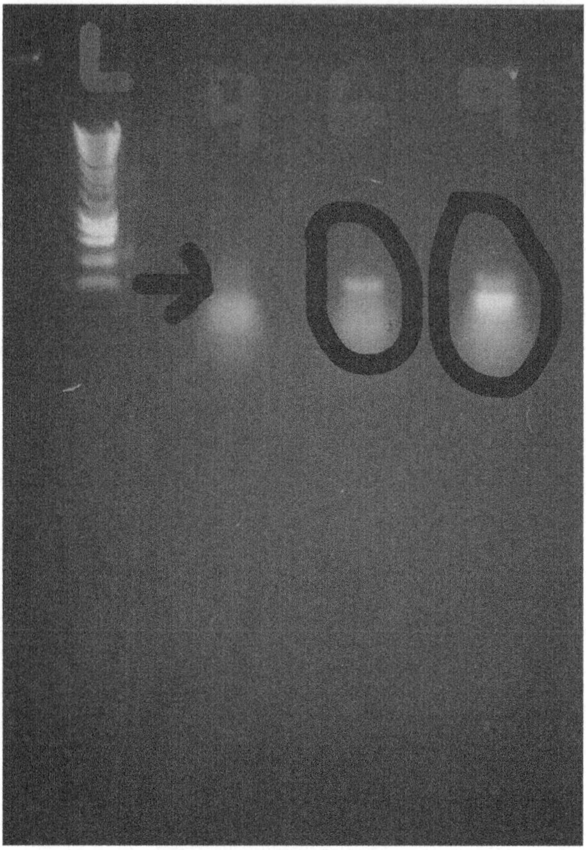

Fig. 19.6 (continued)

but in general the ability to upload files to the ELN will allow additional files to be part of the notebook even if they were not created in the ELN itself.

19.11 Data Sharing and Archiving

The use of the ELN has also facilitated our efforts to teach students some statistics and the power of data by showing them what a change in sample size (N) can do. In 2008, we realized that because of emerging technologies we had an opportunity to easily collect data from an entire class over multiple sections. At that time, we redesigned our labs to facilitate generation by students of data that could be collected in realtime. Data were displayed to the class as they were collected and then used by the students each week to run appropriate statistical tests on their own data and then that of their section and finally the entire class. Analysis of student outcomes showed

Fig. 19.6 (continued)

that this method helped students understand the need for statistics and the impact of sample size on statistical significance (Hunter et al., 2010). Initially, data collection was set up using a personal response system but evolved as technology changed to use other shared forms such as Google Sheets. The ELN can make distribution of links to these shared resources easier but, in this case, the ELN we are using actually

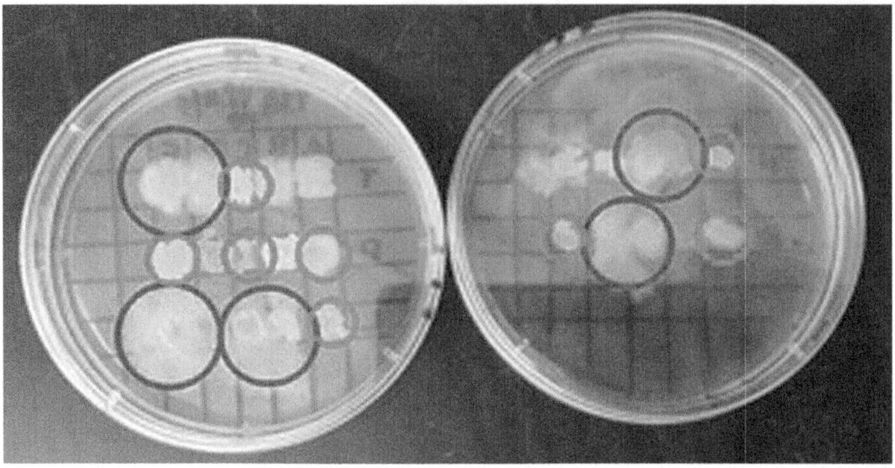

Fig. 19.6 (continued)

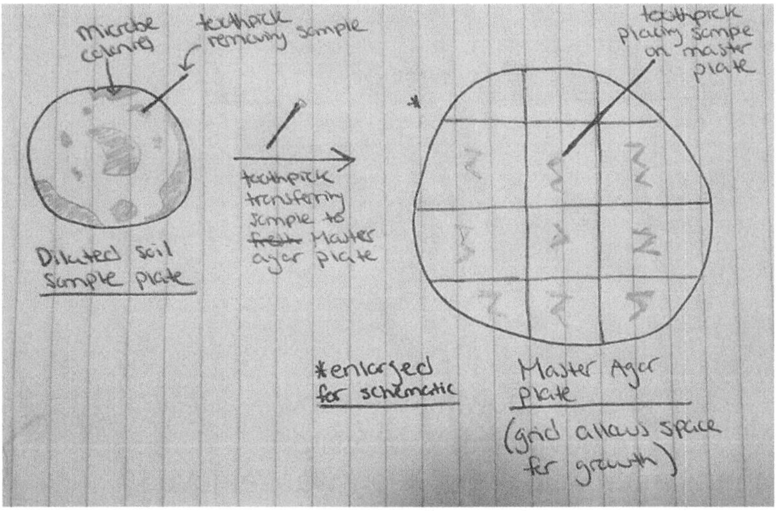

Fig. 19.7 Hand drawn illustration from student ELN, inserted into the ELN as an image. (Used with permission)

allows for sharing an item such as a page or a data collection table within the notebook. It can be shared with users, all of whom have the ability to enter data as edits, just like what can be done using a Google Sheet form. This allows the entire class to upload data in real time that is visible to all. The one caveat is that everyone must be careful not to over write anyone else's data. These data are then available for all to use in performing statistical analysis as we have previously done using other technologies, but now right within the notebook environment. Class experimental

Table 19.2 Data table from student notebook

NA plates	Number of colonies	Percent antibiotic-resistant growth
Control	0	n/a
10^{-1}	235	n/a
10^{-2}	63	n/a
10^{-3}	12	n/a
10^{-4}	0	n/a
10^{-5}	0	n/a
Test 3 plates	**Number of colonies**	**Percent antibiotic-resistant growth**
Control	0	n/a
10^{-1}	82	34%
10^{-2}	8	n/a
10^{-3}	2	n/a
Test 30 plates	**Number of colonies**	**Percent antibiotic-resistant growth**
10^{-1}	14	n/a
10^{-2}	3	n/a
10^{-3}	0	n/a

Used with permission

data can also be archived year to year in the notebook. This can allow students to compare data across years, looking for potential variation, or to examine what happens when the sample size is increased even more. This use of large data sets generated by a number of different students, has allowed us to give our students an early introduction to the use of statistics and to get a feel for what it takes for data to be statistically significant and to reinforce the need for reproducibility. ELN use has facilitated and enhanced what we had been doing and allows us to easily extend it to encompass other concepts.

The ELN environment allows for archiving of all sorts of data that can be readily made available to other students working on similar problems. For example, the course from which most of these examples are taken gives the students an opportunity to get DNA sequence data for the bacteria they are isolating. Samples are sent out for sequencing and the files return from the vendor in electronic form. Students then do alignment searches against the national database to determine isolate identity. All of the sequence from multiple samples over several years are currently archived in the notebook for students to use in practicing their analysis skills on by simply opening the data files linked to the notebook (Fig. 19.8). The availability of these previously analyzed sequences allows them to develop skill before examining their own sequence. It is also an opportunity for those students that are not successful in getting sequence from their own samples to easily have something to explore.

The sharing capabilities of the ELN can be used on a smaller scale as well. The ability of the ELN to facilitate sharing can, and is, utilized by students on a one to one and small group level. Students working on the same experiment can share pages with each other affording work on group projects and experiments while also keeping their own individual notebook. This facilitates collegiality and can promote interaction where students can do some peer teaching.

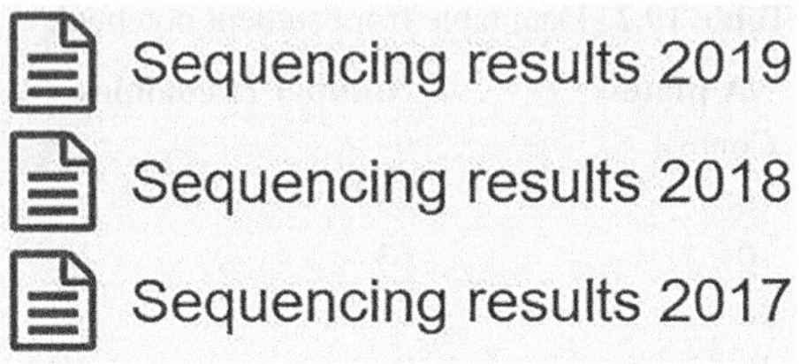

Fig. 19.8 Data files are archived within the notebook environment

A further advantage of archiving notebooks is the students' ability to use the notebook and its data at a later date. For example, at WPI, as a graduation requirement, every student must complete a yearlong senior research project. Since we started using CUREs and developing authentic research labs, several students have chosen to extend aspects of the work they and others have done in these classes in their senior project. Many of them report that they have gone back to their archived notebook to look up procedures they remember doing but don't have the protocol for. They have also related looking at previous data they generated. In many cases a paper version of the notebook, if the students had kept one, might have been lost or destroyed before they got to use it in their project. Sharing archived notebooks has also been useful in another CURE sequence, SEAPHAGES (https://seaphages.org/). In an initial course, students isolate bacteriophage from soil samples, one or two of which are then sent for genomic sequencing. In a separate, subsequent course, students annotate the genomic sequences from the previous course. The student that isolated the phage might not be in the annotation course, but the ELN allows us to share the record of how the phage was isolated because we can archive the notebook of the student who did the isolation. This is easily shared among the annotation student group, something which would not be possible with a paper notebook even if it existed.

In our recent move to remote learning, which included remote laboratory courses, one of our faculty colleagues provided video demonstrations of the "hands-on" experiments the students would have done had they been able to be in the lab. Because the experiments she chose had been used in previous years, she was able to provide the virtual students with real data collected by previous classes and archived in their ELNs. This provided the remote learners an authentic set of data on which to base their analyses and conclusions.

19.12 Other Options

In our initial surveys of students using the notebook, one of the most frequent comments we had from students was that they enjoyed or preferred physically writing in their notebooks and they were less inclined to use an ELN because of this. Sometimes there was also a situation where the student wanted to draw out a diagram or flow chart that was easily sketched into a paper notebook but might take more time or effort to use the built-in sketching program or some other electronic drawing

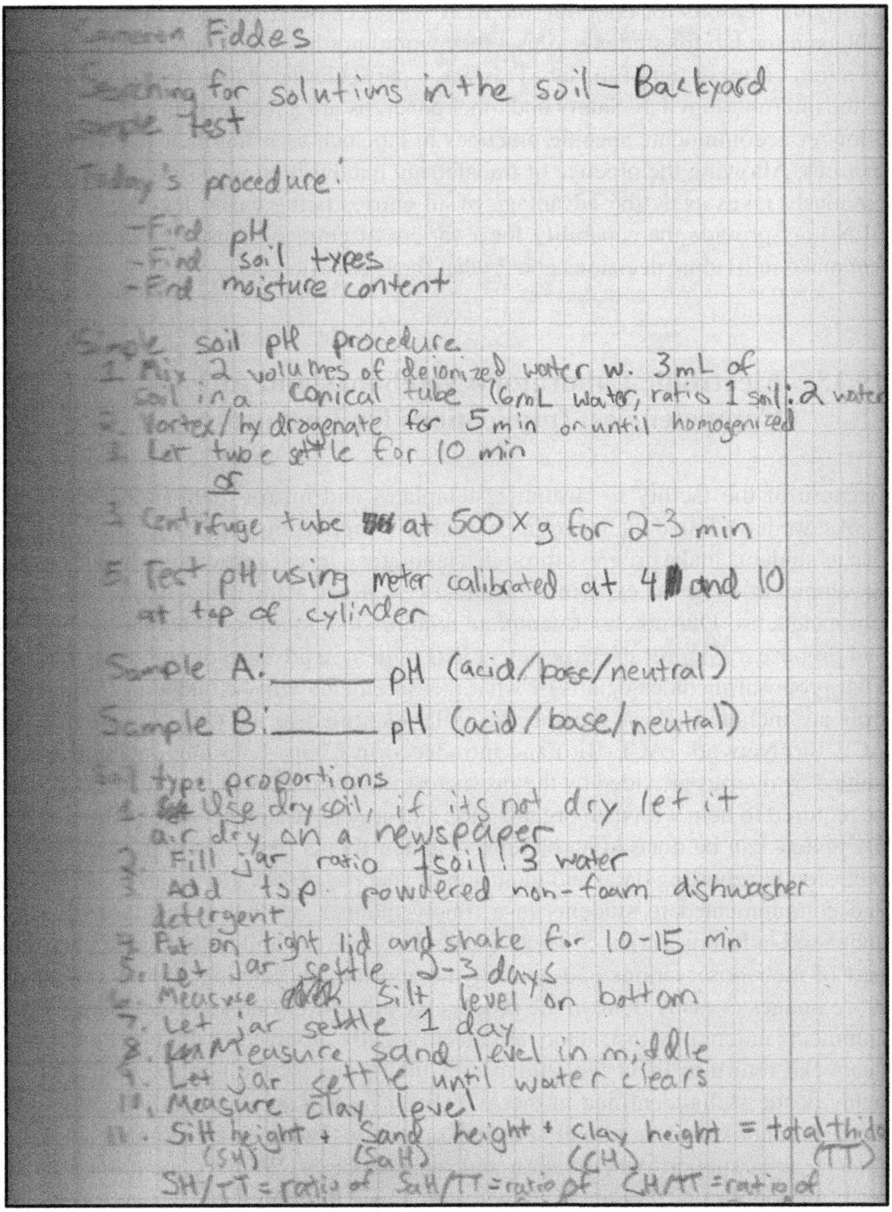

Fiddes
for solutions in the soil – Backyard test
's procedure:
-Find pH
-Find soil types
-Find moisture content
soil pH procedure
1. Mix 2 volumes of deionized water w. 3mL of soil in a conical tube (6mL water, ratio 1 soil : 2 water)
2. Vortex/hydrogenate for 5 min or until homogenized
3. Let tube settle for 10 min
or
3. Centrifuge tube at 500 x g for 2-3 min
5. Test pH using meter calibrated at 4 and 10 at top of cylinder
Sample A: ______ pH (acid/base/neutral)
Sample B: ______ pH (acid/base/neutral)
type proportions
1. Use dry soil, if its not dry let it air dry on a newspaper
2. Fill jar ratio 1 soil : 3 water
3. Add tsp. powdered non-foam dishwasher detergent
4. Put on tight lid and shake for 10-15 min
5. Let jar settle 2-3 days
6. Measure silt level on bottom
7. Let jar settle 1 day
8. Measure sand level in middle
9. Let jar settle until water clears
10. Measure clay level
11. Silt height (SH) + Sand height (SaH) + clay height (CH) = total thick (TT)
SH/TT = ratio of SaH/TT = ratio of CH/TT = ratio of

Fig. 19.9 Image of a student hand-written notebook entry

program. Both are easily accommodated as seen in the example below (Fig. 19.9) of a photo of a student's physical notebook uploaded into the ELN or the hand drawn diagram in the previous section (Fig. 19.7). These formats can present more challenges when trying to make comments particularly if they are photos. As a note one of original reasons for adopting the ELN was because if we have devices such as tablets in the lab for students to use, they would not have to bring anything of their own into or out of out of the lab. This was a particular advantage for us safety wise in the microbiology lab. Safety and other concerns are a factor in how much we can allow or accommodate specific practices in lab such as using their phones to take pictures. Allowing the practice of transferring handwritten work into the ELN also obviously takes away the advantage of all entries being easily legible. While the ELN may provide the capability for a variety of entry options, clearly instructors can make individual decisions about what they will allow.

19.13 Electronic Laboratory Notebooks and ACE-Bio Competencies: Implications for Instructors

Because of the facility to customize templates and mirror them in student notebooks, we have found them extremely useful in supporting good experimental practice as students make progress through a laboratory-based curriculum. In our model, development of good experimental practices aligns with many of the ACE-Bio Competencies. Our notebook template requires that students generate hypotheses and propose a *Plan* for the experiment to test them in advance of coming to the lab. That proposal includes details of what measurements will be taken and what controls are included. These are specific skills identified in the *Question* area by the ACE-Bio Network (ACE-Bio) and introduced in Chap. 1. In lab, data entries are guided by queries provided by the instructor, and formats such as spreadsheets may be required to help with data organization and access. Again, because instructor and TA review can be done in an ongoing manner, things like chart and graph labels, precision in reporting data, and inclusion of specific observations can be addressed and communicated to students in a timely manner, so that expectations can be addressed in future entries. Students particularly like the ability to include photos as part of their observations and guiding the appropriate labeling and annotation of those images is easily done in the product we are using (LabArchives). Because all comments entered by instructors are electronically archived in the notebooks, students can return to them for guidance as they progress through the course. Thus many of the skills identified in the ACE-Bio *Conduct* area are facilitated by electronic notebook use. And finally post lab operations, which are encompassed in the *Analyze* area support data curation and representation. We have mentioned previously the opportunity to teach statistical analysis through data sharing, providing data sets robust enough to address appropriate statistical tests and generate meaningful measures to support hypotheses. While the areas of *Conclude* and *Communicate* encompass skills not directly relevant to notebook keeping, the ability to archive, analyze and curate data certainly makes the use of these data easy in designing oral, written or visual communication genres such as poster presentations or manuscript drafts. Because they can be easily copied to other formats, entries such a graphs, tables and images are a conveniently available for use in other communication venues.

As our students move from the teaching laboratory to individual senior research experiences, their experience guided by formats and templates provide a model of good experimental practice that is transferrable to a less structured teaching environment. The product we have chosen allows students to keep multiple separate notebooks simultaneously and the ability to archive their notebooks allows students to look back at processes, protocols and practices they have used earlier in their laboratory training. To implement this approach, we recommend the following for instructors:

- Review the pros and cons of Electronic Laboratory Notebooks for lab instruction.
- Consider implementing practical strategies to introduce students to data archiving, sharing, and analyses that are now common in professional scientific practice.
- Develop "data documentation" and "data curation" skills by communicating expectations to students and using feedback mechanisms.
- Plan future Design-Based Research (DBR) studies to understand how, when, and why ELNs work better than paper lab notebooks.

Acknowledgments The authors acknowledge and thank Cameron Fiddes, S. Hahn, Lauren Herchenroder, Danielle A. Snider, Jordyn Van Minos and Luis Gregory Gutierrez Zamalloa at Worcester Polytechnic Institute for permission to use their educational material.

References

American Association for the Advancement of Science (AAAS). (2011). Vision and change in undergraduate biology education: A call to action. . Retrieved December 23, 2020 from https://live-visionandchange.pantheonsite.io/wp-content/uploads/2013/11/aaas-VISchange-web1113.pdf

Aoyagi, D. (2018). *No-cost implementation of electronic lab notebooks in an intro engineering design course*. In ASEE annual conference proceedings, Salt Lake City, UT.

Cardenas, M. (2014). *An implementation of electronic laboratory notebooks (ELNs) using a learning management system platform in an undergraduate experimental engineering course*. In ASEE annual conference proceedings, Indianapolis, IN.

Dood, A. J., Johnson, L. M., & Shorb, J. M. (2018). Electronic laboratory notebooks allow for modifications in a general organic, and biochemistry chemistry laboratory to increase authenticity of the student experience. *Journal of Chemical Education, 95*, 1922–1928.

Guerrero, S., Lopez-Cortes, A., Garcia-Cardenas, J. M., Saa, P., Indacochea, A., Armendariz-Castillo, I., et al. (2019). A quick guide for using Microsoft OneNote as an electronic laboratory notebook. *PLoS Computational Biology, 15*(5), e1006918.

Hernandez, S., Tsang, T., Bascom-Slack, C., Broderick, N., & Handelsman, J. (2018). *Tiny Earth, a research guide to studentsourcing antibiotic discovery*. XanEdu Inc.

Hesser, T. L., & Schwartz, P. M. (2013). iPads in the science laboratory: Experiments in designing and implementing a paperless chemistry laboratory course. *Journal of STEM Education, 14*, 5–9.

Hunter, A., Rulfs, J., Caron, J. M., & Buckholt, M. A. (2010). Using a classroom response system for real-time data display and analysis in introductory biology labs. *Journal of College Science Teaching, 40*(2), 19–25.

Johnston, J., Kant, S., Gysbers, V., Hancock, D., & Denyer, G. (2014). Using an ePortfolio system as an electronic laboratory notebook in undergraduate biochemistry and molecular biology practical classes. *Biochemistry and Molecular Biology Education, 42*, 50–57.

Lee, D. B. (2018). Implementation and student perceptions on Google docs as an electronic laboratory notebook in organic chemistry. *Journal of Chemical Education, 95*, 1102–1111.

Meyer, D. G., Johnson, M. C., & Brown, C. M. (2008). *Tablets for timely design documentation*. In ASEE annual conference proceedings, Pittsburgh, PA.

Okon, M. D., & Nocera, T. M. (2017). *The student educational experience with electronic laboratory notebooks*. In ASEE annual conference proceedings, Columbus, OH.

Riley, E. M., Hattaway, H. Z., & Felse, P. A. (2017). Implementation and use of cloud-based electronic lab notebook in a bioprocess engineering teaching laboratory. *Journal of Biological Engineering, 11*, 40.

Van Dyke, A. R., & Smith-Carpenter, J. (2017). Bring your own device: A digital notebook for undergraduate biochemistry laboratory using a free, cross-platform application. *Journal of Chemical Education, 94*, 656–661.

Walsh, E., & Cho, I. (2012). Using Evernote as an electronic lab notebook in a translational science laboratory. *Journal of Laboratory Automation, 18*(3), 229–234.

Chapter 20
Growing Innovation and Collaboration Through Assessment and Feedback: A Toolkit for Assessing and Developing Students' Soft Skills in Biological Experimentation

Sarah M. Beno and Diane C. Tucker

20.1 Introduction

Preparing undergraduate students for their next career steps, including graduate school, professional school, and industry, requires the attainment of both technical skills and soft skills (Scheufele & Krause, 2019). In the biological sciences, necessary skills include the ACE-Bio Basic Competencies of Biological Experimentation (*Identify*, *Question*, *Plan*, *Conduct*, *Analyze*, *Conclude*, and *Communicate*; (Pelaez et al., 2017; Chap. 1 in this volume)) along with leadership skills, creativity, critical thinking skills, teamwork, and innovation. The students involved in this described summer innovation program utilized all seven of the ACE-Bio competencies during their design and prototyping process. We explicitly assessed growth in the ACE-Bio competency, *Communicate*, as well as additional skills that are important to successful biological experimentation and innovation. Individuals in these various situations must be creative and committed to their work. They must also have strong critical thinking skills and be able to communicate their ideas effectively. Importantly, no big project is completed alone; therefore, it is important for students to have good collaboration skills (Jarmai & Vogel-Pöschl, 2020). These innovation skills are important for all career paths and are highly sought-after (Li, 2017). Teamwork is necessary for almost all aspects of life (Salas et al., 2018). These skills

S. M. Beno (✉)
Department of Biology, Birmingham-Southern College, Birmingham, AL, USA
e-mail: smbeno@bsc.edu

D. C. Tucker
UAB Honors College and Department of Psychology, University of Alabama at Birmingham, Birmingham, AL, USA
e-mail: dtucker@uab.edu

N. J. Pelaez et al. (eds.), *Trends in Teaching Experimentation in the Life Sciences*, Contributions from Biology Education Research,
https://doi.org/10.1007/978-3-030-98592-9_20

are also useful in biological experimentation where teams are often essential to project success. Without diligent teamwork, projects are often delayed or abandoned. In research careers in particular, projects require broad skill sets and oftentimes no single person is an expert in all of these areas. Collaboration skills are required in order to design better experiments, ask more refined questions, and conduct more thorough analyses (Bennett & Gadlin, 2012; Cheruvelil et al., 2020). Creativity and critical thinking are important in experimentation, particularly during troubleshooting and experimental design (DeHaan, 2009). These skills are used to help develop solutions after identifying a gap in knowledge or a limitation in previous research. Creativity and critical thinking are also needed to draw inferences and conclusions (DeHaan, 2009). In biological experimentation, having a strong commitment to one's project is critical for success, rather than simply going through the motions. Finally, the ability to communicate one's findings is perhaps the most important piece of experimentation. Scientists present findings in both written and verbal formats at poster sessions, oral presentations, lab meetings, and popular media. Effective communicators can communicate findings to a layperson audience (Brownell et al., 2013; Scheufele & Krause, 2019). In a world of "fake news" and pseudoscience, the ability to communicate findings to diverse audiences is an imperative skill (Scheufele & Krause, 2019). Measuring these innovation and teamwork skills will assist educators to assess student growth in soft skills that lead to success for scientists and entrepreneurs.

Teaching soft skills in classroom settings can be challenging. When done effectively, it requires extensive feedback from both the instructor and from peers. While peer review is implemented in many classroom settings, feedback may not be given in a manner that promotes growth and change (Bailey & Garner, 2010; Poulos & Mahony, 2008). Therefore, improving feedback mechanisms to allow for conversation can also support growth. It is known that students benefit from both giving and receiving feedback (Nicol et al., 2014) so it is important that they are given opportunities to practice this. Understandably, individuals often find feedback uncomfortable (Hattie & Timperley, 2007), but practice in giving and receiving constructive feedback may demonstrate its value.

In higher education settings, teamwork and communication soft skills are often "taught" by using planned activities such as group work and oral presentations. However, methods for assessment of the impact of such activities are limited. One 17-item instrument proposed by Corwin et al. (2015) has been used to assess skills in collaboration, discovery and relevance, and iteration in laboratory courses. In particular, the authors assessed the impact of course-based undergraduate research experiences as a way of growing these skills (Corwin et al., 2015). A recently published rubric was proposed for assessing student critical thinking skills in written assignments, which could be used in evaluating laboratory reports, for example (Reynders et al., 2020). Interdisciplinary or interprofessional teamwork can also be used to expose students to collaboration challenges (Havyer et al., 2014; Morphet et al., 2014). While innovation challenges have been used successfully in many business and entrepreneurship programs (Harkema & Schout, 2008), innovation challenges in health care or other contexts are less frequent and constitute an especially promising way to engage STEM students in collaboration and teamwork (Pellegrini & Jansen, 2013). Systematic assessment of the impact of these experiences is possible, but not the usual practice. Here, we present a template for implementing learning activities and assessing the impact of innovation, problem-based learning, or biological experimentation on the students' collaborative and teamwork skills.

20.2 Assessment Tools

Our proposed toolkit assesses student growth in skills of communication, creativity, critical thinking, and collaboration. This toolkit includes the Qualities of an Innovator survey, behavioral assessment surveys, reflection questions, and semi-structured interview questions. This toolkit was developed to understand student growth during participation in innovation and experimentation projects across disciplines. In our pilot use of the toolkit, we utilized each component at the beginning of the experience and following the experience. Behavioral assessments were used weekly to gauge student perceptions of their own behaviors and those of their teammates. The Qualities of an Innovator survey (Table 20.1) was built to assess a students' inclination for innovation and their confidence in their skills in collaborating, communicating ideas, creating, and thinking critically. This survey is a self-reported indication of confidence in their innovation skill set, rather than a measure of skill attainment. For each skill, a percentage of possible points was computed since not all skills were assessed by the same number of items.

Behavioral assessments were developed to evaluate perceptions of ability to collaborate, create, communicate, and think critically. It also measures commitment to the project. Behavioral assessments can be modified for self-assessment, peer/teammate assessment, and evaluator or instructor assessment (Table 20.2). For modification to a peer or teammate assessment, all "I" statements become "my teammate" statements. To modify to an instructor's assessment, the statements lead with "the student." Behavioral assessments were designed to evaluate these skill sets for an individual and compare self-perceptions with those of others. As with the Qualities of an Innovator survey, scores were expressed as a percentage of possible points for each skill. At the end of the survey, participants were asked to identify their greatest strengths and weaknesses.

Semi-structured interview questions (Tables 20.3, 20.4, and 20.5) explore key constructs in more detail. These interviews help evaluators to assess growth in the innovation skillset, including communication, collaboration, critical thinking, and creativity. The interviews also provide additional insight into challenges students may face as they growing in these skills. A suggested timeline for use of these interview questions is pre- and post- experience, with a more long-term post-experience time point to determine the long-term impact of the program or experience. Examples of student responses to the semi-structured interview questions, collected in a Pilot Study of Clinical Innovation Presidential Fellowship Program participants

Table 20.1 Qualities of an innovator survey. Items A-I were used to assess inclination for in-novation. Items J-L were used to assess communication, items M-P were used to assess creativity, items Q-T were used to assess critical thinking, and items U-X were used to assess collaboration. Item E was reverse scored

How important is it to you...		Not at all	A little	Somewhat	Fairly important	Highly important
A	To have a clear role?	0	1	2	3	4
B	To be sure that your efforts will produce results?	0	1	2	3	4
C	To avoid conflict with others about your ideas or strategies?	0	1	2	3	4
D	To get individual credit for your ideas?	0	1	2	3	4
E	To avoid failure?	0	1	2	3	4
F	To choose your own problems?	0	1	2	3	4
G	To successfully complete a task?	0	1	2	3	4
H	To potentially make a discovery or solve a problem?	0	1	2	3	4
How confident are you that you...		Not at all	A little	Somewhat	Fairly important	Highly important
I	Can tolerate setbacks without giving up?	0	1	2	3	4
J	Can behave professionally in a high stakes situation?	0	1	2	3	4
K	Can present ideas to persons in power?	0	1	2	3	4
L	Can develop a compelling presentation?	0	1	2	3	4
M	Can develop creative solutions?	0	1	2	3	4
N	Can ask questions that lead to examining things in new ways?	0	1	2	3	4
O	Can connect ideas from different contexts?	0	1	2	3	4
P	Can move forward when the path to the solution is not clear?	0	1	2	3	4
Q	Can offer useful ideas for solving problems outside your discipline?	0	1	2	3	4
R	Can offer useful ideas for solving problems in your discipline?	0	1	2	3	4
S	Can develop an effective strategy for approaching a problem?	0	1	2	3	4
T	Can identify problems that need solving?	0	1	2	3	4

(continued)

Table 20.1 (continued)

How important is it to you...		Not at all	A little	Somewhat	Fairly important	Highly important
U	Can effectively redirect a discussion?	0	1	2	3	4
V	Can really listen to the ideas of others?	0	1	2	3	4
W	Can contribute innovative ideas to a team?	0	1	2	3	4
X	Can work effectively as part of a team?	0	1	2	3	4

Table 20.2 Behavioral assessment, in the form of a self-assessment. These assessments can be modified to be used by leadership or by teammates or peers. Items A-C were used to assess communication, items D-G were used to assess collaboration, items H-K were used to assess critical thinking, items L-N were used to assess creativity, and items O-Q were used to assess commitment. Items D and J were reverse scored

Please rate the following, considering your behavior throughout the week		Never	Rarely	Sometimes	Often	Always
A	I contribute substantively to the team discussion	0	1	2	3	4
B	I share ideas with people in positions of power	0	1	2	3	4
C	I adjust my communication to audience or context	0	1	2	3	4
D	I interrupt my teammates	0	1	2	3	4
E	I encourage my teammates	0	1	2	3	4
F	I am respectful of others' ideas	0	1	2	3	4
G	I summarize/paraphrase the comments of others	0	1	2	3	4
H	I troubleshoot effectively	0	1	2	3	4
I	I consider future roadblocks and potential "wins" of others	0	1	2	3	4
J	I am stalled by challenges	0	1	2	3	4
K	I focus on the big picture	0	1	2	3	4
L	I change approaches when stalled	0	1	2	3	4
M	I consider problems from various angles	0	1	2	3	4
N	I organize ideas and information well	0	1	2	3	4
O	I work hours beyond what is required	0	1	2	3	4
P	I bring excitement to the team and project	0	1	2	3	4
Q	I stretch beyond my comfort zone	0	1	2	3	4

Table 20.3 Semi-structured interview questions suggested for an early time point in the project, perhaps pre-innovation or experimentation experience. Questions are organized by theme and common responses from the pilot program are recorded. This interview was conducted in the second week of the pilot program

Theme	Question	Common responses from pilot study
Goals	What do you anticipate will be your next career steps?	Nursing students hoped to pursue advanced degrees in nursing. Neuroscience and engineering students expressed interest in medical school and PhD programs
	What do you hope to get out of this experience?	Students were excited about gaining experience working in interdisciplinary teams, gaining innovation and creativity skills, and to build a network that would benefit their careers
	How does this experience fit in to your career?	Students acknowledged the importance of research, design thinking, and innovation skills for their future careers
Risk-taking	Have you ever started a task and realized it might not be possible? What did you do?	Most could remember a time when this happened. They recalled changing approaches and/or brainstorming with others
	Have you ever decided you are not good at something? How did you reach that conclusion?	Most remembered a time when this happened and that they had tried something a few times but it didn't come easily. Half of the students also mentioned that if it was an important skill or task, they would work hard to become good at it
Brainstorming	How do you approach working on a problem?	Students approached problems by evaluating background research and brainstorming before making a structured plan
	What kinds of tools have helped you in the brainstorming process in the past?	Common brainstorming methods included word dumping, writing everything out, making lists, discussing with others, and sticky notes
Teamwork	What do you anticipate you will bring to the team dynamic?	All students emphasized that teams would not have a leader but that they would hold themselves accountable. Most referred to their organization skills. Some discipline specific skills were also suggested. Neuroscience students felt comfortable with heavy science, engineering students felt good about prototyping, and the nursing students felt their clinical experiences and medical knowledge would benefit their teams
	What roles do you normally take on in a team?	Half of the students tended to lead or organize a team. The others identified themselves as either communicators or supporters
	Do you generally enjoy teamwork? Why or why not?	There were mixed feelings about teamwork. Students had good experiences in teams with set standards and good collaboration. But there were also concerns brought up, including different personality types and differing levels of commitment

(continued)

Table 20.3 (continued)

Theme	Question	Common responses from pilot study
	What do you think is valuable about working in teams?	Students emphasized different perspectives and the ability to talk through problems as beneficial
Overall	What are you most nervous about for this experience?	Nursing students were worried about time management while neuroscience and engineering students were more worried about the final project or final presentation
	What are you most excited about for this experience?	Most were most excited about the projects they would be working on. Other exciting things included networking connections, experience in the hospital, and the ability to work in interdisciplinary teams

at the University of Alabama at Birmingham are provided with the interview questions from each time point: pre-innovation challenge (Table 20.3), post-innovation challenge (Table 20.4), and at a 3-month follow up (Table 20.5). NVivo software was used to help identify common themes in the student responses.

20.3 Feedback and Guided Reflection

Providing structured feedback based upon the behavioral assessments and the Qualities of an Innovator survey can help students to understand their self-perception versus perception of others for effort in these areas. Structured feedback was provided to each of the Clinical Innovation Presidential Fellowship Pro-gram students at weeks 6, 8, and 10 (Beno et al., 2020). This feedback included a compilation of behavioral assessment data from their teammates compared to their self-evaluations. Assessment data was broken down by skill area and provided to students as a percentage of possible points. All strengths and weaknesses comments were shared with the students, as well. Following these feedback sessions, the use of guided reflection questions (Table 20.6) helps encourage student understanding and promote discussion of the feedback that they received. One of the biggest potential benefits of feedback comes from this reflection piece, as individuals gain a stronger sense of understanding, growth, and purpose. Examples of the themes in student responses from the Clinical Innovation Presidential Fellowship Program are provided with the guided reflection questions (Table 20.6).

20.4 The Innovation Toolkit at Work

The University of Alabama at Birmingham Clinical Innovation Presidential Fellowship Program offered 10 interdisciplinary students (traditional STEM and nursing) the opportunity to develop solutions for real-world problems in the hospital

Table 20.4 Semi-structured interview questions for late-stage participation in innovation or experimentation programs with summarized student responses from the Clinical Innovation Presidential Fellowship Program following week 10

Theme	Question	Common responses from pilot study
Overall	What was the most important take-away from this experience?	Students discussed benefits and challenges of working in interdisciplinary teams. They also highlighted the importance of communication in collaboration
	How do you anticipate this summer program will influence your future?	Students talked about newfound passions, their improved understanding of perceptions of others, and new career paths
Brainstorming	What tools or strategies were most useful during the brainstorming process?	One team used a sticky note brainstorming exercise that was influential for all students in that team. The other team dumped many ideas before sorting through them and evaluating the ideas compared to things already available
	Do you feel that your group was able to capture the essence of the sepsis problem while also considering all options?	All students were confident that they had
RIsk-taking & confronting problems	How confident were you in contributing unique ideas or perspectives?	Students were fairly confident. Some started out quiet, but became more comfortable expressing their opinions by the end
	Can you identify a "light bulb moment" during your time working on this project? What happened and how did you get to that moment?	Different students had different lightbulb moments. In general, these lightbulb moments related to the products they were developing, but one student had a lightbulb moment related to team dynamics
	Did you use any strategies in your problem-solving that you did not originally anticipate? What were they?	Students discussed outside collaborations as very important and that they became better communicators both in and out of the team
Teamwork	What were the biggest challenges in working with your team?	Consistent challenges were differences in time spent on the project between nursing and STEM students, communication issues, and personality conflicts
	What were the biggest challenges in working with the other team?	The biggest challenges between teams that were reported were lack of respect for individuals on other teams, communication, and that the teams were often in different stages of the design thinking process during larger group meetings
	What were the benefits of working with students in another discipline?	Neuroscience and engineering students were praised for their technical knowledge, and nursing students for their clinical expertise and background knowledge. Students discussed that working with students outside of their discipline helped them learn to communicate with people who thought differently

(continued)

Table 20.4 (continued)

Theme	Question	Common responses from pilot study
	What were the challenges of working with students in another discipline?	The biggest challenges ultimately had to do with not understanding the needs of students from other disciplines
	What were the best successes?	Teams felt very proud of their final projects and final presentation
	Did you feel valued within your team?	Whether or not students felt valued depended on which project team they were a part of and what discipline they associated with. In general, students felt valued, but the nursing students on the wearable device team felt they were not equals in their team
Feedback	Were the feedback sessions valuable?	Students felt that feedback was very helpful in all forms. The team reflection meeting was mentioned as especially helpful for the team working on the virtual reality project
	What insights did these feedback sessions give you about your role and contribution in your team?	Students saw that everyone had different roles on the team. The feedback allowed students to understand their roles on the team and to work on weaknesses
	Was it challenging to receive feedback from your teammates?	Students said it was challenging at first, but it was important and it got easier
	Was it challenging to receive feedback from the other team?	Students recognized that it was more comfortable to get feedback from their own teammates than from individuals on the other team. This was likely due to trust and intention
	How has your reaction to receiving critical feedback changed over the summer?	Students enjoyed receiving feedback and found it very valuable. They noted that it got easier to give and receive by the end of the summer
Overall	What was the most challenging part of this experience?	Most were challenged by time, communication, and balancing commitments
	What was the most exciting part of this experience?	Most students said the final presentation and seeing their final products were the most exciting part. One mentioned the experience of working with a passionate interdisciplinary team and one talked about the potential that these projects had to make a real difference for sepsis outcomes
	What do you anticipate will be your next career steps?	Many felt more uncertain about career plans because the program opened their eyes to new opportunities
	How did this experience fit into your career path?	The opportunities to immerse in the hospital, learn research techniques, and work in new settings led to more well-rounded individuals who can efficiently work in interdisciplinary teams

Table 20.5 Semi-structured interview questions regarding the long-term impacts of participation in innovation or experimentation programs and summarized student responses from the Clinical Innovation Presidential Fellowship Program 3 months post- program conclusion

Theme	Question	Common responses from pilot study
Team dynamics	Do you feel like your team had a clear leader?	Students reported that teams did not have clear leaders but that STEM students played bigger roles
	What role(s) did each member of your team serve?	Roles were not clearly defined, but they were based on strengths
	How did your team handle the task of splitting up intellectual property?	One team discussed I.P. as a group and split it up based on overall work to the project. The other team's I.P. discussion resulted in a lot of drama and emotions. At this point, a conclusion has not yet been made
	What was the best part about working with your team?	Most students really loved working in their teams and all appreciated the different perspectives from interdisciplinary teammates
	What were the biggest challenges in working with your team?	The biggest challenges were with personality differences, schedule differences, and communication issues within their teams
	If you had the same teammates, but were tasked with the other project, what do you think your summer experience would have been like? (if needed, prompt students to discuss team dynamics)	Students felt team dynamics would have stayed the same but that nursing students would have maybe been more easily connected with the virtual reality-based project
Feedback	How has the feedback that you received influenced you outside of the fellowship? Has it changed the way that you approach group work?	Communication and having the courage to express opinions are important skills. Students say that feedback inspired them to be more confident and more aware of how they are perceived by others
	If you were given the opportunity to participate in a similar program, would you take it? Why or why not?	All students said they would participate again as it gave them a transformative experience in which they matured and grew
Overall	Would you recommend the Clinical Innovation Presidential Fellowship to a friend?	Students said they gained experience in research and enjoyed the opportunity to make real-world connections. They would recommend this fellowship program
	What was the most important takeaway from your summer experience?	Communication is important. Students also felt more confident working in teams with people who think differently than they do

setting over the course of 10 weeks in 2019 (Beno et al., 2020). Fellows were introduced to the design thinking process, which has been presented as a promising method for teaching innovation (Altman et al., 2018). In the specific cases discussed here, teams were tasked with developing solutions to prevalent issues surrounding sepsis. Sepsis is a leading cause of death in hospitals worldwide (Rudd et al., 2020),

Table 20.6 Guided reflection questions and summarized responses from students participating in the Clinical Innovation Presidential Fellowship Program

Week	Questions	Common responses from pilot study
7	What are you working to improve?	Students wanted to improve skills in collaboration and communication
	What are your action items regarding your goals for improvement? For each action item, please elaborate by answering: (A) What is this action item meant to address? (B) What challenges have you experienced/might you anticipate with this? (C) What will you do moving forward to make progress?	Action items varied by student, depending on their goals
9	What are you working to improve or "take to the next level"? (multiple answers are okay and encouraged!)	Students most often wanted to improve their communication and collaboration skills
	What are the steps you have taken to work on this? What challenges are you facing? What successes are you having?	Steps varied depending on the goals of the individual student. In general, students wrote about their efforts and changes being noticed by teammates
	(A) Has the feedback been helpful to you? (B) Will it be helpful as you move forward in your career? Please elaborate	Students said that feedback was helpful for future teamwork situations and for understanding how they are perceived in a group setting

and these teams worked to develop new methods for detection and training to alert for sepsis. While these projects were not traditional biology experiments, they required intensive research analogous to course-based undergraduate research experience or mentored independent research.

20.4.1 Timeline and Methodology of Assessment

For purposes of the fellowship, students were divided into two interdisciplinary teams. These students served as pilot testers for the Innovation Toolkit. Through regular feedback and guided reflection, students identified areas for improvement and discussed important steps for implementing changes that helped them to reach their goals. A timeline of the activities in this fellowship program and assessments used are outlined in Fig. 20.1. In the Clinical Innovation Presidential Fellowship Program, participants spent the first week in a clinical immersion, getting to explore issues related to sepsis in the clinical setting. In week 2, the participants were split into interdisciplinary teams and began to work on problem definition. Ideation and exploration began around week 4 and by week 6 both teams were working on

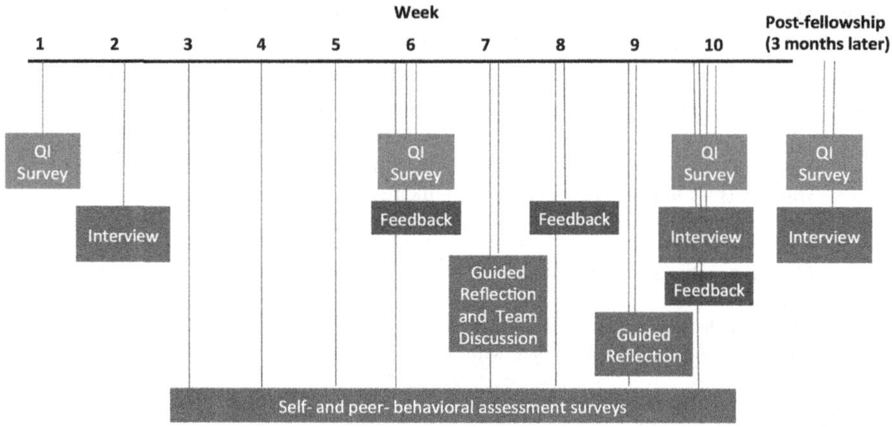

Fig. 20.1 Timeline of activities and assessments during the 10-week Presidential Fellows Clinical Innovation Program. The program included a series of structured feedback and guided reflection opportunities, as well as surveys and semi-structured interviews used to inform the program organizers of shifts in skill development

prototyping their innovative solutions prior to a final presentation in week 10. Throughout the innovation experience, participants worked on building empathy and had weekly structured practices for the final presentation. At week 6, all participants partook in a team reflection session. During these sessions, an evaluator asked students to discuss their structured feedback with their teammates and guided participants through determining action items, changes individuals could make to become more effective teammates. Week 6 was therefore an important evaluation point for the Qualities of an Innovator and Behavioral Assessment surveys.

Students participating in this program spent different amounts of time working on the projects: students pursuing BSN degrees contributed a minimum of 6 h per week to their team projects due to obligations to clinical experiences and required coursework, whereas students from other STEM fields contributed a minimum of 30 h per week to their projects. As such, we hypothesized that STEM students would show greater improvement in their critical thinking, creativity, collaboration, and communication skills as compared to their BSN counterparts. Semi-structured interview data led us to further investigate differences in skills development, particularly in collaboration, within each project team.

Differences in skill perception of STEM and BSN students, as well as between project teams, were assessed using data from the Qualities of an Innovator survey collected at weeks 1, 6, and 10. These were analyzed using independent samples t-tests. Behavioral assessment survey data was evaluating using an average of weeks 3–4 as baseline (early in the team formation), an average of weeks 5–7 as a middle point, and an average of weeks 8–10 as an end point. Self-assessments and peer-evaluations were analyzed separately. Growth was measured using F-tests in a one-way ANOVA. All Qualities of an Innovator survey data were assessed between weeks 1 and 10 and between weeks 6 and 10. Similarly, behavioral assessment data was evaluated using the average of weeks 3–4 compared to the average of weeks 8–10 and the average of weeks 5–7 compared to the average of weeks 8–10. All statistics were run using SPSS.

20.4.2 Pilot Assessment Results

Importantly, all data collected from our pilot study was collected from just ten students, which limits the quantitative conclusions that can be drawn from our study. However, a number of observations were made. All students were highly confident in their innovation skills even at the start of the program. By the end of the program, students reported increased confidence in communicating and presenting ideas. Confidence in creativity also increased from the start to the end of the program, and no differences were observed between disciplines or the two project teams. We observed some behavioral differences between students in traditional STEM disciplines compared to those who were pursuing degrees in nursing. We predicted these differences based on differences in participation requirements as well as discipline-specific stereotypes. Even early in the program, STEM students who were required to spend more time on their projects were perceived as being more committed to their projects ($t(8) = 6.45$, $p < 0.001$) and as having stronger critical thinking skills ($t(8) = 2.39$, $p < 0.05$) by their teammates, as compared to the BSN student team members (Beno et al., 2020). The teammate behavioral assessments revealed that STEM students, were perceived as more effective communicators, being more committed to their projects, and having higher critical thinking skills than their BSN student counterparts from the start of the project to the end of the project ($F(1,8) = 7.90$, $p < 0.03$, $F(1,8) = 16.1$, $p < 0.005$, and $F(1,8) = 33.26$, $p < 0.001$) (Beno et al., 2020).

Interestingly, we also observed a difference in perceived commitment to the projects between teams. The team working on a virtual reality training tool reported increased commitment from the initial to final phases of the project (70% vs 84%) whereas the team working on a wearable device for vital signs self-reported decreased commitment (69% to 60%) from the initial to final phase of the project (Beno et al., 2020). Teammate behavioral assessments corroborated this observation.

Semi-structured interviews helped interpret these observations. The interview responses supported our findings with differences between disciplines and between teams. From the interviews, we discovered that some of the differences in project commitment might have resulted from differences in expertise. STEM students felt valued throughout both the virtual reality and wearable device project design process, but students pursuing BSN degrees reported that while their expertise was needed to develop the virtual reality-training program, they lacked the technical skills needed to participate in the wearable device project as it progressed. Furthermore, participants recognized that their disparate time commitments affected their perceived growth in requisite technical skills. The guided reflection questions indicated students valued the experience of giving and receiving feedback throughout the summer, despite finding it uncomfortable initially. They realized that their teamwork was greatly facilitated by honest interaction about barriers and challenges.

20.4.3 Pilot Observations

Our toolkit provided insight into differences between students from different disciplines and between teams. We found overall increased confidence in presentation and communication skills for all students from the beginning to the end of the fellowship program. Confidence in critical thinking skills and creativity trended higher, but was not statistically significant, likely due to a small sample size. This confidence in these three skill sets was maintained 3 months post-fellowship. Following the guided team reflections, we observed significantly higher confidence in collaboration and creativity skills, suggesting that frequent feedback in both formative and summative instances are crucial for skill development. Future use of this toolkit may give educators further insight into student inclination for innovation.

20.5 Implications of the Toolkit

The toolkit presented here can be used across disciplines to assess student growth from active participation in an innovation challenge, problem-based learning activity, or experimental opportunity. In university classrooms, instructors often teach collaboration using group projects (Beier et al., 2018) and communication through oral presentations (Braun, 2017; Parker et al., 2020). However, these skills are rarely formally assessed. The toolkit can be used to track changes in this skill set over time, to identify time points for instructor innovation, and to understand interdisciplinary team dynamics. This toolkit allows evaluators to determine what changes are maintained throughout the course of a project or experiment, which can create opportunities for intervention to ensure successful teaching of these key skills as opposed to merely waiting for students to develop them.

20.5.1 Toolkit Use for Assessment of Essential Skills in Biological Experimentation

The toolkit assesses collaboration, communication, creativity, and critical thinking and can be utilized at different time points to measure change. This toolkit was designed to measure both confidence in the ability to use this skill set and in the perception of successful use and development of these skills. These skills are essential in biological experimentation and for successful scientists.

Collaboration is often key to a successful experiment. Through collaboration, projects can be completed more efficiently, more quickly, and with greater precision. Most research requires collaboration and relies on technical expertise of different individuals to be done well (Pelaez et al., 2018; Vermeulen et al., 2013). Having the skills to effectively collaborate with others is needed in order to participate in the global scientific culture.

Likewise, scientists must be effective communicators. In the general population, science literacy remains challenging (Rosenthal, 2020; Scheufele & Krause, 2019). Scientists performing biological experimentation therefore must be prepared to communicate their ideas and findings in layman's terms to the appropriate audiences (Brownell et al., 2013). Scientists must also be able to communicate within a project (Pelaez et al., 2017; Chap. 1 in this volume), as highlighted by our pilot study semi-structured interviews. These students often brought up the importance of developing better communication skills.

Of course, creativity, innovation, and critical thinking are also important within biological experimentation. Scientists need to be able to brainstorm new ideas, evaluate existing research, and synthesize new information by designing experiments (DeHaan, 2009). Troubleshooting, which is inevitable in science, requires both critical thinking and creativity. *Communicate* is a core component of the ACE-Bio Competencies (Pelaez et al., 2017; Chap. 1 in this volume) as it is crucial for scientists to convey their research to others. While creativity, critical thinking, and collaboration are not overtly represented in these Basic Competencies of Biological Experimentation, both creativity and critical thinking are represented by the skills outlined in each competency. For example, creativity is needed in *Question*, *Plan*, *Analyze*, and *Conclude*. Likewise, skills outlined in competencies *Identify*, *Question*, *Plan*, *Conduct*, *Analyze*, and *Conclude* are specific examples of critical thinking. While collaboration is not included in the core competencies, it is an essential skill in biological experimentation. The ACE-Bio competencies are complemented by these additional skills, and these skills are required for scientists and relevant in many disciplines.

20.5.2 Toolkit Use in Broad, Interdisciplinary Situations

The piloting of this toolkit was for an interdisciplinary innovation challenge that included students in a nursing program and students from neuroscience and engineering disciplines. Pieces of the toolkit have since been used in courses in various undergraduate disciplines at the University of Alabama at Birmingham. The skill sets assessed in the toolkit presented here are integral to a successful biological experimentation, but they are also important for professional development. Therefore, simple modifications to personalize the toolkit to a specific discipline will help instructors ensure that students are well prepared for success in their careers. The Qualities of an Innovator Survey and self- and teammate- behavioral assessments were recently used in diverse subjects across campus, including a course for education majors planning to teach social studies, a kinesiology course, a nursing course, and another class focused on teaching innovation. Instructors reported compatibility of this toolkit for their various courses and future research will investigate the skill development of students participating in different project-based learning activities.

20.6 Future Directions and Overall Importance

The use of assessment and feedback are of utmost importance for student learning (Andersson & Palm, 2017; Rushton, 2005). A critical point to successful feedback is to keep things constructive. When giving feedback, it helps to focus on something that can be improved (Hardavella et al., 2017). In course design, it is important to include both formative and summative feedback and this principle carries into skill development. In our pilot of this toolkit, students reflected that the consistent use of feedback helped them to recognize how they were perceived and guided methods to improving collaboration and teamwork. Consequently, the implementation of feedback sessions may be highly important in developing these skills quickly.

Successful innovators can be successful in a variety of careers. The skills identified for the innovation toolkit: creativity, collaboration, critical thinking, and communication are all skills that are necessary for success in many careers. In particular, to develop effective scientists through biological experimentation we must also relate these skills to ACE-Bio Basic Competencies of Biological Experimentation outlined by Pelaez et al. (2017 Chap. 1 in this volume). Importantly, it is also critical that students develop competencies in successfully working in diverse and interdisciplinary groups. Data show scientific papers with authors of diverse academic discipline, location, and gender are cited at higher rates (Adams, 2013; AlShebi et al., 2018). While laboratory-based courses often are limited to students within a specific major, instructors can encourage diverse groups by combining students with different career goals or interests. In many careers, project teams are built with experts of different skill sets. The toolkit presented here can be used to better prepare students for their future endeavors.

For instructors and/or education researchers hoping to utilize the Innovation Toolkit, we offer the following recommendations:

- Utilize the Qualities of an Innovator survey at minimum at the beginning and end of the experience.
- Behavioral assessment surveys should be used often to track student self- and peer-perceived growth in the innovation skill areas.
- Midway through the experience, the use of guided discussion following feedback may serve as an intervention. Using a mediator (instructor or evaluator), ask students to elaborate on their greatest strengths and weaknesses.
- Ask students for action items- how will they become more effective experimenters moving forward?
- Provide ample opportunities for self-reflection and for feedback within teams

Acknowledgments We were able to pilot our toolkit with the Presidential Clinical Innovation Summer Fellowship program, which was funded by the Office of the President at the University of Alabama at Birmingham. We would like to thank the students that participated in the program and provided feedback about the implementation of the toolkit. The protocol was reviewed and approved by UAB's Institutional Review Board (protocol #060619003). SMB was supported by the UAB Mentored Experiences in Research, Instruction, and Teaching program, which is funded by a National Institutes for Health grant, GM088010.

References

Adams, J. (2013). The fourth age of research. *Nature, 497*, 557–560.

AlShebi, B. K., Rahwan, T., & Yoon, W. L. (2018). The preeminence of ethnic diversity in scientific collaboration. *Nature Communications, 9*, 5163.

Altman, M., Huang, T. T. K., & Breland, J. Y. (2018). Design thinking in health care. *Preventing Chronic Disease, 15*, E117.

Andersson, C., & Palm, T. (2017). Characteristics of improved formative assessment practice. *Education Inquiry, 8*(2), 104–122.

Bailey, R., & Garner, M. (2010). Is the feedback in higher education assessment worth the paper it is written on? Teachers' reflections on their practices. *Teaching in Higher Education, 15*(2), 187–198.

Beier, M., Kim, M., Saterbak, A., Leautaud, V., Bishnoi, S., & Gilberto, J. (2018). The effect of authentic project-based learning on attitudes and career aspirations in STEM. *Journal of Research in Science Teaching, 56*(1), 3–23.

Bennett, L. M., & Gadlin, H. (2012). Collaboration and team science, from theory to practice. *Journal of Investigative Medicine, 60*(5), 768–775.

Beno, S. M., Wingo, N. P., Berry, J. L., Noles, K., & Tucker, D. C. (2020). Assessment of 21st century skills during an innovation challenge. *The Journal of Higher Education Theory and Practice, 20*(13), 79–108.

Braun, M. (2017). Comparative evaluation of online and in-class student team presentations. *Journal of University Teaching & Learning Practice, 14*(3), 19–40.

Brownell, S. E., Price, J. V., & Steinman, L. (2013). Science communication to the general public: Why we need to teach undergraduate and graduate students this skill as part of their formal training. *Journal of Undergraduate Neuroscience Education, 12*(1), E6–E10.

Cheruvelil, K. S., De Palma-Dow, A., & Smith, K. A. (2020). Strategies to promote effective student research teams in undergraduate biology labs. *The American Biology Teacher, 82*(1), 18–27.

Corwin, L. A., Runyon, C., Robinson, A., & Dolan, E. L. (2015). The laboratory course assessment survey: A tool to measure three dimensions of research-course design. *CBE Life Sciences Education, 14*, 1–11.

DeHaan, R. L. (2009). Teaching creativity and incentive problem solving in science. *CBE Life Sciences Education, 8*(3), 172–181.

Hardavella, G., Aamli-Gaagnat, A., Saad, N., Rousalova, I., & Sreter, K. B. (2017). Incorporating student centered learning in innovation and entrepreneurship education. *European Journal of Education, 43*, 513–526.

Harkema, S., & Schout, H. (2008). Incorporating student-centered learning in innovation and entrepreneuship, education. *European Journal of Education, 43*(4), 513–526.

Hattie, J., & Timperley, H. (2007). The power of feedback. *Review of Educational Research, 77*(1), 81–112.

Havyer, R. D., Wingo, M. T., Comfere, N. I., Nelson, D. R., Halvorsen, A. J., McDonald, F. S., & Reed, D. A. (2014). Teamwork assessment in internal medicine: A systematic review of validity evidence and outcomes. *Journal of General Internal Medicine, 29*, 894–910.

Jarmai, K., & Vogel-Pöschl, H. (2020). Meaningful collaboration for responsible innovation. *Journal of Responsible Innovation, 7*(1), 138–143.

Li, G. (2017). Role of innovation and entrepreneurship education in improving employability of medical university students. *Journal of Mathematics, Science and Technology Education, 13*, 8149–8154.

Morphet, J., Hood, K., Cant, R., Baulch, J., Gibee, A., & Sandry, K. (2014). Teaching teamwork: An evaluation of an interprofessional training ward placement for health care students. *Advances in Medical Education and Practice, 5*, 197–204.

Nicol, D., Thomson, A., & Breslin, C. (2014). Rethinking feedback practices in higher education: A peer review perspective. *Assessment and Evaluation in Higher Education, 39*(1), 102–122.

Parker, A., Lewis, G., & Scott, M. (2020). Improving technical communication with a cue awareness intervention using poster presentations. In *Proceedings of the 4th conference on computing education practice* (Vol. 5, pp. 1–4). Association for Computing Machinery.

Pelaez, N., Anderson, T. R., Gardner, S. M., Yin, Y., Abraham, J. K., Bartlett, E. L., Gormally, C., Hill, J. P., Hoover, M., Hurney, C. A., Long, T. M., Newman, D. L., Sirum, K., & Stevens, M. T. (2017). The basic competencies of biological experimentation: Concept-skill statements. In *PIBERG Instructional Innovation Material Paper 4*. Purdue International Biology Education Research Group.

Pelaez, N., Anderson, T. R., Gardner, S. M., Yin, Y., Abraham, J. K., Bartlett, E. L., Gormally, C., Hurney, C. A., Long, T. M., Newman, D. L., Sirum, K., & Stevens, M. T. (2018). A community-building framework for collaborative research coordination across the education and biology research disciplines. *CBE Life Sciences Education, 17*(2), es2.

Pellegrini, J. J., & Jansen, E. (2013). The Mayo innovation scholars program: Undergraduates explore the science and economics of medical innovations. *Journal of College Science Teaching, 42*(4), 28–32.

Poulos, A., & Mahony, M. J. (2008). Effectiveness of feedback: The students' perspective. *Assessment and Evaluation in Higher Education, 33*(2), 143–154.

Reynders, G., Lantz, J., Ruder, S. M., Stanford, C. L., & Cole, R. S. (2020). Rubrics to assess critical thinking and information processing in undergraduate STEM courses. *International Journal of STEM Education, 7*(1), 9.

Rosenthal, S. (2020). Media literacy, scientific literacy, and science videos on the internet. *Frontiers in Communication, 5*, 581585.

Rudd, K. E., Johnson, S. C., Agesa, K. M., Shackleford, K. A., Kievlan, D. R., Colombara, D. V., Ikuta, K. S., Kissoon, N., Finfer, S., Fleischmann-Struzek, C., Machado, F. R., Reinhart, K. K., Rowan, K., Seymour, C. W., Watson, R. S., West, T. E., Marinho, F., Hay, S. I., Lozano, R., … Naghavi, M. (2020). Global, regional, and national sepsis incidence and mortality, 1990–2017: Analysis for the global burden of disease study. *Lancet, 395*, 200–211.

Rushton, A. (2005). Formative assessment: A key to deep learning? *Medical Teacher, 27*(6), 509–513.

Salas, E., Reyes, D. L., & McDaniel, S. H. (2018). The science of teamwork: Progress, reflections, and the road ahead. *The American Psychologist, 73*(4), 593–600.

Scheufele, D. A., & Krause, N. M. (2019). Science audiences, misinformation, and fake news. *Proceedings of the National Academy of Sciences of the United States of America, 116*(16), 7662–7669.

Vermeulen, N., Parker, J. N., & Penders, B. (2013). Understanding life together: A brief history of collaboration in biology. *Endeavour, 37*(3), 162–171.

Chapter 21
Biological Reasoning According to Members of the Faculty Developer Network for Undergraduate Biology Education: Insights from the Conceptual Analysis of Disciplinary Evidence (CADE) Framework

Chaonan Liu, Nancy J. Pelaez, Shiyao Liu, Ala Samarapungavan, Stephanie M. Gardner, Kari L. Clase, and Deborah Allen

21.1 Background: The Purpose of Undergraduate Biology Education

Undergraduate biology education has been shifting toward helping students to learn science through scientific practice and investigations. The current emphasis on integrating research-based instruction and authentic research experiences into undergraduate biology laboratory courses is meant to improve students' experimentation competence and critical thinking skills in the process of scientific investigation (AAAS, 2011; Laursen, 2019; Pelaez et al., 2017). This is important for a variety of reasons. Already 35 years ago, it was reported that few students had the opportunity to experience a demanding course at the undergraduate level designed to help them

C. Liu (✉) · N. J. Pelaez · S. Liu · S. M. Gardner
Department of Biological Sciences, Purdue University, West Lafayette, IN, USA
e-mail: liu2313@purdue.edu; npelaez@purdue.edu; sgardne@purdue.edu

A. Samarapungavan
Department of Educational Studies, Purdue University, West Lafayette, IN, USA
e-mail: alasam@purdue.edu

K. L. Clase
Department of Agricultural and Biological Engineering, Biotechnology Innovation and Regulatory Science (BIRS) Center, Purdue University, West Lafayette, IN, USA
e-mail: klclase@purdue.edu

D. Allen
Department of Biological Sciences, University of Delaware, Newark, DE, USA
e-mail: deallen@UDel.edu

N. J. Pelaez et al. (eds.), *Trends in Teaching Experimentation in the Life Sciences*, Contributions from Biology Education Research,
https://doi.org/10.1007/978-3-030-98592-9_21

understand the logic of science including the knowledge and methods scientists apply to address major hazards to health such as climate change or disease (Koshland, 1985). More recently, during the COVID-19 pandemic, a science student described problems with understanding the process of science as being "about learning, getting it wrong, and then eventually getting it right" and learning that "when new evidence is constantly being acquired and published… the opinion of the scientific community can change" (Venezia, 2020). Venezia (2020) then pointed out some difficulties and that "making evidence-based decisions is absolutely crucial to an effective pandemic response."

Not all students have the opportunity to learn how disciplinary research techniques produce data that must be appropriately used as evidence, yet there is agreement that helping students learn about the generation and use of scientific evidence to advance scientific knowledge is critical (AAAS, 2011). Seymour and Hunter (2019) reported that changing teaching practices and student support strategies have made a difference according to findings from their sequence of two major studies of science students at the tertiary level in the US, but that variations in educational experience cause some science students to merely "survive" versus others that "thrive." Leaving many at an educational disadvantage. There continue to be reports that students have difficulty understanding the nature, quality, and scope of the evidentiary base that underpins scientific knowledge (Samarapungavan, 2018). Difficulties with understanding scientific evidence may help to explain public misunderstanding of mainstream science, such as vaccine safety. It is time, therefore, for biology educators at the tertiary level to include strategies known to effectively instruct students in the experimentation practices of rigorous research. Among the strategies being studied are undergraduate research experiences designed to induct students into the collaborative practices of science, reported in a number of studies to increase persistence in science and graduation rates for students in groups that have been historically under-represented in science (Seymour & Hunter, 2019). In line with these reports, hard work still needs to be done by biology educators to ensure that society is not left with policy-makers and the general public who are unable to evaluate research-based solutions systematically, which leaves people without protection from being induced to act according to politics or the headlines instead of according to empirical evidence.

For the purposes of this chapter, we argue that an important aim of undergraduate biology education is to train people to understand biology as a research science, to understand the claims that are made based on evidence from modern research, and to evaluate and weigh the importance of those claims. Thus there is a need to teach students to reason with and about evidence upon which scientific claims are made and justified. Examples of difficulties students have with scientific evidentiary reasoning have been reported by others (see examples in Duschl et al., 2007; Ratcliffe & Millar, 2009; Labov et al., 2010). Books that focus on understanding scientific reasoning according to philosophy of science have focused on disciplinary approaches to evidence evaluation and reasoning about causality in cases where causal claims have been established by research in a science discipline (Cartwright, 2007; Giere, 2006; Mayr, 2004). However, although education studies contextualized within biology as the subject matter have revealed the influence of disciplinary knowledge on students' evidentiary reasoning (Lewis & Kattmann, 2004; Pluta et al., 2011), there has been no systematic framework to guide educators who aim to help biology students to develop their evidentiary reasoning abilities until recently. Samarapungavan (2018) developed the *Conceptual Analysis of Disciplinary Evidence (CADE)* framework as a tool for educators by unpacking the notion of biological research evidence and how it is connected to contextual aspects of biology as a discipline.

21.2 The Conceptual Analysis of Disciplinary Evidence (CADE) Framework

In order to identify important practices for helping students understand and use scientific evidence, the *CADE* framework (Samarapungavan, 2018) has been applied as a useful lens to categorize the practices that instructors can focus on scaffolding in order to advance evidentiary reasoning in undergraduate biology class discussions. Derived from philosophy of science ideas about coordination of the models of theories and methodologies, here we use the term scientific evidence to mean the use of data by scientists to evaluate the similarity between scientific theories and the real world (Giere, 2006). The CADE unpacks the notion of evidentiary reasoning, a term we use to refer to the process of generating, using and evaluating evidence to solve problems and make claims. Students need to reason with and about scientific evidence in order to understand the nature, scope and quality of evidence of relevance to substantiate a claim (Samarapungavan, 2018). Because this definition of evidentiary reasoning encompasses the use of evidence at all stages of the research process, the CADE and evidentiary reasoning are more comprehensive than argumentation, which, according to Erduran et al. (2015) refers to the justification of claims through evidence, and evidence-based reasoning according to Brown et al. (2010), which is the use of theoretical statements and scientific evidence to evaluate the quality of a claim. In the science education literature, evidence-based reasoning is not intended to model how scientific knowledge is generated by students or scientists (Brown et al., 2010). In contrast, the *CADE* framework unpacks the complexity of scientific evidence and evidentiary reasoning about evidence in a way that is more comprehensive with four relationships: the *Theory => Evidence* relationship refers to the practice of formulating a research question, testable hypotheses, explanations or the rationale for an investigation; the *Evidence <=> Data* relationship refers to the practice of designing, executing, and analyzing investigations that generate useful data such as biological experiments; the *Evidence => Theory* relationship refers to the analytical processes that lead to inference and critical evaluation of the uncertainty or sufficiency of evidence and the appropriateness of scientific conclusions that are made and justified; and *Social Dimensions* refer to the communication of evidence to the public throughout the research process. Furthermore, for each of these four relationships, the *CADE* framework highlights two essential components of evidentiary reasoning, which are the relevant disciplinary knowledge and epistemic considerations. Disciplinary knowledge provides a foundation for evidentiary reasoning that must build upon a student's prior knowledge, theories and assumptions (Samarapungavan, 2018). It informs decisions about what knowledge is relevant to guide the research, what to choose as evidence and how to interpret the evidence. In parallel, epistemic considerations relate to the logical approaches to reasoning about the nature, scope and the quality of evidence in terms of the sources of such knowledge, its truth, limitations, and uncertainty surrounding the practices applied to generate the evidence for a scientific inference (Sandoval, 2005).

Our chapter examines the CADE framework because it is of relevance to framing the teaching of experimentation. We will show that it is independent of, but complementary to the ACE-Bio Competencies framework and illustrate how both frameworks add different value. Unlike the ACE-Bio framework that focuses on basic competencies for experimentation, the CADE framework primarily targets reasoning with, and about scientific evidence and it is not limited to the area of research that relies on conducting experiments. Although we will map the evidentiary practices under four relationships in CADE to the seven competencies in the ACE-Bio Competencies framework, the CADE relationships are further deconstructed into disciplinary knowledge and epistemic considerations, which are the foundations not only for evidentiary reasoning but also for conducting experiments. By unpacking the concept of evidence, CADE helps to simplify the overlapping and complex integration of the areas covered by the ACE-Bio Competencies to provide guidance for evidentiary reasoning when doing research such as by conducting experiments. Using the CADE framework as a lens in experimentation could provide additional value by emphasizing the understanding and reasoning with and about scientific evidence that scientists practice.

21.3 The Faculty Developer Network for Undergraduate Biology Education (FDN-UBE)

Reform efforts to integrate authentic research with undergraduate biology education are built upon the participation of scientists as instructors or curriculum designers in guiding students' scientific investigations. With their formally trained research experience, scientists can provide students with relevant understanding of disciplinary biological knowledge and also the sophisticated epistemic reasoning applied to experimentation skills essential for developing students' scientific practices. A subset of these expert biology educators have been actively working to create, build, support and sustain a community of practitioners and scholars to advance faculty professional development for undergraduate biology educators. One of us (DA) was Principal Investigator, and two of us (CL and NP) were external evaluators for a project funded by the National Science Foundation that was put together by scientists from different biology subdisciplines who lead faculty professional development to establish a Research Coordination Network, RCN-UBE: Faculty Development Network for Undergraduate Biology (FDN-UB). With Gordon Uno, Karen Sirum, April Maskiewicz, Susan Elrod, and Charlene D'Avanzo as Co-Principal Investigators, since 2014 the project participants have shared biology faculty development resources, mechanisms, and research-based strategies, aiming to improve their delivery of professional development geared toward enhancing teaching and learning of biology across all higher education institutional contexts. Their rich experience in both teaching as well as faculty professional development provides insight about biology education and scientific practices related to improving students' understanding and use of scientific evidence. Implications from project findings provide lessons for young instructors and scientists regarding activities and practices for teaching biology in ways that help students develop evidentiary reasoning.

There is a paucity of reports from a faculty development perspective about what is needed to involve faculty members who are scientists to improve and support students' competence of understanding and using scientific evidence in undergraduate biology education. Thus we conducted an analysis of interview data for a study to document the value of scientists who conduct biology faculty faculty development in terms of their experience and professional perspective. It was found that their knowledge and efforts aligned well with a focus on unpacking evidentiary reasoning in the process of undergraduate biology education in line with the *CADE* framework (Samarapungavan, 2018). Activities and practices that the faculty professional developers mentioned during interviews were analyzed through the *CADE* framework lens to reveal important components that the faculty developers brought to advance undergraduate biology education. Typical quotes from the interviews are presented as examples to reveal insights about important scientific practices for helping students understand and use scientific evidence from the faculty developers' perspectives in terms of their own experience.

Since the *CADE* framework values the role of disciplinary knowledge, it was useful to examine the contributions of FDN-UBE members in order to reflect on ways to develop more sophisticated approaches to teaching and learning of biology by unpacking the notion of evidence according to its disciplinary contexts. In this way, we report on contributions from FDN-UBE members that would not have been possible if professional development had been limited to programs at an institutional Center for Teaching and Learning where the professional development leaders lack the affordances from having a science background.

21.4 Research Method

This study was initially guided by several research questions:

1. What personally motivates/motivated network members to engage in professional development?
2. What professional paths did they take along the way to becoming interested in and effective at professional development for biology faculty?
3. What resources or indicators of success make FDN-UBE members feel qualified to be leading effective faculty development?

Participants drawn from the FDN-UBE membership consisted of 50 individuals who voluntarily responded to online surveys in 2015–2017. FDN-UBE members were asked to take part in this study if they had participated in at least one of the network's activities or attended a network synthesis meeting. They were invited by an e-mail invitation or with flyers at the registration desk at an FDN-UBE meeting for participants to read and determine if they were interested in completing an online study recruitment survey. Survey and interview protocols were approved by the IRBs of Purdue University (protocol # 1510016672, N. Pelaez) and the University of Delaware (protocol # 575674, D. Allen).

Fifty individuals responded to online surveys that were conducted with questions about their motivation for joining the FDN-UBE project and about the major challenges and issues that a network such as the FDN-UBE could address. A final survey question was used for recruiting interview participants and then a stratified representative subset of the participants was selected for oral audiotaped phone interviews. Since interviews were conducted by phone it was possible to select volunteers for interviews to represent different regions in the US, different types of institutions, and a range of different biology sub-disciplines. A representative sample of 18 FDN-UBE network members participated in phone interviews that were recorded for up to 60 min in duration plus a follow-up interview to refine interpretations with assistance from a subset of the original sample who agreed to member-checking so that their words were interpreted as they intended. Results and quotes to illustrate the findings are from the 18 original interviews plus 10 follow-up interviews of representative participants.

The data collected were initially intended to answer research questions 1, 2 and 3 (above), but with our interest in understanding and exploring evidentiary reasoning in undergraduate biology education (Chap. 17 in this volume), an additional research question 4 was explored:

4. What model of professional practice represents the value added and additional potential contributions from FDN-UBE members who are engaged in biology faculty professional development to advance undergraduate biology education?

Guided by this research question, a second tier of data analysis with the CADE as a lens was aimed to reveal instructional practices that help students understand and use scientific evidence from FDN-UBE members' perspective.

21.4.1 Interview Transcription and Coding Methodology

An open coding procedure as characterized by Strauss and Corbin (1990) was based on Khandkar (2009). Interview recordings were transcribed using *Trint.com* online, and then proofread individually. For a subset of the interviews, printed transcripts were cut into pieces for line-by-line coding (by 2 independent coders). Words, sentences or parts of the transcripts were labeled by topic names chosen from within quotes from the transcripts. Coded topic information was categorized based on similarity. After open coding, coded information from the same transcript was put into a spreadsheet to build a coding matrix. Category names were refined and defined by looking for patterns, discovered by comparing coded data from different interviews until saturation was achieved. Following this open coding process, a second tier of coding was conducted according to *CADE* where words, sentences or parts from the interview transcripts were categorized according to the four relationships present in the *CADE* framework as a model of professional practice. Quotes within each relationship were then coded as either disciplinary knowledge or epistemic aspects of reasoning. Any quote of relevance to reasoning about evidence in ways that call on disciplinary science knowledge was coded as *Disciplinary knowledge in biology* and items that relate to the quality of the evidence in terms of good general advice to an investigator regardless of their discipline were coded as *Epistemic considerations*. Finally, the first three authors (CL, NP, and SL) mapped the *CADE* topics from the interviews to the ACE-Bio Competencies (Pelaez et al., 2017; Chap. 1, Tables 1.3–1.9 in this volume) and any *CADE* areas not mentioned in the interviews were also mapped to ACE-Bio Competencies. Before reporting any findings, participants were assigned pseudonyms. Survey and interview protocols were approved by the IRBs of Purdue University (protocol # 1510016672, N. Pelaez) and the University of Delaware (protocol # 575674, D. Allen). No identifying information is reported from the interviews. Summaries of the main points of the interviews are reported in the aggregate.

21.4.2 Selection of FDN-UBE Volunteers for Interviews

Participants who were interviewed were stratified and selected according to biology subdiscipline and to be representative of network subgroups focused in three areas:

Jump-Starting Early Career Faculty in Active Learning (Co-leaders: Mark Connolly & Gili Marbach-Ad). This group used Delphi study methods to develop a consensus among experts on what activities and conditions support adoption of active learning by early-career biology faculty. The study was aimed at producing four prioritized consensus lists: recommendations for individual faculty development in AL strategies; identification of obstacles or barriers; potential sources of support and assistance; and mechanisms that departments, colleges, and universities can adopt to encourage use of active learning approaches.

Inclusive Teaching Practices (Leader: Bryan Dewsbury). This group focused on design of a robust inclusive teaching professional development model to address increasing calls for a transformation of biology classroom culture to support more equitable and inclusive community to welcome students into science. Bryan Dewsbury, the leader of this effort, also successfully garnered support for a scaled-down version of an immersion model from the John Gardner Foundation.

Sustaining Change (Co-leaders: Rachelle Spell, Larry Blumer & Gordon Uno). This group was interested in how to connect faculty development efforts to systemic change initiatives on campuses, and what institutional factors help sustain implementation of best teaching practices learned in faculty development efforts. They developed and implemented a survey of institutional factors in sustainability of best teaching practices for institutions to use to review their efforts.

The experiences of teaching and professional development from the FDN-UBE members of these groups could provide insight of use to others who could include scientific practice and evidentiary reasoning as a focus for undergraduate biology education. By coding according to the Conceptual Analysis of Disciplinary Evidence (*CADE*) framework quotes from interviews of representative FDN-UBE members yielded data for answering the fourth research question. This enabled us to suggest a model of practice that represents value added and additional potential contributions from FDN-UBE members according to their current areas of professional focus. An aim for the future is to extend their current professional practice by identifying areas for potential future development of evidentiary reasoning in undergraduate biology education in areas not yet targeted. A contribution from this work therefore targets future development of new focus areas for faculty professional development aimed at supporting student reasoning with and about evidence to help future students develop abilities to make and critically evaluate the strength of inferences and claims in the biological sciences.

21.5 Findings from the Online Survey of FDN-UBE Members

Survey participants (N=50) reported leading a range of faculty professional development activities, from instructional design, workshop facilitation, education research, program evaluation, consultations at the department or program level, and graduate student/TA training. A relatively large proportion (Fig. 21.1) report performing these activities in a context that is either outside of their institution or not part of their formal roles ("on my own").

In addition to their membership in the FDN-UBE Network, according to their online survey responses, participants in the FDN-UBE network reported participating in the previous 4 years in three or more different types of events (on average) from an impressive array of more than 60 scientific or professional activities, listed alphabetically in Box.

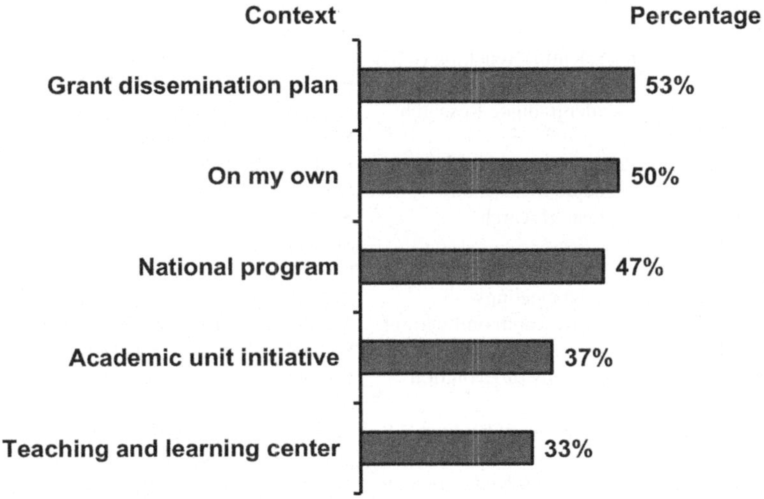

Fig. 21.1 Context of members' past or present biology faculty professional development activities

Achieving the Dream Network
Accelerating Systemic Change Network (ASCN)
Advancement of Competence with Experimentation – Biology (ACE-Bio) Network
American Association for the Advancement of Science (AAAS) events, such as Envisioning the Future of Undergraduate STEM Education (EnFUSE) and Pacific Coast meetings
American Chemical Society (ACS)
American Educational Research Association (AERA)
American Society for Cell Biology (ASCB)
American Physiological Society Institute on Teaching and Learning (APS-ITL)
American Society for Microbiology Conference for Undergraduate Educators (ASMCUE)
Association of American Colleges & Universities (AAC&U) High-Impact Practices (HIPs)
Association of American Colleges & Universities (AAC&U) STEMCentral.net
Association for Biology Laboratory Education (ABLE)
Association of College and University Biology Educators (ACUBE)
Association for Contemplative Mind in Higher Education (ACMHE)
Bio-Link
Biology Teaching Assistant Project (BioTAP)
BioQUEST Curriculum Consortium
Center for the Integration of Research, Teaching, and Learning (CIRTL)
Community College Biology education research (CC-BER)
Community College Biology Faculty Enhancement through Scientific Teaching (CCB FEST)
Community College Undergraduate Research Initiative (CCURI)
Council for Undergraduate Research (CUR)
CUREnet network of people focused on course-based undergraduate research experiences (CUREs)
Ecological Society of America (ESA)
Experimental Biology (EB) meetings
European Association for Research on Learning and Instruction (EARLI)
European Society for the Study of Evolution (SSE)
European Society for Evolutionary Biology (ESEB)
Gordon Research Conference on Undergraduate Biology Education Research
Human Anatomy and Physiology Society (HAPS)
Introductory Biology Project (IBP)
International Conference of the Learning Sciences (ICLS)
Learning Assistant Alliance (LAA)
League for Innovation in the Community College (League)
Life Discovery – Doing Science Biology Education Conference
National Academies Scientific Teaching Alliance (NASTA)
National Association for Biology Teachers (NABT)
National Association for Research in Science Teaching (NARST)
National Centers: NIMBio
National Conference on Race and Ethnicity in Higher Education (NCORE)
National Evolutionary Synthesis Center (NESCent)
National Institute for Staff and Organizational Development (NISOD)
NIMBioS: National Institute for Mathematical and Biological Synthesis
National Science Education Leadership Association (NSELA)
National Science Teachers Association (NSTA)
Network of STEM Education Centers (NSEC)
POD Network: Professional and Organizational Development
Partnership for Undergraduate Life Sciences Education (PULSE) Vision and Change Leadership Fellows
Quantitative Undergraduate Biology Education and Synthesis (QUBES)
RCN-UBE for Visualizations, Interactive Simulations, and Animations for Biology Learning & Instruction
REIL-Biology: Research Experiences in Introductory Laboratory in Biology
Society for the Advancement of Biology Education Research (SABER)
VISABLI: Visualizations, Interactive Simulations, and Animations for Biology Learning & Instruction
ScienceCaseNet: National Center for Case Study Teaching in Science
Science Education for New Civic Engagements and Responsibilities (SENCER) Summer Institutes (SSI)
Society for College Science Teaching
Society for Integrative and Comparative Biology
Society for the Advancement of Chicanos and Native American in Science (SACNAS)
State or regional science education society events such as New England Education Research Organization (NEERO), North East Science Teachers Education Association, NW Biology Instructors Meeting (NWBio), Science Teachers Association of Texas, Wisconsin Society of Science Teachers
Summer Institutes on Scientific Teaching

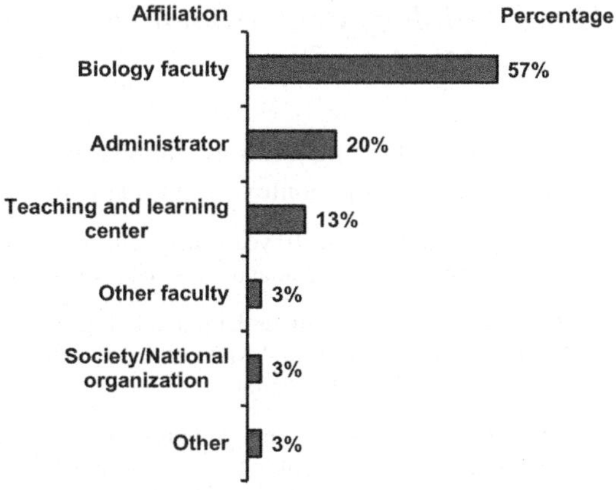

Fig. 21.2 FDN-UBE members' primary professional affiliations

The majority, who are providing professional development expertise to other faculty, are themselves faculty members in departments of biology or other science disciplines, and most were not formally associated with the campus teaching and learning centers whose core mission is professional development (Fig. 21.2).

21.6 Findings from Interviews Reveal Features of Their Biology Faculty Professional Development Interests and Expertise

All original and follow-up interviews were conducted between November, 2016, and January, 2018. Analysis of the interview transcripts in light of the research questions led to identification of several themes. Now, in 2021, a global pandemic has gripped the world and we face important questions about how to incorporate biology as a research science into our collective decisions. Surprisingly, our findings about the role of evidentiary reasoning from the faculty developers' perspective in this study show how well FDN-UBE members are positioned to support other educators in providing students with relevant biology instruction essential for developing students' evidentiary reasoning about biological investigations. As illustrated below by sample quotes from the interview study participants, first we highlight three major themes:

- Faculty professional developers in biology education are visionaries/missionaries.
- The pathways toward education of biology faculty professional developers are unconventional but remain focused on biology as a discipline.
- Knowledge sources include but go beyond the professional development literature.

21.6.1 Biology Professional Developers Are Visionaries/Missionaries

Motivation for leading biology faculty professional development activities was often expressed in the context of an inspiring vision or progressive possibility:

> Phil: "I just think that in 10 years, the ….undergraduate biology education system is gonna look dramatically different, and ….it needs to be driven by faculty…."

> Bill: "When I started as an undergraduate I wanted to change the world, and so, you know, I thought 'so what can I do?' And I thought, well, you know, teaching is a reasonable thing to do and….I think I'd like that."

> John: "I have a very clear and, you know, inspiring vision, to me at least, of what I think a university campus can look like and what it can contribute to a functional society. I think due to no one's particular fault, we have lost the way of it." (and later in the interview) "…but I think I would love that any student walking on any campus feels like this is a place they can come and really grow as a person. And our mission, or our teaching mission, is really dedicated to that."

Interviewees typically expressed almost a moral obligation to lead by providing professional development to biology educators:

> John: "….if you are going to call for changes, you know, and you're going to accuse people of not having the necessary skills to lead that change, you know, part of this is you have a moral obligation to be part of the solution."

> Ellen: "I have no idea what motivates me. It just seems like you reach a point where that's what you should do. It seems that's part of the process of getting more senior in a profession."

> Motivations to engage in this work were often expressed as being intrinsic ones:

> Anna: "…they're coming to activities that I created. This is the first bonus for me."

> Ellen: "….when, for example I have a workshop on writing…. materials and somebody creates something that is just brilliant, completely outshines the 'professor'….when I think I might have had a small part in just helping to create a setting where they feel they could do that….I find that very rewarding."

21.6.2 The Unconventional Pathways of Biology Faculty Professional Developers Remain Focused on Biology as a Discipline

While all of the interviewees received formal graduate education in some aspect of biology, at some point in their career trajectories, they incorporated professional development into their professional roles, not necessarily by design. Teaching and professional development in teaching became a satisfying way to pursue new interests and commitments, without turning their back on biology:

Sarah: "In my experience, the getting into professional development ....is a little bit of a coincidence....it was not necessarily a track that you would say, oh, I would like to do professional development as that goal for me. It was a series of fortunate events."

Anna: "So, they wanted me to continue in biology, but I felt like I love more that I can talk about what I am doing.... I love more science education than science. But it's not... like education as education alone, and science as science alone. I like the energy between science and education."

Jill: "Well, the reason I went to her and said this is something I'm very interested in is it's actually a good percentage of what I do....But I am actually the director of the intro sequence, ...and I took it on as an important service I wanted to provide to help make sure that it was a cohesive, collaborative, 'most effective way as possible' (teaching) group. (later in the interview) ... And because of my experiences and my success with helping faculty develop, I really would like to see my career grow in that way."

John: I have officially left the.... biology [research] world behind. I mean, I loved it, but I just love this more (later in the interview) .... but it requires time, and it requires a different way to think about what a professor's responsibility is.

21.6.3 Knowledge Sources Include But Go Beyond the Professional Development Literature to Include Oral Traditions

In discussing knowledge sources used to acquire expertise in professional development, interviewees (29%) mentioned the literature, but also discussed the importance of oral traditions including personal interactions through attending workshops, conferences, and meetings, and through networks.

Anna: "I'm looking at the literature all the time...."

Ellen: "But we're forced to make people write on pieces of paper or the electronic version of that and create these dead documents and we value that more than this oral tradition that we have that's, that's been so impactful in science education."

John: "...they are the ones who kind of pointed me to a workshop I could go to....to give me the kind of experience that quite frankly I think is what allowed me to get the job I have now. I kind of got into teaching and the scholarship of teaching and learning through the professional development world because they were – these were admin and so the organization was POD, and you know the way they think about scholarship was through professional development."

Bill: "Participants apply to attend these workshops and they actually do the science....and they leave these workshops knowing this is something I can do in my class, I know how to do it and I know that it works. So it's,....it's....to me it was really transformative..... And it really changed my point of view on what works and where our priorities should be."

Sarah: "And also finding other people who were doing similar things and watching how they were doing it and then ultimately getting pulled into projects to work with people and hear how they were doing things."

John: "That's....where I think this becomes a really robust thing, because you're not just meeting to kind of check some ideas back and forth and it is good to see you. And let's talk about what happened at my institution versus yours. No it isn't one particular thing that if we kind of put all our collective intellectual power together we can have a much more powerful paper, a much more powerful workshop, or a much more powerful online training program or assessment scheme and ..."

Analysis of the interview transcripts illustrate that faculty professional developers in biology education are valuable visionaries. Through non-conventional career pathways, they disseminate knowledge through biology faculty professional development, and their knowledge sources are not limited to the professional development literature. In the words of one interviewee:

Sarah: 'today, I do not have, you know, a publication that we can point to, but we're definitely working on formalizing a lot of what we know."

21.7 Interview Findings Through the Conceptual Analysis of Disciplinary Evidence (*CADE*) Lens

The second tier of coding according to the *CADE* framework revealed important practices for helping students understand and use scientific evidence according to the professional developers' responses to a question about what they viewed as an indicator of success in their faculty development work. They often described their success in terms of what they were aiming to accomplish for undergraduate biology education through leading professional development acctivities. The interviews were conducted before the Conceptual Analysis of Disciplinary Evidence (*CADE*) framework was published by Samarapungavan (2018). However, this framework was chosen because it mapped onto ideas about integrating authentic research into biology education for both major and non-major undergraduate students in ways that were reflected as indicators of success according to their professional biology faculty development experiences.

Half of the participants (9/18) explicitly mentioned authentic research practices in undergraduate biology education as having potential for increasing students' interest in learning biology, improving students' biology literacy, and retention to graduation even though they were asked about successful biology faculty development and not about biology instruction. For faculty members who are scientists and have formal scientific research experiences, the important thing is not only to help other educators teach students disciplinary knowledge in biology, but also "we should be teaching them how to do biology and what biologist do."

Julia: "So I think that thinking about that scientific teaching approach … that really emphasizes the use of practices within the disciplinary field and how it is that then improves the way that the students learn all the content and the practices of that discipline as well."

21.7.1 Theory => Evidence Relationship: A Knowledge Foundation for Scientific Research

The *CADE* framework emphasizes the role of disciplinary knowledge to the practice of formulating testable hypotheses, explanations or rationale for an investigation. Decisions about an investigation closely relate to the relevant disciplinary knowledge like theories, mechanisms, and causal relationships, as well as general knowledge about formatting research questions and hypothesis testing process. A Theory => Evidence relationship guides an investigators' decision about what kind of the important unsolved problem to investigate with and what variables are relevant with the investigation (Samarapungavan, 2018). It is important to focus on the knowledge that students need before guiding them through a scientific research experience in biology.

21.7.1.1 A Focus on Conceptual Understanding

Deep conceptual understanding plays a role in evidentiary reasoning and undergraduate biology education. Helping students build meaningful conceptual understanding is one of the important components that participants mentioned that fits a Theory => Evidence relationship according to the *CADE* framework. When students gain deep understanding of the concepts in biology, they become able to organize their biological knowledge and information in a meaningful way. Although concept learning sets the basic foundation for scientific research practice, students have problems in remembering and understanding how to apply concepts and knowledge in biology.

> Steven: "What it is like it's obvious that students, no matter how many times they've learned they won't remember this, because to us, these facts have meaning, like a different molecule has directionality, and the directionality is important. But I think to students they're just random facts."

Clair also mentioned the need for the instructors to provide the knowledge and information in a meaningful way for the students.

> Clair: "They said they are worried about student engagement and their worry about helping to develop... they don't necessarily use the language, but helping students develop certain mental models that organize the information so they have this conflict between a lot of content and being frustrated because it's just develop(ing) the sort of organizational structure they need for it."

Several mentioned a pedagogical way for instructors to improve students' conceptual understanding is to track understanding using a concept inventory.

> Anna: "... one of the things to do is a concept inventory. We use that a lot. We created a concept inventory and we had like five years, maybe more than five now, worth of data from people of... in 8, 9 courses, it caused... interaction. And we gave the concept inventory before and after this course, and we learned that students are not getting it. And now we are working on that with activities."

However, when instructors apply a concept inventory in their teaching, other factors are also carefully considered, like accurate assessment according to the expected and actual student performance level.

> Jill: "… you can't just do course inventory and you can't just concept inventory, because there's no performance parameters associated with that. Those kinds of things only address knowledge maybe in skill, but it doesn't address to what level."

21.7.1.2 Use of Cutting-Edge Research Examples

When talking about activities and scientific practices that relate to the *CADE* Theory=> Evidence relationship, some participants mentioned the importance of including disciplinary knowledge about current science research, especially examples from the instructors' own research experience. Bob shared thoughts about his own teaching experience when he talked about using examples that are more relevant to the students to increase their interest in learning biology.

> Bob: "In bringing other examples, an example with more relevance to students, like examples in Texas, examples of your own work, you know, … whenever I talk about whatever I did, or all my colleague next door did, they just become more interested."

From a professional developer's point of view, Claire also suggested getting research examples into the classroom.

> Claire: "if you're leading a lecturer section right and then what you can do to get research into that classroom is to talk about your own research and they are perfectly comfortable doing that."

Learning about research was suggested to bring about an increase in students' learning motivations and improvement of their learning outcomes.

> Anna: "If the teacher makes students to want to stay in the field and to show that biology is an interesting topic, this is also important, especially in introductory courses".

21.7.2 Evidence <=> Data Relationship: Practice Analysis with Authentic Data

Evidence <=> Data relationships involve the practices of designing and conducting a scientific investigation. Knowledge and practices about data collection and data processing are only relevant when data is considered as useful empirical evidence. Two main themes were identified under the Evidence <=> Data relationship according to findings from the FDN-UBE participant interviews. They indicated advanced research techniques and also mathematical abilities as two types of specific skills or practices to be developed through undergraduate biology education to help students become capable of generating and collecting data as evidence for their scientific investigations.

21.7.2.1 Advanced Research Techniques for Collecting Data

As biology is a rapidly developing discipline, new research techniques and instruments are constantly emerging to meet the changes and challenges in biological investigations. Not only are students facing these challenges, but educators also need help keeping up with advanced techniques.

> Simon: "… think how much is it changed for people particularly at an undergraduate institution or community college, right? They're not getting exposed to modern techniques as frequently. And yet we're expecting that the students are getting exposed. So the faculty need a lot of content help as well."

21.7.2.2 Basic Mathematical Skills for Analyzing Data

Basic mathematical skills, like applying statistical methods and doing calculations, are competencies that enable students to analyze data. Although not formally the focus of what is taught as disciplinary knowledge in biology, mathematical skill, as a component of disciplinary knowledge in biology must be appropriately cultivated rather than being treated as a "weed-out" skill, as it influences the accuracy of the data analyses and provides evidence for validity.

> Emily: "So how many biology programs require students to go through a year or more calculus? And then they just don't make it cause they can't do it. And they go, wow, I didn't pass calculus. It's not that they didn't pass biology. They didn't pass calculus."

> Simon: "So they've (collaborating instructors) created an introductory excel activity because they felt like their students needed some more ramping up before they could analyze the data as it was written in the lab originally."

> Judith: "with my math coworker… she wrote on a board some measurement that we were doing with some milliliters, in liters ....and she'd written some number times ten to the minus seventh (liters), and I went, wow, we would never do that. There's nothing that measures in ten to the minus seventh. We would write, you know 70 microliters or something, ... So we're using microliters."

In this last quote, Judith has recognized the discipline-specific approach to reasoning about measured volumes in biology and that this type of reasoning was not taught by a math coworker.

21.7.3 Evidence => Theory Relationship: Sufficiency of Interpretations

Engaging in practices for interpreting evidence involves considering how to learn from the evidence, whether interpretations are consistent with the totality of disciplinary knowledge available and if any alternate interpretations are compatible with the evidence.

> Ben: We "just give them thousand of photos, say go to look at all these for an hour and say what pattern do you see."
>
> Sarah: They "think about manipulating data and what it tells us."
>
> Anna considered "how to ... interpret trees or how to interpret figures. And then we gave them figures the same in the evolution course as in the genetics course. And every time you go deeper and deeper, understanding and asking more questions. But it's built on the same thing that they saw before."

21.7.4 Social Dimensions: Communication of Evidence to the Public

In motivating students to engage in scientific practice, communicating evidence to the public is an important part of the entire process of scientific investigation. In the interviews, professional developers talked about the practices of collaboration and communication among students, including peer reviews and publication of results from an investigation. According to the professional developers, students who engage in these activities get deep understand of the evidence as they share ideas about scientific evidence with each other. Scientific communication also motivates students to engage in authentic research practice as members of a diverse and welcoming community.

> Ellen: "They (students) like working in groups. They like making their work public. And I think we can harness all those things."
>
> Bill: The students "do peer review, do revision…"
>
> Blake: For "a theoretical client ... they had to develop a sustainable agriculture method you know the principles you know biology or whatever hydroponics or whatever the kids dream of and making it work."

Peer review, revision, use of real-world scenarios, and public presentation of their work are strategies mentioned to support students in communicating their work to stakeholders. These mentions of science communication involve reasoning with and about evidence throughout the research process from the proposal stage to a report of their findings, sometimes with an audience (a theoretical client) identified.

In an attempt to establish a model of professional practice, with the underlined questions in Table 21.1 the quotes above were categorized and posed as faculty develop questions about *Disciplinary knowledge in biology* for the relevant *CADE* relationship, unless the idea related to the quality of the evidence in terms that would apply to an investigator in any discipline, in which case they would be coded as *Epistemic considerations*.

Table 21.1 reveals that *Epistemic considerations* were not mentioned by any of the 18 FDN-UBE members interviewed. To address this gap, the ACE-Bio Competencies includes many science epistemology examples of relevance to reasoning with and about evidence throughout the research process. Note that the examples presented in Table 21.1 are incomplete due to space limitations and some lack of agreement about the categorization. However, according to ideas expressed by FDN-UBE members, CADE relationships can be only partially mapped to the ACE-Bio Competencies as illustrated in Fig. 21.3. As an example of some of the differences, consider Table 1.3 B3 in Chap. 1 of this volume: "*Identify* a problem that is timely relevant, and interesting, and, if addressed, could build on our foundational knowledge of science." For the first parts of the statement, "timely, relevant and interesting" are strongly related to the *Social dimension*. The last part of the statement, "build on our foundational knowledge," is strongly related to the theories within a discipline. This ACE-Bio Competence could be categorized as *Disciplinary knowledge* both in *T=>E* and *Social dimension*. This item is not in Table 21.1. As another example, consider Table 1.4 B2 in Chap. 1 of this volume: "Evaluate ethical, theoretical, practical and cost constraints associated with a research question." Two of us focused on the "ethical and cost" part that relates to *Disciplinary knowledge* in the *Social dimension* category where it is listed in Table 21.1. However, "evaluate theoretical" refers to *Disciplinary knowledge* in that could have been listed as *Disciplinary knowledge* in the T=>E category. This categorization problem reflects an overlapping and complex integration that was simplified to develop the *CADE* framework in order to provide a useful model of practice for reasoning with and about evidence.

Table 21.1 Questions for a professional development model of practice based on analysis of FDN-UBE member interviews according to the Conceptual Analysis of Disciplinary Evidence (CADE) framework's disciplinary knowledge and epistemic science practice relationships[a]

Evidentiary practices (ACE-Bio Competencies)	Questions about relevant disciplinary knowledge in Biology	Questions about epistemic considerations
Theory => Evidence Relationships Articulation of a rationale for an investigation: Formulate a research question, hypotheses or pose explanations *Identify*: relevant background (1.3A) *Identify*: a gap in knowledge (1.3B) *Question*: observation (1.4A) *Question*: hypotheses (1.4D) *Plan*: variables (1.5C)	[b]Is the rationale for the study based on biological knowledge and information organized in a way that is logical and meaningful? (1.3A4) Are observations compared to existing knowledge, models, or theories? (1.4A2) Have multiple explanations that are testable and potentially falsifiable been generated? (1.4D2) Are associations between treatment conditions and outcome variables predicted for the research target? (1.4D3) Are relevant, measurable variables identified for testing the hypothesis (1.5C1)	Has the relevance of information from appropriate sources to the specific research focus been filtered and evaluated? (1.3 A2) Are limits to the background knowledge related to the gap discussed? (1.3B2) What variables are relevant and why is this data appropriate? (1.5C1)

(continued)

Table 21.1 (continued)

Evidentiary practices (ACE-Bio Competencies)	Questions about relevant disciplinary knowledge in Biology	Questions about epistemic considerations
Evidence <=> Data Relationships Designing, executing, and analyzing evidence from investigations *Plan* – Experimental design (1.5 B) *Plan* – variables (1.5C) *Plan* – controls (1.5D) *Plan* – measurement (1.5E) *Plan* – sample (1.5F) *Plan* – Variation (1.5G) *Plan* – Limitations (1.5I) *Conduct* – variable outcomes (1.6B) *Conduct* – Data Documentation (1.6 C) *Analyze* – Statistics (1.7C) *Analyze* – Data Summary (1.7D)	[b]Are relevant research techniques proposed for collecting data? Did the investigator choose the most appropriate design approach to answer the research question(s)? (1.5 B2) Are confounding, and/or covariate variables identified? (1.5C3) Are the limitations of measurement tools/equipment recognized? (1.5E3) Has replication or repeatability needed to quantify variation been determined? (1.5G) [b]Are the approaches to measurement reasonable according to disciplinary norms for biology? Have limitations of methods been evaluated? (1.5I4) Are data recorded with appropriate labels, units of measure, and levels of precision? (1.6D4) [b]What mathematical analysis models are used to organize/compare data? (1.7C1) Are findings displayed with a representation that is effective in summarizing trends or major findings, including illustrating contrasts among categorical groups? (1.7D3)	Does the study propose measurable outcomes that would support or refute hypotheses? (1.5B3) Is the experiment designed with controls to anticipate likely sources of error? (1.5D1) Is the sampling protocol aligned with the research question or hypothesis (1.5F3) Is the sampling strategy designed to expose and account for natural variation and measurement error? (1.5G2) Is there a plan to evaluate uncertainty in protocols (e.g. measured variables) analytical methods (e.g., assumptions of statistical tests), and interpretations of results? (1.5I3) Is the study monitored for unexpected outcomes due to technical errors, equipment failure, subject characteristics, and unplanned factors? (1.6B) Is a written or digital laboratory notebook or field journal maintained that provides a record describing how, when, where, and why data were collected? (1.6C1) Is data recorded in an organized and systematic way (1.6 C3)? Are statistics for a sample planned to summarize and/or describe parameters for a whole population (e.g., mean, median, measures of variance). (1.7C3)

(continued)

Table 21.1 (continued)

Evidentiary practices (ACE-Bio Competencies)	Questions about relevant disciplinary knowledge in Biology	Questions about epistemic considerations
Evidence => Theory Relationships Sufficiency of conclusions; understanding and critically evaluating the appropriateness of scientific claims *Conclude* – Patterns and Relationships (1.8A) *Conclude* – Inferences and conclusions (1.8B) *Communicate* – Limitations (1.9C) *Communicate* – Synthesis and Reflection (1.9D)	Are conclusions aligned with analyses, hypotheses, research question(s), and existing knowledge? (1.8B5) [b]What can be learned from the evidence? Do data support or refute hypotheses and predictions? (1.8B6) Is uncertainty considered in terms of data analysis limitations (sources of error, inaccurate measurement, sample bias, statistical significance vs. biological relevance)? (1.8B7) Are follow up experiments proposed? (1.9D4)	Do the results suggest a causal mechanism beyond simple correlation? (1.8A2) Are results generalized to an appropriate level (more than a single experiment, less than universal)? (1.8B1) Are future directions that will make conclusions more certain suggested? (1.8B8) Are the limitations, unanswered questions, and the tentative nature of results articulated? (1.9C1)
Social Dimensions Communication by reasoning in public with and about biological evidence *Question* – Research Questions (1.4B) *Plan* – Ethics (1.5H) *Analyze* – Data Summary (1.7D) *Communicate* – with Representations (1.9A) *Communicate* – Scientific Communication (1.9B) *Communicate* – Limitations (1.9C)	Have the ethical, theoretical, practical and cost constraints associated with the research been evaluated (1.4B2) [b]Has the work been revised based on peer-review? Was the research plan reviewed by the Institutional Review Board or Animal Care and Use Committee for evaluation, as appropriate? (1.5H2) Are the results distilled into clear numeric and/or graphical forms that are aligned with the experimental objective/ question/hypothesis? (1.9A1)	Are the findings displayed with a representation that is effective in summarizing trends or major findings? (1.7D3) Is the structure and content of a presentation tailored to the probable audience? (1.9B3) Are the limitations, unanswered questions, and the tentative nature of results discussed? (1.9C1) Are justification and counter-justification arguments discussed for each alternative hypothesis? (1.9C3;1.9C4)

[a]Letters and numbers refer to ACE-Bio Competencies in Tables 1.3–1.9 in Chap. 1 of this volume. The ACE-Bio Competence examples were selected based on agreement by at least two authors about the CADE categorizations

[b]Underlined statements reflect ideas from interviews expressed by FDN-UBE members according to Sect. 21.7 above

21.8 Discussion and Implications for Future Direction

By understanding the role of evidentiary reasoning from the faculty developers' perspective in this study, we find that FDN-UBE members are well positioned to support other educators who provide students with relevant biological disciplinary knowledge essential for developing students' scientific research practices. The FDN-UBE network members we interviewed are motivated scientists who hold an

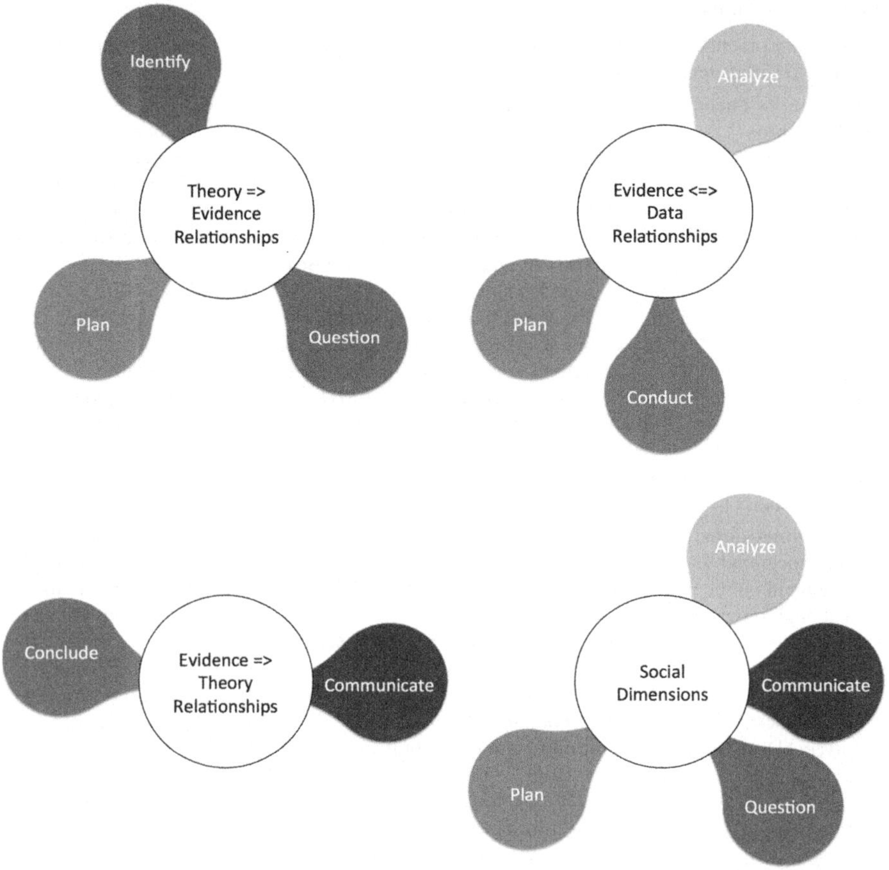

Fig. 21.3 Mapping of CADE relationships to the ACE-Bio Competencies according to ideas expressed by FDN-UBE members in Table 21.1. Each figure panel has a center representing the different evidentiary practices in CADE, including Theory=>Evidence relationships, Evidence <=> Data relationships, Evidence =>Theory relationships and Social Dimensions. For each CADE relationship, there are corresponding ACE-Bio Competencies

inspiring vision of progressive educational possibilities. Their career trajectories were not very conventional, but they found opportunities to inform themselves and pursue their interests in biology with an aim to serve our biology education community. Their sources of knowledge include but go beyond the professional development literature to incorporate learning from meetings that target cutting edge science research, education research or practical pedagogical knowledge applied to higher education. At such meetings or from colleagues they have learned about professional development through oral traditions and from the example of others who are doing similar things, working with faculty to advance biology education.

A real problem most FDN-UBE members we interviewed are working on is the hard work of training undergraduate biology students to accept and deal with the uncertainty inherent to research. In lab instruction, it is easy for students and the instructor to recognize when they learn procedural knowledge, such as the structure of a heart including valves and muscle that propel blood through it, or how to pipette an accurate volume. From our personal experience, we know that students feel a sense of accomplishment when they recognize a well-organized course where they feel they really learned something concrete. However, the abstractions of reasoning about evidence are more difficult to recognize. *CADE* was applied to reveal a value of FDN-UBE member contributions to biology faculty professional development as well as areas that still need to be targeted (*i.e. Epistemic considerations*). As a systematic framework of professional practice, *CADE* can focus students and instructors on their accomplishments as they learn to reason with disciplinary knowledge and consider science epistemology in deciding what counts as evidence and how to use it (Samarapungavan, 2018). Educators could support students by using the CADE framework as a guide for discussions throughout the process of a research study that gives opportunities to investigate a knowledge gap; operationalize relevant treatment, control, and outcome variables for experimental design; apply scientific conventions and standards for precise and accurate measurement; make appropriate decisions about the research subject and sampling; use tools for aggregating and analyzing data such as statistics; and apply science conventions for representing and communicating ideas about evidence throughout the research process, from inception of a study to reporting the findings. In fact, unlike the ACE-Bio Competencies that focus on competence with experimentation, the *CADE* framework is also applicable to research methods such as bioinformatics, structural biology, and evolutionary tree-thinking studies that are not based on experiments, although they do conform to a more generalized consensus research process (Thanukos et al., 2010).

We have also introduced the *CADE* framework as a lens that revealed what the FDN UBE professional developers are doing that cannot be done in a more general way by any leaders in a Center for Teaching and Learning. Such faculty developers lack the required knowledge to incorporate a biological disciplinary perspective needed to ensure society is supported by policy-makers and a general public who can understand and evaluate biological evidence for research-based solutions. In order for biology instructors to guide students' development of reasoning with and about scientific evidence during the process of biological experimentation, instructors need to clarify how evidentiary reasoning happens throughout the entire research process. The *CADE* framework highlights the need to be explicit about disciplinary knowledge, as well as the need to focus on helping students to incorporate more sophisticated epistemic reasoning in their approaches to biological research for success in the shift toward helping students learn the biological sciences through scientific practice and investigations. The FDN-UBE members who are experienced professional developers in this study were trained formally as scientists, they have passions about biology teaching and faculty professional development in biology education as an indispensable part of their career, and they focus on scientific practices related to all four *CADE* components of evidentiary reasoning in biology. Their experience of doing research in biology that they bring to biology education and faculty professional development allows the participants to have insights about the educational reform of learning biology through scientific practices. Their ideas and suggestions about helping students understand and use scientific evidence make them a valuable resource for other faculty who want to engage in biology education and involve evidentiary reasoning in their teaching.

Both the CADE framework and the ACE-Bio framework deconstruct the complexity of scientific research into smaller more manageable components that are organized according to the research process, from forming hypotheses, to generating data as evidence, to basing conclusions on the evidence. On the one hand, both frameworks provide a practical guide for instructors and researchers to design and implement questions that aim at reflecting on details of different aspects of the research process. On the other hand, however, these detail-aiming questions may sometimes shift the instructors' and students' attention to solving procedural problems, such as what the experimental steps are and how techniques and tools should be used, instead of having the whole picture of the research aimed at answering scientific questions. We agree with Manz et al. (2020) who suggest that investigations in the classroom should not be simplified and implemented as practicing procedural skills, and CADE offers a guide to avoid this problem. More importantly, many scientists who focus their expertise on the education of biology students are motivated to help develop inspiring opportunities for students to construct and explore alignments among phenomena, data and explanatory models in meaningful and purposeful ways, since this is what scientists are engaged in doing (Manz et al., 2020).

In summary, four points are offered to improve biology students' opportunities to develop better reasoning about evidence:

- In case there is no faculty professional developer who is a biologist in your institution, attend meetings listed in the findings from this study to benefit from the oral traditions passed on by experts like FDN-UBE members. They are passionately assisting biology educators beyond their own departments and they share innovative ideas about shifting biology lab instruction toward more authentic research experiences that will support students in learning to reason with and about evidence, promoting the application of disciplinary knowledge and science epistemology to biological experimentation.
- Adopt and implement *CADE* as a systematic framework to support a shift in professional practice toward more sophisticated epistemic reasoning in the teaching and learning of biological sciences. By unpacking the multifaceted nature of evidence and the complex integration and coordination of disciplinary knowledge with epistemic considerations, the *CADE* framework can guide evidentiary reasoning education in biology.
- Conduct studies to collect detailed data on how faculty professional developers understand *CADE* and how they integrate *CADE* with their prior experiential pedagogical and disciplinary knowledge in working with undergraduate biology educators to improve evidentiary reasoning in biology.

- Use CADE, which is complementary to the ACE -Bio Competencies framework, to guide the development of more detailed measures of students' reasoning with and about scientific evidence in terms of their integration of science epistemic considerations with disciplinary knowledge for evaluating development of evidentiary reasoning (or lack thereof) as students engage with biological investigations.

Acknowledgments Thanks go to the Faculty Development Network for Undergraduate Biology Education (FDN-UBE) Co-PIs (Gordon Uno, Karen Sirum, April Maskiewicz, Susan Elrod, and Charlene D'Avanzo), and also to the FDN-UBE subgroup leaders (Mark Connolly, Gili Marbach-Ad, Bryan Dewsbury, Rachelle Spell, and Larry Blumer) for discussions and encouragement during the early stages of this study. We also thank Chandrani Mishra, Sharleen Flowers, Nesibe Karakis, and Shuangting Li for discussions about *CADE* coding criteria. This work was supported in part by NSF #1346567 for the RCN-UBE: Advancing Competence with Experimentation in Biology (ACE-Bio) Network, NSF#1346570 RCN-UBE: Faculty Development Network for Undergraduate Biology Education (FDN-UBE), and NSF #1661124 Exploring Biological Evidence (EBE): Helping Students Understand the Richness and Complexity of Research Evidence in Biology. Any opinions, findings, and conclusions or recommendations expressed in this chapter are those of the authors and do not necessarily reflect the views of the National Science Foundation.

References

American Association for the Advancement of Science (AAAS). (2011). *Vision and change in undergraduate biology education*. AAAS. http://visionandchange.org/finalreport

Brown, N. J. S., Furtak, E. M., Timms, M. J., Nagashima, S. O., & Wilson, M. (2010). The evidence-based reasoning framework: Assessing scientific reasoning. *Educational Assessment, 15*(3–4), 123–141.

Cartwright, N. (2007). *Hunting causes and using them: Approaches in philosophy and economics*. Cambridge University Press.

Duschl, R. A., Schweingruber, H. A., & Shouse, A. W. (2007). *Taking science to school: Learning and teaching science in grades K-8*. National Academies Press.

Erduran, S., Ozdem, Y., & Park, J. Y. (2015). Research trends on argumentation in science education: A journal content analysis from 1998–2014. *International Journal of STEM Education, 2*(5). https://doi.org/10.1186/s40594-015-0020-1

Giere, R. N. (2006). *Understanding scientific reasoning* (5th ed.). Thomson/Wadsworth.

Khandkar, S. H. (2009). *Open coding* (p. 23). University of Calgary.

Koshland, D. E. (1985). Scientific literacy. *Science, 230*(4724), 391. https://doi.org/10.1126/science.230.4724.391

Labov, J. B., Reid, A. H., & Yamamoto, K. R. (2010). Integrated biology and undergraduate science education: A new biology education for the twenty-first century? *CBE-Life Sciences Education, 9*(1), 10–16.

Laursen, S. (2019). *Levers for change: An assessment of progress on changing STEM instruction*. American Association for the Advancement of Science. https://www.aaas.org/resources/levers-change-assessment-progress-changing-stem-instruction

Lewis, J., & Kattmann, U. (2004). Traits, genes, particles and information: Re-visiting students' understandings of genetics. *International Journal of Science Education, 26*(2), 195–206.

Manz, E., Lehrer, R., & Schauble, L. (2020). Rethinking the classroom science investigation. *Journal of Research in Science Teaching, 57*(7), 1148–1174. https://doi.org/10.1002/tea.21625

Mayr, E. (2004). *What makes biology unique? Considerations on the autonomy of a scientific discipline*. Cambridge University Press.

Pelaez, N., Anderson, T., Gardner, S., Yin, Y., Abraham, J., Bartlett, E., Gormally, C., Hill, J., Hoover, M., Hurney, C., Long, T., Newman, D., Sirum, K., & Stevens, M. (2017). *The basic competencies of biological experimentation: Concept-skill statements* (PIBERG Instructional Innovation Materials. Paper 4, 2017). http://docs.lib.purdue.edu/pibergiim/4

Pluta, W. J., Chinn, C. A., & Duncan, R. G. (2011). Learners' epistemic criteria for good scientific models. *Journal of Research in Science Teaching, 48*(5), 486–511.

Ratcliffe, M., & Millar, R. (2009). Teaching for understanding of science in context: Evidence from the pilot trials of the twenty first century science courses. *Journal of Research in Science Teaching, 46*(8), 945–959.

Samarapungavan, A. (2018). Construing scientific evidence: The role of disciplinary knowledge in reasoning with and about evidence in scientific practice. In F. Fischer, C. Chinn, K. Engelmann, & J. Osborne (Eds.), *Scientific reasoning and argumentation* (pp. 56–76). Routledge.

Sandoval, W. A. (2005). Understanding students' practical epistemologies and their influence on learning through inquiry. *Science Education, 89*(4), 634–656.

Seymour, E., & Hunter, A.-B. (2019). *Talking about leaving revisited: Persistence, relocation, and loss in undergraduate STEM education*. Springer Nature Switzerland AG. https://doi.org/10.1007/978-3-030-25304-2

Strauss, A., & Corbin, J. (1990). *Basics of qualitative research: Grounded theory procedures and techniques*. Sage.

Thanukos, A., Scotchmoor, J. G., Caldwell, R., & Lindberg, D. R. (2010). The real process of science-understanding science at http://www.understandingscience.org and science 101: Building the foundations for real understanding. *Science, 330*(6012), 1764–1765. https://doi.org/10.1126/science.1186994

Venezia, M. (2020). The American response to COVID-19 illustrates the importance of scientific literacy. *The Statesman COVID-19 Special Issue*. Stony Brook University. https://www.sbstatesman.com/2020/08/27/the-american-response-to-covid-19-illustrates-the-importance-of-scientific-literacy/

Part VI
Approaches to Biological Experimentation Instruction of Relevance to Biology Education Programs in General

Chapter 22
Teaching Successful Student Collaboration Within the Context of Biological Experimentation

Kathryn M. S. Johnson, Heather R. Pelzel, and Namoonga M. Mantina

22.1 Introduction

Successful scientific collaboration results in greater innovation and higher impact research when teams are diverse (AlShebli et al., 2018; Asai & Bauerle, 2016; Figg et al., 2006; Hofstra et al., 2020). Therefore, the ability for scientists to be an effective member of a diverse team, defined as a group of collaborators from various demographics (Johnson et al., 2019), is essential for professional and personal growth (Association of American Colleges & Universities, 2018; Clemmons et al., 2020). As students arrive at an undergraduate institution, it is unlikely they have collaborated with a diverse team, because college is often the first time students engage with people from diverse racial, ethnic, socioeconomic, and religious backgrounds. Collaborative research in an undergraduate course provides a key opportunity for students to learn how to navigate inclusive collaborations successfully.

As biology educators, we often ask undergraduate students to collaborate in the laboratory or field, typically with a partner or in a small group. The reasons for requiring laboratory collaboration vary. Pedagogically, educators may require collaborative, inquiry-based research to mimic the collaborative nature of modern

K. M. S. Johnson (✉)
Trail Build, LLC, East Troy, WI, USA
e-mail: katie.johnson@trailbuildllc.com

H. R. Pelzel
Department of Biological Sciences, University of Wisconsin-Whitewater, Whitewater, WI, USA
e-mail: pelzelh@uww.edu

N. M. Mantina
Mel and Enid Zuckerman College of Public Health, University of Arizona, Tucson, AZ, USA
e-mail: mantinan@email.arizona.edu

N. J. Pelaez et al. (eds.), *Trends in Teaching Experimentation in the Life Sciences*, Contributions from Biology Education Research,
https://doi.org/10.1007/978-3-030-98592-9_22

biological research and to increase student learning gains attributed to collaboration (Dirks, 1999; Fung & Lui, 2016; Hong, 2010; Knight et al., 2015; Strauss et al., 2011). Practically, collaborative student research is likely less expensive than independent projects, as the total number of experiments will be fewer, resulting in a decreased overall cost of supplies and equipment. In addition to coursework, undergraduate students may join faculty-led research groups, requiring collaboration with faculty, staff, and other trainees. Independent of the research context, when students engage in biological experimentation, we must teach them how to become effective scientific collaborators.

Collaborative research also provides an opportunity for students to learn and appreciate the advantages of collaborative work. Many undergraduate students "hate" group work and avoid it, citing its perceived lack of benefit to their own education (Allan, 2016; Le et al., 2018). It is also possible previous science courses reinforced individualistic learning practices, such as memorization, which encourage students to isolate themselves to avoid distraction and prepare for individual evaluation (Salmons, 2019). To counteract these perceptions, explicit explanations and discussions about the benefits of collaboration to their own learning, and the learning of others, are warranted.

In conversations about the benefits of collaboration and how to become a successful scientific collaborator, educators must emphasize the wealth of experience and knowledge of all students. Whether attained through academic, co-curricular, professional, or personal interactions, students bring various assets which can be leveraged to successfully engage biological research (Celedón-Pattichis et al., 2018; Holt et al., 2008; Johnson, 2019). Recognition and appreciation of their own assets, as well as the assets of other stakeholders, is foundational to the success of inclusive research collaborations (Johnson, 2019; Truesdell et al., 2017).

It is essential for educators to provide an intentional and scaffolded framework for students to become successful collaborators, defined as working with others to achieve a common outcome (Salmons, 2019). This framework must include various equitable paths, allowing students to use their own assets to be successful (Johnson et al., 2019). When educators provide little or no guidance to students, as novice scientific collaborators they may create their own strategies which limit their learning and collaborative success (Le et al., 2018). In addition, when working within the context of a primarily white academic institution, these novice strategies have been demonstrated to be more detrimental to Black and Latina/o/x students than white students in diverse student research groups (Johnson et al., 2019; Truesdell et al., 2017).

22.2 Guiding Principles for Collaboration

We aspire for our students to become successful and inclusive collaborators as they learn biological research skills and competencies (Pelaez et al., 2017). Collaboration is an iterative and integrated component of the research process. While not

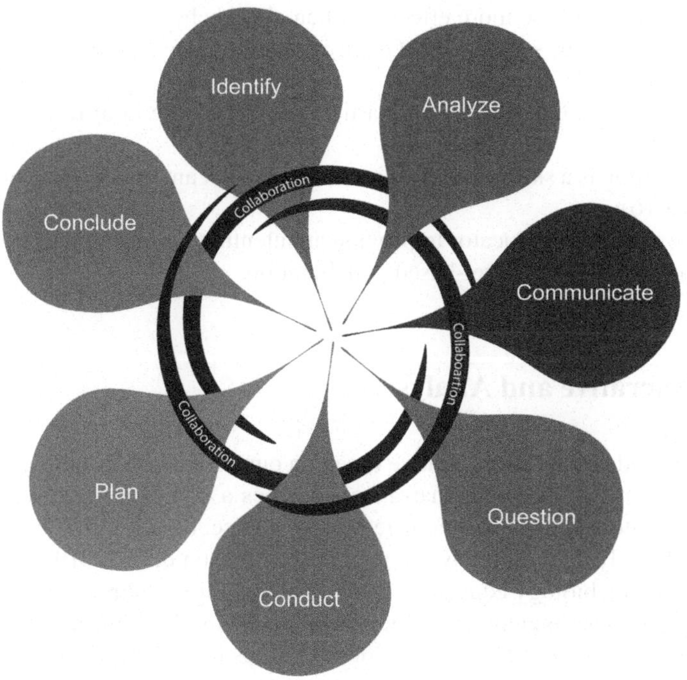

Fig. 22.1 Collaboration throughout biological experimentation. Collaboration is an iterative and integrative process connecting the competencies of biological experimentation. (Modified from Pelaez et al., 2017)

identified as an original component of the ACE-Bio research competencies (Pelaez, et al., 2017), within the context of undergraduate research, collaboration is essential for students as they navigate and learn various research competencies (Fig. 22.1). We propose future iterations of research competencies should investigate and recognize collaboration as a core component of undergraduate research.

The remainder of this chapter is dedicated to an interrupted narrative (White et al., 2009) demonstrating typical challenges which occur during student research collaborations. Each section of the narrative is concluded with an analysis, including practical recommendations to address the challenges. The guiding principles for our recommendations are:

1. Collaborations among individuals with diverse perspectives result in greater innovation and creative problem solving (Barjak & Robinson, 2008; Freeman & Huang, 2014; Freeman et al., 2014; Page, 2008);
2. Outcomes achieved during collaboration cannot be accomplished alone (Fung & Lui, 2016; Salmons, 2019);
3. Successful collaboration enhances learning (Crouch & Mazur, 2001; Johnson et al., 2000; Kilgo et al., 2015; Springer et al., 1999);

4. Collaboration will be more effective when all stakeholders are welcome to contribute their unique set of assets (Green & Haines, 2015; Johnson, 2019; Truesdell et al., 2017);
5. Collaboration is enhanced when participants contribute in different, rather than similar, ways;
6. Collaboration is a skill which requires development and reinforcement (Richards et al., 2016);
7. It is the role of the educator to provide an intentional scaffolding which emphasizes and appreciates asset-based collaboration.

22.3 Narrative and Analysis

Here we provide a five-part narrative based on our experience as biology educators, students, and researchers. The narrative describes a collaborative research project within the context of an undergraduate biology course. It presents scenarios we have observed directly and indirectly in our classes (Johnson et al., 2019) and are likely familiar to many biology educators. The narrative follows four students throughout the process of experimentation, as they navigate various biological experimentation competencies (Pelaez et al., 2017). It demonstrates the overlapping and interactive nature of students engaging the biological competencies (Fig. 22.1), and the ways in which successful collaboration is vital for effective experimentation.

The narrative alternates with analysis of how an educator could guide students to best leverage the assets of the group members and combat common challenges which arise during collaboration. Although provided within the context of a course, these strategies originate from within and beyond academia and are transferable to other types of collaborative work. Although based on real experiences, the characters in this narrative are fictional, and we intentionally did not ascribe them racial, cultural, or gendered identities. However, many of the analyses include how particular challenges may have a disproportionate effect on different students. While processing this narrative and analysis, we ask you to consider and question your own assumptions about each of these characters, including your own mental assignment of gender, race, or other social identities.

22.3.1 NARRATIVE 1: Collaboration on Day One

Students trickle into the classroom, carrying the energy of a new academic year. It is the first day of class, and as students arrive, they find seats at large tables. Some sit by people they know, while others opt for any open seat. By the start of class, Bailey, Devin, Kendall, and Taylor are sharing a table. Bailey and Devin live next door to each other in the residence halls. Devin and Kendall took a class together the previous semester. Taylor does not know anyone in class but found a seat at the table just as the instructor started taking attendance. Much of the first day of class is spent talking through the syllabus. With a few minutes left in the class session, students are asked to select their own groups for a research project. Bailey, Devin, Kendall, and Taylor exchange awkward glances and agree to form a group.

22.3.2 ANALYSIS 1: The First Day of Class

22.3.2.1 Challenge: Setting the Stage for a Long-Term Collaboration

If collaboration is required in a course, it should be intentionally introduced the first day of class. As instructors, we must provide time and guidance to facilitate an effective collaborative process. Lead an interactive class discussion focused on characterizing successful collaboration. Prompt students to offer suggestions or examples from their own experiences as collaborators in various contexts. Ask students to list the traits, experience, and training they could contribute to a collaboration.

Build on examples provided by students with models of effective collaboration from different contexts. For example, First Nations elder epistemology emphasizes a collaborator's responsibility to the community, an imperative to share knowledge, and trust in others (Christensen, 2013). This example may resonate with some students and provide others with a new perspective of collaboration. Highlight the value of diverse contributions rather than identical or individual effort. Early in the semester is also an appropriate time to discuss the dynamic nature of new groups so students expect highs and lows as a normal part of collaborative work (Tuckman, 1965).

22.3.2.2 Challenge: Forming Groups

Forming successful groups for long-term collaboration rarely occurs haphazardly. Some instructors encourage students to form their own groups, relying on previous familiarity to benefit the group (Lacey et al., 2020; Tuckman, 1965). However, these benefits are often only realized if students are familiar with each other in a working context or are already skilled collaborators. For example, it is likely Bailey, Devin, and Kendall, with their previous familiarity, could become a functional group, but this provides a substantial barrier for the inclusion of Taylor, who has never met anyone in the group. While it is not a foregone conclusion that Taylor would be an outsider, this situation is not in Taylor's favor.

There are other disadvantages to group self-selection. When allowed to self-determine groups, students with similar social identities cluster, even though diverse groups enhance the likelihood of finding novel solutions to difficult problems (Barjak & Robinson, 2008; Freeman & Huang, 2014; Freeman et al., 2014; Page, 2008). In addition, the process of self-determining groups is often isolating or anxiety-inducing, particularly for Black, Latina/o/x, Native American, and Pacific Islander students attending predominantly white institutions (Strauss et al., 2011; Strayhorn, 2012). There is substantial evidence that instructor-formed groups, independent of sorting criteria, result in greater student learning gains (Donovan et al., 2018; Jensen & Lawson, 2011).

22.3.3 NARRATIVE 2: Identifying a Research Question

Later that week, the research project assignment is posted on the courses' online learning platform. The assignment asks the student groups to design and implement an experiment to test and compare the microbial abundance of three sampling sites around or beyond campus. To do this, the groups will use swabs to collect samples at various locations, test their samples in the lab, analyze the data, and present their findings to the class.

After receiving the assignment description, Devin sees Bailey in their residence hall, and they chat about the assignment. Bailey suggests sampling the "dirty" restroom at the local gas station where Bailey works. Devin and Bailey quickly agree that collecting the additional samples from the restrooms in their residence hall and in an academic building would make a good experiment.

Devin sees Kendall in the hall before the next class, and Devin fills-in Kendall on the plan to sample three restroom locations. They walk into the room just as class is starting, and find Taylor sitting at their table. Bailey walks in a few minutes late. After general announcements, a teaching assistant begins to describe the swabbing technique for the research project.

22.3.4 ANALYSIS 2: Leveraging the Assets of the Group

22.3.4.1 Challenge: Communication of the Desired Outcome

For any assignment, a clear and direct description of the desired outcome, reiterated over multiple modes of communication, avoids assumptions by either the instructor or students. Assumptions about the expectations of the assignment create a hidden curriculum (Lear & Lear, 2006), disadvantaging students with less working knowledge of academic practices. In this case, posting the assignment and expecting students to proceed with little or no explanation of the expectations requires the students to make assumptions about how to proceed. A more constructive introduction to the assignment, especially one which is a large component of the course, is to post the assignment before class, allow and encourage students to read the assignment, go over the assignment in class, ask for questions, and provide examples which will facilitate a better understanding of the expectations.

Details about grading the assignment should also be explicit and revisited often during the project. One proven strategy is a scaffolded approach, providing assessment and feedback on portions of the project, ultimately building to a final product. Scaffolding assignments enhances student learning, improves the final research product, and alleviates the anxiety and pressure associated with a single, heavily weighted grade for the final product (Burgstahler & Cory, 2010; Frank et al., 2018; Freeman et al., 2011; Hmelo-Silver et al., 2007). For example, the first assignment in a scaffolded student research project could be a brief proposal which outlines a research question and plan, including a timeline. The instructor can review the proposal and provide feedback early in the research process and eliminate hidden assumptions of the research project.

22.3.4.2 Challenge: Engaging the Community

As research-based learning has evolved, more educators are asking students to collaborate with members of the community to collect and analyze data (Anthony & Reagan, 2017; Labov et al., 2019). To maintain working and successful relationships with community stakeholders, communication is key (Brownell et al., 2013). For an assignment, clear instructions on who and how to engage stakeholders is critical. Many academic institutions nurture community relationships. Faculty and staff experienced and tasked with community outreach are valuable resources to implementing best practices. Educators should investigate resources and expertise on their own campus and in the community.

In this example, it seems unlikely the student research group met with appropriate contacts at the gas station or on campus prior to finalizing their research design. Without the appropriate permission, tensions between the student researchers and community members may result. An alternative may have been to direct students to willing and established stakeholders who have been briefed on the objectives of the assignment. This approach increases the likelihood of a positive learning experience for the students and facilitates clear communication of the results to the community stakeholders. In addition, provide and require a signed agreement form to facilitate direct communication between the student researchers and the community members. The added benefits of two-way communication are the knowledge and expertise of the community stakeholders, which will inform the research and enhance the likelihood of using the data to advocate for meaningful change, if warranted.

22.3.4.3 Challenge: Establishing Group Norms to Incorporate the Voices of all Collaborators

This narrative provides a typical but concerning example of how a lack of communication can easily isolate or block the voice of group members. While Devin was eager to start the process of brainstorming for the research project, decisions about shaping the research question were made prior to consulting the entire group.

Moving forward without group consultation limits the assets contributing to the project. It also isolates group members, making them less likely to contribute in the future. Decreased participation in the collaborative process decreases the effects of its learning benefits, and this decrease is more likely to affect females and students of color (Chizhik, 1999). Defined group norms about contributions and group processes are essential for the inclusion of all group members (Feldman, 1984; Rovai & Wighting, 2005; Turpen & Finkelstein, 2010).

Group policies for inclusion and mitigating conflict can be developed within the group or by the entire class, depending on students' experience developing such policies. For example, group members must agree on which decisions will be built by consensus and when individual group members have the autonomy to push the project forward. Instructors should provide guidance and feedback on needed policies. Groups must establish logistical norms, such as agreeing on a shared mode of communication (ex. email, text, or messaging), acceptable response times to group communications, and group attendance policies. Groups must also determine how these policies will be enforced. What happens if a group member moves forward with the project without consulting the group? What process will the group follow if a member is not completing their commitments? Once agreed upon by all group members, students can write these policies into a contract or some other form of agreement, which can be submitted to the instructor for feedback and held to be revisited if issues arise later.

22.3.5 NARRATIVE 3: Conducting the Experiment and Analyzing Data

At the end of class a few weeks later, Devin asks the research group members if they can meet over the weekend to collect samples for the research project. Taylor mentions family commitments. Bailey and Kendall agree to help Devin with the sample collection, but Bailey's availability is limited by basketball practice. Therefore, Bailey volunteers to collect the sample from the gas station restroom when at work. Devin and Kendall agree on a time to meet to collect the other two samples. Devin suggests that Taylor leads the sample analysis, to make-up for missing the sample collection.

The next week, all the group members meet in the lab with the samples. Taylor has read over the very detailed analysis protocol and begins to explain it to the rest of the group. Bailey mentions that basketball practice starts in an hour. Taylor starts to process the samples, and the rest of the group talks about how "dirty" the bathrooms were when they sampled. Devin asks if Taylor has the analysis under control, and Taylor says it is going well. Bailey leaves for basketball practice, and Kendall leaves to study for an exam in another class. While conducting the analysis, Taylor asks Devin about the details of the sampling locations. Devin describes the campus sampling sites, but doesn't know much about the sample from Bailey's workplace.

Before leaving the lab, Devin offers to check the results from the analysis the next day.

The next day, Devin goes alone to check on the samples. To Devin, it appears the analysis didn't go well, and the results are not valid. Devin texts Bailey to collect another swab from work, and Devin does a repeat collection of the other two samples. Devin and Bailey meet the following day to do the laboratory analysis again with new reagents that they make themselves. The new results show that all three bathrooms are extremely "dirty."

22.3.6 ANALYSIS 3: Implementing Clear Communication Strategies

22.3.6.1 Challenge: Establishing Practices for Communication

Unfortunately, all too often, differences in communication are perceived as a lack of engagement or commitment by a group member, resulting in the exclusion of their voice in the research process. In this example, because the group did not establish communication expectations before starting the research project, they defaulted to chance availability and one-on-one communication. One group member, Taylor, has been excluded from most of the decision making.

When research assignments are posted, provide student groups the opportunity to agree on explicit expectations regarding communication. Students may not be aware of group members' individual circumstances, such as caregiving or work schedules, which could change the dynamics of communication. Limited access to the internet also changes the ability for group members to communicate (Aguilar, 2020). In addition, patterns of communication differ among students of various gender, cultural, and socioeconomic identities (Hargittai, 2008; Junco et al., 2010). While explaining or justifying life circumstances or preferences is not necessary, group agreement about what will and what will not work for group communication is essential.

Apart from social media and email, students may not be familiar with collaborative communication tools and will benefit from instructor provided examples. Direct students to institutional sharing platforms (e.g. Google or Office 365) or to tools embedded in the campus learning management system (e.g. Canvas or Blackboard). These tools can be used for group chats and calls, as well as real-time collaborative work on documents and presentations.

22.3.6.2 Challenge: Group Trust

Devin and Bailey repeated the collection and analysis of the samples without consulting the entire group. This is a clear violation of the group's informal agreement of Taylor's role in leading the sample analysis, and it sends a strong message that

Devin did not trust the initial analysis by Taylor. Although Devin had the best of intentions to move the research project forward, it was inappropriate for Devin to proceed without consulting Taylor about the preliminary findings.

Trust motivates group members to work together, and it is arguably one of the most important components of a successful collaboration (Dirks, 1999; Salmons, 2019). This group has failed to build and communicate trust. Devin's demonstrated distrust of Taylor could initiate further feelings of distrust within the group (Nguyen et al., 2010). It is likely Taylor will now have decreased trust in other group members, the group as a whole, and the work of the team (Alesina & La Ferrara, 2002; Costa, 2003; Smith, 2010). Feelings of distrust lead to a decreased sense of belonging and self-efficacy, determinants of student success and retention in science, which also disproportionately affect women and people of color (Trujillo & Tanner, 2014). Therefore, a well-meaning attempt by Devin to complete the research project may result in a barrier to the success of other students. Explicit discussions of how to foster trust through communication will improve student learning and research outcomes.

22.3.6.3 Challenge: Group Roles and Responsibilities

In this narrative, group roles were assigned hastily. Without planning, casual delegation of duties can result in inequitable differences in time commitment. Poorly designated or unheeded roles can also result in conflicts with other individual responsibilities. Clearly defined roles in laboratory group work increase student learning and collaboration productivity (Hunnicutt et al., 2015; Moog et al., 2006).

Provide student research groups time and guidance to discuss and define their group roles and responsibilities with input from all group members early in the process. For example, if a group member has a heavier course load later in the semester, that group member could volunteer to do more of the work earlier in the project. If another group member has family commitments in the evenings, perhaps they can volunteer to stay on campus during the day to complete laboratory tasks or attend meetings. In addition, encouraging student research groups to use the scheduled laboratory period, rather than meet on the evening or on the weekend, for the research projects provides all students the opportunity to fulfill their roles more fully and equitably.

22.3.7 *NARRATIVE 4: Drawing Conclusions*

The next day before class starts, Devin tells the group about the new results. Taylor looks confused and asks about why the samples needed to be rerun. Devin mentions that something didn't seem right with the first analysis, so Devin and Bailey saved everyone some time and redid the experiment. Taylor wants to ask more about what didn't "look right", but class is starting, and Taylor must leave right after class to arrive on time for another class across campus.

Later that week, Devin approaches the teaching assistant (TA) to schedule a time to discuss the results and get tips on preparing for the required presentation. The presentation is scheduled for the following week, and Devin is getting nervous because the group has not started preparing. The TA suggests attending the scheduled TA hours on Sunday afternoon. Devin texts the information to the group. Bailey and Kendall can make the meeting, but Taylor cannot.

On Sunday afternoon, Devin, Bailey, and Kendall meet at the TA session, and the room is crowded with other groups looking to prepare for their research presentation in two days. The TA is working with one group over in the corner and ignores Devin's frantic waves. Frustrated, Devin, Bailey, and Kendall begin to design the presentation. They decide that each person should be responsible for a section of the talk. After some discussion, they decide Bailey will present the background information, Taylor the methods, Devin the results, and Kendall will conclude the presentation. Devin, Bailey, and Kendall agree to add their parts to the online presentation file and practice on their own. Devin emails Taylor about their plan for the presentation.

Later that evening at work, Bailey is chatting with a coworker at the gas station and mentions the research project, including the results indicating the gas station restrooms were really dirty. Totally grossed-out, the coworker goes to the manager and demands better cleaning practices for the gas station. An extremely angry gas station manager approaches Bailey about the research project.

22.3.8 ANALYSIS 4: Addressing Group Issues

22.3.8.1 Challenge: An Inclusive Group Dynamic

As unexperienced researchers, undergraduate students often prioritize the research result over the process. In this case, Devin did not consider how repeating the sample analysis in this manner could impact Taylor or the group dynamic. In fact, Devin would likely complain to the instructor about the lack of Taylor's time commitment to the project, rather than realizing Devin's own role in the group dynamic.

To illuminate the dynamic of the research group, educators can implement mid-project peer-evaluations. Students can score each member's contributions to the collaboration, including themselves, and provide a short justification for each score. TA observations of group interactions, weekly student reflections about group progress, or notes from group meetings could provide additional insight. The instructor can review this information and provide constructive feedback on the group process, without revealing detailed criticisms of individuals. Feedback could direct groups to revisit group norms and defined group roles. Working through a case study of a collaborative process, such as this narrative, may also help students examine their own assumptions about collaborative work.

As a follow-up to these discussions, continue to emphasize the benefits and importance of utilizing the assets of all group members. Allow time for conversations about different cultural expectations of collaboration and the impact of these differences on the collaborative process. For example, students with an individualistic culture focus will approach problem solving differently than those from collectivist cultures (Popov et al., 2012). Remind students that incorporating different perspectives may take more time to discuss ideas and make decisions, but ultimately result in a much more innovative product and better learning for all students (Asai & Bauerle, 2016; Crouch & Mazur, 2001; Johnson et al., 2000; Kilgo et al., 2015; Page, 2008; Springer et al., 1999).

22.3.8.2 Challenge: Navigating Scheduling Conflicts

For students who typically have numerous academic, professional, and personal commitments, meeting outside of class is a challenge and is likely to create a barrier to the equitable success of all students. It is tempting for educators to leave scheduling details up to the student research groups, but as new researchers likely collaborating with a diverse student group for the first time, they need guidance establishing equitable group guidelines. In addition, if group research projects and collaborative work are objectives for the course, substantial class time should be dedicated to group planning.

Group meetings during class time provide a more inclusive approach, rather than privileging students with fewer commitments, such as caregiving and provider obligations. Even with class time, groups should consider what work requires all group members and what can be done by one or a few individuals. For example, groups should prioritize planning, communication, and decision-making during meetings when all members are present. Teams should consider saving menial tasks, such as labeling test tubes or making photocopies, for individual team members to complete at their convenience. Collaborators should keep in mind that members of highly effective teams rarely all complete the same role or are all present for every group task.

22.3.8.3 Challenge: Preparing TAs to Facilitate Inclusive Collaboration

Teaching Assistants (TAs), typically advanced undergraduates or graduate students, are a valuable resource for student learning in research-focused science courses (Knight et al., 2015; Talbot et al., 2015). Instructors, especially of courses with larger enrollments, are not always aware of research group dynamics, but TAs can gain insight into group issues. TAs may be more approachable for students than instructors and are often available to students at times when instructors are not, such as evenings or weekends. These unique aspects of student-TA relationships provide an opportunity to engage TAs as facilitators in the collaborative research process.

However, TAs should have appropriate training and be provided with strategies to help assist student collaboration (Pentecost et al., 2012; Ruder & Stanford, 2018).

When tasked with assisting undergraduate group research projects, TAs should do so with the intent of promoting inclusive, asset-based collaborations. TA training must include discussions of strategies to address barriers to collaboration, such as those described in this narrative. TAs should be familiar with group norms regarding collaborative communication and conflict resolution. They should also consider how their interactions inhibit or facilitate student collaboration. In this example, the TA was a barrier to effective collaboration. A scheduled group meeting time with the TA to provide direct feedback about the group's research project and presentation would have been more productive than a large open TA session. Communication among the instructor and TAs is essential so TAs can receive the support and feedback they need to facilitate student learning and group collaboration.

22.3.8.4 Challenge: Communication to the Public

Scientists must effectively communicate their work to all research stakeholders so they can understand the findings and make informed decisions (Brownell et al., 2013). Thoughtful communication must respect and acknowledge the assets and interests of the community collaborators. Referring to the analysis findings as "dirty" establishes a negative connotation of the sample site. This language could easily be miscommunicated, cause long-term detrimental effects to the community stakeholders, and damage the relationship between the academic institution and the community (Strand et al., 2003). In this case, ineffective communication resulted in a conflict between Bailey and the gas station manager. If trust among researchers and community partners is broken, it is difficult to work together to enact change. Thoughtful consideration, discussion, and practice for how and when students should engage community members is necessary (Brownell et al., 2013).

22.3.9 NARRATIVE 5: The Final Presentation

The day of the final research presentation, Devin, Bailey, Kendall, and Taylor are sitting at their table making last minute preparations for their talk. All of the group members completed their portions of the talk late last night. Devin compiled the presentation and brought it to class. When combining the different portions of the presentation, Devin grew anxious because the methodology described by Taylor was different than the protocol used by Devin and Bailey to reanalyze the samples. There was no time to consult with Taylor prior to the presentation, so Devin and the rest of the group make their way to the front of the room.

Bailey leads off with the background, including details about collecting samples from the restroom at the gas station, but leaving out the angry confrontation with the gas station manager. As Taylor explains the methods, Devin's suspicions are confirmed. This was not the technique used to reanalyze the samples. Nevertheless, Devin takes a deep breath, and describes the results of the reanalyzed samples, hoping no one will notice the discrepancy. Kendall concludes the presentation with an emphasis on how dirty all of the restrooms are on campus and at the gas station. The presentation comes across as disjointed and unrehearsed. At the end of the class, the student groups are asked to complete and submit peer-evaluations for their research projects. Devin, Kendall, and Bailey all rate each other highly, but give low scores to Taylor. Taylor gives everyone low scores.

After noticing some irregularities in their data, the instructor asks the group to stay after class for a few minutes to discuss their findings. The instructor states the findings presented during the talk seem very unlikely and asks the group to describe their sample analysis. Devin explains that the group actually collected two sets of samples and conducted the analysis twice. After Devin describes the second analysis, the instructor explains the error in the analysis resulted in a much higher level of contamination reported than what was actually in the sample. Kendall and Bailey look embarrassed about the error. Taylor looks mad.

22.3.10 ANALYSIS 5: Improving the Collaboration Process

22.3.10.1 Challenge: Normalizing and Addressing Research Difficulties

As new researchers, undergraduate students may not consider how to recognize and respond to problems with their experiments. When teaching students to engage in research, missteps and difficulties should be normalized. Strategies to navigate research issues should be discussed prior to the start of the research. In this example, the group members had no plan to recognize or address their initial research issue, a confusion about the results of their first sample analysis. Instead, Devin and Bailey moved forward to reanalyze new samples, which was detrimental to the research project and the group members.

Providing students with iterative feedback teaches students how to monitor their own progress and increases learning gains (Morrell, 2019; Sadler, 1989), with greatest gains observed by students identifying as Black, Latina/o/x, and Native American (Theobald et al., 2020). In the context of a group research project, the larger research assignment can be scaffolded into smaller assignments, allowing for instructor feedback throughout the process. A required group meeting with the instructor or TA at various checkpoints during the research also facilitates timely feedback. In this example, if the student group had discussed the results of their original sample analysis with the instructor, they would have understood their results and avoided further issues.

22.3.10.2 Challenge: Time Management

This research group's presentation was compiled at the last minute, was unrehearsed as a group, and described conflicting information. Their unfamiliarity with successful research timelines provided a barrier to their success. Time management training with practical advice and tactics can improve student time management (Häfner et al., 2014; Nadinloyi et al., 2013). However, the causes and effects of procrastination are complex, and anxiety around procrastination can serve as a major stressor for students (Sirois, 2014). Therefore, unfamiliarity with developing an appropriate research timeline could lead to unintentional procrastination, likely contributing to student stress. This stress disproportionately affects students of color compared to white students, contributing to decreases in persistence (Arbona et al., 2018; Johnson et al., 2014; Wei et al., 2011). Campus resources, with skilled staff, may be available to provide training and advice to students around issues of time management and stress.

As part of assignment scaffolding within the course, require student research groups to submit a work plan, including scheduled meetings and protocols, prior to beginning their experiment. Feedback on this work plan facilitates conversations about realistic timelines. Assigning progress reports, possibly modelling typical research group lab meetings, encourages research groups to discuss their findings prior to the final presentation.

22.3.10.3 Challenge: Assessing Gains in Students' Ability to Collaborate

There are various formative and summative assessments to evaluate the ability of students to collaborate (Macdonald, 2003; Salmons, 2019). The outcome of the research project, per se, and the individual contributions to the project can be assessed for scientific impact and accuracy. Comparisons between preliminary research proposals and the final research project can provide evidence of growth in collaborative skills. An innovative group research project and successful presentation represent a successful collaborative process. Adaptation of collaboration and teamwork skills rubrics can assist in this assessment (Aguirre et al., 2013; Firdausa & Istiyono, 2019; Herro et al., 2017).

Final peer- and self-evaluations provide insight into each group member's collaborative effort and growth. Prompts for written reflections could ask students to consider how the actions of each group member affected the work of the group or how each group member gained collaborative skills throughout the project. Other valuable assessment information could include student descriptions of the advantages of completing this research project collaboratively.

22.4 Conclusions & Recommendations

Engaging students in biological research provides a venue to demonstrate and reinforce inclusive collaboration practices within the context of the discipline. It is the responsibility of educators to provide a framework to prioritize the development of skills needed for successful collaboration through explicit directions, discussions, and assignments. When students recognize and appreciate the value of collaboration, their own value, and the value of other collaborators, the collaboration will become more inclusive, innovative, and productive. This framework for learning and engaging in the biological sciences pushes away historically individualistic norms and replaces them with inclusive and collaborative practices.

Educators seeking to teach successful and inclusive student collaboration within the context of biological experimentation are encouraged to consider these recommendations:

- Do the following during the development of courses involving long-term collaborative group work:
  - Plan a scaffolded approach with small assignments and group check-ins throughout the project;
  - Investigate and prepare campus resources for group communication (e.g. Microsoft Teams, Webex Teams, Blackboard Collaborate);
  - Determine an inclusive process for group selection;
  - Schedule class time in the syllabus for specific discussions about collaboration and provide ample class time to conduct the research project (see below);
  - Train TAs to facilitate productive group work.
- Plan and facilitate class discussions about the following:
  - The characteristics of successful collaboration;
  - Asset-based inclusive collaborative work which relies on various types of contributions from different collaborators;
  - Establishing group norms, policies, and the roles of each group member;
  - How actions of group members could damage the trust of various stakeholders and impact the learning of others.
- To facilitate effective research collaborations, provide class time for students to do the following:
  - Create contracts about group norms for communication, responsibilities, and roles;
  - Plan and complete the project;
  - Meet with each other and the instructor to receive feedback throughout the project.

Acknowledgments We thank the members of the ACE-Bio Network and the reviewers for their thoughtful and productive feedback on this work. Thank you to Dr. Trevor Anderson, Dr. Stephanie

Gardner, and Dr. Nancy Pelaez for sharing your vision and inviting us to contribute. ACE-Bio Network is funded by the National Science Foundation (NSF#1346567). Author Statement: Kathryn M.S. Johnson: Conceptualization, Formal Analysis, Writing – Original Draft, Writing – Review & Editing, Supervision, Project Administration; Heather R. Pelzel: Conceptualization, Formal Analysis, Writing – Original Draft, Writing – Review & Editing; Namoonga M. Mantina: Conceptualization, Formal Analysis, Writing – Original Draft, Writing – Review & Editing, Visualization.

References

Aguilar, S. J. (2020). Guidelines and tools for promoting digital equity. *Information and Learning Sciences, 121*(5/6), 285–299. https://doi.org/10.1108/ILS-04-2020-0084

Aguirre, K. M., Balser, T. C., Jack, T., Marley, K. E., Miller, K. G., Osgood, M. P., Pape-Lindstrom, P. A., & Romano, S. L. (2013). PULSE vision & change rubrics. *CBE—Life Sciences Education, 12*(4), 579–581. https://doi.org/10.1187/cbe.13-09-0183

Alesina, A., & La Ferrara, E. (2002). Who trusts others? *Journal of Public Economics, 85*(2), 207–234. https://doi.org/10.1016/S0047-2727(01)00084-6

Allan, E. G. (2016). "I hate group work!": Addressing students' concerns about small-group learning. *InSight: A Journal of Scholarly Teaching, 11*, 81–89.

AlShebli, B. K., Rahwan, T., & Woon, W. L. (2018). The preeminence of ethnic diversity in scientific collaboration. *Nature Communications, 9*(1), 5163. https://doi.org/10.1038/s41467-018-07634-8

Anthony, B. M., & Reagan, K. M. (2017). Community-engaged projects in operations research. *Science Education and Civic Engagement: An International Journal, 8*. http://new.seceij.net/articletype/research/community-engaged-projects-in-operations-research/

Arbona, C., Fan, W., & Olvera, N. (2018). College stress, minority status stress, depression, grades, and persistence intentions among Hispanic female students: A mediation model. *Hispanic Journal of Behavioral Sciences, 40*(4), 414–430. https://doi.org/10.1177/0739986318799077

Asai, D. J., & Bauerle, C. (2016). From HHMI: Doubling down on diversity. *CBE-Life Sciences Education, 15*(3), fe6. https://doi.org/10.1187/cbe.16-01-0018

Association of American Colleges & Universities. (2018, August 27). *Fulfilling the American dream: Liberal education and the future of work (text)*. Association of American Colleges & Universities. https://www.aacu.org/research/2018-future-of-work

Barjak, F., & Robinson, S. (2008). International collaboration, mobility and team diversity in the life sciences: Impact on research performance. *Social Geography, 3*(1), 23–36. https://doi.org/10.5194/sg-3-23-2008

Brownell, S. E., Price, J. V., & Steinman, L. (2013). Science communication to the general public: Why we need to teach undergraduate and graduate students this skill as part of their formal scientific training. *Journal of Undergraduate Neuroscience Education, 12*(1), E6–E10.

Burgstahler, S. E., & Cory, R. C. (2010). *Universal design in higher education: From principles to practice*. Harvard Education Press.

Celedón-Pattichis, S., Borden, L. L., Pape, S. J., Clements, D. H., Peters, S. A., Males, J. R., Chapman, O., & Leonard, J. (2018). Asset-based approaches to equitable mathematics education research and practice. *Journal for Research in Mathematics Education, 49*(4), 373–389. https://doi.org/10.5951/jresematheduc.49.4.0373

Chizhik, A. W. (1999). Can students work together equitably? An analysis of task effects in collaborative group work. *Social Psychology of Education, 3*(1), 63–79. https://doi.org/10.1023/A:1009611032434

Christensen, R. A., & author. (2013). *Connective pedagogy: Elder epistemology, oral tradition and community*. Aboriginal Issues Press. https://search.library.wisc.edu/catalog/9910219229802121

Clemmons, A., Timbrook, J., Herron, J., & Crowe, A. (2020). BioSkills guide: Develop-ment and national validation of a tool for interpreting the vision and change core competencies. *CBE—Life Sciences Education, 19*, 4. https://doi.org/10.1187/cbe.19-11-0259

Costa, A. C. (2003). Work team trust and effectiveness. *Personnel Review, 32*(5), 605–622. https://doi.org/10.1108/00483480310488360

Crouch, C. H., & Mazur, E. (2001). Peer instruction: Ten years of experience and results. *American Journal of Physics, 69*(9), 970–977. https://doi.org/10.1119/1.1374249

Dirks, K. T. (1999). The effects of interpersonal trust on work group performance. *Journal of Applied Psychology, 84*(3), 445.

Donovan, D. A., Connell, G. L., & Grunspan, D. Z. (2018). Student learning outcomes and attitudes using three methods of group formation in a nonmajors biology class. *CBE—Life Sciences Education, 17*(4) https://www.lifescied.org/doi/full/10.1187/cbe.17-12-0283

Feldman, D. C. (1984). The development and enforcement of group norms. *Academy of Management Review, 9*(1), 47–53. https://doi.org/10.5465/amr.1984.4277934

Figg, W. D., Dunn, L., Liewehr, D. J., Steinberg, S. M., Thurman, P. W., Barrett, J. C., & Birkinshaw, J. (2006). Scientific collaboration results in higher citation rates of published articles. *Pharmacotherapy, 26*(6), 759–767. https://doi.org/10.1592/phco.26.6.759

Firdausa, A. R., & Istiyono, E. (2019). Developing instrument for assessing student collaboration in vocational high schools. *Journal of Physics: Conference Series, 1273*, 012043. https://doi.org/10.1088/1742-6596/1273/1/012043

Frank, B., Simper, N., & Kaupp, J. (2018). Formative feedback and scaffolding for developing complex problem solving and modeling outcomes. *European Journal of Engineering Education, 43*(4), 552–568. https://doi.org/10.1080/03043797.2017.1299692

Freeman, R. B., & Huang, W. (2014). Collaboration: Strength in diversity. *Nature News, 513*(7518), 305. https://doi.org/10.1038/513305a

Freeman, S., Eddy, S. L., McDonough, M., Smith, M. K., Okoroafor, N., Jordt, H., & Wenderoth, M. P. (2014). Active learning increases student performance in science, engineering, and mathematics. *Proceedings of the National Academy of Sciences, 111*(23), 8410–8415. https://doi.org/10.1073/pnas.1319030111

Freeman, S., Haak, D., & Wenderoth, M. P. (2011). Increased course structure improves performance in introductory biology. *CBE—Life Sciences Education, 10*(2), 175–186. https://doi.org/10.1187/cbe.10-08-0105

Fung, D., & Lui, W. (2016). Individual to collaborative: Guided group work and the role of teachers in junior secondary science classrooms. *International Journal of Science Education, 38*(7), 1057–1076. https://doi.org/10.1080/09500693.2016.1177777

Green, G. P., & Haines, A. (2015). *Asset building & community development*. SAGE.

Häfner, A., Oberst, V., & Stock, A. (2014). Avoiding procrastination through time management: An experimental intervention study. *Educational Studies, 40*(3), 352–360. https://doi.org/10.1080/03055698.2014.899487

Hargittai, E. (2008). The digital reproduction of inequality. In D. Grusky (Ed.), *Social stratification* (pp. 936–944). Westview Press.

Herro, D., Quigley, C., Andrews, J., & Delacruz, G. (2017). Co-measure: Developing an assessment for student collaboration in STEAM activities. *International Journal of STEM Education, 4*(1), 26. https://doi.org/10.1186/s40594-017-0094-z

Hmelo-Silver, C. E., Duncan, R. G., & Chinn, C. A. (2007). Scaffolding and achievement in problem-based and inquiry learning: A response to Kirschner, Sweller, and Clark. *Educational Psychologist, 42*(2), 99–107. https://doi.org/10.1080/00461520701263368

Hofstra, B., Kulkarni, V. V., Galvez, S. M.-N., He, B., Jurafsky, D., & McFarland, D. A. (2020). The diversity–innovation paradox in science. *Proceedings of the National Academy of Sciences, 117*(17), 9284–9291. https://doi.org/10.1073/pnas.1915378117

Holt, N. L., Tink, L. N., Mandigo, J. L., & Fox, K. R. (2008). Do youth learn life skills through their involvement in high school sport? A case study. *Canadian Journal of Education/Revue Canadienne de l'éducation, 31*(2), 281–304 JSTOR. https://doi.org/10.2307/20466702

Hong, Z. (2010). Effects of a collaborative science intervention on high achieving students learning anxiety and attitutdes toward science. *International Journal of Science Education, 32*(15), 197–1988. https://doi.org/10.1080/09500690903229304

Hunnicutt, S. S., Grushow, A., & Whitnell, R. (2015). Guided-inquiry experiments for physical chemistry: The POGIL-PCL model. *Journal of Chemical Education, 92*(2), 262–268. https://doi.org/10.1021/ed5003916

Jensen, J., & Lawson, A. (2011). Effects of collaborative group composition and inquiry instruction on reasoning gains and achievement in undergraduate biology. *CBE Life Sciences Education, 10*, 64–73. https://doi.org/10.1187/cbe.10-07-0089

Johnson, D. R., Wasserman, T. H., Yildirim, N., & Yonai, B. A. (2014). Examining the effects of stress and campus climate on the persistence of students of color and White students: An application of bean and Eaton's psychological model of retention. *Research in Higher Education, 55*(1), 75–100. https://doi.org/10.1007/s11162-013-9304-9

Johnson, D. W., Johnson, R. T., & Stanne, M. B. (2000). *Cooperative learning methods: A meta-analysis*. University of Minnesota. https://www.researchgate.net/publication/220040324_Cooperative_learning_methods_A_meta-analysis

Johnson, K. M. S. (2019). Implementing inclusive practices in an active learning STEM classroom. *Advances in Physiology Education, 43*(2), 207–210. https://doi.org/10.1152/advan.00045.2019

Johnson, K. M. S., Briggs, A., Hawn, C., Mantina, N., & Woods, B. C. (2019). Inclusive practices for diverse student populations: Experimental biology 2017. *Advances in Physiology Education, 43*(3), 365–372. https://doi.org/10.1152/advan.00201.2018

Junco, R., Merson, D., & Salter, D. W. (2010). The effect of gender, ethnicity, and income on college students' use of communication technologies. *Cyberpsychology, Behavior and Social Networking, 13*(6), 619–627. https://doi.org/10.1089/cyber.2009.0357

Kilgo, C. A., Ezell Sheets, J. K., & Pascarella, E. T. (2015). The link between high-impact practices and student learning: Some longitudinal evidence. *Higher Education, 69*(4), 509–525. https://doi.org/10.1007/s10734-014-9788-z

Knight, J. K., Wise, S. B., Rentsch, J., & Furtak, E. M. (2015). Cues matter: Learning assistants influence introductory biology student interactions during clicker-question discussions. *CBE—Life Sciences Education, 14*(4), ar41. https://doi.org/10.1187/cbe.15-04-0093

Labov, J., Brenner, K. A., & Middlecamp, C. H. (2019). Integrating undergraduate research in STEM with civic engagement. *Science Education and Civic Engagement: An International Journal, 11*(1) http://new.seceij.net/articletype/review/integrating-undergraduate-research-in-stem-with-civic-engagement/

Lacey, M. M., Campbell, S. G., Shaw, H., & Smith, D. P. (2020). Self-selecting peer groups formed within the laboratory environment have a lasting effect on individual student attainment and working practices. *FEBS Open Bio, 10*(7), 1194–1209. https://doi.org/10.1002/2211-5463.12902

Le, H., Janssen, J., & Wubbels, T. (2018). Collaborative learning practices: Teacher and student perceived obstacles to effective student collaboration. *Cambridge Journal of Education, 48*(1), 103–122. https://doi.org/10.1080/0305764X.2016.1259389

Lear, J., & Lear, J. U. N. D. S. P. in the C. on S. T. and the D. of P. J. (2006). *Radical Hope: Ethics in the face of cultural devastation*. Harvard University Press.

Macdonald, J. (2003). Assessing online collaborative learning: Process and product. *Computers & Education, 40*(4), 377–391. https://doi.org/10.1016/S0360-1315(02)00168-9

Moog, R. S., Spencer, J. N., & Straumanis, A. R. (2006). Process-oriented guided inquiry learning: POGIL and the POGIL project. *Metropolitan Universities, 17*(4), 41–52.

Morrell, L. J. (2019). Iterated assessment and feedback improves student outcomes. *Studies in Higher Education*. https://doi.org/10.1080/03075079.2019.1643301

Nadinloyi, K. B., Hajloo, N., Garamaleki, N. S., & Sadeghi, H. (2013). The study efficacy of time management training on increase academic time management of students. *Procedia – Social and Behavioral Sciences, 84*, 134–138. https://doi.org/10.1016/j.sbspro.2013.06.523

Nguyen, V.-A., Lim, E.-P., Tan, H.-H., Jiang, J., & Sun, A. (2010). Do you trust to get trust? A study of trust reciprocity behaviors and reciprocal trust prediction. In *Proceedings of the 2010 SIAM international conference on data mining* (Vol. 1–0, pp. 72–83). https://doi.org/10.1137/1.9781611972801.7

Page, S. E. (2008). *The difference: How the power of diversity creates better groups, firms, schools, and societies*. Princeton University Press.

Pelaez, N., Anderson, T., Gardner, S. M., Yin, Y., Abraham, J. K., Bartlett, E., Gormally, C., Hill, J. P., Hoover, M., Hurney, C., Long, T., Newman, D. L., Sirum, K., & Stevens, M. (2017). *The basic competencies of biological experimentation: Concept-skill statements*. PIBERG Instructional Innovation Materials (Paper 4). https://docs.lib.purdue.edu/pibergiim/4/

Pentecost, T. C., Langdon, L. S., Asirvatham, M., Robus, H., & Parson, R. (2012). Graduate teaching assistant training that fosters student-centered instruction and professional development. *Journal of College Science Teaching, 41*(6), 68–75.

Popov, V., Brinkman, D., Biemans, H. J. A., Mulder, M., Kuznetsov, A., & Noroozi, O. (2012). Multicultural student group work in higher education: An explorative case study on challenges as perceived by students. *International Journal of Intercultural Relations, 36*(2), 302–317. https://doi.org/10.1016/j.ijintrel.2011.09.004

Richards, S. B., Frank, C. L., Sableski, M.-K., & Arnold, J. M. (2016). *Collaboration among professionals, students, families, and communities: Effective teaming for student learning*. Routledge.

Rovai, A. P., & Wighting, M. J. (2005). Feelings of alienation and community among higher education students in a virtual classroom. *The Internet and Higher Education, 8*(2), 97–110. https://doi.org/10.1016/j.iheduc.2005.03.001

Ruder, S. M., & Stanford, C. (2018). Strategies for training undergraduate teaching assistants to facilitate large active-learning classrooms. *Journal of Chemical Education, 95*(12), 2126–2133. https://doi.org/10.1021/acs.jchemed.8b00167

Sadler, D. R. (1989). Formative assessment and the design of instructional systems. *Instructional Science, 18*(2), 119–144. https://doi.org/10.1007/BF00117714

Salmons, J. (2019). *Learning to collaborate, collaborating to learn: Engaging students in the classroom and online*. Stylus Publishing, LLC.

Sirois, F. M. (2014). Procrastination and stress: Exploring the role of self-compassion. *Self and Identity, 13*(2), 128–145. https://doi.org/10.1080/15298868.2013.763404

Smith, S. S. (2010). Race and trust. *Annual Review of Sociology, 36*(1), 453–475. https://doi.org/10.1146/annurev.soc.012809.102526

Springer, L., Stanne, M. E., & Donovan, S. S. (1999). Effects of small-group learning on undergraduates in science, mathematics, engineering, and technology: A meta-analysis. *Review of Educational Research, 69*(1), 21–51. https://doi.org/10.2307/1170643

Strand, K. J., Cutforth, N., Stoecker, R., Marullo, S., & Donohue, P. (2003). *Community-based research and higher education: Principles and practices*. Wiley.

Strauss, P., & U, A., & Young, S. (2011). 'I know the type of people I work well with': Student anxiety in multicultural group projects. *Studies in Higher Education, 36*(7), 815–829. https://doi.org/10.1080/03075079.2010.488720

Strayhorn, T. L. (2012). *College students' sense of belonging: A key to educational success for all students*. Routledge.

Talbot, R. M., Hartley, L. M., Marzetta, K., & Wee, B. S. (2015). Transforming undergraduate science education with learning assistants: Student satisfaction in large-enrollment courses. *Journal of College Science Teaching, 44*(5), 24–30.

Theobald, E. J., Hill, M. J., Tran, E., Agrawal, S., Arroyo, E. N., Behling, S., Chambwe, N., Cintrón, D. L., Cooper, J. D., Dunster, G., Grummer, J. A., Hennessey, K., Hsiao, J., Iranon, N., Jones, L., Jordt, H., Keller, M., Lacey, M. E., Littlefield, C. E., … Freeman, S. (2020). Active learning narrows achievement gaps for underrepresented students in undergraduate science, technology, engineering, and math. *Proceedings of the National Academy of Sciences, 117*(12), 6476–6483. https://doi.org/10.1073/pnas.1916903117

Truesdell, N., Carr, J., & Orr, C. M. (2017). The role of Combahee in anti-diversity work. *Souls, 19*(3), 359–376. https://doi.org/10.1080/10999949.2017.1389632

Trujillo, G., & Tanner, K. D. (2014). Considering the role of affect in learning: Monitoring students' self-efficacy, sense of belonging, and science identity. *CBE-Life Sciences Education, 13*(1), 6–15. https://doi.org/10.1187/cbe.13-12-0241

Tuckman, B. W. (1965). Developmental sequence in small groups. *Psychological Bulletin, 63*(6), 384–399. https://doi.org/10.1037/h0022100

Turpen, C., & Finkelstein, N. D. (2010). The construction of different classroom norms during Peer Instruction: Students perceive differences. *Physical Review Special Topics – Physics Education Research, 6*(2), 020123. https://doi.org/10.1103/PhysRevSTPER.6.020123

Wei, M., Ku, T.-Y., & Liao, K. Y.-H. (2011). Minority stress and college persistence attitudes among African American, Asian American, and Latino students: Perception of university environment as a mediator. *Cultural Diversity and Ethnic Minority Psychology, 17*(2), 195–203. https://doi.org/10.1037/a0023359

White, T. K., Whitaker, P., Gonya, T., Hein, R., Kroening, D., Lee, K., Lee, L., Lukowiak, A., & Hayes, E. (2009). The use of interrupted case studies to enhance critical thinking skills in biology. *Journal of Microbiology & Biology Education, 10*(1), 25–31.

Chapter 23
Biochemistry and Art: Incorporating Drawings, Paintings, Music, and Media into Teaching Biological Science

Latisha R. Jefferies and Shanae S. Jefferies

23.1 Introduction

Biochemistry, the study of chemistry in biological systems, has a reputation of being a fundamentally difficult course to learn and thus teach (Vella, 1990; Wood, 1990). The vast amount of information involved in biochemistry combined with the exponential growth of knowledge, the constant development of new technology (i.e. bioinformatics), and the dependency of many highly complex subjects (i.e. biology, chemistry and physics) contributes to the difficulty of teaching biochemistry in an enjoyable manner. Additionally, student's first impressions tend to be that the subject requires a lot of memorization and is mundane (Valenti et al., 2016). This impression is heavily tied to the popular teaching method of traditional lecture. Understandably, lecture-style teaching is popular in biochemistry because it is germane to large class sizes. However, pure lecture courses provide little contribution to stimulate creative thinking, improve student's attitude toward the subject, or strengthen student problem solving development (Adkins et al., 2017). As Generation Alpha (born after 2015) enters grade school and Generation Z (born between 1995 and 2015) spans through post-secondary education, classrooms are filled with students raised on the internet and social media. Accordingly, educators are experimenting with alternative teaching methods to critically engage evolving

L. R. Jefferies (✉)
Department of Chemistry, United States Air Force Academy, USAFA,
Colorado Springs, CO, USA
e-mail: latisha.jefferies@usafa.edu

S. S. Jefferies
Department of Sociology, Collin College, McKinney, TX, USA
e-mail: SJefferies@collin.edu

N. J. Pelaez et al. (eds.), *Trends in Teaching Experimentation in the Life Sciences*, Contributions from Biology Education Research,
https://doi.org/10.1007/978-3-030-98592-9_23

student populations. One successful approach is to combine the rigors of science with the creativity of art (Colucci-Gray et al., 2017).

Merging arts and science has shown to improve cognitive scores, increase innovation, and motivate students to further pursue scientific studies (Colucci-Gray et al., 2017). Expressing a biochemical concept through artistic representation gives students the opportunity to actively learn and breaks the mundane experience of traditional lecture. Active learning is "*the process of having students engage in some activity that forces them to reflect upon ideas and upon how they are using those ideas*" (Freeman et al., 2014; Hanauer et al., 2016; Wiggins et al., 2017). Active learning methods shift the student's mindsets from brute memorization (or simply "Googling" the answer) and allows the students to exercise critical thinking skills. Although not the driver, the seven ACE-Bio Experimentation Competencies (*Identify*, *Analyze*, *Communicate*, *Question*, *Conduct*, *Plan*, and *Conclude*) (Pelaez et al., 2017) were used to evaluate the strength of an existing activity or assessment relative to foundational skills used in biologic experimentation. All of these works, viewed through this lens, provide guidance in design or implementation of activities that support the basic competencies of Biologic Experimentation. This chapter also explores the many artistic approaches educators have successfully experimented in using (drawings, paintings, music and media) to teach and/or reinforce biochemical concepts through expressions, as well as define areas in which improvement is still needed.

23.2 Drawing

The most historic connection of art and science comes from drawing. Prior to the ability to quickly snap a photo with a digital camera (or now-a-days a smart phone), biologists depended on hand-drawn representations to capture their ideas and subjects. In fact, if you look throughout history you will see many scientific advancements were stimulated by interaction with the arts. For example, motivated by his interest in science, Leonardo da Vinci drew some of the first realistic representations of the human body (Atalay, 2011) as shown in the highly publicized *Fetus in the Womb* (Fig. 23.1).

Considering today's technological advancements, it is important not to lose sight of the benefits of handmade drawings. Aninsworth et al. (Ainsworth et al., 2011) investigated an array of drawing programs used to help students learn science and concluded that drawing engages motor and visual learners while acting as a simple, low-cost method to foster science literacy and engage science curricula. Ainsworth et al. (2011) noted five reasons why drawing should be a key element in science education:

1. Drawing enhances engagement
2. Drawing helps students learn to conceptualize science
3. Drawing helps to reason in science
4. Drawing can be used as a learning strategy
5. Drawing helps students communicate

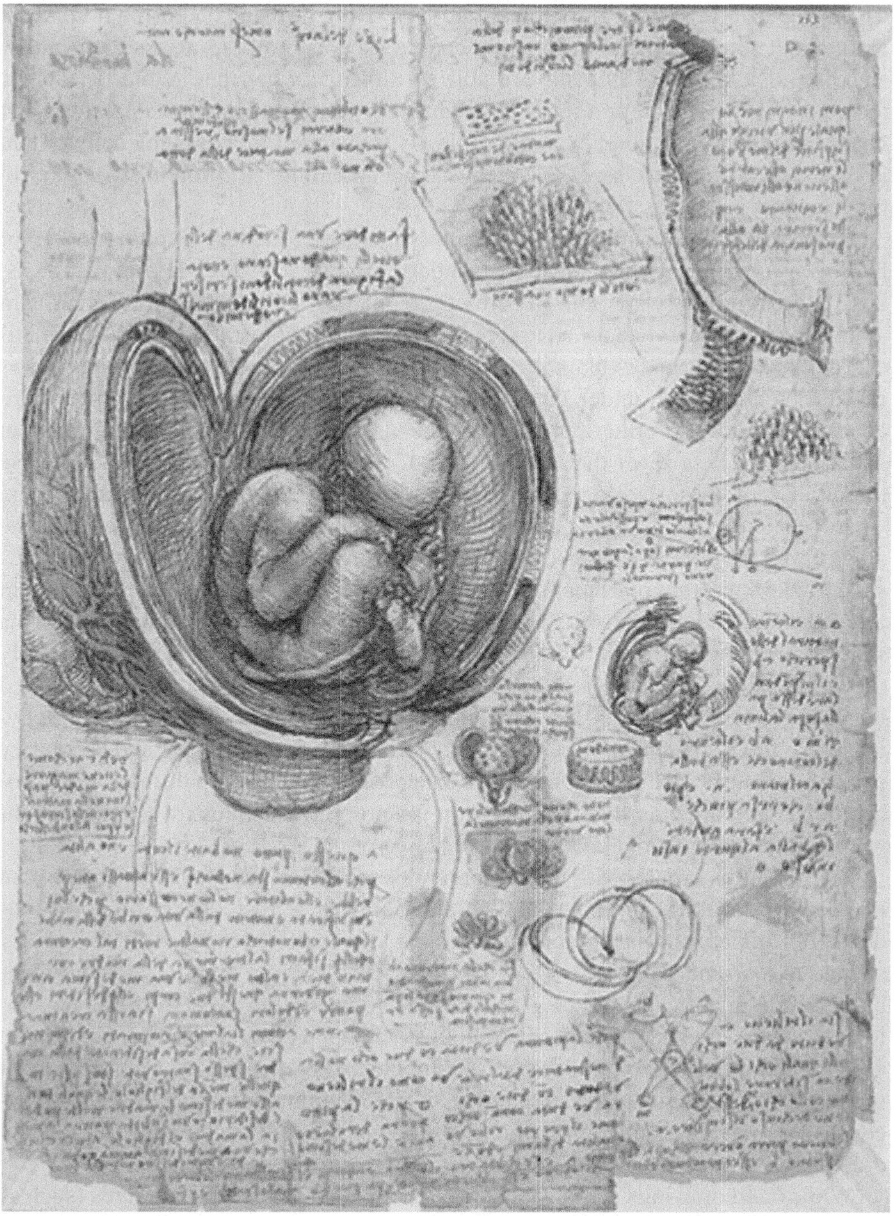

Fig. 23.1 ***Fetus in the Womb*** by Leonardo da Vinci in 1511. Image is in the public domain in the United States

This analysis allows educators to explore how using drawings helps their students learn about science and how drawings can be used as a tool to assess their understanding of a concept. Students are able to illustrate what they retained, and the instructor has a visual representation of what may have been misunderstood or what needs clarification. A good example of this is showcased in a drawing activity that is used as a tool to assess student's understanding of protein structure in biochemistry.

Harle and Towns conducted a study investigating 21 students' understanding of hydrogen-bonding in primary and secondary protein structures (Harle & Towns, 2013). The students were able to correctly recognize and identify primary protein structures; however when asked to draw secondary protein structures for alpha helices and beta sheets and explain the role of hydrogen bonding, their drawings showcased a breakdown in understanding the role of hydrogen bonding in these more complex structures. Furthermore, when students were asked to draw a portion of an alpha helix or beta sheet that included the hydrogen bonds-while articulating the importance of these hydrogen bonds, the results revealed a disconnect between what the students were saying and what they were drawing. For the drawings of the alpha helix, the drawings ranged from no hydrogen bonds shown to hydrogen bonds shown with incorrect placement to hydrogen bonds shown with correct placement and atoms involved (Harle & Towns, 2013). A similar result was seen for the beta sheet as well, ranging from strands without hydrogen bond interaction and correctly drawn with two strands of hydrogen bonding interactions and atoms involved labeled (Harle & Towns, 2013). The authors concluded from this study that drawings gave them insight into the student's understanding of primary and secondary protein structure. The student's lack of coherence between verbally saying hydrogen bonds and drawing them correctly on the protein structures revealed a disconnection in their fundamental knowledge.

Based on this article, with the visual representation of drawings, instructors can identify a breakdown in understanding and reintroduce the concept on terms students can understand. This approach allows students and instructors to actively engage in learning while also allowing instructors to provide more directed feedback. While historically physical drawings have been effective, given the technological advancements of today, generations Z and Alpha may also be accustomed to drawing through sketch technology (i.e computer programs such as Microsoft Paint or electronic drawing tablets). Regardless of the medium, one could expect the outcome to be the same since in both cases students are actively engaged in manual drawing.

While hand drawings are advantageous for smaller molecular structures (such as single units of proteins, nucleic acids, carbohydrates and lipids), larger and more complex systems in biochemistry (such as macromolecules and metabolic pathways) can be more challenging. To overcome this challenge, drawing or sketch technology can still be used as an effective avenue to help aid in a student's understanding of a biochemical concept. Becker and Rojas developed an algorithm for drawing cyclic or partially cyclic pathways to represent a combination of complex pathways such as glycolysis, the citric acid cycle and the urea cycle (Becker &

Rojas, 2001). Currently, many biochemistry educators have their students draw (or construct) their own metabolic pathways in attempts to help the students draw a parallel between the different metabolic pathways.

In summary, drawings are a useful tool for biochemistry education as a means to improve student learning and represent challenging concepts (Krajcik & Sutherland, 2010). Educators can use hand drawings as a means for quick feedback (Forbus et al., 2011) or a combination of drawing and technology to assess how students are seeing and understanding complex visualization environments (Zhang, 2010). Furthermore, as technology advances, technology will play a big role in the broadening of the concept of drawing. Even in the growing field of educators using drawings to help students learn science (particularly in biochemistry (Digby, 2017)) there are still areas of research that should be explored. One of the more active areas is the exploration of how combining drawing and new technologies can benefit education. In fact, Harle and Towns mention researchers should consider using the LiveScribe Smartpen, a tool used for collecting data that synchronizes student audio with drawings (Harle & Towns, 2012, 2013; Livescirbe). This device can greatly benefit those designing research and collecting data, especially from an assessment standpoint. Additional foundational research involving drawing in biochemistry should also be explored to include topics such as these:

- Is there a best drawing platform (free draw, computer programs, etc.)?
- Does the complexity of some topics make drawing counterproductive?
- What factors (gender, race, location) effect student's ability to learn from drawing or communicate their ideas?

These three research questions focus on the competencies of *Identify* (help students better understand a molecule and the molecule in a pathway which could in turn lead them to *Identify* a potential research question) and *Communicate* (increases the ability and capacity for students to discuss biochemistry to others, including in layman's term). Exploring these questions could aid in the development of better communication between students and the instructor, as well as recognize gaps where miscommunication commonly occurs. Drawing is a continually growing resource used for biochemistry education and should be considered as an artistic avenue in biochemistry courses.

23.3 Painting

The interdisciplinary activity of painting and scientific learning is one that is highly used in all levels of education. While painting is much more time consuming than drawing, painting allows for a more versatile range of expression. While typical artists like to use easels and paints as their tools, in the realm of biology, a recent implementation involves painting agar with bacteria, this is known as Agar Art (Fig. 23.2).

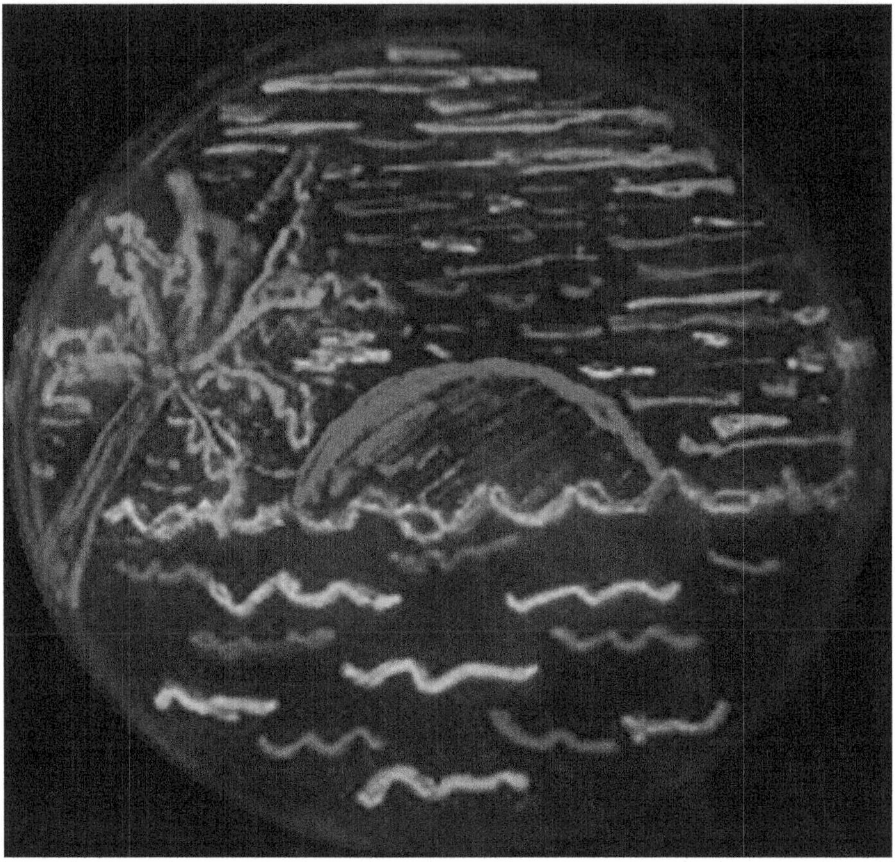

Fig. 23.2 Picture of Agar Art created by Nathan Shaner using fluorescent proteins (Cressey, 2016). Image is in the public domain in the United States

Agar Art allows students to create "living" works of art on agar plates by "painting" with E. coli that express visible or fluorescent proteins of various colors. Painting with bacteria allows students to learn the nuances of gene expression (when a fluorescent proteins had been inserted into a gene of interest so that the presence and level of the fluorescence gives a readout of the activation of that gene) while also engaging the students in a fun, expressive, and creative activity. In fact, it is so popular there is a worldwide annual Agar Art competition hosted by the American Society for Microbiology (Art). Agar Art is easily executed in primary and secondary education as a fun activity to get students interested in the sciences while introducing them to the basics of bacteria and gene expression; such is seen in Wu's activity for DNA and gene expression for grades 8–12 (Wu et al., 2018). However, the ability to use this activity as an educational tool to discuss some of the more advanced concepts, such as seen in biochemistry, is much more of a challenge. While there are no specific examples posed yet for biochemistry courses, adaptation from activities used in other biology courses described below could help create an avenue to use paintings in biochemistry teachings.

One example is seen by Morris et al. where they describe the use of a Course-based Undergraduate Research Experience (CURE) in an introductory undergraduate microbiology class (Adkins et al., 2017). Their CURE allows students to develop their own research question based on the observations they made during the production of their bioart. The students began the experience by isolating bacteria from a soil at a nearby park. They then used the genetic and biochemical methods they learned to grow and isolate select colonies based on the characteristics they hypothesized would provide colorful artwork for the agar. Students then explored their creativity by designing their own templates to use for the paintings. They painted their work on two different types of media to study how the nutritional quality of their agar affected the growth of their bacteria art. The independent research of this activity began when students had to form a hypothesis of the ecological relationship between the bacteria and the respective media that could be tested using standard laboratory techniques, creating an opportunity for the students to participate in active learning. While this is a semester long project, parts of this could be used to use as stand-alone activities in lab courses.

Due to the time constraints of painting activities, there are few examples of how to incorporate actual painting into biochemistry lecture. However, one idea can be borrowed from Dolev et al. (Dolev et al., 2001) who utilized paintings as a teaching tool for medical students at Yale University School of Medicine. The purpose of this study was to introduce an avenue within medical students' education that allowed them to develop human interactions skills necessary to be efficient and respectful communicators to their future patients. The students were separated into two groups: (1) the control and (2) an art intervention group. While the control group of medical students continued with traditional clinical tutorials, the intervention group traveled to an art center and was asked to study a painting for 10 min, then describe to four of their peers the details of what they saw. When all students were asked to describe a photo of someone with a medical disorder in 3 min with as much detail as possible, those in the intervention group scored significantly higher than the control group in their observatory skill by being able to give greater detail and depth to their descriptions. As a result, the medical students who participated in the intervention group could be better communicators with their patients when discussing the factors associated with their health. This study suggests it may be beneficial to have students describe their interpretation of biological structures as a way to participate in active learning of the material. While instructors may not always have the luxury of taking students to art centers or museums, instructors could show pieces of art in the class and have students do the same activity.

Painting and biochemistry education, while developing, is still largely underexplored. As of now, the only easily implemented expression of painting relatable to biochemistry is using Agar Art as a way to supplement teaching about bacterial transformation and gene expression. The current delivery of biochemistry content makes it difficult to commit the amount of time needed to complete a painting activity outside of the lab portion of the course. Experiments such as CUREs or pure research projects can be incorporated into biochemistry lab courses to allow the fusion of painting and biochemistry to take place. Additionally, entire curriculums can be created around this theme as seen by Dr. Cole who created a "Biology Through Art" course, where students create their own art in a biology laboratory course (Dybas, 2018).

Further research for painting in biochemistry could include:

- What other ways can we use paintings to help students ask research questions?
- Could having students explain pictures in lecture lead to better communication/understanding?
- What other painting mediums can be used outside of Agar and bacteria?

These questions focus on the competencies of *Analyze*, *Question*, and *Plan*. Just like drawing, painting or explaining pictures could help students *Identify* research questions. These research proposals will aid in the development of new designs and techniques that may allow painting to span from being used in lecture courses all the way to entire courses being built around this artistic technique. The versatility of paintings allow them to be utilized in various levels of STEM education.

23.4 Music

Educational curriculum is currently structured to initially develop student's minds and learning abilities through music (such as the alphabet song); however, music is uncommonly used in high school and college-level courses, likely due to the perception that educational songs are only tools for rote memorization. Many of the songs found today in the realm of biochemistry are used as an aid to memorize specific metabolic pathways. Some of the earliest songs associated with biochemistry come from Dr. Weiss in 1971 dealing with protein synthesis using a "free love" theme and from the book, "A Biochemists' Songbook," by Harold Baurm, originally released in 1982. The songs in "A Biochemists' Songbook" cover an array of biochemistry topics such as "Fatty Acid Biosynthesis", "The Pentose Phosphate Shunt" and "We're Here Because Urea". The songs of this edition and Baum's second edition (released in 1995), are publicly available online and are still being used in classrooms today. Music videos by Glen Wolkenfield (YouTube) on metabolism and other biological process are also highly popular and a common source used in aiding science learning.

In efforts to take further advantage of what music can offer to biochemistry educators, instructors usually turn to music as a way to help students feel relaxed, more engaged and welcomed in unsettling environments (Crowther, 2012). Students are more inclined to learn in environments where they feel comfortable and represented (Rainey et al., 2018; Stump et al., 2011). Traditionally STEM has been dominated by white men (O'Dea et al., 2018; Rainey et al., 2018); however, as the field grows with more scholars of color and women, the infusion of culture can help add clarity to the concepts that were previously only created for the dominant group. Music is extremely popular and important to cultures around the world (Franklin, 2016; Herndon & McLeod, 1981; Long et al., 2017). Furthermore, music is a relatively easy way to give students the liberty to explore concepts on their own cultural terms or preferences; some of the most popular tunes can be used to help students, regardless of color and gender, find familiarity in elaborate concepts.

Heineman (2018) has used songs as a comedic tool to review material recently covered in class. Cleverly changing the words of the chorus from a popular Adele song "Hello", Heineman wrote lyrics to the song, which expound the "complicated" relationship between cancer cells and healthy cells. The chorus of the song entitled "Hello, I've Metastasized" is as follows:

Hello from the other side
I've divided a thousand times
But when you call and say that I've gotta slow down
I don't listen 'cause my genes have had a breakdown.

Hello, I've metastasized
Now I refuse to suicide
I'm immortal and I grow in other tissue
I bring in new blood and I'll happily grow over you, I don't care....

The students enjoyed hearing their instructor sing songs and gave it positive reviews; over 90% of the students stated the songs made them more attentive, engaged in the context, and when incorporated into lectures helped them to refocus their attention on the matter at hand (Heineman, 2017). This example showcases a real-time opportunity where science can be fun, especially when the songs are correlated to popular songs students are familiar with. Some instructors have encouraged students to sing along in order to reinforce the concepts, others have even created entire assignments around the idea-allowing students to create their own songs using the biochemical concepts they know.

Crowther and Davis (2013) developed a sing-along activity for the subject of protein structures which they call "Amino Acid Jazz". Comparable to the way music is composed using the letters A–G, Crowther and Davis used the letter abbreviations of the 20 amino acids to create a new musical scale, which in turn is used to build musical polypeptides. For amino acids that raise the pH, these are sung in a higher octave, while those that lower the pH are sung at a lower octave. This activity allows students to reinforce their knowledge of the amino acids because their names are the lyrics to the song and the characteristics effect their position on the musical scale. Additionally, the same way changing an amino acid in a polypeptide can affect the way the protein functions, changing the amino acid in the song can change the entire melody. These correlations stimulate metacognition, thus helping students develop a deeper connection with the material. General reviews from students and teachers showed positive results with over half of the students seeing this as an enjoyable exercise.

After an in-depth look on the interdisciplinary approach of using music in science education, Crowther (2012) identified some of the challenges, to include:

- Comfortability of the student and instructor with musical composition
- Being vulnerable enough to sing off-key if not musically savvy
- Finding a music genre all the students would like

A way to overcome these challenges may be to incorporate some of the method of Heineman (2017, 2018) by using popular songs or giving students the freedom to incorporate their own preferred musical choice into the assignment. Also, students that are shy could record their submissions before class or as an added exercise, the class could sing each students submission.

A common theme seen in these examples is that musical exercises tend to be utilized after the information has been taught in a traditional lecture setting. It is important to note that music should be used as a supplementary tool and not used to replace the teaching of content (Crowther & Davis, 2013). Of the different artistic teaching tools discussed in this section, creating music to integrate into traditional biochemistry courses could be the most difficult, however, as showcased in the examples above it can be very rewarding.

Music can be further incorporated into biochemistry by exploring the following research questions:

- Do students grasp concepts better by writing their own songs?
- Which topics can benefit most from music?

These research interest focus on the competency of *Questions*. Gaining more background information and insight into the use of music in biological science will create a platform to further expand when and where to use music in teaching so that it is beneficial to all.

23.5 Media

In this digital age, media (video, animation, and film) is the most connected art form to our technology driven world. Students have greater access to creating media, ranging from formal oral presentations to full on theatrical play performance. Using media as an artistic art form extends the unique ability for students to showcase their technical abilities, expand their creative thinking, and gives them an opportunity to convey their own cultural and personal uniqueness; all of these benefits encourage a higher level of personal investment in student's work, which leads to greater success (Adkins et al., 2017).

Bruna (2013) performed a case study called Art and Biochemistry to motivate first-year veterinary medicine students to learn biochemical concepts. The course was taught in three main sections (molecular biology, protein structure and enzymology, and metabolism), students formed small teams and randomly selected the section they would present for their project. The various media representations covered several biochemical concepts, as shown in Table 23.1.

Table 23.1 Examples of media representations and corresponding biological concepts created by students from the study Art and Biochemistry

Media representation	Biochemical concept
Animated cartoon	Composition of straight and curly hair
Animated cartoon	Oxidative phosphorylation
Animation using clay figurine	Eukaryotic replication
Educational video	Effect of the brown recluse spider
Educational video	The cloning of Dolly
Silent film	Beer and nutrition
Video interview	Protein synthesis

Bruna (2013)

Students stated the project motivated them, helped them understand the concepts better, and should be included in future versions of the course. However, one of the drawbacks of this project was the grading system, some students felt bias because the evaluation of their work was based on their peers and a panel of other professors within the university. It is important to note that while creativity opens the door for a wider range of expression, assigning unbiased grades can be challenging. Some ways to address this challenge is to methodically assign points for portions of the project including creativity and amount of effort or labor students needed to complete the project. Another option is to create a panel of unbiased individuals from various academic backgrounds to assess all student's work. In addition, as students are challenged to think in more creative manners, instructors will also be challenged to think broadly about concepts and outside the realm they have been traditionally applied. As we encourage more diverse populations of students to enter STEM, we have to remember to be inclusive of their perspectives and input.

Now more than ever, with the events of COVID-19 forcing many instructors to teach courses remotely, diverse presentations of concepts could help students tangibly grasp content in a digital environment. Media is likely a viable option for alternative testing and other traditional in-person assignments. More so, students can create their own media based in any of the other forms of expression discussed in this chapter. Through animation of paintings or drawings in conjunction with music, students have the opportunity to capitalize on artistic expression and create resources to help them, or others, learn and study.

Research involving media in biochemistry should also be explored to include topics such as:

- What is the best way to assess the data of student's understanding upon the conclusion of these activities?
- Is there a common artistic media form students prefer for biochemistry?
- Can these avenues be used to teach biochemistry to non-majors or the lay audience?

These research questions focus on the competency of <u>*Communicate*</u>. Exploring these questions could aid in the development of better communication between scientist and non-scientist (whether it be in academia or to the general public), as well as identify the best ways to assess the effectiveness of these types of activities on student understanding/learning of concepts. Through more and diverse levels of understanding, the general uneducated public could understand complex concepts of biochemistry that will improve their everyday life.

23.6 Conclusion and Implications for Instructors

It is certain the combination of art and biochemistry education helps students relate to the content and reduce monotony of standard oral lectures while increasing the motivation for students to want to learn biochemistry. Considering the upcoming generations and their familiarity with technology, developments and changes in the standard way of teaching need to be adjusted to meet the demands and expectations of learning. Drawing can be used as a form of assessment and is a quick and easy way to incorporate artistic representation into a lecture. Paintings, while not highly explored in the realm of biochemistry education, could be a nice inclusion for CUREs, research projects, or as a topic for major or elective curriculum course. Music should be used as a fun way to reinforce the information and keep students engaged. Media, the one format most likely to be explored in the future, is a great way to motivate students to integrate learning and engage in a manner they feel most comfortable. Finally, as we include more avenues of learning for students, we should include more avenues of teaching for instructors. Exploration into the research questions posed for the different artistic avenues under the direction of the seven ACE-Bio Experimentation Competencies (Pelaez et al., 2017) provides an opportunity to develop novel approaches to the incorporation of art in biochemistry and other biological science courses. This chapter serves as a guide to help instructors create more inclusive content to reach the more diverse students we serve/target using artistic representation as the vehicle.

To assist instructors who are new to this area, we offer the following guidelines for teaching:

- Look for opportunities within the curriculum to utilize artistic representation in teaching. Drawing, painting, music, and media representation are a good start.
- Creative projects involving artistic exploration can be more reliable to assess students' knowledge when compared to traditional testing. Including opportunities for creative projects and active learning also broadens the appeal of the field to students who traditionally have been excluded.
- We encourage instructors and students to tie in current topics and social trends in the classroom, when applicable, to increase student engagement and content retention.
- Be inclusive of student creativity as a tool of education. The projects students submit could be useful to help educate future students of the course or inspire research questions for further exploration.

- We suggest instructors be creative in their assignment creation and grading criteria. When grading, keep in mind the various backgrounds and learning styles of students in the classroom; there is more than one-way to understand a concept.

Acknowledgments We would like to acknowledge and thank the US Air Force Office of Scientific Research and Collin College. This chapter has received public approvial by the US Air Force Office of Scientific Research under PA#: USAFA-DF-2021-83. We also thank Stephanie Gardner and Jenean O'Brien for their helpful discussions and edits of the manuscript.

References

Adkins, S. J., Rock, R. K., & Morris, J. J. (2017). Interdisciplinary STEM education reform: Dishing out art in a microbiology laboratory. *FEMS Microbiology Letters, 365*(1), 1–8.

Ainsworth, S., Prain, V., & Tytler, R. (2011). Drawing to learn in science. *Science, 333*(6046), 1096–1097.

Art, A. Retrieved from https://asm.org/Events/2019-ASM-Agar-Art-Contest/Home

Atalay, B. (2011). *Math and the Mona Lisa: The art and science of Leonardo da Vinci*. Smithsonian Institution.

Baurm. Retrieved from https://web.csulb.edu/~cohlberg/songbook.html

Becker, M. Y., & Rojas, I. (2001). A graph layout algorithm for drawing metabolic pathways. *Bioinformatics, 17*(5), 461–467.

Bruna, C. (2013). Motivating active learning of biochemistry through artistic representation of scientific concepts. *Journal of Biological Education, 47*(1), 46–51.

Colucci-Gray, L., Burnard, P., Cooke, C., Davies, R., Gray, D., & Trowsdale, J. (2017). Reviewing the potential and challenges of developing STEAM education through creative pedagogies for 21st learning: How can school curricula be broadened towards a more responsive, dynamic, and inclusive form of education? *BERA*. Retrieved from http://oro.open.ac.uk/62156/

Cressey, D. (2016). Roger Tsien's legacy: The creations that lit up biology. *Nature*. Retrieved from https://doi.org/10.1038/nature.2016.20532

Crowther, G. (2012). Using science songs to enhance learning: An interdisciplinary approach. *CBE—Life Sciences Education, 11*(1), 26–30.

Crowther, G. J., & Davis, K. (2013). Amino acid jazz: Amplifying biochemistry concepts with content-rich music. *Journal of Chemical Education, 90*(11), 1479–1483.

Digby, A. K. (2017). Using drawing-to-learn, argumentation, and 3-D physical models to study non-covalent. *Interactions for Biochemistry Education, 47*. Retrieved from http://www.lib.ncsu.edu/resolver/1840.20/34755

Dolev, J. C., Friedlaender, L. K., & Braverman, I. M. (2001). Use of fine art to enhance visual diagnostic skills. *JAMA, 286*(9), 1020–1021.

Dybas, C. L. (2018). Meeting of the minds: Biology and art find common ground. *Bioscience, 68*(12), 933–939.

Forbus, K., Usher, J., Lovett, A., Lockwood, K., & Wetzel, J. (2011). CogSketch: Sketch understanding for cognitive science research and for education. *Topics in Cognitive Science, 3*(4), 648–666.

Franklin, M. I. (2016). *Resounding international relations: On music, culture, and politics*. Palgrave Macmillan/Springer Nature.

Freeman, S., Eddy, S. L., McDonough, M., Smith, M. K., Okoroafor, N., Jordt, H., & Wenderoth, M. P. (2014). Active learning increases student performance in science, engineering, and mathematics. *Proceedings of the National Academy of Sciences, 111*(23), 8410–8415.

Hanauer, D. I., Graham, M. J., & Hatfull, G. F. (2016). A measure of college student persistence in the sciences (PITS). *CBE—Life Sciences Education, 15*(4), ar54.

Harle, M., & Towns, M. H. (2012). Students' understanding of external representations of the potassium ion channel protein part II: Structure–function relationships and fragmented knowledge. *Biochemistry and Molecular Biology Education, 40*(6), 357–363.

Harle, M., & Towns, M. H. (2013). Students' understanding of primary and secondary protein structure: Drawing secondary protein structure reveals student understanding better than simple recognition of structures. *Biochemistry and Molecular Biology Education, 41*(6), 369–376.

Heineman, R. H. (2017). Engaging college students by singing the science. *Journal of Microbiology & Biology Education, 18*(1).

Heineman, R. H. (2018). Songs about cancer, gene expression, and the biochemistry of photosynthesis. *Biochemistry and Molecular Biology Education, 46*(1), 98–99.

Herndon, M., & McLeod, N. (1981). *Music as culture*. Norwood. Retrieved from https://www.hugoribeiro.com.br/biblioteca-digital/Herndon_Mcleod-Music_a_culture.pdf

Krajcik, J. S., & Sutherland, L. M. (2010). Supporting students in developing literacy in science. *Science, 328*(5977), 456–459.

Livescribe. Retrieved from http://www.livescribe.com/en-us/

Long, P., Baker, S., Istvandity, L., & Collins, J. (2017). A labour of love: The affective archives of popular music culture. *Archives and Records, 38*(1), 61–79.

O'Dea, R. E., Lagisz, M., Jennions, M. D., & Nakagawa, S. (2018). Gender differences in individual variation in academic grades fail to fit expected patterns for STEM. *Nature Communications, 9*(1), 1–8.

Pelaez N, Anderson TR, Gardner SM, Yin Y, Abraham JK Bartlett EL, Gormally C, Hill JP, Hoover M, Hurney CA, Long TM, Newman DL, Sirum K, Stevens MT (2017). The basic competencies of biological experimentation: Concept-skill statements. *PIBERG Instructional Innovation Material Paper 4*. https://docs.lib.purdue.edu/pibergiim/4/. Accessed 8 Nov 2021.

Rainey, K., Dancy, M., Mickelson, R., Stearns, E., & Moller, S. (2018). Race and gender differences in how sense of belonging influences decisions to major in STEM. *International Journal of STEM Education, 5*(1), 10.

Stump, G. S., Hilpert, J. C., Husman, J., Chung, W. t., & Kim, W. (2011). Collaborative learning in engineering students: Gender and achievement. *Journal of Engineering Education, 100*(3), 475–497.

Valenti, S., Masnick, A., Cox, B., & Osman, C. (2016). Adolescents' and emerging adults' implicit attitudes about STEM careers: "Science is not creative". *Science Education International, 27*(1), 40–58.

Vella, F. (1990). Difficulties in learning and teaching of biochemistry. *Biochemical Education, 18*(1), 6–8. https://doi.org/10.1016/0307-4412(90)90004-8

Weiss. Retrieved from https://www.youtube.com/watch?v=u9dhO0iCLww

Wiggins, B. L., Eddy, S. L., Wener-Fligner, L., Freisem, K., Grunspan, D. Z., Theobald, E. J., & Crowe, A. J. (2017). ASPECT: A survey to assess student perspective of engagement in an active-learning classroom. *CBE—Life Sciences Education, 16*(2), ar32.

Wolkenfield, G. Retrieved from https://www.youtube.com/user/sciencemusicvideos

Wood, E. (1990). Biochemistry is a difficult subject for both student and teacher. *Biochemical Education, 18*(4), 170–172.

Wu, R., Brinkema, C., Peterson, M., Waltzer, A., & Chowning, J. (2018). STEAM connections: Painting with bacteria. *The American Biology Teacher, 80*(4), 305–307.

Zhang, H. Z. (2010). Exploring drawing and critique to enhance learning from visualizations. *Proceedings of the 9th International Conference of the Learning Sciences*, ICLS 2010, Chicago, IL, USA, June 29 – July 2, 2010, Volume 2.

Chapter 24
Strategies for Targeting the Learning of Complex Skills Like Experimentation to Different Student Levels: The Intermediate Constraint Hypothesis

Eli Meir

24.1 Introduction

Students enter biology classes at different skill levels, and this affects how prepared they are to understand difficult concepts or use complex skills, such as the experimental process. We know students learn better if they are actively engaged in their learning through student-centered activities, whether that be through worksheets and other questions, exploring through open-ended wet labs or simulations, discussions, self-guided research, or many other modes of activity (Freeman et al., 2014). But an activity that works well for one student or population of students will be too easy or too difficult for other populations. Instructors must write and/or choose activities at an appropriate level for their students in order to keep the students learning optimally. This can be thought of in terms of Vygotsky's idea of keeping a student within their zone of proximal development (discussed, for instance, in Chaiklin, 2003) – that degree of challenge where the student is pushed to develop to a new level of understanding, but not challenged so much that they are lost. More simply, I think of maximizing the efficiency with which a student learns new material.

One way to tune activities to different student populations is through changing how the activity is presented to the student, often called scaffolding. A number of studies have examined the effects of differing amounts of scaffolding on learning. Under the name levels of inquiry, several papers (e.g. Bell et al., 2005; Buck et al., 2008; Weaver et al., 2008) have tried to classify degrees of scaffolding or amount of student-directed inquiry: from highly scaffolded activities that provide students very directive instructions including the expected result ("Confirmation inquiry"); providing a question but leaving the student to design the experiment to answer it

E. Meir (✉)
SimBiotic Software, Inc, Missoula, MT, USA
e-mail: eli@simbio.com

N. J. Pelaez et al. (eds.), *Trends in Teaching Experimentation in the Life Sciences*, Contributions from Biology Education Research,
https://doi.org/10.1007/978-3-030-98592-9_24

("Guided inquiry"); and at the most constructivist end asking students to both think of a question and answer it themselves ("Authentic inquiry") with other levels in-between. Many studies have investigated both whether inquiry itself improves learning, and what level of inquiry is best, at different ages. While there is a consensus within the college science education community that some amount of inquiry, or student-directed learning, is beneficial (for instance Freeman et al., 2014), there is much debate over when inquiry is beneficial and how highly scaffolded activities should be to aid learning (e.g. Jerrim et al., 2020 and many others).

Thinking in terms of levels of inquiry/scaffolding, though, ignores related but separate axes along which instructional activities can vary. In authoring simulation-based and other interactive exercises for teaching biology, my colleagues and I consider a trinity of intertwined strategies for adjusting the level of an activity to maximize learning in a given student population. In addition to scaffolding, we also consider when and how to provide feedback, and the degree of constraint on an activity (Table 24.1). Feedback means giving students some guidance DURING and/or AFTER they have done some activity, based on their actions, as opposed to before (guidance beforehand is scaffolding). Constraint, as described further below, refers to the amount of freedom to make choices that is provided to the student by the environment in which the activity takes place.

The thesis of this chapter is that thinking explicitly and even quantitatively about scaffolding, feedback, and constraint may provide practical strategies to tune learning activities. I'll consider these strategies in the context of teaching and measuring skills involved in the experimental process, as delineated by the ACE-Bio framework (Pelaez et al., 2017). The framework details the muti-layered and complex nature of successfully doing experiments, making it clearer why these are such difficult skills for students to learn. But the framework, on its own, does not provide strategies for how to teach these types of skills.

This chapter complements the many specific demonstrations throughout the book by contributing a more general framework for designing tools to aid students learning to do their own experiments. Constraint is an often overlooked but potentially easily measured and manipulated axis for tuning activities. I propose the "Intermediate Constraint Hypothesis" – that considering and tuning the degree of constraint in teaching tools for difficult skills like experimentation, especially within an intermediate range of constraint, may be a useful approach in designing activities for particular students or populations. To reach that hypothesis, I will first describe what I mean by constraint and provide a number of examples of using

Table 24.1 Three axes along which educational activities can vary

Axis type	Description
Scaffolding	Instruction or other guidance prior to a student conducting an activity, and/or choosing activities for a student based on students prior work
Feedback	Guidance during and after a student completes an activity based on their actions
Constraint	Limits on the meaningful choices a student has in the environment where they are conducting an activity

constraint, feedback, and scaffolding to adjust educational activities. I'll then give some data on the effect of constraint and feedback on learning, and finish with a description of the intermediate constraint hypothesis and how it might be of use in both learning and assessment activities.

24.2 What Is Constraint?

Intuitively, one way to adjust the level of teaching activities is to provide fewer or more choices within the activities given to students. A novice on a skill may do best with an activity that has just a few, basic choices. This may help the student focus on the most relevant concepts and ideas. By contrast, a more advanced practitioner of that same skill may learn best when challenged with a more open activity with more choices available, exercising a larger set of practices within the skill. We can consider the number of choices given to a student as a level of constraint on the activity. A highly constrained activity has few choices for the student, while a lightly constrained activity has many to infinite numbers of choices. This is not the same thing as scaffolding. An activity can be highly scaffolded – i.e. the student is told very explicitly what to do – whether the environment within which the activity takes place is highly constrained (few choices available) or lightly constrained (many choices available). Similarly, in principal the feedback given to students can be specified independently of the level of scaffolding as well as the level of constraint. One could imagine a highly scaffolded, low constraint activity with no feedback, or a lightly scaffolded, high constraint activity with lots of specific feedback, or any other point within that three-dimensional space.

Scalise and Gifford (2006) discuss this in the context of questions. A multiple-choice question is highly constrained, while an essay question is very lightly constrained. Other question types can fall in between (Fig. 24.1). For instance, my colleagues and I have explored intermediate constraint question types that we call LabLibs (similar formats have been used by others, e.g., Biswas et al., 2016) and WordBytes (Kim et al., 2017; Meir et al., 2019). One place we used a LabLibs question is to elicit a hypothesis and prediction (Pelaez et al., 2017: ACE-Bio competency area of *Plan*) in a lab on predator-prey dynamics (Isle Royale, Meir et al., 2009). The students are asked to consider how predators may affect the health of their prey as paraphrased below:

> Construct a hypothesis for how a predator might affect prey health and predict the pattern you'll see if your hypothesis is correct. There is more than one good hypothesis-prediction combination.
>
> Answer: I hypothesize that prey will be **[more | less | equally]** healthy when predators are present because **[with predation there will be fewer prey and each prey will have more access to food | avoiding predation takes time away from eating | the carrying capacity for prey will be lower, so food availability will change]**. If the hypothesis is true, I predict prey will have **[higher | lower | about the same]** health with and without predators.

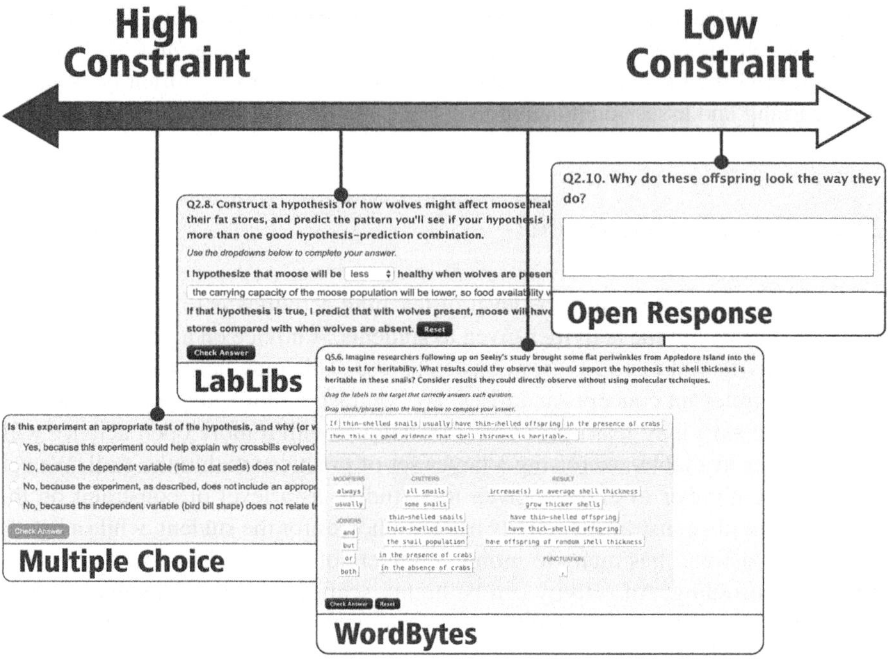

Fig. 24.1 Examples of question types along the axis of constraint. LabLibs and WordBytes are both intermediate constraint questions, with multiple choice representing a high constraint question and short-answer a lower-constraint question

The items within [brackets] are choices that the student must make from pull-down menus to fill in blanks in the paragraph. In total, there are 27 ways to answer the question, several of which are reasonable hypothesis/prediction combinations. The student is thus constructing their own answer to some extent as they have many more choices, or less constraint, than with a typical 4 option multiple choice question. But the constraint is still much higher than if the student was provided a blank text box in which to write their answer with their own words.

A WordByte is an example of reducing constraint even further. Figure 24.1 shows a WordByte asking students to design an experiment to test for heritability of shell thickness in snails. Students construct a sentence by dragging pre-written tiles with words or phrases into an answer area. Here students have a very large, but not infinite, number of possible choices of response, and in practice we see hundreds of different responses to these questions among students (Kim et al., 2017). As Scalise and Gifford (2006) outlined, one could place many other question types along this axis of constraint – drag-n-drop labeling questions might fall near the LabLibs side of the constraint axis for instance. While particularly useful for questions asked through a computer, questions asked in other environments, including paper and pencil, can also be thought of in terms of degree of constraint.

Constraint is not only set by the question type, but also by the number of choices given the student within that question. One can write a LabLibs, for instance, with a single blank to fill in the answer and four choices for that blank. This would be no different in constraint than a multiple-choice question. Conversely, one could add additional options to the predator-prey LabLibs above that would reduce the constraint (increase the available choices), assuming the options were written based on common student responses to such questions or to capture known student confusions as is common practice for creating the options in multiple choice assessments (e.g., concept inventories). Thus, the total constraint for a given question is a function of both the type of question and the number of possible answers.

24.3 Constrained Simulated Experiments

While the amount of choice vs. constraint is easy to see in question formats, that same axis also exists in many other activities. In simulation-based labs I've helped write, we constantly make decisions about how much of a simulation to expose to a student and how much freedom of choice to give for each exposed parameter. As an example, we have two versions of a lab on action potentials, one more in-depth than the other (Action Potentials Explored/Extended: Kim & Meir, 2015; see also Meir, 2022), both of which are effective at helping students learn core neurophysiology concepts (Cerchiaria et al., 2019). Both versions include an exercise where students play with membrane channel dynamics to re-create an action potential. In the shorter Explored version, we provide binary controls to open or close both Na^+ and K^+ channels (Fig. 24.2 Explored). In the Extended version (Fig. 24.2 Extended), we provide sliders for opening and closing Na^+ and K^+ channels (more choices of values) as well as exposing Na^+ and K^+ concentrations inside and outside the membrane (more parameters). The additional parameters and value choices reduce the constraints on students using the Extended version compared to Explored.

While the Extended version has less constraint than Explored, both would count as having intermediate levels of constraint on a constraint axis for simulation-based learning activities, similar, perhaps, to a LabLibs or WordBytes amongst different question types. The Explored version only has four possible parameter combinations (each channel open or closed) but because students can choose the timing of opening and closing, the space of possible manipulations is larger.

HHMI has a very different neurobiology simulation that, while complex in its own way, might be considered a higher constraint activity (HHMI, 1997). Their simulation tries to hue more closely to actual experimental techniques – students are presented with a portion of a nervous system from a leech and have to identify types of cells based on their response to three different stimuli, as well as the shapes of the neurons. Students can pick from a couple dozen neurons to probe, but for each neuron the choices in manipulation are limited to three possible stimuli, akin to a multiple-choice question. The complexity in this simulation is in the data interpretation rather than the experimental manipulations. Simulations that try to recreate the

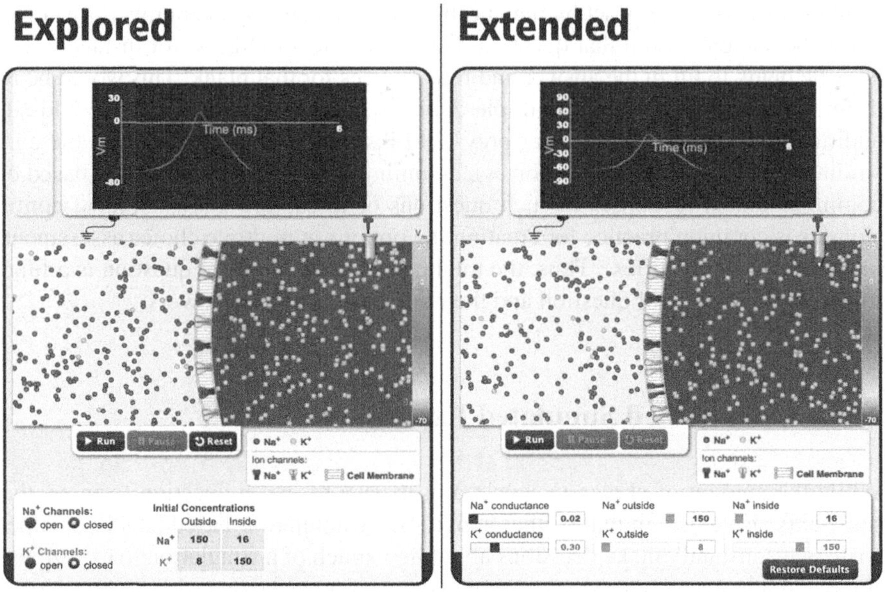

Fig. 24.2 Changing constraint in a simulation as displayed in Action Potentials Explored versus Extended simulation-based labs. Note the sliders for Na^+ and K^+ channel density in Extended compared to the binary option in Explored, as well as the additional concentration parameters exposed in Extended

look and feel of real experiments tend to be higher constraint, as showing realistic images and manipulations is difficult to do in a more open-ended simulation. This is not always the case, though, as shown in the chapter in this volume on a virtual microscope simulator, which includes most of the controls found in a real microscope (Casali et al., Chap. 14).

By contrast, a low-constraint activity for teaching about action potentials could be designed in programs such as StarLogo (Klopfer et al., 2009) and similar modeling environments where students can build their own simulations. Rather than providing students with a few parameters to manipulate in a premade simulation of diffusion, a lab in StarLogo might provide them with programmatic building blocks to place molecules and barriers, provide each molecule with a velocity, detect and respond to collisions, etc. Students would then need to build their own simulation of molecular diffusion to understand, or show their understanding, of electrochemical gradients. By analogy to question types this might be the simulation-based activity equivalent of an essay question.

To show how this idea of constraint intersects with feedback and scaffolding to aid in building activities, I'll first give a qualitative example of how we used constraint to refine an exercise on experimental design. Then I'll discuss more quantitative measures of the effects of constraint. My examples are all drawn from simulation-based labs, but I believe the same principles would apply to many types of activities, both on and off the computer.

24.4 Changing Constraint, Feedback, and Scaffolding in a Virtual Lab to Improve Learning

The trajectory of a design research project for a popular virtual lab called Darwinian Snails illustrates how explicitly thinking about constraint, together with feedback and scaffolding, can help improve science learning activities. Darwinian Snails explores the conditions that lead to natural selection using populations of snails whose shell thickness evolves in response to predation by a crab. The original version of this lab (Herron et al., 2005), which I'll call "Original Darwinian Snails", has a series of simulations on screen, accompanied by a 20-page PDF "workbook". The workbook contains instructions as well as many short-answer questions and a couple of graphing exercises where students handwrite or type their answers into the workbook. Because of this format, it's biased towards low-constraint question types, with no immediate feedback.

The last section in the lab asks students to do their own experiments to show whether a population of snails is likely to have evolved due to natural selection. Figure 24.3a shows a screenshot from the experimental design section of the lab. Students are given two natural populations of snails, predators of the snails, and four experimental tanks where they can establish lab populations. They can drag snails and predators between populations, remove them from populations, and measure relevant data. As shown, this simulation activity is towards the lower constraint end. For instance, there are no limits placed on moving snails or crabs from one

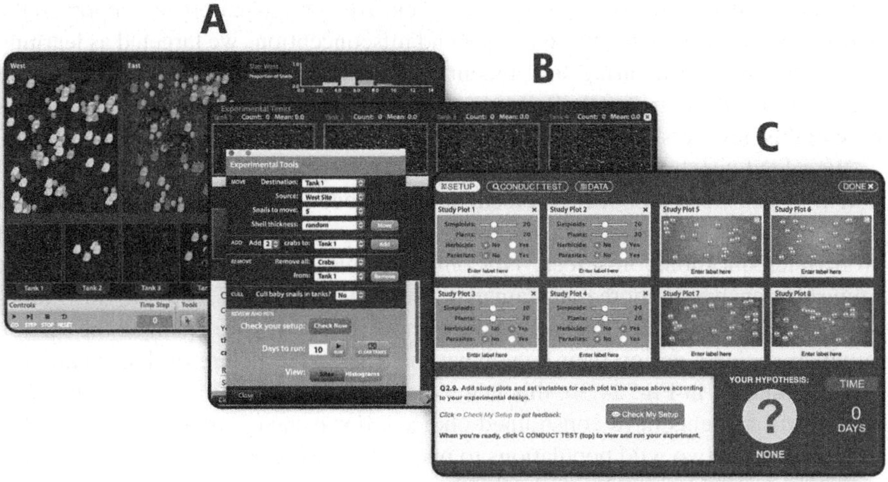

Fig. 24.3 Design-based research iterations of an experimental design activity. (**a**) Original Darwinian Snails has a low-constraint simulation where students design and carry out their experiment by moving individual creatures between study areas. (**b**) Darwinian Snails added some constraint in the form of a window, shown on the left, where students specify how many creatures to move from one site to another and other features of the experiment through limited menus, as well as providing some feedback. (**c**) Understanding Experimental Design further constrained students to manipulating only four parameters, and provided more feedback

population to another – students are free to move as few or many as they'd like, at whatever time points they would like, with whatever traits they choose. The instructions are similarly scaffolded to help students see this as an open-ended exercise. Rather than telling students the experiments to do, the instructions provide questions for students to answer and then describe how to use each tool at the student's disposal. For instance, sets of instructions are labeled "Snail Tanks" on how to use the experimental tanks, "Gathering Populations of Snails" on how to move snails, etc. The idea is that these instructions have the effect of broadening student's horizon of possibilities without guiding them to particular experimental designs. This section is scaffolded to be less guided, with low constraint and no immediate feedback.

Our measured learning outcomes in our studies of the Original Darwinian Snails lab were all around understanding natural selection. As measured using pre-post testing from 637 students attending 25 classes in U.S. colleges and universities, this first version of the lab is effective at teaching some, but not all, of the targeted evolution learning outcomes (Abraham et al., 2009). One common misconception on directional mutations stands out as not improving in that version of the lab. Though we do not have quantitative data on student learning about experimentation from the lab, from watching and interviewing students we saw a wide range of abilities to design reasonable experiments within this environment.

In 2013, as part of a research study, we had the opportunity to re-imagine this lab using a design-based research approach (DBR Collective, 2003). In doing so, we explicitly set about adding constraints to both the questions and the experimental design activity to see if we could improve learning gains. To facilitate this, we moved the instructions and questions onscreen. We both increased and improved the scaffolding by refining the key concepts and misconceptions we targeted as learning outcomes, and restructuring the student task sequence addressing the directional mutation misconception (Clarke-Midura et al., 2018). In a nod to the new format, we call this new version "Darwinian Snails Tutorial-Style" (Fig. 24.3b).

We added constraint in two ways. First, over three iterations involving internal rewriting followed by both expert feedback and student interviews, we converted all the short-answer questions in the workbook to multiple choice, LabLibs, and WordBytes questions on-screen. Not only did this increase constraint, the move from short-answer to more constrained question types enabled us to provide immediate formative feedback to students throughout. Secondly, we changed the section in which students design their own experiment to add structure to their manipulations. As shown in Fig. 24.3b, this new version has an "Experimental Tools" window where students have constrained choices. For instance, they can move snails from one of the two wild populations to one of their experimental tanks, but cannot move snails between tanks, or back to wild populations, and are only given the choice to move 5 or 10 snails at a time. Other actions were similarly more constrained, although in one respect there was a bit less constraint – students now had up to eight experimental tanks available rather than four in the original version.

As the project continued, we thought that the experimental design activity could be further improved with yet another new version where it was separated into its own lab, now called Understanding Experimental Design (Pope et al., 2016). Unlike

Darwinian Snails, we used an imaginary species in this lab ("Simploids") and story (Simploids are getting sick – figure out why) as we noticed that students preconceived ideas about natural selection, snails, and crabs interfered with our ability to assess their experimental design skills. We also, in think-aloud interviews of about 30 students who went through Darwinian Snails Tutorial-Style, noticed that many students have trouble articulating the building blocks of good experiments, but that such difficulty is not always correlated with their ability to plan a well-designed experiment (in prep). This led us to split the experimental design activity in two, with a first section introducing the language and core building blocks of good experimental design (independent and dependent variables, controls, replication, etc.), and a second section where they apply those to design an experiment.

Once settled on those changes, we used the idea of constraint to guide our rewrite. We did this in the context of Bloom's taxonomy (overview in Crowe et al., 2017), a commonly used framework which separates types of knowledge into levels. Remembering and understanding knowledge are at lower levels of the taxonomy, while applying, analyzing, evaluating and creating knowledge are higher level (or higher-order) skills. We hypothesized that learning the language of experimental design was at a lower Blooms level, and that high constraint activities were sufficient for learning these lower order skills. I say "sufficient" rather than better because we wanted to limit the time students spent on this first section, and in our experience highly constrained activities tend to take less time to cover the same material. Thus the first section of the new lab introducing the language is highly-scaffolded and only has highly constrained activities, including 15 of the 18 questions in multiple choice format and no exploratory simulations.

We also hypothesized that less constrained activities would be needed to learn experimental design at higher Blooms levels. Thus, we transferred the student time-on-task we saved in the first section to a second section where students have to <u>*Conduct*</u> their own experiments. This second section includes 7 multiple-choice questions, 2 check-all-that-apply, and 7 LabLibs questions, skewing much farther to the intermediate part of the constraint axis. It also includes two central experimental design activities that, while still quite open-ended, are more constrained than what was in Darwinian Snails Tutorial-Style (Fig. 24.3c). Rather than a window of actions to place things in sites, we put four controls below each experimental site. Two sliders allow students to add Simploids and Simploid food (plants) to the site in increments of 10 each. Two sets of radio buttons allow students to specify the presence or absence in each site of the two factors that might be responsible for the Simploid disease. There is still a huge variety of experimental designs that students can make, but their choices are more constrained than in Darwinian Snails Tutorial-Style, and much more constrained than in Original Darwinian Snails.

The constraints we put into the experimental activities in Darwinian Snails Tutorial-Style and Understanding Experimental Design had an additional benefit. As with questions, constraining the simulation activity allowed us to algorithmically capture common student confusions in experimental design and provide students with immediate, specific feedback on their confusions. For instance, we were able to identify if students were systematically varying variables, had appropriate controls and replication, and if their design matched their hypothesis (Pelaez et al., 2017: ACE-BIO *Plan*). We were also able to capture some aspects of whether students conducted their experiments in accordance with the given natural history about the Simploids, such as whether they ran the experiment for an appropriate length of time (Pelaez et al., 2017: ACE-BIO *Conduct; Chap. 1 in this volume*). We used that algorithm to provide optional feedback to students on their design through a "Check My Setup" button.

Does this focus on constraint, and constraint-enabled feedback, help students? It appears to. We compared Original Darwinian Snails to Darwinian Snails Tutorial-Style in two studies. The first used shared multiple choice quiz questions at the end of each lab. The second added a pre and post-test in Darwinian Snails Tutorial-Style that included both multiple choice and an essay question drawn from the EvoGrader collection (Moharreri et al., 2014). The pre/post test in the latter study allowed us to compare learning gains in Darwinian Snails Tutorial-Style to an earlier study on Original Darwinian Snails (Abraham et al., 2009). Again, we conducted the studies on U.S. college students taking biology classes, in this case with 2082 students in 33 classes across the two studies. We can't tease apart the effects of constraint, feedback, scaffolding, and other changes, but overall, these changes led to increased learning about natural selection among students using Darwinian Snails Tutorial-Style (Clarke-Midura et al., 2018). Among the significant learning differences are that students using Darwinian Snails Tutorial- Style now improve on the directional mutation misconception, perhaps in part because of two LabLibs questions we added on the topic.

In newer research involving Understanding Experimental Design (in prep), we used both a multiple-choice exam as a pre-post test around the first, highly constrained section of Understanding Experimental Design, and a paper and pencil design exercise as pre-post around that whole lab. With a sample of 42 students recruited from Boston-area college biology classes, we saw very little improvement in experimental design skills after students complete the high constraint first section, but large improvements in experimental design skill after the second, intermediate-constraint section of the lab. Digging deeper within the quiz students took after completing the first section, we saw small but significant improvement (Cohen's $d = 0.27$, $p << 0.01$ on 2-tailed paired t-test) on lower-order Blooms questions, but not on higher-order Blooms questions, supporting our hypothesis that this more highly constrained section could help students with lower-order Blooms portions of experimental design, but that to learn the higher-order parts of experimental design students need less constrained activities.

24.5 Constraint Can Affect Student Learning

The above stories show how my colleagues and I have used the idea of constraint and constraint-enabled feedback to guide development of learning activities that address or involve the experimental process. I talked about indirect evidence that constraint and feedback may be helpful. What I haven't presented so far is direct evidence on the role of constraint in aiding learning, separate from other factors. Although the idea of constraint was crystallized over a decade ago (Scalise & Gifford, 2006), it is difficult to find studies that explicitly vary constraint in a controlled way to test its effects. With colleagues I've helped perform two studies recently where we've found evidence that properly tuning constraint can sometimes improve learning outcomes.

In one study, we wrote a lab where secondary school students use models within the educational programming environment StarLogo to do experiments on natural selection (in this case using StarLogo to run pre-built models, not building their own). We wrote three versions of this lab, which differed only in the format of two embedded questions asking students for conclusions from their experiments. In one version, those two questions used the LabLibs format. In a second, they used WordBytes. In a third, they were formatted as a text box where students typed their own response. By having 319 students across 17 different high school biology classes answer an essay question before and after the lab in a pre/post design, we were able to measure learning gains of students using each version.

The results supported an effect of constraint on learning. Students who completed a version of the lab with LabLibs or WordBytes questions learned more on some key learning outcomes than those who did the same lab with the short answer format (Meir et al., 2019). Answering two questions in an intermediate constraint rather than short-answer format, absent any other changes in a lab that typically took two class periods, was enough to have a measurable impact on learning.

In a separate study, we tested fine gradations of variation in constraint in the experimental design activity in Understanding Experimental Design. One version was as described above with limited choices for what went into each study site (e.g., herbicide could either be included or excluded in each site, but nothing in between) and immediate feedback to students on their designs. Another version relaxed some of the constraints in the design activity. Students were able to drag and drop items into each study site as in Original Darwinian Snails, allowing them to add any number of each item rather than being constrained to a binary choice (such as with herbicide) or specific values. Although in a broad sense the difference in constraint is not huge between these two versions, the less-constrained version was different enough that it was too complex for us to provide automated feedback. We compared these in two independent studies, one with the aforementioned 42 Boston-area college students and the other with 129 students in an introductory biology class at a state college in the Western U.S. where class sections were randomly assigned to different versions. In both cases, we found that students using the more constrained, with feedback version were later able to design better experiments than those using the less constrained, no feedback version (in prep).

24.6 Skilled Students May Be Better Challenged in Lower Constraint Activities

What I have talked about so far is how high and intermediate constraint activities may benefit students. High constraint can be more efficient for student learning on lower Blooms levels. Intermediate constraint can benefit student learning on higher-order skills such as many of those involved in the experimental process. But lower constraint activities also likely have benefits, especially for students whose skill level is higher coming in. Lower constraint activities may better challenge those students and allow them to exercise more aspects of a skill like experimental design.

Many instructors use low constraint activities to have students design and carry out their own experiments based on what they already know about experimentation. This lets students use and solidify experimental process skills they have, while learning more from guidance and feedback given by an instructor. As one example, Jon Herron, a co-author of a number of SimBio's labs including Darwinian Snails, uses lower constraint, more lightly-scaffolded activities to teach the experimental process in his evolution classes at University of Washington. He gives students a choice of pre-made simulation labs and asks them to come up with a *Question* they want to answer using one of those simulations. He then works with them throughout the term to come up with a hypothesis, design how they will test it within the simulation, carry out the experiment, *Analyze* the data, and *Communicate* their conclusions (Herron, 2020). Some CURE's and other structured undergraduate research experiences are organized in the same way to give students the opportunity to learn the many subtler aspects of conducting informative experiments by carrying out the whole process themselves (Cole & Beck, Chap. 2; Gouvea et al., Chap. 17).

Our research on Understanding Experimental Design had hints of data that supported this benefit of lower constraint activities. There were two potential causal variables students could manipulate. In the intermediate constraint versions of the lab, 34% of students in the introductory biology class attempted experiments that manipulated both of those simultaneously. In the lower constraint version, a higher 41% tried manipulating both simultaneously (N = 170 across the treatments). We saw this same pattern in the Boston-area college students. However, neither was statistically significant so we cannot draw any firm conclusion.

24.7 Quantifying Constraint as a Way of Informing Designs

In many of the examples above, we can quantify how constrained a question or other activity is. One simple way to do this is to add up the number of possible answers, or in more open-ended activities, the number of distinct answers typically seen from students. To make this concrete, a typical multiple-choice question has 3–5 answer choices. The Isle Royale LabLibs where students are asked to make a hypothesis and prediction about predator-prey relationships has three blanks that students must fill in with 3 × 3 × 3 = 27 total possible choices, or 5–9× as many choices as a typical multiple-choice question. With the WordBytes question on heritability (Fig. 24.1c) we could quantify number of choices in at least two ways: through the number of tiles students have available to use in their answer; or through counting the number of distinct answers given by students. The former has the advantage that it can be done prior to collecting any student data, but the latter is possible to do not just for WordBytes, but for almost any type of question or activity, including those done on paper rather than computer.

A more refined way of quantifying relies on having data on student answers and calculates the informational entropy of those data (Meir et al., 2019). This calculation measures how student answers to a question vary, with more variation corresponding to a greater entropy. In other words, entropy measures how much variety is there in actual student answers as opposed to measuring the total of possible answers. In the Isle Royale LabLibs, if in practice students only answered using 4 of the 27 possible answer choices, the entropy would be much less than if student answers were spread evenly across all 27 possible choices. Using this measure, we find that LabLibs questions within our body of work are equivalent in entropy to 1–3 multiple choice questions, while WordBytes are equivalent to 2–4 multiple choice questions (Meir et al., 2019). We haven't done so, but in principle this calculation could be done for essay questions and many other types of activities, though for less constrained activities the entropy value would depend on how one defined answers as being equivalently the same.

Quantifying the degree of constraint allows us to better explore the effects of constraint. Figure 24.4 shows question difficulty (percent of students answering wrong on the first try) as a function of degree of constraint across a number of short-answer essay, LabLibs, WordBytes, and multiple-choice questions from four different SimBio labs (N = approx. 6500 students from 87 U.S. college biology classes). Whether measured by number of choices (Fig. 24.4a) or entropy (Fig. 24.4b), there is a clear pattern of less constrained questions being harder to answer. Assuming the choices are meaningfully capturing different student thinking about the question topic, this indicates that constraint affects question difficulty. This could happen simply because with more choices, the possibility of randomly choosing a wrong answer goes up. While this is undoubtedly part of the explanation, we have data that students are not just guessing at answers (Meir et al., 2019), so I would argue that the increased difficulty is at least in part due to the questions challenging students to use greater knowledge or skill in their answers.

We see a similar pattern in simulation-based activities. We randomly assigned 108 students in the introductory biology class mentioned above to a version of Understanding Experimental Design with intermediate constraint (but in this case, without any feedback) or to the lower constraint version. While there was no learning difference ultimately between those treatments, 48% of students using the intermediate constraint version had a good experimental design on the very first experiment they ran, even without receiving any feedback, while only 34% of those in the low constraint version had a good first design. As with questions, less constraint makes the activity more challenging.

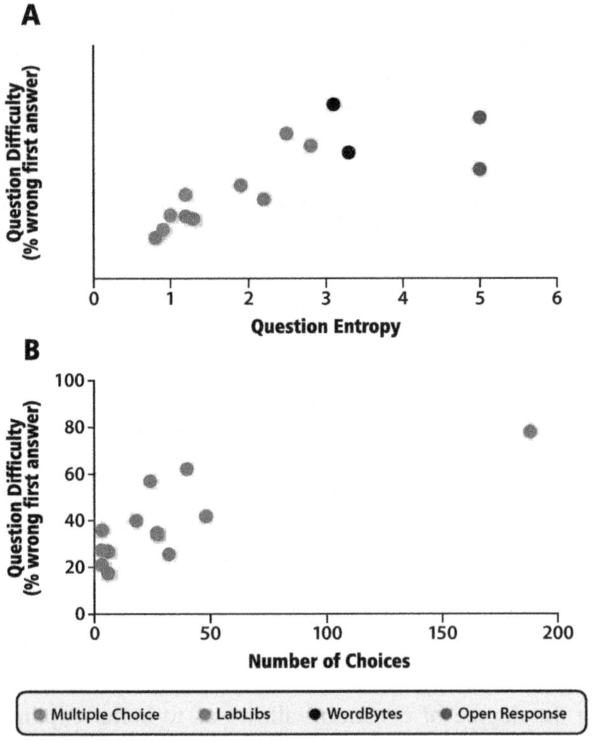

Fig. 24.4 Question difficulty (as measured by percent of students answering wrong on first attempt) versus number of choices (**a**) and question entropy (**b**). Questions include multiple choice (averages from 3 labs), LabLibs (10 questions), WordBytes (2 questions), and short-answer (2 questions) formats. Question format of each point is shown by color. The points representing multiple choice questions are averages for all multiple choice across a lab. For technical reasons WordBytes and short-answer questions are only included in (**b**), and some LabLibs questions are only included in (**a**). Both graphs show question difficulty increasing as constraint decreases

24.8 The Interplay of Constraint, Feedback, and Scaffolding

The data presented above often conflates constraint, scaffolding and specific immediate feedback, with only a couple of studies that manage to tease those apart. There is a reason for this. Constraint enables better feedback, and both constraint and feedback can be used in place of instructional scaffolding.

A great deal of research shows that fast and specific feedback is important to learning (reviewed in Shute, 2008). It is easy to write feedback for multiple choice questions. One can provide a feedback item specific to each of the choices. By contrast, while there is progress being made (Urbain Lurain et al., 2015), it is still very difficult to provide specific automated feedback in low constraint activities such as essay questions or very open-ended simulations. But as I show above, adding some intermediate level of constraint can often make an activity more amenable to providing feedback. Constraint enables feedback.

It is not only computer-based automated feedback that benefits from some constraint. Many instructors have trouble effectively implementing active learning strategies (Andrews et al., 2017; Ebert-May et al., 2011). Adding constraints to activities, on or off the computer, may make it easier for instructors, particularly those not used to student-centered teaching techniques, to successfully implement student-centered teaching strategies, including providing productive and specific feedback, by limiting the number of possible outcomes to which they need to respond in the moment.

Another, more subtle benefit of constraint and feedback is that they can provide implicit rather than explicit scaffolding. If you only give students productive choices to make in an activity by constraining the range of choices they have available, this can replace scaffolding such as a set of instructions guiding students towards what to do. A potential advantage of constraint over guidance is that it encourages students to think for themselves a bit more. In a simulation, for instance, they have to make decisions about what to vary among the given parameters, rather than being told "try changing X".

Feedback can do the same thing. In the LabLibs about predator effects on prey, we do not say "make sure to match your prediction to your hypothesis". Instead, if the student chooses an answer that does not match prediction to hypothesis, we provide feedback pointing that out. Again, this may encourage more thought on the part of the student – now they are realizing they did something poorly, and are more engaged to understand what their mistake was and learn from it.

24.9 The Intermediate Constraint Hypothesis: Tuning Constraint to Match the Student

Different student populations, and individual students within those populations, arrive to learning situations with differing levels of skill and experience on higher-order skills such as experimentation (e.g. Deane et al., 2014). The data we've accumulated, both qualitative through design-research practices and quantitative, suggests to me that between scaffolding, feedback, and constraint, focusing on degree of constraint may be an efficient way for a teacher or educational materials author to tune a particular exercise to different student levels.

The intermediate constraint hypothesis that I propose here is summarized in Fig. 24.5. The X-axis is the degree of constraint in an activity, where I am assuming some amount of active learning in the activity. The active learning could be answering conceptual questions, designing an experiment, or anything else where the student has to put some of their own discipline-based thought into completing the activity. The Y-axis shows how efficiently a particular student learns from that activity. By efficiency, I mean how much learning gain they have for a unit of effort put

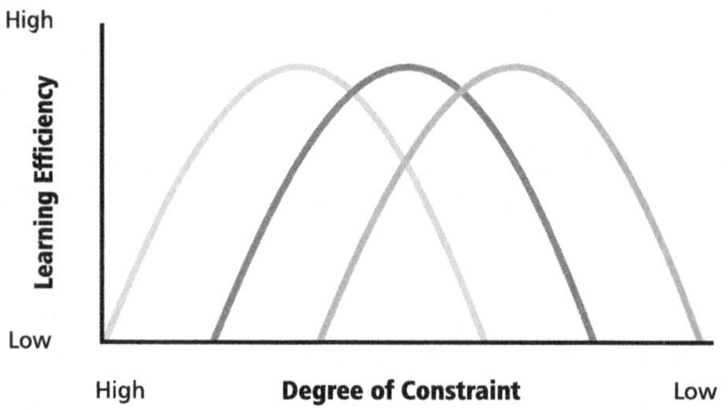

Fig. 24.5 The Intermediate Constraint Hypothesis. Curves represent learning efficiency of individual students or populations of students. The hypothesis is that for any given student (population), there will be a degree of constraint in an activity which maximizes learning efficiency, but this point will be different for different students. Thus, manipulating constraint can be used to tune an activity to different students

in, where that unit could be time spent, brain juice used up, etc. I hypothesize that if the activity is too constrained for that student, they may expend little effort but will have very low learning gains for a low efficiency of learning. If the activity has too little constraint, the student may become lost, or may not have enough direction to gain all the learning available from the activity. They may expend a lot of effort without a lot of gain, again giving a low efficiency of learning.

Somewhere in the middle, there is an intermediate degree of constraint that will challenge the student to think about the subject in a productive way that pushes the edge of their skill level, leading to learning. That point will maximize their efficiency of learning. Further, as that student becomes more knowledgeable in the topic, their optimal level of constraint may shift towards lower constraint (right on the diagram).

Factoring in feedback may shift the curves in 24.5 to the left. A higher constraint activity with fast, specific feedback may provide higher learning efficiency than a lower constraint activity without such feedback, even if the lower constraint is, on its own, better suited to the current knowledge base of the student. This has increasingly become the assumption of my colleagues and I as we author simulation-based labs. We tend to err on the side of higher constraint for the level of student we are targeting, and including some form of instant, specific feedback to students. We also tend, where possible, to use constraint and feedback rather than instructions to scaffold activities.

This is a hypothesis rather than a theory, in that the data I presented above suggests the possibility that this idea is helpful, but is far from proving it. Stronger evidence would come from studies that systematically vary constraint in an exercise and test those levels of constraint on populations of students with differing initial skill levels. The shape of the learning efficiency curve is likely more complex than in Fig. 24.5, and may depend on the topic, type of activity, student background and so on. As long as there is an optimum in the middle, the hypothesis would hold.

Why would this hypothesis be useful, if true? As I showed above, constraint can be relatively easy to manipulate. In some cases, such as certain question formats, we can quantify it. We can then alter the amount of constraint by, for instance, removing or adding additional (meaningful) answer choices. In other cases, quantifying may be more trouble than it's worth, but it is often relatively easy to envision how to qualitatively alter the level of constraint. In a simulation, we can expose or hide additional parameters, or change how many settings are available per parameter. Changing constraint is thus a very concrete focal point for adjusting the challenge of an activity to the needs of a student or the average incoming knowledge in a population of students.

Thinking in terms of constraint could be particularly useful for designers of digital educational tools. A designer can choose from many different question formats with different degrees of constraint (see summary of question formats and constraint in Scalise and Gifford (2006) for ideas) and constraint might inform the choice of which format to use in a particular situation. In an interactive, the designer can consider constraint in choosing the number and flexibility of controls to provide the student. Constraint might also be useful in non-digital activities. As an example, if students were struggling to learn how to use a microscope, the instructor might set the slide in place, choose the magnification, adjust the lighting, and then cover all those controls, only exposing the focus knob for the student. Or in an ecological sampling exercise, an instructor might choose whether to provide students with tools to build quadrats of different sizes, conduct transects, etc., or whether to provide only a single fixed size of quadrat. These are choices that we as instructors make intuitively all the time, but considering them in terms of constraint may help make such choices more deliberate.

24.10 Considering Degree of Constraint Can Improve Summative Assessment of Skills

I've couched the Intermediate Constraint Hypothesis in terms of learning, but the relation of constraint to feedback hints at another key area of education where focusing on constraint may be useful. Assessment is critical to all efforts to improve learning of complex skills such as those involved in experimentation, as detailed in Part 2 of this book (Chaps. 6, 7, 8, 9 and 10). Providing feedback to students necessarily involves assessing their work. In online activities, this assessment is done algorithmically. The same or similar algorithms may be used to provide feedback to instructors on the amount of learning taking place due to their instruction. I believe these summative assessments can also benefit from a consideration of degree of constraint, whether algorithmically-provided online or provided through a rubric offline. This is especially true for performance-based assessments where the student's skill level is deduced by watching them perform the skill. I'll provide two brief examples of how constraint affects assessment to close out this chapter.

An important aspect of the experimental process is presenting data, and for many sciences this is often done in graphs (Pelaez et al., 2017, ACE-Bio *Communicate*). Students have many challenges with graph construction, and it would be useful to have an automated way of documenting these challenges in order to measure the effect of different learning activities. As part of a team building an online assessment of graph construction ability among undergraduates, called GraphSmarts (Suazo-Flores et al., 2018), we needed to consider how to capture the various practices that, when done well together, lead to clear and informative graphs.

We referred first to prior automated graph construction assessments we and others had written. The ones we found were all quite constrained, albeit in different ways. For instance, in our Isle Royale lab, the predator-prey LabLibs discussed above is followed by an exercise where students collect data to test the prediction, and graph the data. There are three variables available for students to measure, and three choices they must make in their graphing (variables to graph on each axis; using field or summarized data; graph type). While many undergraduate students make non-optimal graphs in this interface, there is a sharp limit to information we are able to extract about their graphing skills because they simply don't have enough freedom to show many of the common graphing mistakes (e.g., selecting graph type) (Meir et al., 2022). On the other hand, when we ask students to draw graphs by hand on paper, we get a plethora of interesting information about how they are confused in thinking about data and data presentation, but it is time-consuming to score and not possible to automate scoring (Angra & Gardner, 2017; Gardner et al., 2021).

As we iterated towards our latest version of GraphSmarts, we honed in on a level of constraint intermediate between those. The current version (Fig. 24.6), which has been successfully class tested and proved useful to instructors, has seven possible variables and six choices/functions (e.g., variables to plot, graph type, scales, etc.) for students to use in their graphing (Gardner et al., 2021; Meir et al., 2022). Here students have a much broader ability to show both good graphing skill and problems they have constructing graphs (albeit not as many as on paper), and we have been able to capture a much fuller picture of their graphing ability.

As a second example, in the Understanding Experimental Design project, we needed a summative assessment of experimental design skill to test our lab. After looking at several published assessments, we favored one called the Experimental Design Ability Test (EDAT: Sirum & Humburg, 2011) which presents students with a hypothesis and asks them to write down (on paper) an experimental design to test it. Because the EDAT as published uses human examples, and we worried about

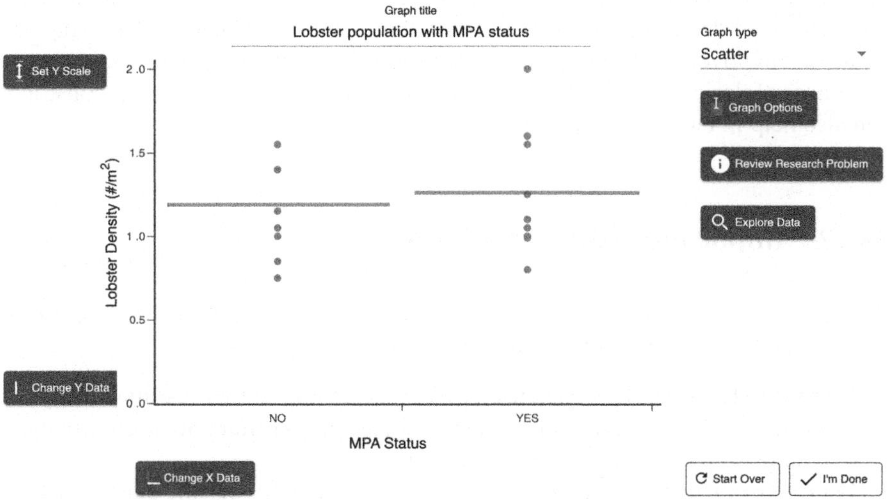

Fig. 24.6 The GraphSmarts interface for assessing students' graphing skills. Students construct a graph onscreen using the controls shown to test a given prediction

confounding human-specific ideas such as placebos with more general experimental design practices, we wrote a very parallel assessment that used other species. Initially, in parallel to the EDAT, we wrote a prompt that discussed a single causal variable. For instance, one prompt talked about a "new anti-rabbit spray RabbitRepel" which is supposed to prevent rabbits from eating your flowers, and asked students to design an experiment to test that claim.

The assessment worked at eliciting experimental design practices from students, but it was too easy for the undergraduate population we were testing. Most of them produced a good design. In terms of constraint, the assessment was too highly constrained with respect to number of variables to be informative. We then wrote new prompts that included three possible causal variables as well as another variable that was a distractor. The number of choices the student had to make increased severalfold, and this less constrained assessment was much better tuned to differentiate students within the target population (Meir et al., in prep). As these two examples show, the intermediate constraint hypothesis may apply not only to efficient learning, but also to efficient assessment.

24.11 Conclusion and Implications for Instructors

Scaffolding, feedback, and constraint are all important ways of tuning learning activities to the current skill level of different students. Such tuning is especially useful for higher-order skills such as experimentation where there is a large difference between novices and experts at the skill. Explicitly thinking about these components of design (Table 24.1) can help those building educational activities do so more effectively. The Intermediate Constraint Hypothesis states that of those three axes of design, constraint is often the most easily tuned. A focus on constraint can also help in building appropriate summative assessments.

24.12 Implications for Instructors

Below are recommendations and avenues for future research which follow out of the Intermediate Constraint Hypothesis:

- **Deliberately consider constraint.** When designing active learning exercises and/or environments, explicitly consider degree of constraint in the environment in relation to the target student population.
- **Quantify constraint.** Where possible, consider quantifying the degree of constraint in an activity (either numerically, or by placing along an axis from low to high) to help in tuning the activity to target population(s).
- **Constraint and feedback in place of instructional scaffolding.** Instructional scaffolding that guides students in what to do can in many cases be replaced by constraining the choices available, and/or providing specific feedback as students make choices. Using feedback and constraint in place of instructional scaffolding may lead to more student learning.
- **Constrain assessments to match population.** Especially when using performance-based assessments, too much constraint does not reveal a full range of student skill; too little constraint makes interpreting results difficult and can hide more basic skills.
- **Vary constraint in research**. Studies varying level of constraint in both educational activities and assessments could prove very interesting both theoretically and practically.

While I couched much of this chapter in digital questions and simulation-based activities – my own specialty – I believe the same ideas would apply to a wide range of learning materials from instructors writing worksheets for their own class to all manner of activities built for wider distribution. Of course, there are many other important aspects of making a good learning activity but constraint should take its place among the principles of good educational design.

Acknowledgments While I alone take the blame for the intermediate constraint hypothesis and other potentially naïve ideas introduced here, I built the ideas and narrative upon research and discussions shared with many collaborators. At SimBiotic Software, thanks to Susan Maruca, Kerry Kim, and all my other colleagues who have participated in our research activities. The Understanding Experimental Design project involved many collaborators including Denise Pope, now at Univ. Massachusetts Amherst, Joel Abraham at California State Univ, Fullerton, Jody

Clarke-Midura at Utah State, Daniel Wendell, Ling Hsao, and others from the lab of Eric Klopfer at Massachusetts Institute of Technology, and many people at SimBio including Susan Maruca, Kerry Kim, Jennifer Palacio, Jenna Conversano, and numerous content and software developers. We also received help from a great advisory board of Ross Nehm, Kathryn Perez, and Ryan Baker. The GraphSmarts project was led by Stephanie Gardner at Purdue University and involved several researchers in her lab, most prominently Elizabeth Suazo-Flores. Also participating were Joel Abraham, and many of the same research team at SimBio. Thanks to Isobel Buck for making the figures prettier. Thanks to the many instructors and students who agreed to allow us access to their classes and data. Finally, thanks to the organizers of this volume for inviting my contribution. This material is based upon work supported in part by the National Science Foundation under Grants No. 1227245 and 1726180. Thanks to all of those parties for the ideas, data, and funding upon which this chapter is built.

References

Abraham, J. K., Meir, E., Perry, J., Herron, J. C., Maruca, S., & Stal, D. (2009). Addressing undergraduate student misconceptions about natural selection with an interactive simulated laboratory. *Evolution: Education and Outreach, 2*(3), 393–404.

Andrews, T. M., Leonard, M. J., Colgrove, C. A., & Kalinowski, S. T. (2017). Active learning *not* associated with student learning in a random sample of college biology courses. *CBE Life Sciences Education, 10*(4). https://doi.org/10.1187/cbe.11-07-0061

Angra, A., & Gardner, S. M. (2017). Reflecting on graphs: Attributes of graph choice and construction practices in biology. *CBE Life Sciences Education, 16*(3), 1–15. https://doi.org/10.1187/cbe.16-08-0245

Bell, R. L., Smetana, L., & Binns, I. (2005). Simplifying inquiry instruction. *The Science Teacher, 72*(7), 30–33.

Biswas, G., Segedy, J. R., & Bunchongchit, K. (2016). From design to implementation to practice a learning by teaching system: Betty's brain. *International Journal of Artificial Intelligence in Education, 26*, 350–364. https://doi.org/10.1007/s40593-015-0057-9

Buck, L. B., Bretz, S. L., & Towns, M. H. (2008). Characterizing the level of inquiry in the undergraduate laboratory. *Journal of College Science Teaching, 38*, 52–58.

Cerchiara, J. A., Kim, K. J., Meir, E., Wenderoth, M. P., & Doherty, J. H. (2019). A new assessment to monitor student performance in introductory neurophysiology: Electrochemical gradients assessment device. *Advances in Physiology Education, 43*, 211–220. https://doi.org/10.1152/advan.00209.2018

Chaiklin, S. (2003). The zone of proximal development in Vygotsky's analysis of learning and instruction. In A. Kozulin, B. Gindis, V. Ageyev, & S. Miller (Eds.), *Vygotsky's educational theory and practice in cultural context*. Cambridge University Press.

Clark-Midura, J., Pope, D. S., Maruca, S., Abraham, J. K., & Meir, E. (2018). Iterative design of a simulation-based module for teaching evolution by natural selection. *Evolution: Education and Outreach, 11*(4). https://doi.org/10.1186/s12052-018-0078-6

Collective, D. B. R. (2003). Design-based research collective. Design-based research: An emerging paradigm for educational inquiry. *Educational Research, 1*(32), 5–8.

Crowe, A., Dirks, C., & Wenderoth, M. P. (2017). Biology in bloom: Implementing Bloom's taxonomy to enhance student learning in biology. *CBE Life Sciences Education, 7*(4). https://doi.org/10.1187/cbe.08-05-0024

Deane, T., Nomme, K., Jeffery, E., Pollock, C., & Birol, G. (2014). Development of the biological experimental design concept inventory (BEDCI). *CBE Life Sciences Education, 13*(3), 540–551. https://doi.org/10.1187/cbe.13-11-0218

Ebert-May, D., Derting, T. L., Hodder, J., Momsen, J. L., Long, T. M., & Jardeleza, S. E. (2011). What we say is not what we do: Effective evaluation of faculty professional development programs. *Bioscience, 61*(7), 550–558. https://doi.org/10.1525/bio.2011.61.7.9

Freeman, S., Eddy, S. L., McDonough, M., Smith, M. K., Okoroafor, N., Jordt, H., & Wenderoth, M. P. (2014). Active learning increases student performance in science, engineering, and mathematics. *Proceedings of the National Academy of Sciences of the United States, 111*, 8410–8415.

Gardner, S. M., Suazo-Flores, E., Maruca, S., Abraham, J. K., Karippadath, A., & Meir, E. (2021). Biology undergraduate students' graphing practices in digital versus pen and paper graphing environments. *Journal of Science Education & Technology, 30*, 431–446.

Herron, J. (2020). *Teaching with SimBio's evolution modules*. https://vimeo.com/simbio/jon-herron-evolution. Accessed 4 Aug 2020.

Herron, J. C., Maruca, S., Stal, D., Kingsolver, J., & Meir, E. (2005). *Darwinian snails*. SimBiotic Software.

HHMI. (1997). https://www.biointeractive.org/classroom-resources/neurophysiology-virtual-lab. Accessed 30 July 2020.

Jerrim, J., Oliver, M., & Sims, S. G. (2020). The relationship between inquiry-based teaching and students' achievement. New evidence from a longitudinal PISA study in England. *Learning and Instruction, 61*, 35–44. https://doi.org/10.1016/j.learninstruc.2018.12.004

Kim, K., & Meir, E. (2015). *Action potentials extended*. SimBiotic Software.

Kim, K. J., Pope, D. S., Wendel, D., & Meir, E. (2017). WordBytes: Exploring an intermediate constraint format for rapid classification of student answers on constructed response assessments. *Journal of Educational Data Mining, 9*(2), 45–71. https://doi.org/10.5281/zenodo.3554721

Klopfer, E., Scheintaub, H., Huang, W., Wendal, D., & Roque, R. (2009). The simulation cycle: Combining games, simulations, engineering and science using StarLogo TNG. *e-Learning, 6*(1), 71–96.

Meir, E. 2022. Designing a simulation lab: The process that led to action potentials explored and extended, two simulation-based neurobiology labs. *Journal of Undergraduate Neuroscience Education (JUNE), 20*(2): in press. https://www.funjournal.org/current-issue/

Meir, E., Steinberg, E. K., & Maruca, S. (2009). *Isle Royale*. SimBiotic Software.

Meir, E., Wendel, D., Pope, D. S., Hsiao, L., Chen, D., & Kim, K. J. (2019). Are intermediate constraint question formats useful for evaluating student thinking and promoting learning in formative assessments? *Computers & Education, 142*. https://doi.org/10.1016/j.compedu.2019.103606

Meir, E., Gardner, S.M., Maruca, S., Suazo-Flores, E., & Abraham, J. K. (2022). Can My Students Make A Graph? Using Evidence-Based Design to Build an Auto-Scored Performance-Based Assessment of Graph Construction. In J.M. Spector, B. B. Lockee, & M.D. Childress (Eds.) *Learning, Design, and Technology. An International Compendium of Theory, Research, Practice, and Policy*. Springer Nature, New Dehli, India.

Moharreri, K., Ha, M., & Nehm, R. H. (2014). EvoGrader: An online formative assessment tool for automatically evaluating written evolutionary explanations. *Evolution: Education and Outreach, 7*(15). https://doi.org/10.1186/s12052-014-0015-2

Pelaez, N., Anderson, T., Gardner, S. M., Yin, Y., Abraham, J. K., Bartlett, E., Gormally, C., Hill, J. P., Hoover, M., Hurney, C., Long, T., Newman, D.L., Sirum, K., & Stevens, M. (2017). *The basic competencies of biological experimentation: Concept-skill statements*. PIBERG Instructional Innovation Materials. Paper 4. http://docs.lib.purdue.edu/pibergiim/4

Pope, D. S., Maruca, S., Palacio, J., Meir, E., & Herron, J. (2016). *Understanding experimental design*. SimBiotic Software.

Scalise, K., & Gifford, B. (2006). Computer-based assessment in e-learning: A framework for constructing "intermediate constraint" questions and tasks for technology platforms. *The Journal of Technology, Learning, and Assessment*, *4*(6), Retrieved from http://files.eric.ed.gov/fulltext/EJ843857.pdf

Shute, V. J. (2008). Focus on formative feedback. *Review of Educational Research, 78*(1), 153–189. https://doi.org/10.3102/0034654307313795

Sirum, K., & Humburg, J. (2011). The experimental design ability test (EDAT). *Bioscene, 37*, 8–16.

Suazo-Flores, E., Allison-Bunnell, S., Maruca, S., Quick, J., Abraham, J. K., Meir, E., & Gardner, S. M. (2018). *Developing a digital tool to evaluate and teach graphing in introductory biology*. SABER Annual Meeting 2018 Poster Session, Minneapolis MN.

Urban-Lurain, M., Cooper, M. M., Haudek, K. C., Kaplan, J. J., Knight, J. K., Lemons, P. P., et al. (2015). Expanding a national network for automated analysis of constructed response assessments to reveal student thinking in STEM. *Computers in Education Journal, 6*(1), 65–81.

Weaver, G. C., Russell, C. B., & Wink, D. J. (2008). Inquiry-based and research-based laboratory pedagogies in undergraduate science. *Nature Chemical Biology, 4*(10), 577–580.

Chapter 25
Implementing Innovations in Undergraduate Biology Experimentation Education

Trevor R. Anderson and Nancy J. Pelaez

25.1 Introduction

There is a major move in undergraduate biology education to develop curricula aimed at advancing students' competence to do authentic biology research, the major focus of this book. To make studies like those presented in this book worth their time, energy, intellectual input and financial investment, it is clearly highly desirable to disseminate the innovations produced to as wide an audience as possible and, where relevant, to ensure that they are effectively and efficiently implemented in teaching practice. Usually, as demonstrated by the success stories documented in this book and other similar educational literature, it is relatively easy for educational researchers to incorporate their own innovations into their courses. In contrast, for a wide range of reasons, it can be a major challenge for educational specialists to harness support from other course instructors, program administrators and other stakeholders to agree to make any such significant curricular changes. This is because numerous barriers can be encountered that need to be addressed and overcome before a new teaching innovation targeting undergraduate research in biology can be implemented.

Thus, the first issue addressed in this chapter focuses on the process of innovative change and the importance of considering the influence of contextual factors when attempting to introduce an innovation. Secondly, we highlight the importance of establishing the feasibility for introducing an innovation by determining the

T. R. Anderson (✉)
Department of Chemistry, Purdue University, West Lafayette, IN, USA
e-mail: ander333@purdue.edu

N. J. Pelaez
Department of Biological Sciences, Purdue University, West Lafayette, IN, USA
e-mail: npelaez@purdue.edu

N. J. Pelaez et al. (eds.), *Trends in Teaching Experimentation in the Life Sciences*, Contributions from Biology Education Research,
https://doi.org/10.1007/978-3-030-98592-9_25

availability of key resources, as well as gauging the level of tolerance of various stakeholders to curricular changes by considering contextual forces that might support or oppose the changes. Finally, we end this chapter by illustrating how various published change strategies could be used to overcome such opposing forces so that an innovation can be successfully translated and implemented into the teaching, learning and assessment of students about biological research, so that more biology students will understand the importance of validating all science through rigorous studies to validate its findings.

25.2 The Process of Innovative Change: Consider the Potential Influence of Contextual Factors on Innovative Change

Before attempting to implement a teaching innovation, including those in this book focused on teaching how to do biological research, it is essential to consider any contextual factors that might influence the chances of a successful implementation process. Figure 25.1 shows a modified version of a model published by Rogan and Anderson (2011) illustrating the overall relationships between the key contextual factors and forces influencing curriculum change and the need for strategies to overcome them so that innovative curriculum change can be successful. In this section, we briefly address these contextual factors, suggesting how to respond to them where relevant at your institution.

In 2011, Anderson and Rogan (2011) discussed the important elements of sound curriculum design and development and identified key contextual factors that may

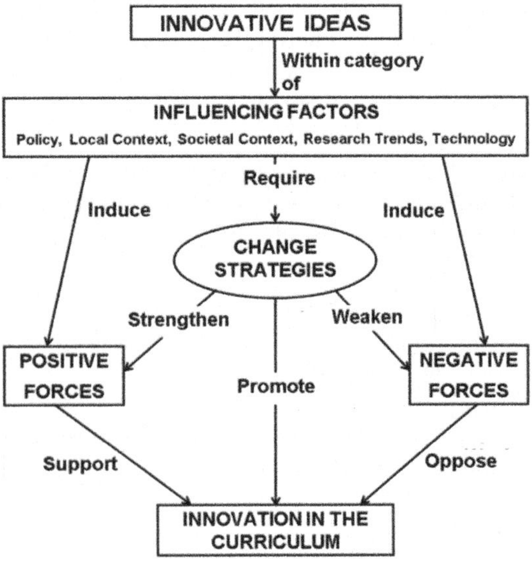

Fig. 25.1 Relationships between influencing factors, change forces, and strategies for promoting innovative change in the curriculum. (Modified from Rogan & Anderson, 2011)

influence curriculum change, including educational policy, the local context, societal expectations, research trends, and technology. Educational policy (Fig. 25.1) could, for example, stem from national policy statements like the *Vision and Change in Undergraduate Biology Education* report (AAAS, 2011), as well as more local policy guidelines from university provosts, college deans and departmental education committees. Ideally, such policy would not be unduly restrictive of your academic freedom to be innovative but would instead foster any curricular innovation change in a positive manner (Boyer et al., 1994; Altbach, 2001). For example, where you might encounter individuals who favor maintaining the status quo and oppose your efforts, policy could help pressure them to be more adventurous and open-minded to testing your innovation.

Addressing the influence of local context (Fig. 25.1) on your goals to implement your innovation requires knowledge of student, instructor, and scientific context. Regarding student context, modern universities need to meet the needs of an ever-changing student body in terms of cultural and language diversity, level of educational preparedness and socio-economic backgrounds. For example, some students may lack experience working in a lab using equipment, performing techniques, solving problems and processing and interpreting data. In this case, the innovation would need to be matched to the educational level of the student and either modified or replaced with a more manageable exercise so that the students feel confident and competent to perform the required tasks. In addition, instructors would also need to feel confident that they know the subject matter and can competently teach a particular lab about experimentation. If this presents a problem, then appropriate instructor training may be necessary before you consider introducing your innovation. Regarding the influence of societal context (Fig. 25.1), universities always need to be accountable to the public, funding bodies, professional societies, activist groups, the government and industry for the work they do in educating people. Thus, when you consider modifying a curriculum to incorporate an innovation, you will need to consider any stakeholder expectations that might affect your efforts. For example, funders might expect certain defined course objectives or learning outcomes while certain industry, as a major employer of your students, may expect graduates to develop specific knowledge and skills.

If in the views of your colleagues, your planned innovation qualifies as a modern research trend (Fig. 25.1), this could help you gain acceptance for its implementation. Supportive colleagues could include instructors and program administrators who favor keeping undergraduate curricula up to date with the latest cutting-edge research trends in science and educational theory and practice. Colleagues implementing some of the innovations from this book will have an advantage in that they focus on some of the latest trends, but the onus will still be on you to convince any stakeholders of the importance of the proposed innovation to graduates' repertoire of knowledge and skills. There will also likely be concerns about how the new material will be taught- i.e., will the pedagogical approach be appropriate for the students and material being taught? For example, in the current context of biology research education, the success of teaching an innovation will be dependent on how the competencies are taught, how learning is assessed, and how students respond to feedback from formative assessment. Technology (Fig. 25.1) is also a key factor influencing curricular change, especially regarding visualization and online learning. For this reason, this book has chapters that address teaching (Sect. 25.3) assessment (Sect. 25.4) and that illustrate various ways to incorporate technology (Chaps. 7, 10, 18, 19, and 24) dedicated to these topics.

25.3 Establishing Feasibility and Tolerance: Contextual Forces Supporting and Opposing the Innovation

When considering whether to introduce a new teaching innovation into the curriculum, it is important to establish what stakeholders, especially those most affected by the changes, will consider feasible and what they will tolerate. These two issues of feasibility and tolerance are discussed in more detail in Rogan & Anderson (2011) as the concepts of the Zone of Feasible Innovation (or ZFI) and the Zone of Tolerance (ZoT). Briefly, ZFI is especially about whether a teaching innovation, while going beyond current practice, could still be implemented with current resources. In the context of biology research education, such resources would include lab space, relevant equipment and consumables or funding to purchase them, space in the teaching timetable, as well as human resources such as instructors, teaching assistants and technicians to support lab setup and testing. The ZoT, originally introduced by McGivney and Moynihan (1972), is about the extent to which stakeholders, impacted in various ways by the innovation, can tolerate its introduction. It is possible that on paper and considering all required resources that introducing the innovation is feasible but still may not be tolerated by stakeholders because of a number of factors such as those listed in Box 25.1. For example, colleagues may be resistant to any change and wishing to maintain the status quo, or they may feel insecure about their competence to learn something new and then to implement and teach the innovation. Ideally, an institution that fully supports innovation in its teaching program would always consider what is feasible as being acceptable.

Key factors that impact the ZFI and ZoT are the various negative and positive forces that might respectively oppose and support introduction of the teaching innovation (Fig. 25.1). Selected examples of such forces are listed in Boxes 25.1 and 25.2 while a more comprehensive list can be found in Rogan & Anderson (2011). The literature contains extensive examples of negative forces (Fig. 25.1) or barriers that could oppose the introduction of your innovation. When attempting to introduce an innovation, the best is to investigate what barriers may be a problem in your specific context. Box 25.1 lists a few of these potential barriers.

Regarding the final bullet in Box 25.1, student resistance must be recognized and addressed when teaching them to understand and do the hard work of biological research. Based on their prior experiences with biology labs, undergraduate students may be accustomed to learning concrete procedural knowledge where

Box 25.1: Selected Examples of Contextual Forces Opposing innovation (Rogan & Anderson, 2011)

- Change agent and faculty at variance
- Others are comfortable with status quo
- Historical and psychological barriers
- Complexity of jargon-laden ideas
- Faculty competence, expertise, ideas
- Resources- faculty, facilities, technical, financial, time, logistical
- Others do not believe change will add value
- Innovation overload
- Social, cultural, philosophical issues
- Student resistance

mastery is easily recognized. In contrast, the experimental sciences are a messy human enterprise, requiring students to do more cognitively demanding tasks such as those in the list of competences documented by Pelaez et al. (2017) and in the tables of Chap. 1 in this volume. For example, when students are resistant to dealing with the feelings of discomfort that reflect uncertainty, they need explicit instruction about science epistemology and the joys of searching for refutation or confirmatory evidence (and not just a "right" answer). Educators can call on examples from this book to motivate the hard work needed for students to ***Identify*** a gap in the literature that is worth investigating, to ***Plan*** what data to collect and how to use it as evidence, and to justify inferences using data from their experiments and other studies. That is, as discussed by Anderson (2007) and many others over the years, there is a need to bridge the gap between research and its application in teaching and learning.

When you attempt to implement an innovation for teaching experimentation in biology, you will be viewed as the so-called 'change agent' (Box 25.1) with the task of convincing other faculty and students of the importance of your innovation. If you are a scientist, then gathering support for your innovation may be easier if you can bring your own research into the classroom. In cases where you feel you are at variance with other faculty, we suggest you identify a 'champion of your cause' (Box 25.3), a faculty member or administrator who supports your innovation and enjoys wide support from key stakeholders. This will promote tolerance as per the concept of ZoT. Box 25.1 lists various common problems that may also cause faculty to oppose your innovation attempts. In summary, many programs and curricula have been well established over many years and faculty and technicians are loath to make any drastic changes, which may put them out of their comfort zone and disrupt their desire for *status quo*. Related to this, often psychological and historical barriers to change exist where faculty, teaching assistants and technicians may feel insecure about tackling something new that may require them to develop new competences. They would favor historical habit. Furthermore, there will be knowledgeable colleagues who simply disagree that the innovation will add value to the program – it does not fit with their educational philosophy. Finally, some colleagues may already be suffering from innovation overload- too many new ideas to get their head around- or concerns about the availability of resources to run the innovation, even if you have already checked on these when considering feasibility (ZFI).

The good news is that there will always be some positive forces that you could consider that will support the implementation (Box 25.2) and in some cases counteract the negative forces. One such force could be positive pressure on you, or a participating colleague, to satisfy management expectations and to demonstrate scholarship in your/their teaching with the promise of tenure or promotion. Alternatively, there may be pressure to meet accreditation guidelines from a national society that you could use to motivate for your innovation. Endorsement of a novel approach by policymakers or external private and government funding agencies can motivate efforts to incorporate more authentic research practices such as experimentation into biology instruction. Recommendations from policymakers to teach about biological research, the last bullet listed in Box 25.2, include consensus reports from the National Research Council (NRC, 1999, 2003) and the American Association for the Advancement of Science (AAAS, 2011). Even incentives from higher administration at the institution, such as release time from teaching or professional development funds, can leverage efforts to influence colleagues to move beyond their comfort zone. External funding has driven change, but so have institutional priorities, as illustrated by the number of chapters in this book that were written by scientists who learned from the literature or professional development in biology education but who did most of their work without external support. Since students should always be considered a crucial part of the equation, their expectations and any pressure from peer-, course- and lecturer evaluations should also be taken cognizance of when implementing and testing a newly introduced innovation. Moreover, related to this, future employers of your students may want a voice in what is introduced. At the end of the day, one would hope that better educated undergraduate students would lead to more competent researchers and people who support more funding for scientific research endeavors.

Box 25.2: Selected Examples of Contextual Forces Supporting innovation (Modified from Rogan & Anderson, 2011)

- Pressure to demonstrate scholarship
- Promise of tenure or promotion
- Shift in time commitments
- Pressure of quality evaluation
- Student/employer expectations
- More competent research students
- Competition for recruiting the best students
- Pressure of accreditation
- Recommendations from policy-makers

25.4 Potential Strategies for Implementing an Innovation: Examples

By considering the above-mentioned positive and negative forces we recommend you identify potential strategies for implementation (Fig. 25.1), which promise to support the positives and oppose the negatives. Box 25.3 contains a far-from-exhaustive list of some such strategies which could, if needed, be tested. Obviously, your academic freedom to be innovative should be an expectation, which you should address if you feel you are being unduly restricted in your position. And you should be prepared to motivate why your innovation of interest would add value to the program and not simply be change for the sake of change. Where feasible you should also not try to throw out another course and replace it with yours. Numerous courses have been developed over many years with a substantial amount of investment put into them so, if possible, adding your innovation to an existing course could be more desirable, especially in the short term until your innovation has been fully tested. This is in line with the well-established curriculum practice of phasing in new developments through an "evolution not revolution" approach. Also, in line with the tenets of sound curriculum theory (Anderson & Rogan, 2011), it will be advantageous to use a backward curriculum design strategy (Wiggins & McTighe, 2005, 2011) by aligning class activities with learning outcomes and assessments that test the achievement of such learning outcomes. In cases where the innovation you wish to introduce will require faculty/TA development, you might consider establishing a Learning Community for all stakeholders (Cox, 2003) who can share knowledge and expertise in a collegial manner. In order to do this successfully, we need to optimize collaboration between participants and to deeply integrate the knowledge across disciplinary boundaries. According to recommendations by Pelaez et al., (2018), it is also important that all participants share a commitment to address a problem; focus on collegiality and mutual respect; communicate by using inclusive language and avoiding jargon; and achieve consensus with equitable processes of collective decision making about roles, tasks, and processes for achieving a useful product, and then to plan for project continuity. This will reduce the chances of anyone feeling threatened by the change. In some cases, it may also be useful for the change agent to work one-on-one with stakeholders at the 'coalface' so to speak, so that any specific training and mentoring can be performed, and any questions answered, or concerns resolved. In our experience, involving colleagues and staff in such a way that they feel empowered to take joint ownership of the innovation can provide an advantage.

Box 25.3: Selected Examples of Strategies Promoting Change (Modified from Rogan & Anderson, 2011 and Pelaez et al., 2018)

- 'Hands-on' at the coalface
- Mutual respect between change agent and educators
- Driven by 'champions of the cause'
- Faculty empowered to take ownership
- Align plans with a backward design strategy
- Emphasize faculty development via Learning Communities
- Operate with mutual respect and value expertise of team members
- Communicate without jargon and with inclusive language
- Academic freedom to be innovative
- Adopt an equitable process of collective decision making
- No change just for the sake of change
- Evolution not revolution
- Addition not replacement
- Plan for continuity to continue learning from the experience of others

Chapters in this book provide practical examples to give change agents and innovative educators a source of ideas. In summary, as you consider how to introduce an innovation presented in this book, we recommend you consider the following questions (modified from Rogan & Anderson, 2011):

- Into which category of influencing factor (see Fig. 25.1) does your innovation fit and how will that factor affect the decision you make about your approach to implementing the innovation?
- Do you think that the implementation of the innovation will be feasible (i.e. ZFI)?
- Will it be acceptable to colleagues and students (i.e., ZoT)?
- Which contextual forces might support the implementation?
- Which contextual forces might hinder the implementation?
- Which strategies, therefore, could strengthen or weaken the supporting and opposing contextual forces?

Acknowledgments This work was funded in part by the National Science Foundation (NSF) grant #1346567 for the Advancing Competencies in Experimentation–Biology (ACE-Bio) Network, as well as by the National Science Foundation (NSF), Division of Undergraduate Education (DUE) grants 1503798 and 1710051 for the Biochemistry Authentic Scientific Inquiry Lab (BASIL) project. Any opinions, findings, and conclusions or recommendations expressed are those of the authors and do not necessarily reflect the views of the NSF.

References

Altbach, P. G. (2001). Academic freedom: International realities and challenges. *Higher Education, 41*, 205–219.

American Association for the Advancement of Science (AAAS). (2011). *Vision and change in undergraduate biology education: A call to action.* Retrieved from Washington, DC: https://live-visionandchange.pantheonsite.io/wp-content/uploads/2013/11/aaas-VISchange-web1113.pdf. Accessed 5 Jan 2021.

Anderson, T. R. (2007). The importance of bridging the gap between science education research and its application in biochemistry teaching and learning: Barriers and strategies. *Biochemistry and Molecular Biology Education, 35*(6), 465–470.

Anderson, T. R., & Rogan, J. M. (2011). Bridging the educational research-teaching practice gap: Curriculum development, part 1: Components of the curriculum and influences on the process of curriculum design. *Biochemistry and Molecular Biology Education, 39*(1), 68–76.

Boyer, E. L., Altbach, P. G., & Whitelaw, M. J. (1994). *The academic profession: An international perspective.* Carnegie Foundation for the Advancement of Teaching.

Cox, M. D. (2003). Proven faculty development tools that foster the scholarship of teaching in faculty learning communities. In C. Wehlburg & S. Chadwick-Blossey (Eds.), *To improve the academy: Resources for faculty, instructional, and organizational development* (Vol. 21, pp. 109–142). Anker.

McGivney, J. H., & Moynihan, W. (1972). School and community. *Teachers College Record, 74*, 209–224.

National Research Council (NRC). (1999). *Transforming undergraduate education in science, mathematics, engineering, and technology.* The National Academies Press.

National Research Council (NRC). (2003). *BIO 2010. Transforming undergraduate education for future research biologists.* National Academy Press.

Pelaez, N. J., Anderson, T. R., Gardner, S. M., Yin, Y., Abraham, J. K., Bartlett, E., Gormally, C., Hill, J. P., Hoover, M., Hurney, C., Long, T., Newman, D. L., Sirum, K., & Stevens, M. (2017). *The basic competencies of biological experimentation: Concept-skill statements.* Retrieved from http://docs.lib.purdue.edu/pibergiim/4. Accessed 5 Jan 2021.

Pelaez, N. J., Anderson, T. R., Gardner, S. M., Yin, Y., Abraham, J. K., Bartlett, E., Gormally, C., Hurney, C., Long, T., Newman, D. L., Sirum, K., & Stevens, M. (2018). A community-building framework for collaborative research coordination across the education and biology research disciplines. *CBE-Life Science Education, 17*, es2. https://doi.org/10.1187/cbe.17-04-0060

Rogan, J. M., & Anderson, T. R. (2011). Bridging the educational research-teaching practice gap: Curriculum development, part 2: Becoming an agent of change. *Biochemistry and Molecular Biology Education, 39*(3), 233–241.

Wiggins, G., & McTighe, J. (2005). *Understanding by Design.* Association for Supervision and Curriculum Development (ASCD).

Wiggins, G., & McTighe, J. (2011). *The understanding by design guide to creating high-quality units.* Association for Supervision and Curriculum Development (ASCD).

Index

N. J. Pelaez et al. (eds.), *Trends in Teaching Experimentation in the Life Sciences*, Contributions from Biology Education Research,
https://doi.org/10.1007/978-3-030-98592-9

D

E

The manufacturer's authorised representative in the EU is Springer Nature Customer Service Centre GmbH, Europaplatz 3, 69115 Heidelberg, Germany. If you have any concerns regarding our products, please contact ProductSafety@springernature.com

Printed and bound by CPI Group (UK) Ltd, Croydon, CR0 4YY

30/09/2026

02225497-0020